DYNAMICAL SYMMETRY BREAKING IN QUANTUM FIELD THEORIES

DYNAMICAL SYMMETRY BREAKING IN QUANTUM FIELD THEORIES

V A Miransky
Institute for Theoretical Physics
Academy of Sciences of Ukraine
Kiev, Ukraine

World Scientific
Singapore • New Jersey • London • Hong Kong

Published by

World Scientific Publishing Co. Pte. Ltd.

P O Box 128, Farrer Road, Singapore 9128

USA office: Suite 1B, 1060 Main Street, River Edge, NJ 07661

UK office: 73 Lynton Mead, Totteridge, London N20 8DH

DYNAMICAL SYMMETRY BREAKING IN QUANTUM FIELD THEORIES

ISBN 981-02-1558-4

Printed in Singapore by JBW Printers & Binders Pte. Ltd.

Preface

Dynamical symmetry breaking is one of the deepest, most important and most beautiful phenomena in physics.

In condensed matter physics, the phenomenon of dynamical symmetry breaking is synonymous with superconductivity. In particle physics, dynamical chiral symmetry breaking is the main source of hadron masses and governs the low energy hadron dynamics. All, or almost all, high energy physicists believe that the $W^{\pm}$ and Z vector bosons are massive because of spontaneous electroweak symmetry breaking, and some of us believe that this breakdown may be dynamical in origin.

The first physical theory in which dynamical symmetry breaking was encountered was the Bardeen-Cooper-Schrieffer (BCS) theory of supreconductivity. The BCS theory not only explained a marvelous phenomenon in nature, but also introduced a new concept in physics: the dynamical rearrangement of the ground state (vacuum) which results from fermion pairing.

The universality of this concept was first recognized by Nambu who interpreted the fermion pairing as generating a fermion mass. The universal definition of the latter (valid both in relativistic and nonrelativistic physics) is as a gap in a fermion spectrum. Nambu then introduced the radical hypothesis that the pairing mechanism is responsible for generating the masses of elementary particles. At present, it is difficult for us to assess how revolutionary this idea was in 1960. One should recall that one of the paradigms in particle physics at that time was that the vacuum in relativistic quantum field theory is unique. Since vacuum degeneracy is inevitable in systems with dynamical (and, more generally, spontaneous) symmetry breaking, the accepted wisdom therefore excluded the existence of such a phenomenon in relativistic physics.

The first model with dynamical symmetry breaking in relativistic quantum field theory was constructed by Nambu and Jona-Lasinio. Their model still remains a basic example of model building in particle physics where dynamical symmetry breaking is now one of the central concepts.

Hopefully, this short introduction has convinced the readers that the phenomenon of dynamical symmetry breaking deserves its own book ! At the beginning of this project,

I had decided that it should be really a book and not just a review. Shortly after starting, I recognized that writing such a book is a rather difficult task. The point is that dynamical symmetry breaking is not a closed, isolated, subject. It is a part of quantum field theory and is, despite its beauty, rather technical. Without knowing about such technical machinery as the equations for Green's functions on the nonperturbative vacuum, the Bethe-Salpeter equation for bound states, the effective action for composite operators and so on, it is impossible to work in this area. Therefore I decided to discuss this material in detail. I hope, however, that the technical content does not overwhelm the rest of the book.

In order to understand better the phenomenon of dynamical (and, more generally, spontaneous) symmetry breaking in relativistic physics, it is extremely useful, I believe, to consider first the realization of this phenomenon in condensed matter physics. The condensed matter examples provide a clear and rather general physical picture which is very useful in the discussion of their much less visual analogues in particle physics. I have therefore devoted one chapter of the book to superfluidity and superconductivity.

The description of dynamical symmetry breaking can be divided into two parts: the first part is the description of the microscopic dynamics (as the BCS theory in superconductivity), the second part is the derivation of the low energy effective action (as the Ginzburg-Landau action in superconductivity). I consider both these approaches in detail.

The last four chapters in the book are devoted to the consideration of dynamical symmetry breaking in such realistic theories as quantum chromodynamics (QCD) and electroweak theory: spontaneous chiral symmetry breaking in QCD and electroweak symmetry breaking. At present, there exist whole "industries" studying spontaneous chiral symmetry breaking in hadronic physics and electroweak symmetry breaking in electroweak dynamics. The aim of the present book is not to review all of the directions in these two subjects but rather to give a detailed treatment of the key physical ideas.

I hope that my initial goal, to write a book and not a review, has been achieved at least in part: but above all my great hope is that the book will be useful.

I am indebted to many colleagues who over the years have been a source of stimulation and encouragement. I gratefully acknowledge the long and fruitful collaboration with my Kiev colleagues P. I. Fomin, V. P. Gusynin and Yu. A. Sitenko. I am grateful to W. A. Bardeen, S. T. Love, R. R. Mendel, S. Peris, S. Raby, M. Tanabashi and K. Yamawaki for enjoyable collaboration which has increased my depth of understanding of this subject. I would also like to thank T. W. Appelquist, D. Atkinson, M. Bando, S. R. Chivukula, R. E. Cutkosky, M. B. Einhorn, V. Elias, A. T. Filippov, K. Higashijima, C. T. Hill, B. Holdom, P. W. Johnson, J. B. Kogut, T. Kugo, M. G. Mitchard, T. Muta, Y. Nambu, K. Nishijima, J. Shigemitsu and L. C. R. Wijewardhana for numerous discussions which certainly influenced this book.

I also would like to express my thanks to V. Elias, M. Harada, A. Kovner, T. Kugo, M. G. Mitchard and B. Rosenstein for their remarks and suggestions concerning the book.

This book has been written during the 1992-93 academic year while I was at the Yukawa Institute for Theoretical Physics at Kyoto University. I am grateful to Professor Y. Nagaoka and all the members of the Institute for their hospitality. Special thanks to Dr. M. Toya and the members of the Institute Secretariat for their help in preparing the manuscript. In particular, I want to express my gratitute to Ms. M. Imanishi and Ms. A. Yamaguchi for patiently typing the manuscript in TEX.

The book is essentially an extended version of the course of lectures that I gave to graduate students of the Department of Physics of Kiev State University, the Department of Applied Mathematics of the University of Western Ontario, and the Department of Physics of Ohio State University. I thank all the students involved for asking good questions.

I also would like to thank the students of the Department of Physics of Kyoto University T. Hamazaki, K.-I. Izawa, T. Kawano, S. Miyazaki, J. Sato, M. Sato, T. Takahashi and Y. Taniguchi for reading the manuscript, finding typos and asking good questions.

Finally I thank my mother and my family for their support, although from a long way off.

Vladimir Miransky

Kyoto, September 1993

Contents

DYNAMICAL SYMMETRY BREAKING IN QUANTUM FIELD THEORIES

Chapter 1

The Theory of Free Fields and Nonequivalent Representations of the Canonical Commutation Relations in Field Theories

1.1. Introduction

One of the differences between field theoretical and mechanical systems is that field theories have an infinite number of degrees of freedom. This is important because, as we shall see, in most cases, spontaneous symmetry breaking can only occur in systems with an infinite number of degrees of freedom.

A local field $\varphi(\mathbf{x}, x_0)$ describes the dynamics of degrees of freedom labeled by the coordinates $\mathbf{x}$. Because of the complication of dealing with infinite degrees of freedom, it is often convenient to define the field theoretical system as a limit of systems with finite degrees of freedom. The usual construction has two steps: we start by enclosing the system in a finite rectangular box of size V and assume that $\varphi(\mathbf{x}, x_0)$ satisfies periodic boundary conditions. In this finite volume, momenta can take only discrete values: when we take the limit $V \to \infty$ we have the equivalence

$$V^{-1} \sum_{\mathbf{k}} \longrightarrow (2\pi)^{-3} \int d^3\mathbf{k} \,. \tag{1.1}$$

The next step is to introduce an ultraviolet cutoff Λ in momentum space; there now remain only a finite number of momentum modes, and so a finite number of degrees of freedom, in the system. We recover the field theory by taking the limits $V \to \infty$ and $\Lambda \to \infty$. The introduction of the volume V and the cutoff Λ are both examples of regularization in quantum field theory. In many cases, particularly those involving nonperturbative dynamics, regularization is not just convenient but in fact necessary in order to define properly field theoretical systems.

Some peculiarities in the behavior of field theoretical systems as the regulators are removed are evident in free field theories. In this chapter we give a brief summary of the theory of free fields, and using free field theory as an example, we then discuss one particular property of field theory that will give us physical insight into the phenomenon of spontaneous symmetry breaking: the existence of nonequivalent representations of canonical commutation relations. The demonstration of this property is the central point of the present chapter.

1.2. The Klein-Gordon Field

In order to introduce our notation we briefly recall the properties of the Klein-Gordon field. The lagrangian density of a real free scalar field $\varphi(x)$ takes the form

$$\mathcal{L} = \frac{1}{2}\left(\partial_\mu\varphi\partial^\mu\varphi - m^2\varphi^2\right) . \tag{1.2}$$

The equation of motion for $\varphi(x)$ corresponding to this lagrangian is the free Klein-Gordon equation

$$\left(\Box + m^2\right)\varphi(x) = 0 \tag{1.3}$$

where the operator $\Box$ is

$$\Box = \partial^\mu\partial_\mu = \frac{\partial}{\partial x^0}\frac{\partial}{\partial x^0} - \frac{\partial}{\partial x^i}\frac{\partial}{\partial x^i} .$$

An arbitrary solution of this equation may be expanded as a Fourier integral over elementary plane-wave solutions

$$\varphi(x) = \int \frac{d^3\mathbf{k}}{\sqrt{(2\pi)^3 2\omega_k}}\left[a(k)e^{-ikx} + a^\dagger(k)e^{ikx}\right] , \tag{1.4}$$

where

$$k^0 = \omega_k = \sqrt{\mathbf{k}^2 + m^2} \quad \text{and} \quad kx = k^\mu x_\mu . \tag{1.5}$$

If we follow the canonical quantization procedure, we find that $a(k)$ and $a^\dagger(k)$ satisfy the following commutation relations

$$\left[a(k),\, a(k')\right] = \left[a^\dagger(k),\, a^\dagger(k')\right] = 0 \quad \text{and} \quad \left[a(k),\, a^\dagger(k')\right] = \delta^3(\mathbf{k} - \mathbf{k}') . \tag{1.6}$$

The total energy-momentum vector for the free Klein-Gordon field takes the form

$$\begin{aligned} P^\mu &= \frac{1}{2}\int d^3\mathbf{k}\, k^\mu \left[a^\dagger(k)a(k) + a(k)a^\dagger(k)\right] \\ &= \int d^3\mathbf{k}\, k^\mu a^\dagger(k)a(k) + \frac{1}{2}\delta^3(0)\int d^3\mathbf{k}\, k^\mu . \end{aligned} \tag{1.7}$$

The meaning of the second term in Eq.(1.7) becomes clear if the scalar field is considered in a finite volume V. Then, since

$$\delta^3(\mathbf{k}) = \frac{1}{(2\pi)^3}\int d^3\mathbf{x}\, e^{i\mathbf{k}\cdot\mathbf{x}} , \tag{1.8}$$

we obtain the following (somewhat heuristic) relation

$$\delta^3(\mathbf{k}-\mathbf{k}') \longrightarrow \frac{V\delta_{\mathbf{k}\mathbf{k}'}}{(2\pi)^3} , \tag{1.9}$$

valid in the limit of large volume V. If we further assume that the field $\varphi(x)$ obeys periodic boundary conditions on the surfaces the volume V, we find that the number of states in the interval $d^3\mathbf{k}$ around the momentum $\mathbf{k}$ is

$$dn_k = \frac{V d^3\mathbf{k}}{(2\pi)^3} . \tag{1.10}$$

Therefore in a large but finite volume V, the expression for P^μ can be rewritten as

$$P^\mu = \sum_{\mathbf{k}} k^\mu a_k^\dagger a_k + \frac{1}{2}\sum_{\mathbf{k}} k^\mu , \tag{1.11}$$

where the operators a_k and $a_k^\dagger$ defined by

$$a_k = (2\pi)^{3/2} V^{-1/2} a(k) \ , \quad a_k^\dagger = (2\pi)^{3/2} V^{-\frac{1}{2}} a^\dagger(k) , \tag{1.12}$$

satisfy the commutation relations

$$\begin{aligned} \left[a_k, a_{k'}\right] &= \left[a_k^\dagger, a_{k'}^\dagger\right] = 0 \\ \left[a_k, a_{k'}^\dagger\right] &= \delta_{\mathbf{k}\mathbf{k}'} . \end{aligned} \tag{1.13}$$

The expression (1.11) corresponds to a system of the harmonic oscillators in the volume V; the second term in the sum (1.11),

$$\frac{1}{2}\sum_{\mathbf{k}} k^\mu = \frac{1}{2}\delta_{\mu 0}\sum_{\mathbf{k}} \omega_k ,$$

represents the zero-point energy of the system.

The energy operator P^0 is positive definite and the ground state or vacuum is by definition a state with the lowest energy. In this case, the vacuum is the state devoid of excitations created by the operators $a_{\mathbf{k}}^\dagger$, that is, the state annihilated by all operators $a_{\mathbf{k}}$. Its energy is just the zero-point energy $\frac{1}{2}\sum_{\mathbf{k}} \omega_{\mathbf{k}}$ and is infinite.

One can easily extend this discussion to a model with a complex field $\varphi(x)$. This field can be written as

$$\varphi = \frac{1}{\sqrt{2}}(\varphi_1 + i\varphi_2) , \tag{1.14}$$

where φ_1 and φ_2 are real fields. The lagrangian density for the complex field φ is

$$\mathcal{L} = \partial^\mu \varphi^\dagger \partial_\mu \varphi - m^2 \varphi^\dagger \varphi . \tag{1.15}$$

We also recall that the propagator of a free scalar field is

$$D(x) = i\langle 0|T\varphi(x)\varphi^\dagger(0)|0\rangle = \frac{1}{(2\pi)^4}\int d^4k \frac{1}{m^2 - k^2 - i\epsilon} e^{-ikx} , \tag{1.16}$$

where the time-ordered product $T\varphi(x)\varphi^\dagger(0)$ is

$$T\varphi(x)\varphi^\dagger(0) = \theta(x^0)\varphi(x)\varphi^\dagger(0) + \theta(-x^0)\varphi^\dagger(0)\varphi(x) . \tag{1.17}$$

1.3. The Spin-$\frac{1}{2}$ Field

As is well known [1], for a spin-$\frac{1}{2}$ field there are two inequivalent representations of $SL(2,C)$, the double cover of the Lorentz group: $\xi^\alpha \in (\frac{1}{2},\ 0)$ with $\alpha = 1,2$ and $\xi^{\dot\alpha} \in (0,\ \frac{1}{2})$ with $\dot\alpha = 1,2$.

Under

$$\begin{pmatrix} \alpha & \beta \\ \gamma & \delta \end{pmatrix} \in SL(2,C) \ , \ \alpha\delta - \gamma\beta = 1 ,$$

these spinors transform as

$$\begin{aligned} \xi^{1'} &= \alpha\xi^1 + \beta\xi^2 , \\ \xi^{2'} &= \gamma\xi^1 + \delta\xi^2 , \end{aligned} \tag{1.18}$$

and

$$\begin{aligned} \xi^{\dot{1}'} &= \alpha^* \xi^{\dot{1}} + \beta^* \xi^{\dot{2}} , \\ \xi^{\dot{2}'} &= \gamma^* \xi^{\dot{1}} + \delta^* \xi^{\dot{2}} , \end{aligned} \tag{1.19}$$

where α^* denotes the complex conjugate of α, and so on.

The metric spinor of $SL(2,C)$ is $\epsilon_{\alpha\beta}$ with $\epsilon_{12} = -\epsilon_{21} = 1$, $\epsilon_{11} = \epsilon_{22} = 0$, and $\xi^\alpha \epsilon_{\alpha\beta} \eta^\beta$ is a $SL(2,C)$ invariant. The spinors ξ^α and $\xi_\alpha = \epsilon_{\alpha\beta}\xi^\beta$ are referred to as contravariant and covariant spinors respectively. For the dotted representation the metric spinor $\epsilon_{\dot{\alpha}\dot{\beta}} = \epsilon_{\alpha\beta}$, and we define $\xi_{\dot{\alpha}} = \epsilon_{\dot{\alpha}\dot{\beta}}\eta^{\dot{\beta}}$.

With the respect to the subgroup of rotations $SU(2)$, we note that these representations are unitary, so that the combination $\xi^1\xi^{1*} + \xi^2\xi^{2*}$ is invariant under rotations. Under this subgroup, the conjugate $\xi^{\alpha *}$ transforms as $\epsilon_{\alpha\beta}\xi^\beta = \xi_\alpha$ (that is, they belong to the same representation which we will denote by $\xi^{\alpha *} \sim \xi_\alpha$). However under the full $SL(2,C)$ group the conjugate $\xi^{\alpha *}$ is assigned to the dotted representation $(0,\, \frac{1}{2})$, $\xi^{\alpha *} \sim \xi^{\dot{\alpha}}$.

A four-dimensional vector a^μ is assigned to the representation $(\frac{1}{2},\, \frac{1}{2})$, and can be represented explicitly as

$$\chi^{\alpha\dot{\beta}} = (\mathbf{a}\cdot\boldsymbol{\sigma})^{\alpha\dot{\beta}} + a^0 \sigma_0^{\alpha\dot{\beta}} , \tag{1.20}$$

where σ_1, σ_2 and σ_3 are the Pauli matrices and σ_0 is the identity matrix:

$$\sigma_1 = \begin{pmatrix} 0 & 1 \\ 1 & 0 \end{pmatrix} , \quad \sigma_2 = \begin{pmatrix} 0 & -i \\ i & 0 \end{pmatrix} , \quad \sigma_3 = \begin{pmatrix} 1 & 0 \\ 0 & -1 \end{pmatrix} , \quad \sigma_0 = \begin{pmatrix} 1 & 0 \\ 0 & 1 \end{pmatrix} . \tag{1.21}$$

The Dirac equation for a particle of mass m can be written in momentum space in terms of two spinors $\eta_{\dot{\beta}} \in (0,\, \frac{1}{2})$ and $\xi^\alpha \in (\frac{1}{2},\, 0)$ as

$$\begin{aligned} p^{\alpha\dot{\beta}} \eta_{\dot{\beta}} &= m\xi^\alpha , \\ p_{\dot{\beta}\alpha} \xi^\alpha &= m\eta_{\dot{\beta}} \end{aligned} \tag{1.22}$$

with

$$p^{\alpha\dot{\beta}} = (\mathbf{p}\cdot\boldsymbol{\sigma} + p_0)^{\alpha\dot{\beta}} \quad , \quad p_{\dot{\beta}\alpha} = (-\mathbf{p}\cdot\boldsymbol{\sigma} + p_0)_{\dot{\beta}\alpha} . \tag{1.23}$$

This formalism will be useful when we later discuss chiral symmetry (see Chapter 4), since the equations (1.22) decouple when $m = 0$. The Dirac equation may be rewritten in the standard form by introducing the Dirac matrices

$$\gamma^0 = \begin{pmatrix} 0 & \sigma_0 \\ \sigma_0 & 0 \end{pmatrix}, \; \boldsymbol{\gamma} = \begin{pmatrix} 0 & -\boldsymbol{\sigma} \\ \boldsymbol{\sigma} & 0 \end{pmatrix}, \tag{1.24}$$

and the bispinor

$$\psi = \begin{pmatrix} \xi^\alpha \\ \eta_{\dot\beta} \end{pmatrix}, \tag{1.25}$$

which transforms in the reducible representation $\left(\frac{1}{2},\, 0\right) \oplus \left(0,\, \frac{1}{2}\right)$. Eq.(1.22) now becomes

$$(\gamma^\mu p_\mu - m)\,\psi = 0 \; . \tag{1.26}$$

The Dirac equation (1.26) can be derived from the lagrangian density

$$\mathcal{L} = i\bar\psi\gamma^\mu\partial_\mu\psi - m\bar\psi\psi \; , \tag{1.27}$$

where the Dirac adjoint $\bar\psi$ is defined by $\bar\psi = \psi^\dagger\gamma_0$. This lagrangian density is invariant under the $U(1)$ transformation

$$\psi \longrightarrow e^{i\alpha}\psi \;\; , \quad \bar\psi \longrightarrow e^{-i\alpha}\bar\psi \; . \tag{1.28}$$

Because of the Noether theorem, this invariance leads to the conservation of the current $\jmath^\mu(x) = \bar\psi(x)\gamma^\mu\psi(x), \partial_\mu \jmath^\mu(x) = 0$, i.e. to the conservation of the fermion number (the Noether theorem is discussed in Sec. 4.2).

We emphasize that in relativistic physics, a mass term which conserves fermion number can be constructed only when there is a bispinor ψ assigned to the reducible representation $\left(\frac{1}{2},\, 0\right) \oplus \left(0,\, \frac{1}{2}\right)$. For example, the combination $\xi^{\alpha^*}\xi^\alpha$ transforms as the p_0-component of a vector under Lorentz transformations (although it is invariant under $SU(2)$ transformations) so that it cannot be used as a mass term. However if one does not require fermion number conservation, one can also construct the following $SL(2, C)$-invariant mass term

$$m\xi^\alpha\xi_\alpha \;+\; H.c. \tag{1.29}$$

where $H.c.$ denotes the hermitean conjugated term. This mass term is not invariant under the $U(1)$ rotation Eq.(1.28) and so violates fermion number conservation. Such a term is called a Majorana mass term. We note that since $\xi^\alpha\xi_\alpha = \xi^\beta\epsilon_{\beta\alpha}\xi^\alpha$, the Majorana mass can appear only due to the anticommutativity of ξ^α ; $\xi^\beta\xi^\alpha = -\xi^\alpha\xi^\beta$.

The general solution of the Dirac equation

$$i\gamma^\mu \partial_\mu \psi - m\psi = 0 \tag{1.30}$$

in coordinate space takes the form

$$\begin{aligned}\psi(x) =& \frac{1}{(2\pi)^{3/2}} \sum_{s=\pm1} \int d^3\mathbf{p} \left[b(p,s)u(p,s)e^{-ipx} + d^\dagger(p,s)v(p,s)e^{ipx} \right] ,\\ \bar{\psi}(x) =& \frac{1}{(2\pi)^{3/2}} \sum_{s=\pm1} \int d^3\mathbf{p} \left[b^\dagger(p,s)\bar{u}(p,s)e^{ipx} + d(p,s)\bar{v}(p,s)e^{-ipx} \right] ,\end{aligned} \tag{1.31}$$

where, as usual, $p^0 = \sqrt{\mathbf{p}^2 + m^2}$, and spinors $u(p,s)$ and $v(p,s)$ are taken to be helicity eigenstates:

$$\frac{\mathbf{\Sigma} \cdot \mathbf{p}}{|\mathbf{p}|} u(p,s) = s u(p,s) \ , \quad \frac{\mathbf{\Sigma} \cdot \mathbf{p}}{|\mathbf{p}|} v(p,s) = -s v(p,s) \quad \text{for} \quad s = \pm 1 \ , \tag{1.32}$$

where

$$\mathbf{\Sigma} = \begin{pmatrix} \sigma & 0 \\ 0 & \sigma \end{pmatrix} . \tag{1.33}$$

Properties of these spinors are considered in Appendix A.

They individually satisfy the Dirac equations

$$\begin{aligned}(\gamma^\mu p_\mu - m)\, u(p,s) =& 0 \ ,\\ (\gamma^\mu p_\mu + m)\, v(p,s) =& 0 \ ,\end{aligned} \tag{1.34}$$

which in terms of the adjoints are

$$\begin{aligned}\bar{u}(p,s)\, (\gamma^\mu p_\mu - m) &= 0 \ ,\\ \bar{v}(p,s)\, (\gamma^\mu p_\mu + m) &= 0 \ .\end{aligned} \tag{1.35}$$

We will use a normalisation which leads to the orthogonality relations

$$\begin{aligned}u^\dagger(p,s)u(p,s') = v^\dagger(p,s)v(p,s') &= \delta_{ss'}\\ \bar{v}(p,s)u(p,s) = v^\dagger(p,s)u(-p,s') &= 0\end{aligned} \tag{1.36}$$

with $u(-p,s) \equiv u(p_0, -\mathbf{p}, s)$.

To quantize, we impose the canonical equal time anticommutation relations

$$\begin{aligned} \{\psi(x), \bar{\psi}(y)\}|_{x_0=y_0} =& \gamma_0 \delta^3(\mathbf{x}-\mathbf{y}) , \\ \{\psi(x), \psi(y)\}|_{x_0=y_0} =& \{\bar{\psi}(x), \bar{\psi}(y)\}|_{x_0=y_0} = 0 . \end{aligned} \tag{1.37}$$

This leads to the following anticommutation relations for b and d

$$\{b(p,s), b^\dagger(p',s')\} = \{d(p,s), d^\dagger(p',s')\} = \delta_{ss'}\delta^3(\mathbf{p}-\mathbf{p}\,') , \tag{1.38}$$

with all other anticommutators being zero. The operators $b^\dagger(b)$, $d^\dagger(d)$ are creation (annihilation) operators of fermions and antifermions respectively.

The energy-momentum vector of the free Dirac field is

$$P^\mu = \sum_{s=\pm 1} \int d^3\mathbf{p}\, p^\mu \left[b^\dagger(p,s)b(p,s) + d^\dagger(p,s)d(p,s) \right] - 2\delta^3(0) \int d^3\mathbf{p}\, p^\mu . \tag{1.39}$$

If we consider the system in a large volume V, we can rewrite P^μ as

$$P^\mu = \sum_{s=\pm 1} \sum_{\mathbf{p}} p^\mu \left[b_p^\dagger(s) b_p(s) + d_p^\dagger(s) d_p(s) \right] - 2 \sum_{\mathbf{p}} p^\mu , \tag{1.40}$$

where we have defined

$$\begin{aligned} b_p(s) =& (2\pi)^{3/2} V^{-1/2} b(p,s) , \\ d_p(s) =& (2\pi)^{3/2} V^{-1/2} d(p,s) , \end{aligned} \tag{1.41}$$

with similar definitions for $b_p^\dagger(s)$ and $d_p^\dagger(s)$, so that

$$\{b_p(s), b_{p'}^\dagger(s')\} = \{d_p(s), d_{p'}^\dagger(s')\} = \delta_{ss'}\delta_{\mathbf{p}\mathbf{p}'} , \tag{1.42}$$

The last term on the right hand side of Eq.(1.40) represents the negative zero-point energy of the fermion system, in contrast to the expression for the scalar field in Eq.(1.11).

We recall that the Dirac propagator is

$$S(x) = -i\langle 0|T\psi(x)\bar{\psi}(0)|0\rangle = \frac{1}{(2\pi)^4} \int d^4p \frac{\hat{p}+m}{p^2-m^2+i\epsilon}\, e^{-ipx} , \tag{1.43}$$

where $\hat{p} = p_\mu \gamma^\mu$, and the time-ordered product $T\psi(x)\bar{\psi}(0)$ is

$$T\psi(x)\bar{\psi}(0) = \theta(x^0)\psi(x)\bar{\psi}(0) - \theta(-x^0)\bar{\psi}(0)\psi(x) . \tag{1.44}$$

1.4. The notion of the vacuum. Nonequivalent representations of the canonical commutation relations in quantum field theory

A universal definition of the vacuum for any hamiltonian system is that the vacuum is the state with the lowest energy.

In the case of hamiltonian systems with a finite number of degrees of freedom (quantum mechanical systems), different hamiltonians posses different vacua, however all these vacua lie in the same Hilbert space (in a sense which we will discuss later). As we shall show below, this is not true for field theoretical systems, or even for free field theories.

In a more formal language, this difference can be characterized in the following way. For a system with a finite number of creation and annihilation operators, all representations of their commutation relations are unitarily equivalent. However for field theoretical systems, there are an infinite number of unitary inequivalent representations of the commutation relations. In this section we consider two examples which illustrate this point.

Let us consider two real scalar fields φ_1 and φ_2 with different masses, μ_1 and μ_2, defined in finite volume V

$$\left(\Box + \mu_1^2\right) \varphi_1(x) = 0 \ , \ \left(\Box + \mu_2^2\right) \varphi_2(x) = 0 \ . \tag{1.45}$$

Since these differential equations are second order, we can subject them to the $x_0 = 0$ boundary conditions

$$\varphi_1(\mathbf{x}, 0) = \varphi_2(\mathbf{x}, 0) \quad , \quad \dot{\varphi}_1(\mathbf{x}, 0) = \dot{\varphi}_2(\mathbf{x}, 0) \ , \tag{1.46}$$

where $\dot{\varphi} \equiv \frac{\partial}{\partial t}\varphi$. Using the Fourier expansions for φ_1 and φ_2,

$$\varphi_i(x) = V^{-1/2} \sum_{\mathbf{k}} (2\omega_{ik})^{-1/2} \left[a_{ik} e^{-ikx} + a^\dagger_{ik} e^{ikx}\right] \ , \tag{1.47}$$

we find the relation

$$a_{1k} = (4\omega_{1k}\omega_{2k})^{-1/2} \left[(\omega_{1k} + \omega_{2k}) a_{2k} + a^\dagger_{2(-k)}(\omega_{1k} - \omega_{2k})\right] \ , \tag{1.48}$$

where $a^\dagger_{(-k)} \equiv a^\dagger_{(k^0, -\mathbf{k})}$.

The vacua $|0_1\rangle$ and $|0_2\rangle$ satisfy the conditions

$$a_{1k}|0_1\rangle = a_{2k}|0_2\rangle = 0\ , \tag{1.49}$$

and are normalized so that

$$\langle 0_1|0_1\rangle = \langle 0_2|0_2\rangle = 1\ . \tag{1.50}$$

Taking into account the relation (1.48), one can check that

$$|0_1\rangle = A|0_2\rangle\ , \tag{1.51}$$

where

$$A = K \exp\left[\sum_{\mathbf{k}} \frac{(\omega_{2k} - \omega_{1k})}{2(\omega_{2k} + \omega_{1k})} a^{\dagger}_{2k} a^{\dagger}_{2(-k)}\right] \tag{1.52}$$

with the normalization factor

$$K = \exp\left[\frac{1}{4}\sum_{\mathbf{k}} \ln\left(1 - \frac{(\omega_{2k} - \omega_{1k})^2}{(\omega_{2k} + \omega_{1k})^2}\right)\right]\ . \tag{1.53}$$

To check these equations, it is convenient to use the relation

$$\begin{aligned} a_k \exp\left(\alpha a^{\dagger}_k a^{\dagger}_{-k}\right)|0\rangle &= \alpha a^{\dagger}_{-k} \exp\left(\alpha a^{\dagger}_k a^{\dagger}_{-k}\right)|0\rangle\ ;\ k \neq 0, \\ a_0 \exp\left[\alpha \left(a^{\dagger}_0\right)^2\right]|0\rangle &= 2\alpha a^{\dagger}_0 \exp\left[\alpha \left(a^{\dagger}_0\right)^2\right]|0\rangle \end{aligned} \tag{1.54}$$

with α an arbitrary c-number.

In the large volume limit, $V \to \infty$, the normalization factor is (see Eq.(1.1))

$$K \simeq \exp\left[\frac{V}{4(2\pi)^3}\int d^3\mathbf{k} \ln\left(1 - \frac{(\omega_{2k} - \omega_{1k})^2}{(\omega_{2k} + \omega_{1k})^2}\right)\right]\ . \tag{1.55}$$

The crucial point is the appearance of the factor V in the exponent; as a result K rapidly goes to zero as $V \to \infty$. Eqs.(1.51) – (1.53) imply that the scalar product

$$\langle 0_1|0_2\rangle = K \tag{1.56}$$

vanishes in the infinite volume limit. Moreover, using Eqs.(1.48) and (1.51), one can check that $\langle n_1|m_2\rangle = 0$ at $V = \infty$ for any pair $|n_1\rangle$ and $|m_2\rangle$ of vectors from the Fock spaces F_1 and F_2 corresponding to the fields φ_1 and φ_2 respectively (recall that the Fock spaces F_i are built on a basis of eigenstates of the number operators $N_i = \sum_{\mathbf{k}} a^\dagger_{ik} a_{ik}$ with finite eigenvalues).

What is the physical meaning of this somewhat mysterious result? In the case of quantum mechanical systems, eigenvalues of different hamiltonians lie in the same Hilbert space, the space of square integrable functions L_2. For example, although the vacuum of the hamiltonian of the quantum harmonic oscillator $|0\rangle$ is an excited state with respect to the hamiltonian of the hydrogen atom H_h, the value $\langle 0|H_h|0\rangle$ is finite. A similar situation holds for the φ_1 and φ_2 systems with ultraviolet momentum cutoff in a finite volume. The transformation (1.48) preserves the canonical commutation relations and is unitary. One can expand the $|n_1\rangle$ states in the $|m_2\rangle$ basis and vice versa. In particular, $\langle 0_1|0_2\rangle = K \neq 0$ in finite volume. What happens in the limit $V \to \infty$ is the following: although the coefficients $\langle n_1|m_2\rangle$ in the expansion

$$\langle n_1| = \sum_{m_2} \langle n_1|m_2\rangle\langle m_2|$$

go to zero as $V \to \infty$, the norm

$$\langle n_1|n_1\rangle = \sum_{m_2} \langle n_1|m_2\rangle\langle m_2|n_1\rangle \equiv 1 \; ,$$

independent of the value of V. This implies that the larger V, the more terms with a large particle number dominate in the sum $\sum_{m_2}$. In other words, as $V \to \infty$, the number of φ_2 particles in the $|n_1\rangle$ states goes to infinity. Let us make this concrete: defining $\bar{N}_2$ by

$$\bar{N}_2(|n_1\rangle) \equiv \langle n_1|N_2|n_1\rangle \; , \tag{1.57}$$

where $N_2 = \sum_{\mathbf{k}} a^\dagger_{2k} a_{2k}$ is the number operator for φ_2 particles, and using Eqs.(1.51) and (1.52), we find that

$$\bar{N}_2(|0_1\rangle) = \sum_{\mathbf{k}} \frac{(\omega_{2k} - \omega_{1k})^2}{8\omega_{1k}\omega_{2k}} \simeq \frac{V}{64\pi^3} \int d^3\mathbf{k} \frac{(\omega_{2k} - \omega_{1k})^2}{\omega_{1k}\omega_{2k}} \; . \tag{1.58}$$

As $V \to \infty$, $\bar{N}_2 \to \infty$ and so the vectors $|n_1\rangle$ dissapear from the space F_2 (and vice versa).*)

We note also the following interesting point. If we define the average energy $\bar{E}_2$ to be

$$\bar{E}_2(|n_1\rangle) \equiv \langle n_1|H_2|n_1\rangle - \langle 0_2|H_2|0_2\rangle ,$$

we find

$$\bar{E}_2(|0_1\rangle) = \sum_{\mathbf{k}} \frac{(\omega_{2k} - \omega_{1k})^2}{8\omega_{1k}} \simeq \frac{V}{64\pi^3} \int d^3\mathbf{k} \frac{(\omega_{2k} - \omega_{1k})^2}{\omega_{1k}} \simeq \frac{V}{64\pi^2} (\mu_2^2 - \mu_1^2)^2 \ln \frac{\Lambda}{\mu_1} ,$$

where Λ is the ultraviolet momentum cutoff which we introduce in order to define the divergent integral which appears. Now as $\Lambda \to \infty$, the energy $\bar{E}_2(|0_1\rangle)$ is infinite even when the volume V is finite. From the formal point of view this means that although at finite V, the state $|0_1\rangle$ is in the Fock space F_2 (since $\bar{N}_2$ is finite), it is out of the domain $D(H_2)$ of the hamiltonian H_2 (that is, the action of H_2 on $|0_1\rangle$ is not properly defined).

This example shows that, besides volume divergences, there is another source of infinities in field theory, ultraviolet divergences connected with the high momentum region. This point will become clearer in the next example.

Let us consider now two free fermion fields ψ_1 and ψ_2 with different masses m_1 and m_2. We choose $m_1 = 0$ and $m_2 = m$. This example will be very useful when we discuss dynamical chiral symmetry breaking in Chapter 6. The Dirac equations for ψ_1 and ψ_2 take the form.

$$i\gamma^\mu \partial_\mu \psi_1 = 0 \ , \quad (i\gamma^\mu \partial_\mu - m)\, \psi_2 = 0 \ . \tag{1.59}$$

We subject the fields to the following boundary condition at $x_0 = 0$:

$$\psi_1(\mathbf{x}, 0) = \psi_2(\mathbf{x}, 0) \ . \tag{1.60}$$

---

*) This example gives a simple illustration of the Haag theorem [2] asserting that in relativistic field theory there is no unitary operator transforming a local theory with interaction into a free theory. Indeed, in this example, one may choose the theory with the mass μ_1 as a "genuine" free theory; then, the theory with the mass μ_2 can be considered as a theory with the superrenormalizable "interaction" $\mathcal{L}_{int} = (\mu_1^2 - \mu_2^2)\varphi^2$.

The Fourier expansion for the fields takes the form (see Eqs.(1.31) and (1.41)):

$$\psi_{1,2}(\mathbf{x},0) = V^{-1/2} \sum_{\mathbf{p},\, s=\pm 1} \left[b_{1,2p}(s) u_{1,2}(p,s) e^{i\mathbf{p}\cdot\mathbf{x}} + d^\dagger_{1,2p}(s) v_{1,2}(p,s) e^{-i\mathbf{px}} \right] . \tag{1.61}$$

This equation and Eq.(1.60) imply

$$\begin{aligned} &\sum_{s=\pm 1} \left[b_{1p}(s) u_1(p,s) + d^\dagger_{1(-p)}(s) v_1(-p,s) \right] \\ &= \sum_{s=\pm 1} \left[b_{2p}(s) u_2(p,s) + d^\dagger_{2(-p)}(s) v_2(-p,s) \right] . \end{aligned} \tag{1.62}$$

Taking into account the explicit form of the spinors $u_i(p,s)$ and $v_i(p,s)$, (see Eq.(A.15) in Appendix A), we find the relations

$$\begin{aligned} b_{2p}(s) &= \alpha_p b_{1p}(s) + \beta_p d^\dagger_{1(-p)}(s) , \\ d_{2p}(s) &= \alpha_p d_{1p}(s) - \beta_p b^\dagger_{1(-p)}(s) , \end{aligned} \tag{1.63}$$

where

$$\alpha_p = \left(\frac{\omega + p}{2\omega} \right)^{1/2} , \quad \beta_p = \left(\frac{\omega - p}{2\omega} \right)^{1/2} , \quad \omega = \left(\mathbf{p}^2 + m^2 \right)^{1/2} , \quad p = |\mathbf{p}\,| . \tag{1.64}$$

The vacua $|0_1\rangle$ and $|0_2\rangle$ are defined by the relations

$$\begin{aligned} b_{1p}(s)|0_1\rangle &= d_{1p}(s)|0_1\rangle = 0 , \\ b_{2p}(s)|0_2\rangle &= d_{2p}(s)|0_2\rangle = 0 . \end{aligned} \tag{1.65}$$

Using Eq.(1.63), we can check that

$$|0_2\rangle = \prod_{\mathbf{p},\, s=\pm 1} \left[\alpha_p - \beta_p b^\dagger_{1p}(s) d^\dagger_{1(-p)}(s) \right] |0_1\rangle , \tag{1.66}$$

and so we find that the scalar product $\langle 0_1|0_2\rangle$ is

$$\langle 0_1|0_2\rangle = \prod_{\mathbf{p},\, s=\pm 1} \alpha_p = \exp\left(2 \sum_{\mathbf{p}} \ln \alpha_p \right) \simeq \exp\left[2V \int \frac{d^3\mathbf{p}}{(2\pi)^3} \ln \alpha_p \right] . \tag{1.67}$$

Since $\ln \alpha_p \sim -m^2/8p^2$ as $p \to \infty$, we find that

$$\langle 0_1|0_2\rangle \simeq \exp\left[-V\Lambda m^2/8\pi^2 \right] , \tag{1.68}$$

where Λ is the ultraviolet momentum cutoff. Thus the scalar product $\langle 0_1|0_2\rangle$ goes to zero both in the infinite volume limit and in the limit $\Lambda \to \infty$.

One can check that the scalar product $\langle n_1|m_2\rangle$ of any pair of vectors from the Fock spaces F_1 and F_2 goes to zero with both $\Lambda \to \infty$ and $V \to \infty$.

This example is even more striking than the previous one. We see how the two sources of the infinite number of degrees of freedom (infinite volume and infinite ultraviolet cutoff) lead to the appearance of nonequivalent representations of the canonical commutation relations for fermion fields.

The main goal of the present chapter was the demonstration of the existence of nonequivalent representations of the canonical commutation relations in field theories. As will be shown in the next chapters, this point has a direct relation to the phenomenon of spontaneous symmetry breaking. However its manifestation is much more sophisticated in that case. In the case of spontaneous symmetry breaking, different representations of the canonical commutation relations relate to the same dynamical system. Moreover, the energies of the vacua from different representations are equal, a manifestation of the phenomenon of the vacuum degeneracy for such systems.

Chapter 2

Spontaneous Symmetry Breaking in Condensed Matter. Superfluidity and Superconductivity

2.1. Introduction

In this chapter, we consider the phenomenon of spontaneous symmetry breaking in condensed matter physics. We shall discuss two remarkable examples, superfluidity and superconductivity, which are particularly important because of their relativistic analogues in particle physics. The condensed matter examples provide a clear and rather general physical picture which will be very useful in our later discussions of their much less visual analogues in particle physics.

2.2. Ideal Bose Gas

The lagrangian density of an ideal Bose gas of particles of mass m is

$$\mathcal{L} = i\varphi^\dagger \frac{\partial}{\partial t}\varphi - \frac{1}{2m}\frac{\partial \varphi^\dagger}{\partial x^i}\frac{\partial \varphi}{\partial x^i} \, . \tag{2.1}$$

The field φ satisfies the Schrödinger equation

$$i\frac{\partial}{\partial t}\varphi + \frac{1}{2m}\Delta\varphi = 0 \quad \text{where } \Delta = \frac{\partial^2}{\partial x_i \partial x_i} \, . \tag{2.2}$$

The general solution of this equation in a finite volume V can be written as

$$\varphi(\mathbf{x}, t) = V^{-1/2}\sum_{\mathbf{k}} a_{\mathbf{k}} e^{-(i\epsilon(k)t - i\mathbf{k}\cdot\mathbf{x})} \quad \text{with} \quad \epsilon(k) = \frac{\mathbf{k}^2}{2m} \equiv \frac{k^2}{2m} \, , \tag{2.3}$$

where the operators $a_{\mathbf{k}}$ and $a^\dagger_{\mathbf{k}}$ satisfy the commutation relations $\left[a_{\mathbf{k}}, a^\dagger_{\mathbf{k}'}\right] = \delta_{\mathbf{k}\mathbf{k}'}$.

The hamiltonian H and the particle number operator $\hat{N}$ are then

$$H = \sum_{\mathbf{k}} \frac{\mathbf{k}^2}{2m} a^\dagger_{\mathbf{k}} a_{\mathbf{k}} \, , \quad \hat{N} = \sum_{\mathbf{k}} a^\dagger_{\mathbf{k}} a_{\mathbf{k}} \, . \tag{2.4}$$

Since $[H, \hat{N}] = 0$, particle number is conserved. Therefore one can consider the system with a fixed particle number

$$\hat{N} = N . \tag{2.5}$$

The ground state (which we call the vacuum) of this N-particle system is

$$|N\rangle = \frac{(a_0^\dagger)^N}{\sqrt{N!}}|0_{\text{bare}}\rangle \quad \text{where} \quad a_0^\dagger \equiv a_{\mathbf{k}}^\dagger|_{\mathbf{k}=0} , \tag{2.6}$$

where the "bare" vacuum satisfies the condition

$$a_{\mathbf{k}}|0_{\text{bare}}\rangle = 0 . \tag{2.7}$$

Because of the constraint Eq.(2.5), the vacuum, Eq.(2.6), is unique.

Up to now we have considered the system in a finite volume V. Let us show that the situation changes drastically when $V \to \infty$.

2.3. Spontaneous Symmetry Breaking in an Ideal Bose Gas

In the infinite volume limit, $V \to \infty$, it is appropriate to consider not the particle number but the particle density, n, defined by

$$n = \frac{N}{V} . \tag{2.8}$$

As $V \to \infty$ with fixed n, N goes to infinity. We will show that in this limit the ideal Bose gas provides an example of a system with spontaneous symmetry breaking. The symmetry which is broken is associated with particle number conservation,

$$\varphi \to e^{-i\theta}\varphi , \quad \varphi^\dagger \to e^{i\theta}\varphi^\dagger . \tag{2.9}$$

The lagrangian density (2.1) is of course invariant under this transformation.

A general discussion of spontaneous symmetry breaking will be given in the next chapter. Here we need to know only that spontaneous symmetry breaking occurs when the action of a dynamical system is invariant under some symmetry transformation but the ground state or vacuum of the system is not assigned to a singlet representation of the symmetry group. Since in the case of spontaneous symmetry breaking, the vacuum is assigned to a nonsinglet representation of a group connected with the symmetry, and since the group generators commute with the hamiltonian, a vacuum degeneracy takes place in such field systems.

The ground state $|N\rangle$ (see Eq.(2.6)) is an eigenvector of the operator $\hat{N}$, and therefore in the ideal Bose gas with this vacuum the $U(1)$ symmetry (2.9) is explicitly realized. We will now show that as $V \to \infty$ with fixed particle density n, the $U(1)$ symmetry (2.9) is spontaneously broken in the ideal Bose gas.

We begin by considering the coherent states defined by

$$|\theta\rangle_0 \equiv \exp(-N/2)\exp(N^{1/2}e^{i\theta}a_0^\dagger)|0_{\text{bare}}\rangle\,. \tag{2.10}$$

Under the transformation (2.9), $|\theta\rangle_0$ transforms as

$$|\theta\rangle_0 \longrightarrow U_{\theta'}|\theta\rangle_0 = |\theta+\theta'\rangle_0\,, \tag{2.11}$$

where the unitary operator

$$U_\theta = \exp(i\theta\hat{N})\,. \tag{2.12}$$

Since $H|\theta\rangle_0 = 0$, all these states have the same energy $E = 0$. However, although these states are eigenvectors of the operators $a_{\mathbf{k}}$,

$$a_{\mathbf{k}}|\theta\rangle_0 = \delta_{\mathbf{k}0}N^{1/2}e^{i\theta}|\theta\rangle_0\,, \tag{2.13}$$

they are *not* eigenvectors of the operators $\hat{N}$. On the other hand, as Eq. (2.13) implies, the average number of particles for all $|\theta\rangle_0$ is equal to N :

$${}_0\langle\theta|\hat{N}|\theta\rangle_0 = N\,. \tag{2.14}$$

Let us consider now the limit $V \to \infty$. First of all, using the Baker - Hausdorf formula,

$$\exp(\alpha a)\exp(\beta a^\dagger) = \exp(\beta a^\dagger)\exp(\alpha a)\exp(\alpha\beta)\,, \tag{2.15}$$

we find that

$${}_0\langle\theta'|\theta\rangle_0 = \exp\{N[\cos(\theta-\theta') - 1 + i\sin(\theta-\theta')]\} \underset{N\to\infty}{\longrightarrow} \delta_{\theta\theta'}\,, \tag{2.16}$$

i.e. since N/V^0 is fixed, $|\theta\rangle_0$-states become orthonormal as $V \to \infty$. Moreover, one can check that all vectors from the Fock space F_θ with the vacuum $|\theta\rangle_0$ are orthogonal to vectors from the space $F_{\theta'}$, $\theta' \neq \theta$, as $V \to \infty$.

In a finite volume, the states $|\theta\rangle_0$ are not eigenvectors of the operator $\hat{N}$ as we noted above. However, as $V \to \infty$, the appropriate operator is not $\hat{N}$ but the density operator $\hat{n} = \lim_{V\to\infty} \hat{N}/V$. Let us calculate the dispersion of the density for the states $|\theta\rangle_0$:

$$
\begin{aligned}
D =& [{}_0\langle\theta|\hat{n}^2|\theta\rangle_0 - ({}_0\langle\theta|\hat{n}|\theta\rangle_0)^2\,]^{1/2} \\
=& V^{-1}[{}_0\langle\theta|(a_0^\dagger a_0)^2|\theta\rangle_0 - ({}_0\langle\theta|a_0^\dagger a_0|\theta\rangle_0)^2]^{1/2} \\
=& V^{-1}N^{1/2} = nN^{-1/2} \underset{V\to\infty}{\longrightarrow} 0\,.
\end{aligned}
\tag{2.17}
$$

Thus, with $V \to \infty$, the fluctuations of the density in these states are completely suppressed. What have happened looks as a miracle : as $N \to \infty (V \to \infty)$, we find a continuum set of different states with the same $n = \lim_{V\to\infty} N/V$ and the same energy $E = 0$!

What we have shown is that the Bose - Einstein condensate gives a realization (apparently the simplest one) of the phenomenon of spontaneous symmetry breaking: while the lagrangian density (2.1) is invariant under the transformations (2.9), the vacua $|\theta\rangle_0$ of the system are not (see Eq. (2.11)).

We would like to point out the following characteristic features of the phenomenon:

a) The degeneracy of the vacua $|\theta\rangle_0$ reflects the presence of the "excitations " (created by the operator $a_0^\dagger$) with zero energy. This in turn connected with a specific dispersion law for the ideal Bose gas : $\epsilon(k) = \frac{k^2}{2m}$, and $\epsilon(k) \to 0$ with $k \to 0$. This point is a manifestation of a general phenomenon known as the Goldstone theorem in particle physics: spontaneous breakdown of a continuous symmetry leads to the appearance of an excitation branch with $\epsilon(k) \to 0$ as $k \to 0$ (a gapless mode). The situation in particle physics is in a sense even simpler : since the Poincare group fixes the form of the dispersion law for one particle states as $\epsilon(k) = (k^2 + m^2)^{1/2}$, the gapless mode corresponds to a massless particle with $\epsilon(k) = k$ (see Chapter 3 for a more detail);

b) Why should we consider the vacua $|\theta\rangle_0$ instead the vacuum $|N\rangle$ (2.6) in the infinite volume ? The latter choice would give us a system without spontaneous breakdown of the $U(1)$ symmetry.

The answer to this question is connected with the so called clusterization property of the vacuum. Let us consider a Green's function

$$G^{(n+k)} = \langle 0|T\prod_{i=1}^{n} A_i(x_i)\prod_{j=1}^{k} B_j(y_j)|0\rangle \ , \tag{2.18}$$

where $A_i(x_i)$, $B_j(y_j)$ are some local operators . The clusterization property implies that when $r_{ij}^2 \to \infty \left(r_{ij}^2 = (\mathbf{x}_i - \mathbf{y}_j)^2 \right)$ for all i and j, the Green's function then factorizes as follows:

$$G^{(n+k)} \to \langle 0|T\prod_{i=1}^{n} A_i(x_i)|0\rangle\langle 0|T\prod_{j=1}^{k} B_j(y_j)|0\rangle \,. \tag{2.19}$$

The physical meaning of this property is clear: clusterization implies the absence of instantaneous long-range correlations in the system, so that the dynamics in two distant spatially-separated regions are independent.

Let us now consider the Green's function

$$G^{(2)} = \langle N|T\varphi(x)\varphi^{\dagger}(0)|N\rangle \,. \tag{2.20}$$

The clusterization decomposition would imply that

$$\lim_{\mathbf{x}^2\to\infty} G^{(2)}(x) = \langle N|\varphi(x)|N\rangle\langle N|\varphi^{\dagger}(0)|N\rangle = 0 \tag{2.21}$$

in the infinite volume. However, since

$$\langle N|\varphi(x)|N+1\rangle \neq 0, \ \langle N+1|\varphi^{\dagger}(0)|N\rangle \neq 0 \ , \tag{2.22}$$

we see that $G^{(2)}(x)$ does not vanish as $\mathbf{x}^2 \to \infty$. (We note that in the infinite volume, both the state $|N\rangle$and the state $|N+1\rangle$ have the same density n, therefore we *must* take into account the contribution of the $|N+1\rangle$-state).

Thus the clusterization property does not take place for the $|N\rangle$ - vacuum.

On the other hand, since all vectors from the Fock space F_θ built on the vacuum $|\theta\rangle_0$ are orthogonal to the vectors from $F_{\theta'}$, $\theta' \neq \theta$, the clusterization property does take place for the $|\theta\rangle_0$ - vacua;

c) Although all these statements are strictly valid only at $V = \infty$, the physical picture remains practically the same in a finite but sufficiently large volume (though $1/N$ corrections can be interesting for some purposes, they are practically irrelevant for most physical quantities).

2.4. Superfluidity. Landau Criterion

Superfluidity is the property of a liquid of flowing through capillaries without exhibiting viscosity.

Superfluidity is a quantum phenomenon. In classical physics, interactions of a liquid with capillary walls will always cause the liquid velocity to decrease:

$$E = \frac{Mv^2}{2} \rightarrow E' = \frac{Mv'^2}{2} < E \text{ if } v' < v, \tag{2.23}$$

where $E(E'), v(v')$ are initial (final) energy and velocity of a liquid (with a mass M), respectively. In classical physics, such a process is never forbidden by the energy-momentum conservation.

In quantum physics, as it was shown by Landau, the situation is essentially different [1, 2]. Following Landau, let us consider the flow in a coordinate system moving with the liquid. The key point in the Landau consideration was that the interactions of the liquid with the walls of the capillary (moving with velocity - $\mathbf{v}$ in this frame) cannot initiate movement of the liquid as a whole. The interactions initially lead to the appearance of elementary excitations in the liquid.

An elementary excitation appears with momentum $\mathbf{k}$ and energy $\epsilon(k)$. Therefore the energy E'_0 of the liquid becomes equal to $\epsilon(k)$ and its momentum $\mathbf{P}'_0 = \mathbf{k}$. Transforming the liquid to the coordinate system where the capillary is at rest, one finds:

$$\begin{aligned} E' =& E'_0 + \mathbf{P}'_0\mathbf{v} + \frac{Mv^2}{2} = \epsilon(k) + \mathbf{k}\cdot\mathbf{v} + E, \\ \mathbf{P}' =& \mathbf{k} + M\mathbf{v}\,. \end{aligned} \tag{2.24}$$

Since the energy of the liquid must decrease as a result of the appearance of the elementary excitation, we come to the condition:

$$\epsilon(k) + \mathbf{k}\cdot\mathbf{v} < 0\,. \tag{2.25}$$

Since minimum of the combination $\epsilon(k) + \mathbf{k}\cdot\mathbf{v}$ appears with antiparallel $\mathbf{k}$ and $\mathbf{v}$, we find from Eq. (2.25):

$$v > \epsilon(k)/k\,. \tag{2.26}$$

If this criterion is not upheld at any k, superfluidity takes place (the Landau criterion).

We can now understand impossibility of superfluidity in classical physics in a somewhat different way. As " elementary excitations " in a classical liquid, one can consider small pieces of the liquid with masses $m \ll M$. Then the energy $\epsilon(k)$ is $\epsilon(k) = k^2/2m$. We find that the Landau criterion (2.26) is upheld at any v provided $k < 2mv$, consequently superfluidity is impossible in a classical liquid.

Since the energy of the excitations in an ideal Bose gas is also $\epsilon(k) = k^2/2m$, we conclude that superfluidity is impossible for this case as well.

In the next section, we consider a soluble model (an almost ideal Bose gas) in which superfluidity is realized.

2.5. Superfluidity in an Almost Ideal Bose Gas

We will discuss here the model of an almost ideal Bose gas first considered and solved by Bogolyubov [3, 2].

The hamiltonian of the model takes the form:

$$\hat{H} = \sum_{\mathbf{k}} \frac{k^2}{2m} a^\dagger_{\mathbf{k}} a_{\mathbf{k}} + \frac{U_0}{2V} \sum_{\substack{\mathbf{k}_1, \mathbf{k}_2, \\ \mathbf{k}'_1, \mathbf{k}'_2}} \delta_{(\mathbf{k}_1+\mathbf{k}_2),(\mathbf{k}'_1+\mathbf{k}'_2)} a^\dagger_{\mathbf{k}'_1} a^\dagger_{\mathbf{k}'_2} a_{\mathbf{k}_2} a_{\mathbf{k}_1} \quad . \tag{2.27}$$

One assumes that the hamiltonian describes a weakly interacting gas, i.e. that $a \cdot n^{1/3} \ll 1$, where a is the scattering length (see below).

Because of this assumption, the vacua $|\theta\rangle_0$ of the ideal Bose gas should be good zero approximations for the vacua $|\theta\rangle$ of the hamiltonian (2.27). In particular one can expect that

$$\langle\theta|a_0^\dagger a_0|\theta \simeq {}_0\langle\theta|a_0^\dagger a_0|\theta\rangle_0 = N_0 \sim N \gg \langle\theta|a^\dagger_{\mathbf{k}} a_{\mathbf{k}}|\theta\rangle, k \neq 0 \,. \tag{2.28}$$

The operator a_0 is nearly classical, since while

$$a_0 \sim N^{1/2} e^{i\theta} , \tag{2.29}$$

the commutator $[a_0, a_0^\dagger]$ is

$$[a_0, a_0^\dagger] = 1 \ll N \,. \tag{2.30}$$

The next step is to simplify the hamiltonian by retaining only the terms $(a_0^\dagger)^2(a_0)^2$,

$(a_0^\dagger)^2 a_{\mathbf{k}} a_{-\mathbf{k}}$ and $a_{\mathbf{k}}^\dagger a_{-\mathbf{k}}^\dagger a_0^2$ in it. One can use the constraint

$$\hat{N} = a_0^\dagger a_0 + \sum_{\mathbf{k}\neq 0} a_{\mathbf{k}}^\dagger a_{\mathbf{k}} = N \tag{2.31}$$

to express $a_0^\dagger a_0$ in terms of $a_{\mathbf{k}}^\dagger a_{\mathbf{k}}$ ($\mathbf{k} \neq 0$) in the leading $(a_0^\dagger)^2 (a_0)^2$ term. After this substitution, the remaining a_0 and $(a_0^\dagger)$ are approximated by

$$a_0 = N^{1/2} e^{i\theta}, \ a_0^\dagger = N^{1/2} e^{-i\theta}, \tag{2.32}$$

and the hamiltonian takes the form:

$$\hat{H}_{app} = \sum_{\mathbf{k}} \frac{k^2}{2m} a_{\mathbf{k}}^\dagger a_{\mathbf{k}} + \frac{N^2}{2V} U_0 + \frac{N}{2V} U_0 \sum_{\mathbf{k}\neq 0} \left[e^{2i\theta} a_{\mathbf{k}}^\dagger a_{-\mathbf{k}}^\dagger + e^{-2i\theta} a_{\mathbf{k}} a_{-\mathbf{k}} + 2 a_{\mathbf{k}}^\dagger a_{\mathbf{k}} \right]. \tag{2.33}$$

Following Bogolyubov, we use the transformation

$$a_{\mathbf{k}} = u_k b_{\mathbf{k}} + v_k b_{-\mathbf{k}}^\dagger, a_{\mathbf{k}}^\dagger = u_k^* b_{\mathbf{k}}^\dagger + v_k^* b_{-\mathbf{k}} \tag{2.34}$$

to diagonalize this hamiltonian. The coefficients u_k and v_k satisfy the condition

$$u_k^* u_k - v_k^* v_k = 1 \tag{2.35}$$

in order to retain the canonical commutation relations for $b_{\mathbf{k}}, b_{\mathbf{k}}^\dagger$. Consequently, the transformation (2.34) can be rewritten in the form :

$$a_{\mathbf{k}} = e^{i\theta} \frac{b_{\mathbf{k}} + L_k b_{-\mathbf{k}}^\dagger}{\sqrt{1 - L_k^2}}, a_{\mathbf{k}}^\dagger = e^{-i\theta} \frac{b_{\mathbf{k}}^\dagger + L_k b_{-\mathbf{k}}}{\sqrt{1 - L_k^2}}, \tag{2.36}$$

where the quantity L_k is constructed to eliminate nondiagonal terms like $b_{\mathbf{k}} b_{-\mathbf{k}}$ and $b_{\mathbf{k}}^\dagger b_{-\mathbf{k}}^\dagger$ in the hamiltonian:

$$L_k = \frac{1}{mu^2} \left[\epsilon(k) - \frac{k^2}{2m} - mu^2 \right] \tag{2.37}$$

with

$$\begin{aligned} \epsilon(k) &= \left[u^2 k^2 + \left(\frac{k^2}{2m} \right)^2 \right]^{1/2}, \\ u &= \left(\frac{4\pi a N}{m^2 V} \right)^{1/2}. \end{aligned} \tag{2.38}$$

The scattering length a is $a = \frac{m}{4\pi} U_0$.

The hamiltonian now takes the form:

$$\hat{H}_{app} = E_0 + \sum_{\mathbf{k}\neq 0} \epsilon\,(k) b_{\mathbf{k}}^{\dagger} b_{\mathbf{k}} \tag{2.39}$$

with $\epsilon(k)$ given by Eq.(2.38) and

$$E_0 = \frac{1}{2} N m u^2 + \frac{1}{2} \sum_{\mathbf{k}\neq 0} [\epsilon(k) - \frac{k^2}{2m} - m u^2 + \frac{m^3 u^4}{k^2}] \ . \tag{2.40}$$

Thus $\epsilon(k)$ determines the dispersion law for the excitations in this system. The crucial point is that

$$\epsilon(k) \simeq uk \tag{2.41}$$

(and not $\epsilon \sim k^2$) as $k \to 0$. Comparing this relation with the Landau criterion (2.26), one finds that when the velocity v is smaller than u, superfluidity occurs in the system.

Let us consider the content of the vacua $|\theta\rangle$ in the terms of the initial particles connected with the operators $a_{\mathbf{k}}, a_{\mathbf{k}}^{\dagger}$.

Since $b_{\mathbf{k}}|\theta\rangle = 0$, the relation (2.36) implies that

$$N_{\mathbf{k}} \equiv \langle\theta|a_{\mathbf{k}}^{\dagger} a_{\mathbf{k}}|\theta\rangle = \frac{L_k^2}{1 - L_k^2}; k \neq 0 \ . \tag{2.42}$$

In particular the density $n_{\mathbf{k}} = N_{\mathbf{k}}/V$ goes to zero as $V \to \infty$. However the density $n_0 = N_0/V$ is finite at $V = \infty$. In fact we have

$$\begin{aligned} N_0 =& N - \frac{V}{(2\pi)^3} \int d^3k N_{\mathbf{k}} \\ =& N \left[1 - \frac{8}{3} \left(\frac{N a^3}{\pi V} \right)^{1/2} \right] \ . \end{aligned} \tag{2.43}$$

By using the same approach as in the section 1.4, we find the explicit representation of the vacua $|\theta\rangle$:

$$|\theta\rangle = \exp(-N/2) \exp(N_0^{1/2} e^{i\theta} a_0^{\dagger}) \prod_{\mathbf{k}\neq 0} (1 - L_k^2)^{1/4} \cdot \exp\left(\frac{1}{2} e^{2i\theta} L_k a_{\mathbf{k}}^{\dagger} a_{-\mathbf{k}}^{\dagger}\right) |0_{bare}\rangle \tag{2.44}$$

(compare with Eq. (1.51) and Eq.(2.10)). Under the $U(1)$ transformations (2.9), the vacua transform as

$$|\theta\rangle \to U_{\theta'}|\theta\rangle = |\theta + \theta'\rangle\,. \tag{2.45}$$

Therefore the spontaneous breakdown of the $U(1)$ symmetry takes place in the system. Since U_θ commutes with the hamiltonian, the vacua $|\theta\rangle$ have the same energy. The degeneracy of the vacuum in this case is connected with the presence of a gapless mode with $\epsilon(k) \simeq uk$ at small k (phonon excitations).

It is convenient to introduce a wave function for the condensate $\varphi_c(\mathbf{x}, t)$. In the ground state,

$$\varphi_c \equiv \langle\theta|\varphi(x)|\theta\rangle = n_0^{1/2} e^{i\theta} \tag{2.46}$$

is independent of $\mathbf{x}$ and t. In the general case (for a moving liquid, for example), φ_c can depend on $\mathbf{x}$ and t : $\varphi_c(\mathbf{x}, t) = n_0^{1/2}(\mathbf{x}, t)e^{i\theta(\mathbf{x},t)}$. The current density of the superfluid condensate is

$$j_c^i = \frac{i}{2m}\left[\varphi_c \partial^i \varphi_c^* - \varphi_c^* \partial^i \varphi_c\right] = \frac{1}{m} n_0 \partial^i \theta\,. \tag{2.47}$$

2.6. Superconductivity. General Remarks

The phenomenon of superconductivity can be considered as superfluidity of a charged fermion gas (or liquid). Therefore the Landau criterion (2.26) of superfluidity is applicable to superconductivity as well.

In an ideal Fermi gas, with the dispersion law $\epsilon(p) = \frac{p^2}{2m}$, superconductivity is impossible, as is evident from this criterion. Therefore the phenomenon of superconductivity implies an essential alteration of the form of the spectrum in a superfluid Fermi gas.

A theory of superconductivity was elaborated by Bardeen, Cooper and Schrieffer (the BCS theory)[4]. The key physical phenomenon underlying the BCS theory is the creation of an energy gap near the Fermi surface in the spectrum of one fermion excitations. The gap appears as a result of the interactions of electrons with phonons.

We emphasize however that although the presence of the gap implies the validity of the Landau criterion for fermion excitations (if the velocity v of the fermion gas is sufficiently low), it itself is not sufficient to realize superconductivity. In fact, an energy gap is present in the electron spectrum of an insulator in which there is no even electrical conductivity ! The essence of superconductivity is that the generation of the gap in superconductors is related to the spontaneous (dynamical) breakdown of the $U(1)$ gauge symmetry that generates the superconductivity current analogous to the superfluid current (2.47).

Our aim here is to discuss superconductivity from the field theoretical point of view in order to exhibit its deep connection with the phenomenon of dynamical generation of fermion masses in particle physics. In particular, we will show that the BCS theory can be regarded as a theory of dynamical generation of a nonrelativistic Majorana mass for electrons in metals.

2.7. Ideal Fermi Gas. Fermi Gas with the Majorana Mass

The lagrangian density of an ideal Fermi gas is

$$\mathcal{L} = i\psi^\dagger \frac{\partial}{\partial t}\psi + \psi^\dagger \left(\frac{\Delta}{2m} + \mu\right)\psi \, , \tag{2.48}$$

where ψ is a two component fermion field, and μ is the chemical potential which is equal to the energy on the Fermi surface :

$$\mu = \frac{p_F^2}{2m} \, . \tag{2.49}$$

The Fermi momentum p_F is expressed through the gas density [2] :

$$n = p_F^3/3\pi^2 \, . \tag{2.50}$$

The lagrangian density (2.48) implies the following equation of motion :

$$i\frac{\partial\psi}{\partial t} + \left(\frac{\Delta}{2m} + \mu\right)\psi = 0 \, . \tag{2.51}$$

Its general solution takes the form :

$$\psi = \frac{1}{(2\pi)^3}\sum_{s=\pm 1}\int d^3p e^{-i(\epsilon(p)t - \mathbf{p}\cdot\mathbf{x})} a_{\mathbf{p}}(s) v(s), \tag{2.52}$$

where $\epsilon(p) = \frac{p^2}{2m} - \mu$, and the spinors $v(s)$ are

$$v(+1) = \begin{pmatrix}1\\0\end{pmatrix}, \; v(-1) = \begin{pmatrix}0\\1\end{pmatrix} \, . \tag{2.53}$$

The ground state of an ideal Fermi gas is the state in which all levels with $|\mathbf{p}| \leq p_F$

are filled by fermions. Let us define $E_p \equiv |\epsilon(p)|$, i.e. $E_p = -\epsilon(p)$ at $|\mathbf{p}| \leq p_F$, and $E_p = \epsilon(p)$ at $|\mathbf{p}| > p_F$. Let us also define $b_{\mathbf{p}}^{\dagger}(s) \equiv a_{-\mathbf{p}}(-s)$ at $|\mathbf{p}| \leq p_F$. Then the general solution (2.52) can be rewritten in a form similar to that of the general solution of the Dirac equation (see Eq.(1.31)) :

$$\psi = \frac{1}{(2\pi)^3} \sum_{s=\pm 1} \int d^3p \Big[e^{-i(E_p t - \mathbf{p}\cdot\mathbf{x})} \theta(|\mathbf{p}| - p_F) a_{\mathbf{p}}(s) v(s) \\ + e^{i(E_p t - \mathbf{p}\cdot\mathbf{x})} \theta(p_F - |\mathbf{p}|) b_{\mathbf{p}}^{\dagger}(s) v(-s) \Big] . \tag{2.54}$$

The ground state satisfies the condition :

$$a_{\mathbf{p}}(s)|0\rangle = b_{\mathbf{p}}(s)|0\rangle = 0 . \tag{2.55}$$

The operators $a_{\mathbf{p}}^{\dagger}(s)$ and $b_{\mathbf{p}}^{\dagger}(s)$ create the fermion (electron) and hole excitations, respectively.

Let us now alter the lagrangian density (2.48) :

$$\tilde{\mathcal{L}} = i\psi^{\dagger} \frac{\partial}{\partial t} \psi + \psi^{\dagger} \left(\frac{\Delta}{2m} + \mu \right) \psi + \frac{M^*}{2} \psi_{\alpha} \epsilon_{\alpha\beta} \psi_{\beta} + \frac{M}{2} \psi_{\alpha}^{\dagger} \epsilon_{\alpha\beta} \psi_{\beta}^{\dagger} . \tag{2.56}$$

The new term here is a Majorana mass term (compare with Eq.(1.29)). It is invariant with respect to the $SO(3)$ (rotational) group, although it violates the conservation of the fermion number connected with the $U(1)$ transformations :

$$\psi \to e^{-i\theta} \psi, \ \psi^{\dagger} \to e^{i\theta} \psi^{\dagger} . \tag{2.57}$$

As we shall see in the next section, this model is very useful in studying the BCS model of superconductivity.

We introduce a bispinor

$$\psi_M \equiv \begin{pmatrix} \psi \\ \psi^{\dagger} \end{pmatrix} = \frac{1}{(2\pi)^4} \int d\omega d^3p e^{-i(\omega t - \mathbf{p}\cdot\mathbf{x})} . \begin{pmatrix} u(\omega, \mathbf{p}) \\ u^{\dagger}(-\omega, -\mathbf{p}) \end{pmatrix} . \tag{2.58}$$

Then the equations of motion corresponding to the lagrangian density (2.56) take the following form in the momentum space :

$$\omega u_\alpha(\omega, \mathbf{p}) + (\mu - \frac{p^2}{2m})u_\alpha(\omega, \mathbf{p}) + M\epsilon_{\alpha\beta}u^\dagger_\beta(-\omega, -\mathbf{p}) = 0, \tag{2.59}$$

$$-\omega u^\dagger_\alpha(-\omega, -\mathbf{p}) + (\mu - \frac{p^2}{2m})u^\dagger_\alpha(-\omega, -\mathbf{p}) + M^*\epsilon_{\alpha\beta}u_\beta(\omega, \mathbf{p}) = 0 \cdot \tag{2.60}$$

The dispersion law implied by there equations is

$$\omega = \pm E_p = \pm(\epsilon^2(p) + |M|^2)^{1/2} \tag{2.61}$$

with $|M| = (MM^*)^{1/2}$ and $\epsilon(p) = \frac{p^2}{2m} - \mu$.

After rather long but routine algebraic manipulations, we obtain the following general solution for ψ_M :

$$\psi_M = \frac{1}{(2\pi)^3} \sum_{s=\pm 1} \int d^3p\{ e^{-i(E_pt - \mathbf{p}\cdot\mathbf{x})} [\, \theta(|\mathbf{p}| - p_F)c_\mathbf{p}(s)u_p(s) + \theta(p_F - |\mathbf{p}|)d_\mathbf{p}(s)\tilde{v}_p(s) \,]$$
$$+ e^{i(E_pt - \mathbf{p}\cdot\mathbf{x})} \Big[\, \theta(|\mathbf{p}| - p_F)c^\dagger_\mathbf{p}(s)\tilde{u}_p(s) + \theta(p_F - |\mathbf{p}|)d^\dagger_\mathbf{p}(s)v_p(s) \,\Big] \} , \tag{2.62}$$

where the four component bispinors are :

$$u_p(s) = \left(\frac{E_p + \epsilon(p)}{2E_p}\right)^{1/2} \begin{pmatrix} v(s) \\ \frac{(-1)^{\frac{s+1}{2}} M^*}{\epsilon(p)+E_p} v(-s) \end{pmatrix}, \tag{2.63}$$

$$\tilde{u}_p(s) = \left(\frac{E_p + \epsilon(p)}{2E_p}\right)^{1/2} \begin{pmatrix} \frac{(-1)^{\frac{s+1}{2}} M}{\epsilon(p)+E_p} v(-s) \\ v(s) \end{pmatrix}, \tag{2.64}$$

$$v_p(s) = \left(\frac{E_p - \epsilon(p)}{2E_p}\right)^{1/2} \begin{pmatrix} v(-s) \\ \frac{(-1)^{\frac{1-s}{2}} M^*}{\epsilon(p)-E_p} v(s) \end{pmatrix}, \tag{2.65}$$

$$\tilde{v}_p(s) = \left(\frac{E_p - \epsilon(p)}{2E_p}\right)^{1/2} \begin{pmatrix} \frac{(-1)^{\frac{1-s}{2}} M}{\epsilon(p)-E_p} v(s) \\ v(-s) \end{pmatrix}. \tag{2.66}$$

The ground state satisfies the condition :

$$c_\mathbf{p}(s)|0_M\rangle = d_\mathbf{p}(s)|0_M\rangle = 0 \, . \tag{2.67}$$

The operators $c^\dagger_\mathbf{p}(s), d^\dagger_\mathbf{p}(s)$ create rather unusual excitations which are a coherent mixture of electron and hole excitations of an ideal Fermi gas. Fermion number becomes a bad quantum number in the system.

Let us consider in more detail the structure of the ground state $|0_M\rangle$ in the terms of the dynamics of the usual ideal Fermi gas. We shall see that $|0_M\rangle$ coincides in fact with the ground state in the BCS model.

To obtain the structure $|0_M\rangle$, we will follow the methodology of the analysis of the ground state of the massive Dirac field (see Sec.1.4). As in that case, we utilize the boundary condition

$$\psi_M(0, \mathbf{x}) = \psi_{M=0}(0, \mathbf{x}) \ , \tag{2.68}$$

where $\psi_{M=0} = \begin{pmatrix} \psi \\ \psi^\dagger \end{pmatrix}$ with ψ being given by Eq. (2.54) .

Eqs. (2.54), (2.62) and (2.68) yield the relations:

$$\begin{aligned} d^\dagger_{\mathbf{p}}(s) &= \alpha(p) b^\dagger_{\mathbf{p}}(s) + e^{2i\theta}\beta_s(p) b_{-\mathbf{p}}(-s) \ , \\ d_{\mathbf{p}}(s) &= \alpha(p) b_{\mathbf{p}}(s) + e^{-2i\theta}\beta_s(p) b^\dagger_{-\mathbf{p}}(-s) \ , \\ c^\dagger_{\mathbf{p}}(s) &= \tilde{\alpha}(p) a^\dagger_{\mathbf{p}}(s) + e^{-2i\theta}\tilde{\beta}_s(p) a_{-\mathbf{p}}(-s) \ , \\ c_{\mathbf{p}}(s) &= \tilde{\alpha}(p) a_{\mathbf{p}}(s) + e^{2i\theta}\tilde{\beta}_s(p) a^\dagger_{-\mathbf{p}}(-s) \ , \end{aligned} \tag{2.69}$$

where $M = |M|e^{2i\theta}$ and where

$$\begin{aligned} \alpha(p) &= \left[\frac{1}{2}\left(1 - \frac{\epsilon(p)}{E_p}\right)\right]^{1/2} , \\ \beta_s(p) &= (-1)^{\frac{1-s}{2}} \left[\frac{1}{2}\left(1 + \frac{\epsilon(p)}{E_P}\right)\right]^{1/2} , \\ \tilde{\alpha}(p) &= \left[\frac{1}{2}\left(1 + \frac{\epsilon(p)}{E_p}\right)\right]^{1/2} , \\ \tilde{\beta}_s(p) &= (-1)^{\frac{1+s}{2}} \left[\frac{1}{2}\left(1 - \frac{\epsilon(p)}{E_p}\right)\right]^{1/2} . \end{aligned} \tag{2.70}$$

We note that

$$\alpha^2(p) + \beta_s^2(p) = 1, \tilde{\alpha}^2(p) + \tilde{\beta}_s^2 = 1 \ .$$

Eqs. (2.67) and (2.69) imply that

$$|0_M\rangle = \prod'_{\substack{|\mathbf{p}| \leq p_F \\ s=\pm 1}} \left(\alpha(p) - e^{-2i\theta}\beta_s(p) b^\dagger_{\mathbf{p}}(s) b^\dagger_{-\mathbf{p}}(-s)\right) \cdot \prod'_{\substack{|\mathbf{p}| > p_F \\ s=\pm 1}} \left(\tilde{\alpha}(p) - e^{2i\theta}\tilde{\beta}_s(p) a^\dagger_{\mathbf{p}}(s) a^\dagger_{-\mathbf{p}}(-s)\right) |0\rangle \,, \tag{2.71}$$

where $|0\rangle = |0_{M=0}\rangle$ is the ground state of the ideal Fermi gas. The symbol Π' means that in order to avoid the double counting, we consider here the vectors $\mathbf{n} = \mathbf{P}/|\mathbf{P}|$ lying in the upper semisphere ($n_z > 0$) only.

Thus $|0_M\rangle$ is a superposition of states with differing eigenvalues for fermion number.

Let us find the average of the fermion number operator

$$\hat{N} = \sum_{\substack{|\mathbf{p}| \leq p_F \\ s=\pm 1}} b^\dagger_{\mathbf{p}}(s) b_{\mathbf{p}}(s) + \sum_{\substack{|\mathbf{p}| > p_F \\ s=\pm 1}} a^\dagger_{\mathbf{p}}(s) a_{\mathbf{p}}(s) \tag{2.72}$$

in $|0_M\rangle$:

$$\begin{aligned} \bar{N} =& \langle 0_M|\hat{N}|0_M\rangle = \sum_{\substack{|\mathbf{p}| \leq p_F \\ s=\pm 1}} \beta^2_s(p) + \sum_{\substack{|\mathbf{p}| > p_F \\ s=\pm 1}} \tilde{\beta}^2_s(p) \\ =& \frac{V}{(2\pi)^2} \int p^2 dp \left[1 + \theta(p_F - |\mathbf{p}|)\frac{\epsilon(p)}{E_p} - \theta(|\mathbf{p}| - p_F)\frac{\epsilon(p)}{E_p}\right] \,. \end{aligned} \tag{2.73}$$

We see that $\bar{N}$ goes to infinity with $V \to \infty$, so that the state $|0_M\rangle$ lies outside of the Fock space of the ideal Fermi gas $F_M|_{M=0}$ in the infinite volume. The same is true for any state with a finite number of c and d excitations possessing the Majorana mass M. This of course implies that the Fock spaces $F_{M=0}$ and F_M with $M \neq 0$ correspond to two inequivalent representations of the canonical commutation relations for the field $\psi_M(0, \mathbf{x}) = \psi_{M=0}(0, \mathbf{x})$.

Let us now calculate the fermion propagator

$$G(x) = -i\langle 0_M|T\left(\psi_M(x)\psi_M^\dagger(0)\right)|0_M\rangle \ . \tag{2.74}$$

We rewrite the equation for ψ_M in the matrix form :

$$i\Gamma_0\frac{\partial}{\partial t}\psi_M + \left(\frac{\Delta}{2m}+\mu\right)\psi_M + |M|\Gamma_2\psi_M = 0 \ , \tag{2.75}$$

where

$$\Gamma_0 = \begin{pmatrix} I & 0 \\ 0 & -I \end{pmatrix} , \quad \Gamma_2 = \begin{pmatrix} 0 & e^{2i\theta}i\sigma_2 \\ e^{-2i\theta}i\sigma_2 & 0 \end{pmatrix} . \tag{2.76}$$

The equation for the propagator is

$$\left[i\Gamma_0\frac{\partial}{\partial t} + \left(\frac{\Delta}{2m}+\mu\right) + |M|\Gamma_2\right]G(x) = \Gamma_0\delta^4(x) \ , \tag{2.77}$$

or

$$\left[\Gamma_0\omega + \left(\mu - \frac{p^2}{2m}\right) + |M|\Gamma_2\right]G(\omega,\mathbf{p}) = \Gamma_0 \tag{2.78}$$

with

$$G(\omega,\mathbf{p}) = \int d^4x e^{i(\omega t - \mathbf{p}\cdot\mathbf{x})}G(x) \ . \tag{2.79}$$

The solution of the equation (2.78) is

$$G(\omega,\mathbf{p}) = \frac{\Gamma_0\omega + \left(\frac{p^2}{2m}-\mu\right) + |M|\Gamma_2}{\omega^2 - |M|^2 - \left(\frac{p^2}{2m}-\mu\right)^2 + i0}\Gamma_0 \ . \tag{2.80}$$

Rewriting the propagator G(x) as

$$\begin{aligned} G_{\alpha\beta}(x) &= -\,i\begin{pmatrix} \langle 0_M|T\left(\psi_\alpha(x)\psi_\beta^\dagger(0)\right)|0_M\rangle & \langle 0_M|T\left(\psi_\alpha(x)\psi_\beta(0)\right)|0_M\rangle \\ \langle 0_M|T\left(\psi_\alpha^\dagger(x)\psi_\beta^\dagger(0)\right)|0_M\rangle & \langle 0_M|T\left(\psi_\alpha^\dagger(x)\psi_\beta(0)\right)|0_M\rangle \end{pmatrix} \\ &\equiv \begin{pmatrix} D_{\alpha\beta}(x) & F_{\alpha\beta}(x) \\ F_{\alpha\beta}^\dagger(x) & D_{\alpha\beta}^\dagger(x) \end{pmatrix} , \end{aligned} \tag{2.81}$$

we find that

$$G_{\alpha\beta}(\omega, \mathbf{p}) = \begin{pmatrix} D_{\alpha\beta}(\omega, \mathbf{p}) & F_{\alpha\beta}(\omega, \mathbf{p}) \\ F^{\dagger}_{\alpha\beta}(-\omega, -\mathbf{p}) & D^{\dagger}_{\alpha\beta}(-\omega, -\mathbf{p}) \end{pmatrix} \tag{2.82}$$

with

$$D_{\alpha\beta}(\omega, \mathbf{p}) = \frac{\left[\omega + \left(\frac{p^2}{2m} - \mu\right)\right]\delta_{\alpha\beta}}{\omega^2 - |M|^2 - \left(\frac{p^2}{2m} - \mu\right)^2 + i0} \;, \tag{2.83}$$

$$\begin{aligned} F_{\alpha\beta}(\omega, \mathbf{p}) = & \frac{-M\epsilon_{\alpha\beta}}{\omega^2 - |M|^2 - \left(\frac{p^2}{2m} - \mu\right)^2 + i0} \\ \equiv & F(\omega, \mathbf{p})\epsilon_{\alpha\beta} \,. \end{aligned} \tag{2.84}$$

The function $F_{\alpha\beta}$ describes the "anomalous" part of the propagator connected with the Majorana mass M and, consequently, with the violation of the fermion number conservation. As we shall see in the next section, the fermion propagator in the BCS model has exactly the same structure as $G_{\alpha\beta}(\omega, \mathbf{p})$ in Eq. (2.82).

2.8. BCS Model of Superconductivity

We are ready now to discuss the BCS model. We will do this comparing BCS dynamics with those of the Fermi gas with a Majorana mass.

The BCS lagrangian density takes the form

$$\mathcal{L}_{BCS} = i\psi^{\dagger}\frac{\partial}{\partial t}\psi + \psi^{\dagger}\left(\frac{\Delta}{2m} + \mu\right)\psi + \frac{g}{2}\psi^{\dagger}_{\beta}(x)\psi^{\dagger}_{\alpha}(x)\psi_{\alpha}(x)\psi_{\beta}(x) \,. \tag{2.85}$$

Thus, instead of a Majorana term, a four-fermion term appears in L_{BCS}. This interaction, unlike the Majorana mass term, is invariant under the $U(1)$ transformations (2.57).

The four-fermion term of (2.85) describes the interactions of electrons with phonons near the Fermi surface. The BCS model can be considered as an effective theory one obtains after integrating out the phonon degrees of freedom in the path integral of a more fundamental theory.

The equation of motion in the BCS model is

$$i\frac{\partial}{\partial t}\psi_\alpha(x) + \left(\frac{\Delta}{2m} + \mu\right)\psi_\alpha(x) + g\psi_\beta^\dagger(x)\psi_\beta(x)\psi_\alpha(x) = 0 \ , \tag{2.86}$$

a rather complicated nonlinear equation. The BCS approximation to its solution occurs upon making the following substitution into (2.86):

$$\begin{aligned} \psi_\beta(x)\psi_\alpha(x) &\longrightarrow \langle 0|\psi_\beta(x)\psi_\alpha(x)|0\rangle \\ &= \lim_{x\to y}\langle 0|T\left(\psi_\beta(x)\psi_\alpha(y)\right)|0\rangle \equiv iF(0)\epsilon_{\beta\alpha} \ , \end{aligned} \tag{2.87}$$

where $|0\rangle$ is the ground state of the model, and $F(x-y)\epsilon_{\beta\alpha}$ is the anomalous part of the fermion propagator (compare with Eq.(2.81)).

This substitution reflects the essence of the dynamics in the BCS model. The composite operator $\psi_\beta(x)\psi_\alpha(x)$ describes a condensate of electrons. The substitution corresponds to the approximation in which this operator is treated as the c-number $iF(0)\epsilon_{\alpha\beta}$, i.e. it is assumed that its classical part dominates. One can see a striking similarity of the BCS approximation with the approximation used before in the problem of a nonideal Bose gas (see Sec.2.5). In that case the condensate was described by the elementary field $\varphi(x)$; the condensate led to the spontaneous breakdown of particle number conservation. In the BCS model the condensate described by the composite operator $\psi_\beta(x)\psi_\alpha(x)$ leads to the spontaneous breakdown of the $U(1)$ symmetry connected with the conservation of the electric charge.

With this substitution, equation (2.86) simplifies to become the equation

$$i\frac{\partial}{\partial t}\psi_\alpha(x) + \left(\frac{\Delta}{2m} + \mu\right)\psi_\alpha(x) + M_d\epsilon_{\alpha\gamma}\psi_\gamma^\dagger(x) = 0 \ , \tag{2.88}$$

where $M_d \equiv -igF(0)$ can be considered as a *dynamical* Majorana mass. Indeed, if we replace M_d with M , this equation exactly coincides with that of the Fermi gas with a Majorana mass. Therefore we can use freely the results of the preceding section.

In particular the fermion propagator (see Eqs. (2.80) and (2.84)) is given by

$$G(\omega, \mathbf{p}) = \frac{\Gamma_0\omega + \epsilon(p) + |M_d|\Gamma_2}{\omega^2 - |M_d|^2 - \epsilon^2(p) + i0}\Gamma_0 \ , \tag{2.89}$$

$$F(\omega,\mathbf{p}) = \frac{-M_d}{\omega^2 - |M_d|^2 - \epsilon^2(p) + i0} \ . \tag{2.90}$$

However we now have an additional constraint for M_d :

$$M_d = -igF(x)|_{x=0} = \frac{g}{(2\pi)^4 i}\int d\omega d^3pF(\omega,\mathbf{p}) \ . \tag{2.91}$$

Substituting the expression (2.90) for $F(\omega,\mathbf{p})$ in the equation (2.91), we find

$$1 = \frac{g}{2(2\pi)^3}\int d^3p(\epsilon^2(p) + |M_d|^2)^{-1/2} \ . \tag{2.92}$$

Thus, M_d is a dynamical quantity and not a free parameter like the Majorana mass M of the preceding section.

To solve the equation (2.92), we need to know the following additional information. The BCS model describes the interactions of electrons near the Fermi surface. It means that the quantity $\bar{E} \equiv \frac{p^2}{2m}$ is in the interval

$$|\bar{E} - \bar{E}_F| \leq \frac{\omega_D}{2} \ , \tag{2.93}$$

where $\bar{E}_F = \frac{p_F^2}{2m}$ and ω_D is the Debye frequency which is an important (and measurable) property of crystals. Thus ω_D plays the role of an ultraviolet cutoff in the problem.

Therefore equation (2.92) becomes

$$1 = \frac{2\pi gm}{(2\pi)^3}\int_{\bar{E}_F-\frac{\omega_D}{2}}^{\bar{E}_F+\frac{\omega_D}{2}} d\bar{E}\frac{(2m\bar{E})^{1/2}}{\sqrt{(\bar{E}-\bar{E}_F)^2 + |M_d|^2}} \ . \tag{2.94}$$

The solution of this equation is

$$|M_d| \simeq \omega_D \exp(-2/g\nu_F) \ , \tag{2.95}$$

where ν_F is the energy density of the number of states of electrons on the Fermi surface $\left(dN_F = \frac{8\pi V p_F^2}{(2\pi)^3}dp\right)$,

$$\nu_F \equiv V^{-1}\frac{dN_F}{d\bar{E}}\bigg|_{\bar{E}=\bar{E}_F} = mp_F/\pi^2 \ . \tag{2.96}$$

Thus because of the presence of the Fermi surface ($\nu_F > 0$), the mass (gap) M_d appears for any positive value of the coupling constant g($g > 0$ corresponds to the attraction interactions between electrons).

A characteristic point of the expression (2.95) for $|M_d|$ is an essential singularity at $g = 0$. Another characteristic point is that equation (2.94) determines the absolute value but not the phase of M_d $\left(M_d = |M_d|e^{2i\theta}\right)$. This is a signature of the vacuum degeneracy in the BCS model associated in turn with the spontaneous breakdown of the $U(1)$ gauge symmetry. We shall discuss this point in more detail in the next section.

2.9. Dynamical $U(1)$ Gauge Symmetry Breaking in the BCS Model

Dynamical symmetry breaking is a special case of spontaneous symmetry breaking in which the condensate (order parameter) is given by the vacuum average of a composite (and not elementary) operator.

In the case of the BCS model, such an operator is $\psi_\beta(x)\psi_\alpha(x)$, and the order parameter is

$$\langle 0|\psi_\beta(x)\psi_\alpha(x)|0\rangle = iF(0)\epsilon_{\beta\alpha} = -g^{-1}M_d\epsilon_{\beta\alpha} \; . \tag{2.97}$$

It is clear that $M_d \neq 0$ implies noninvariance of the vacuum with respect to the gauge $U(1)$ transformations (2.57). Indeed, if $U_\theta|0\rangle = |0\rangle$, then

$$\begin{aligned}\langle 0|\psi_\beta(x)\psi_\alpha(x)|0\rangle =&\langle 0|U_\theta^{-1}\psi_\beta(x)U_\theta U_\theta^{-1}\psi_\alpha(x)U_\theta|0\rangle \\ =&e^{-2i\theta}\langle 0|\psi_\beta(x)\psi_\alpha(x)|0\rangle\end{aligned} \tag{2.98}$$

in which case M_d is necessarily zero.

Let us consider the structure of the BCS ground state.

First of all, let us recall that in the case of the Fermi gas with a Majorana mass $M = |M|e^{2i\theta}$, the vacuum state is given by the expression (2.71). Since in the BCS approximation, the equation of motion (2.88) coincides with that of the Fermi gas with a Majorana mass $M_d = |M_d|e^{2i\theta}$, we can write immediately the ground state :

$$\begin{aligned}|\theta\rangle = &\prod_{\substack{|\mathbf{p}| \leq p_F \\ s=\pm 1}}{}' \left(\alpha(p) - e^{-2i\theta}\beta_s(p)b^\dagger_{\mathbf{p}}(s)b^\dagger_{-\mathbf{p}}(-s)\right) \\ &\cdot \prod_{\substack{|\mathbf{p}| > p_F \\ s=\pm 1}}{}' \left(\tilde\alpha(p) - e^{2i\theta}\tilde\beta_s(p)a^\dagger_{\mathbf{p}}(s)a^\dagger_{-\mathbf{p}}(-s)\right)|0\rangle \; ,\end{aligned} \tag{2.99}$$

with $\alpha(p), \beta_s(p), \tilde\alpha(p), \tilde\beta_s(p)$ are given in Eq.(2.70).

However there is a crucial difference between the dynamics of these two cases. In the case of the Fermi gas, there is one-to-one correspondence between the vacua $|\theta\rangle$ and the lagrangians with different $M = |M|e^{2i\theta}$. In the case of the BCS model, the equations of motion do not determine the phase θ ; therefore all vacua $|\theta\rangle$ correspond to the same lagrangian density (2.85).

Under the transformations (2.57), the vacuum $|\theta\rangle$ transforms as

$$U_{\theta'}|\theta\rangle = |\theta + \theta'\rangle \ . \tag{2.100}$$

Since the operators U_θ commute with the BCS hamiltonian, Eq.(2.100) implies spontaneous breakdown of the $U(1)$ gauge symmetry and the degeneracy of the vacua $|\theta\rangle$.

Although in the present approximation, the equation for the fermion propagator in the BCS model coincides with that in a Fermi gas with a Majorana mass, the physical content of the two cases is essentially different. In the Fermi gas, the spectrum of the hamiltonian is very simple (the excitations created by the operators $c^\dagger_{\mathbf{p}}(s), d^\dagger_{\mathbf{p}}(s)$ with $E_p = \left(\epsilon^2(p) + |M|^2\right)^{1/2}$). By contrast, the vacuum degeneracy in the BCS model implies that these excitations are accompanied by some collective excitations, such as those in the superfluid Bose gas.

The study of the spectrum of excitations in a system with dynamical symmetry breaking is a rather complicated problem equivalent to that of bound states (electron - electron bound states in the case of the BCS model)(see Sec.9.5). In the next section, we will use a rather heuristic approach to address this problem.

2.10. The Ginzburg-Landau Effective Action. The Meissner Effect

The characteristic quantity in superconductivity is the condensate

$$\langle\theta|\psi_\alpha(x)\psi_\beta(x)|\theta\rangle = i\epsilon_{\alpha\beta}F(x,x) \ . \tag{2.101}$$

In the absence of external fields, $F(x,x)$ is independent of x ($F(x,x) = F(0)$); in general, however, it depends on x. This function describes the dynamics of superconducting electron pairs (Cooper pairs) in a superconductor. In particular, collective excitations can be described as small variations (fluctuations) of the condensate near the ground state with $F(x,x) = \bar{F}(0)$.

The form of the effective action describing such low frequency fluctuations was established by Ginzburg and Landau on the basis of the general Landau theory of the second order phase transitions [5,2]. In the case of superconductivity, the phase transition is a transition in temperature. At $T < T_c$, the wave function

$$\varphi_c(x) \equiv iF(x,x) \tag{2.102}$$

is nonzero, and therefore the $U(1)$ gauge symmetry is spontaneously broken. At $T > T_c$, the function $\varphi_c(x) = 0$, and the symmetry is restored.

In the absence of an external electromagnetic field, the Ginzburg-Landau (GL) effective action at zero temperature takes the form:

$$S_{GL} = c\int dtd^3x \left[d\partial_t\varphi_c^\dagger\partial_t\varphi_c - \frac{1}{4m}\partial_i\varphi_c^\dagger\partial_i\varphi_c - a\varphi_c^\dagger\varphi_c - \frac{1}{2}b(\varphi_c^\dagger\varphi_c)^2\right] \ , \tag{2.103}$$

where a, b, c, d are some parameters; the coefficient $\frac{1}{4m}$ is fixed so that the mass of a Cooper pair is $2m$.

Because of the action S_{GL} describes low energy fluctuations, all terms with higher derivatives are neglected (there are some general arguments in support of the absence of terms with one derivative in S_{GL} [2]). The Landau theory implies that $a < 0$ and $b > 0$ at $T < T_c$; this provides the spontaneous breakdown of the $U(1)$ symmetry.

In principle, S_{GL} can be derived from the BCS action [6] (see Sec. 11.3), however some important information can be obtained directly from S_{GL}.

The value of $\varphi_c^\dagger\varphi_c$ corresponding to the ground state ($\varphi_c^\dagger\varphi_c = \bar\varphi_c^\dagger\bar\varphi_c$) is independent of t and x, and therefore it is given by the minimum of the potential

$$V_{GL} = c\int d^3x \left[a(\varphi_c^\dagger\varphi_c) + \frac{1}{2}b(\varphi_c^\dagger\varphi_c)^2\right] \ . \tag{2.104}$$

We find

$$\bar\varphi_c^\dagger\bar\varphi_c = -a/b > 0 \ . \tag{2.105}$$

Let us rewrite $\varphi_c(x)$ as

$$\varphi_c(x) = n_s^{1/2}(x)\exp\left(2i\theta(x)\right) \ , \tag{2.106}$$

where $n_s(x)$ can be treated as the density of superconducting electron pairs.

The dispersion law for the excitations connected with the fluctuations of the density $n_s(x)$ and the phase $\theta(x)$ can be found from S_{GL}. Since the phase $\theta(x)$ appears only in the terms with derivatives, it is associated with a gapless mode in which

$$\epsilon_\theta(k) \sim k \tag{2.107}$$

as $k \to 0$. On the other hand, there is a gap in the spectrum of the excitations connected with $n_s(x)$:

$$\epsilon_{n_s}(k) \to \mu \tag{2.108}$$

as $k \to 0$. In fact one can show that $\mu = 2|M_d|$ in this approximation (see Sec. 9.5).

The gapless mode (2.107) is intimately connected with spontaneous $U(1)$ symmetry breaking (see Sec. 3.4). However, as shown by Anderson and Nambu [7, 8,], the character of the spectrum changes dramatically in the presence of an external electromagnetic field. This can be shown in GL framework in the following way.

In the presence of an external electromagnetic field, the GL action is

$$S_{GL} = \int dt d^3x \Big\{ -\frac{1}{4}F_{\mu\nu}F^{\mu\nu} + c\Big[d\left(\partial_t + 2ieA_0\right)\varphi_c^\dagger\left(\partial_t - 2ieA_0\right)\varphi_c - \frac{1}{4m}\left(\partial_i + 2ieA_i\right)\varphi_c^\dagger\left(\partial_i - 2ieA_i\right)\varphi_c - a{\varphi_c}^\dagger\varphi_c - \frac{1}{2}b\left({\varphi_c}^\dagger\varphi_c\right)^2\Big]\Big\}; \tag{2.109}$$

$$F_{\mu\nu} = \partial_\mu A_\nu - \partial_\nu A_\mu \,.$$

The appearance of $2e$ follows from having the charge of a Cooper pair be equal to $2e$.

The remarkable point of the action (2.109) is that the phase $\theta(x)$ appears only in the combination $\partial_\mu\theta - eA_\mu$. Therefore it can be removed by the local gauge transformation $\partial_\mu\theta - eA_\mu \longrightarrow -eA_\mu$. Thus we conclude that the phase $\theta(x)$ disappears from the dynamics.

What happens to the excitations associated with $\theta(x)$? Let us consider the Maxwell equations for A_μ :

$$\partial^\mu F_{\mu\nu} = \jmath_\nu \,, \tag{2.110}$$

where, as it follows from Eq.(2.109), the current $\jmath_\nu$ is

$$\begin{aligned} \jmath_0 &= -2iecd\left[\varphi_c^\dagger\partial_t\varphi_c - \varphi_c\partial_t\varphi_c^\dagger - 4ie\varphi_c^\dagger\varphi_c A_0\right] \,, \\ \jmath_i &= -\frac{iec}{2m}\left[\varphi_c^\dagger\partial_i\varphi_c - \varphi_c\partial_i\varphi_c^\dagger - 4ie\varphi_c^\dagger\varphi_c A_i\right] \,. \end{aligned} \tag{2.111}$$

Let us consider a rather weak external field when $\varphi_c^\dagger \varphi_c \simeq \bar{\varphi}_c^\dagger \bar{\varphi}_c = \bar{n}_s$, where $\bar{n}_s$ is the density of the Cooper pairs in the ground state. Now one can see that the terms $8e^2 cd\bar{n}_s A_0^2$ and $2e^2 c\bar{n}_s A_i^2$ in the equation (2.109) reveal themselves to be mass terms for A_0 and A_i, respectively.

Let us consider the case of an external, time-independent, magnetic field $\mathbf{B} = \text{curl}\mathbf{A}$. The Maxwell equations (2.110) for $\mathbf{B}$ are

$$\text{curl}\mathbf{B} = \jmath \ , \tag{2.112}$$

$$\text{div}\mathbf{B} = 0 \ . \tag{2.113}$$

On the other hand, Eq.(2.111) implies that

$$\text{curl}\jmath = -\frac{2e^2 c\bar{n}_s}{m}\mathbf{B} \ . \tag{2.114}$$

This is the famous London equation. Substituting $\jmath$ from Eq.(2.112) in Eq.(2.114) and taking into account that, because of Eq.(2.113), curl curl $\mathbf{B} = -\Delta\mathbf{B}$, we obtain the equation

$$\Delta\mathbf{B} = \mathbf{B}/\delta^2 \tag{2.115}$$

with $\delta = m/(2e^2 c\bar{n}_s)$.

Let us consider a superconductor with the surface in xy - plane, and an external $\mathbf{B}$ field depending only on the coordinate z. Then Eqs. (2.113) and (2.115) imply that $B_z = 0$. The equation (2.115) now takes the simple form:

$$\frac{d^2\mathbf{B}}{dz^2} = \mathbf{B}/\delta^2 \ . \tag{2.116}$$

Its solution is

$$\mathbf{B}(z) = \mathbf{B}_0 \exp(-z/\delta) \ , \tag{2.117}$$

where $\mathbf{B}_0 = (B_{0x}, B_{0y}, 0)$ is the field on the surface of the superconductor.

Thus the magnetic field decreases exponentially into the superconductor. This phenomenon is known as the Meissner effect.

Now we can answer the question concerning the fate of the gapless mode θ in the external field. As we already noted, the Meissner effect can be treated as the appearance of a mass for the electromagnetic field. The massive vector field has not only transverse components but also a longitudinal one. This degree of freedom is just the transformation of the gapless mode in the electromagnetic field (the plasmon excitation).

The Meissner effect is a nonrelativistic analog of the Higgs mechanism of generating the mass for vector fields in particle physics (see Chapter 3).

In this chapter, we considered two remarkable phenomena in condensed matter physics, superfluidity and superconductivity. These phenomena reveal all essential features of the spontaneous symmetry breaking and dynamical symmetry breaking mechanisms. In particular they can serve as laboratories for studying these phenomena in particle physics. Being much more visual, they will help us to obtain clearer insight in the physical content of these phenomena.

Chapter 3

Spontaneous Symmetry Breaking in Relativistic Quantum Field Theory. General Consideration

3.1. Introduction

In this chapter, we consider general properties of the phenomenon of spontaneous symmetry breaking in relativistic quantum field theory.

We shall consider relativistic field systems at zero temperature and zero particle density. Our everyday intuition suggests that the vacuum of such systems should be an "empty" state containing "nothing". This is of course connected with the point that the vacuum must be invariant under the Lorentz transformations (if one does not consider the possibility of spontaneous Lorentz symmetry breaking). It is very difficult to imagine another medium than "nothing" being the same in all Lorentz frames.

On the other hand, by definition, in the case of spontaneous symmetry breaking, an invariance of the action of a system is not an invariance of its vacuum. It is difficult to imagine how "empty" state ("nothing") can have a lower symmetry than the action.

This point was a severe psychological obstacle for introducing the concept of spontaneous symmetry breaking in particle physics and was overcome only in the sixties. The concept was introduced in the classical papers of Nambu, Nambu and Jona-Lasinio, and Goldstone [1, 2, 3]. Remarkably, Nambu suggested the first and most important example of a spontaneously broken symmetry in particle physics — the chiral symmetry, intimately connected with the dynamics generating masses of elementary particles.

One should also indicate the important paper of Vaks and Larkin [4], unfortunately almost unknown outside of the former USSR. They also considered the dynamics generating fermion masses. It is interesting that in their paper, Vaks and Larkin discussed a model in which a Majorana mass was dynamically generated (see Sec. 6.4).

Spontaneous (dynamic) chiral symmetry breaking is the main subject of the present book. In this chapter, we will discuss general characteristics of spontaneous symmetry breaking phenomenon.

3.2. The Goldstone Models

As we already know, spontaneous symmetry breaking corresponds to the situation in which the symmetry of the action of a system is higher than the symmetry of its vacuum. The simplest examples of this phenomenon are given by the Goldstone models [3].

The first Goldstone model is the model of a real scalar field with the lagrangian density

$$\mathcal{L} = \frac{1}{2}\left(\partial_\mu \varphi \partial^\mu \varphi\right) - U(\varphi)\ , \tag{3.1}$$

where the potential U is

$$U(\varphi) = \frac{\mu^2}{2}\varphi^2 + \frac{\lambda}{4!}\varphi^4\ ;\ \lambda > 0\ . \tag{3.2}$$

The lagrangian density is invariant under the discrete transformation

$$\varphi \longrightarrow -\varphi\ . \tag{3.3}$$

The energy density of the system is

$$\mathcal{H} = \frac{1}{2}\left(\partial_0 \varphi\right)^2 + \frac{1}{2}\left(\nabla \varphi\right)^2 + U(\varphi)\ . \tag{3.4}$$

Let us first consider the classical limit in which the field $\varphi(x)$ is a c-number field. Eq.(3.4) implies that a solution with the lowest energy density is x-independent, i.e. $\varphi = const.$. In this case

$$\mathcal{H} = U(\varphi)\ , \tag{3.5}$$

and φ satisfies the equation:

$$\frac{dU}{d\varphi} = \mu^2\varphi + \frac{\lambda}{6}\varphi^3 = 0\ . \tag{3.6}$$

When $\mu^2 > 0$, the potential $U(\varphi)$ has only one minimum at $\varphi = 0$ (see Fig.3.1). However, when $\mu^2 < 0$, the form of the potential changes dramatically (see Fig.3.2):

now the trivial solution $\varphi = 0$ corresponds to its maximum, and there are two nontrivial solutions corresponding to the two minima:

$$\varphi = \pm \left(\frac{-6\mu^2}{\lambda} \right)^{1/2} \equiv \pm v \,. \tag{3.7}$$

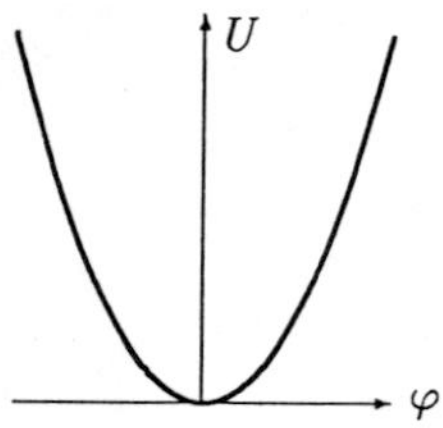

Fig. 3.1

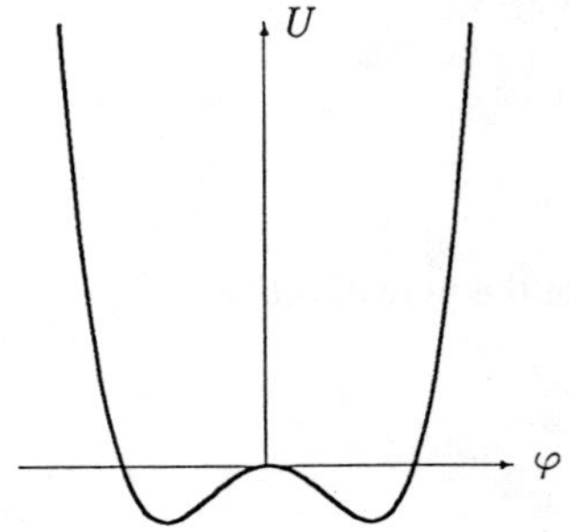

Fig. 3.2

The trivial solution is invariant under the transformation (3.3), so that when $\mu^2 > 0$, this discrete symmetry is the symmetry both of the action and of the vacuum solution. However, when $\mu^2 < 0$, the two solutions (3.7) are noninvariant under this transformation:

$$\varphi_+ \equiv \left(-\frac{6\mu^2}{\lambda} \right)^{1/2} \longleftrightarrow \varphi_- \equiv -\left(-\frac{6\mu^2}{\lambda} \right)^{1/2} . \tag{3.8}$$

In this case the symmetry of the vacuum solutions is lower than the symmetry of the action. Therefore the spontaneous breakdown of the discrete symmetry (3.3) takes place in the system.

Since the hamiltonian is invariant under this transformation, the energy density of the solutions $\varphi_\pm$ is the same

$$\mathcal{H}\,(\varphi_\pm) = -3\mu^4/2\lambda \,. \tag{3.9}$$

Thus the vacuum degeneracy takes place in the system.

In the case with $\mu^2 > 0$, μ is the mass of the field φ. To determine its mass at $\mu^2 < 0$, it is convenient to introduce a field

$$\varphi' = \varphi - (\pm v) \tag{3.10}$$

(one can choose either sign here; for the sake of definiteness, we shall use $+v$). Then one finds:

$$U(\varphi) = \frac{\lambda v^2}{6}\varphi'^2 + \frac{\lambda v}{6}\varphi'^3 + \frac{\lambda}{4!}\varphi'^4 - \frac{\lambda}{24}v^4 \,. \tag{3.11}$$

Therefore the parameter $\frac{\lambda v^2}{3}$ is the mass square of the field φ'.

Let us make the following remark. Although we used the classical approximation, the treatment of the phenomenon is in fact quantum: the Planck relation was used to express the mass (the rest energy) through the frequency of the classical field.

Beyond the classical (tree) approximation the interpretation will be slightly changed in that the field φ is an operator, and one should consider its matrix element

$$\varphi_c \equiv \langle 0|\varphi|0\rangle \tag{3.12}$$

(see Chapter 7).

One can see that $\varphi_c \neq 0$ is a signature of the spontaneous breakdown of the $P-$ symmetry, $\varphi \to P^{-1}\varphi P = -\varphi$. Indeed, the invariance of the vacuum under P, $P|0\rangle = |0\rangle$, implies that

$$\varphi_c = \langle 0|\varphi|0\rangle = \langle 0|P^{-1}\varphi P|0\rangle = -\varphi_c \,, \tag{3.13}$$

i.e. $\varphi_c = 0$. The matrix element $\varphi_c = \langle 0|\varphi|0\rangle$ is called the order parameter. It defines to what extent the symmetry is spontaneously broken in a system.*)

Let us now consider the second Goldstone model giving an example of the spontaneous breakdown of a continuous symmetry. It is a model of a complex scalar field with

---

*) The general definition of an order parameter of a spontaneously broken symmetry is the following: an order parameter is a non-vanishing vacuum expectation value of some local field which transforms non-trivially under the symmetry group.

the lagrangian density

$$\mathcal{L} = \partial_\mu \varphi^\dagger \partial^\mu \varphi - U(\varphi^\dagger \varphi) \,, \tag{3.14}$$

where

$$U = \mu^2 \varphi^\dagger \varphi + \frac{\lambda}{3!} \left(\varphi^\dagger \varphi\right)^2 \,. \tag{3.15}$$

This lagrangian density is invariant under the continuous $U(1)$ - transformations

$$\varphi \longrightarrow \exp(i\omega)\varphi, \varphi^\dagger \longrightarrow \exp(-i\omega)\varphi^\dagger \,. \tag{3.16}$$

At $\mu^2 > 0$, the minimum of the potential is at $\varphi = 0$; however, at $\mu^2 < 0$, the minima lie on the circle

$$2\varphi^\dagger \varphi = \varphi_1^2 + \varphi_2^2 = v^2 = -6\mu^2/\lambda \,, \tag{3.17}$$

where $\varphi = \frac{1}{\sqrt{2}}(\varphi_1 + i\varphi_2)$. These minima correspond to a continuum set of the degenerate vacua.

Choosing the minimum with

$$\varphi_1 = v, \;\; \varphi_2 = 0 \tag{3.18}$$

and shifting the fields,

$$\varphi' = \varphi_1 - v, \;\; \varphi_2' = \varphi_2 \,, \tag{3.19}$$

we find that the potential is (up to some inessential constant)

$$U(\varphi_1, \varphi_2) = \frac{\lambda}{4!} \left(\varphi_1'^2 + \varphi_2'^2 + 2v\varphi_1'\right)^2 \,. \tag{3.20}$$

The dynamics describes two real scalar fields: the massive φ_1' field with the mass $\mu_1 = \left(\frac{\lambda v^2}{3}\right)^{1/2}$ and the massless field φ_2' .

That the presence of the massless field is not an accident can be seen from the following consideration. Let us introduce new (angular) variables:

$$\varphi_1 = \rho\cos\theta, \varphi_2 = \rho\sin\theta \ . \tag{3.21}$$

The transformation (3.16) for them is

$$\rho(x) \longrightarrow \rho(x), \theta(x) \longrightarrow \theta(x) + \omega \ . \tag{3.22}$$

Since $\mathcal{L}$ is invariant under this transformation, the potential U is independent of the angular variable $\theta(x)$. Therefore $\theta(x)$ appears in $\mathcal{L}$ only through the derivative $\partial_\mu\theta$. This implies that the excitation connected with $\theta(x)$ is massless.

As will be shown in the section 3.4, the existence of such massless particles in relativistic theories with spontaneous breakdown of a continuous symmetry follows from very general principles. They are called Nambu-Goldstone (NG) particles.

Let us show, following Ref.[5], how NG particles appear in a system with spontaneous breakdown of an arbitrary continuous symmetry including K real scalar fields assigned to a representation $D^{(k)}$ of the group.

Let us write an element g of the group in the $D^{(k)}$ representation as

$$g = \exp\left[iQ^\alpha\omega_\alpha\right] \ , \tag{3.23}$$

where N matrices Q^α are generators of the corresponding Lie algebra.

Under the group transformations,

$$\varphi \longrightarrow \exp\left[iQ^\alpha\omega_\alpha\right]\varphi \ , \tag{3.24}$$

the lagrangian density does not change, and therefore

$$U(\varphi) = U\left(\exp\left[iQ^\alpha\omega_\alpha\right]\varphi\right) \ . \tag{3.25}$$

Let us assume that a solution corresponding to the minimum of the potential U is a K-dimensional vector v. Let us also assume that the subgroup with elements

$$\bar{g} = \exp\left[i\bar{Q}^\alpha\omega_\alpha\right] \tag{3.26}$$

with respect to which the vector v is invariant ($\bar{g}v = v$) contain $\bar{N}$ generators. Then group elements corresponding to the remaining $N - \bar{N}$ generators change the vector

v (in the example with the lagrangian density (3.14), N is $N = 1$ and $\bar{N}$ is $\bar{N} = 0$). Therefore there is an $(N - \bar{N})$-dimensional surface corresponding to the same value of U but different values of the vector v. This implies that the potential U is independent of these $(N - \bar{N})$ variables, and there are $(N - \bar{N})$ NG massless scalar particles in the system.

We shall consider a more rigorous proof of this fact known as the Goldstone theorem in the section 3.4. In the next section, we shall discuss the role of the thermodynamic limit $V \longrightarrow \infty$ in spontaneous symmetry breaking phenomenon in relativistic systems.

3.3. Why Do We Need the Infinite Volume for Spontaneous Symmetry Breaking ?

In the previous chapter, it was shown that the spontaneous breakdown of the $U(1)$ symmetry in superfluidity and superconductivity can be realized only in the infinite volume. Let us consider the role of V in relativistic systems. For simplicity, we shall consider only the lagrangian density (3.14) with the $U(1)$ symmetry.

As it was shown above, there is a NG field $\varphi \equiv \varphi_2'$ when the spontaneous breakdown of this symmetry takes place. Let us consider the mean-field approximation in which the interactions between the fields φ_1' and $\varphi_2' = \varphi$ are neglected. Then the lagrangian density for φ is that of massless free field,

$$\mathcal{L} = \frac{1}{2}\partial^\mu\varphi\partial_\mu\varphi\ , \tag{3.27}$$

invariant under the transformations $\varphi(x) \longrightarrow \varphi(x) + const.$ The conserved current connected with these transformations is

$$\jmath_\mu = \partial_\mu\varphi\ ;\ \partial^\mu\jmath_\mu = 0\ . \tag{3.28}$$

The commutator $[\partial_0\varphi(x), \varphi(y)]|_{x_0=y_0}$ is

$$[\partial_0\varphi(x), \varphi(y)]|_{x_0=y_0} = i^{-1}\delta^3(\mathbf{x} - \mathbf{y})\ . \tag{3.29}$$

Therefore

$$i[Q, \varphi(x)] = 1\ , \tag{3.30}$$

where the conserved charge Q is

$$Q = \int d^3x \partial_0 \varphi(x) . \tag{3.31}$$

Let us consider the system in a finite volume. Then (see Sec. 1.2)

$$\varphi(x) = 2^{-1/2} V^{-1/2} \sum_{\mathbf{k}} k^{-1/2} \left(a_k e^{-ikx} + a_k^\dagger e^{ikx} \right) , \tag{3.32}$$

$$\partial_0 \varphi(x) = 2^{-1/2} V^{-1/2} i \sum_{\mathbf{k}} k^{1/2} \left(a_k^\dagger e^{ikx} - a_k e^{-ikx} \right) , \tag{3.33}$$

$$Q = \int d^3x \partial_0 \varphi(x) = 2^{-1/2} V^{1/2} i \lim_{k \to 0} k^{1/2} \left(a_k^\dagger - a_k \right) . \tag{3.34}$$

One can see that

$$[Q, \hat{q}] = \frac{1}{i} , \tag{3.35}$$

where the collective coordinate

$$\hat{q} = \varphi_k |_{k=0} = 2^{-1/2} V^{-1/2} \lim_{k \to 0} k^{-1/2} \left(a_k e^{-ikx} + a_k^\dagger e^{ikx} \right) \tag{3.36}$$

is the coordinate of the centre of the mass of the system in the volume V.

Identifying Q with the momentum $\hat{p}$, $[\hat{p}, \hat{q}] = 1/i$, one might think that what we have encountered is the dynamics of the harmonic oscillator,

$$H = \frac{\hat{p}^2}{2m} + \frac{m\omega^2 \hat{q}^2}{2} \tag{3.37}$$

with the "mass" $m = V$ and the frequency $\omega = k \to 0$. Naively one might expect that as $k \to 0$, there is an infinite number of degenerate vacua, $\frac{(a_0^\dagger)^N}{\sqrt{N!}} |0\rangle$. However, as $k \to 0$, the hamiltonian H (3.37) reduces to that of a free particle, $H = \frac{\hat{p}^2}{2V} = -\frac{1}{2V} \frac{\partial^2}{\partial q^2}$, with eigenvectors

$$\psi_p(q) = \exp(ipq); H\psi_p(q) = \frac{p^2}{2V} \psi_p(q) . \tag{3.38}$$

Therefore there is no vacuum degeneracy in a finite volume; the unique vacuum (ground state) is the state $\psi_{p=0}(q) = 1$.

Thus we conclude that there is no spontaneous breakdown of the $U(1)$ symmetry in a finite volume. We note, however, that as the volume V grows, we have "almost" vacuum degeneracy in the system: all eigenvectors $\psi_p(q)$ with $p^2V^{-1} \ll v$ practically have the same energy (v is the characteristic energy of the system (see Eq. (3.17)).

As $V \longrightarrow \infty$, we get a continuum set of degenerate vacua $|c\rangle$ characterized by $\langle c|\varphi|c\rangle = c$.

Actually, we have somewhat simplified the situation. The symmetry of the lagrangian density (3.27) is not $U(1)$ but $R(1)$ symmetry under which $\varphi \longrightarrow \varphi + c, -\infty < c < +\infty$. The lagrangian density for the NG field corresponding to the spontaneous breakdown of the $U(1)$ symmetry is (in the mean field approximation)

$$\mathcal{L} = \rho^2 \partial^\mu \theta \partial_\mu \theta \tag{3.39}$$

with $0 \leq \theta < 2\pi$ and $\rho \equiv v$ (see Eq.(3.21)), i.e. with ρ being equal the value of the field at the minimum of the potential.

The hamiltonian for the collective coordinate $\hat{q}$ (with $k \rightarrow 0$) has the same form as before,

$$H = \hat{p}^2/2V \ , \tag{3.40}$$

however, since q is now the subject of the constraint $0 \leq q < 2\pi v$, the wave functions are periodic functions ($\psi(q + 2\pi v) = \psi(q)$). Therefore the eigenvectors of H are:

$$\psi_n(q) = \exp(inq/v), n = 0, \pm 1, 2.... \tag{3.41}$$

The spectrum of H coincides with the spectrum of the hamiltonian of the quantum rotator:

$$E_n = n^2/_{2v^2V}, n = 0, \pm 1, \pm 2, ... \tag{3.42}$$

It is clear that when $q \ll 2\pi v$, this dynamical system is practically identical to the system with the plane wave eigenvectors (3.38).

This simple example clearly illustrates the role of the limit $V \longrightarrow \infty$ in spontaneous symmetry breaking phenomenon. When V is finite, there is no vacuum degeneracy; one must consider the collective coordinate $\hat{q}$ describing the movement of the box containing the whole volume V. In this case, there is the unique vacuum with the momentum $p = 0$. As $V \longrightarrow \infty$, the states with such p^2 that $p^2V^{-1} \longrightarrow 0$ become degenerate, and the phenomenon of spontaneous symmetry breaking is realized.

We emphasize once again that in a finite but rather large volume the states with $p^2V^{-1} \ll v$ are almost degenerate. It implies that the conception of spontaneous symmetry breaking is practically applicable for such systems too.*)

3.4. The Coleman Theorem and the Goldstone Theorem

The vacuum plays a very special role in describing the character of the realization of invariance in quantum field theory. In this section, we consider two theorems establishing this special role of the vacuum in an explicit form.

Let us first consider the Coleman theorem [7].

Theorem. If the vacuum in a Lorentz and translational invariant local theory is invariant under the group generated by the space integral of the time component of a local current $\jmath_\mu$, then the current satisfies the conservation law $\partial^\mu \jmath_\mu(x) = 0$.

Coleman also expressed this theorem as the statement that "the invariance of the vacuum is the invariance of the World".

Below, we shall present the original proof of the theorem given by Coleman. Although, strictly speaking, this proof is not rigorous (see [7] and below), it expresses the physical essence of the theorem in the best way. For a rigorous proof, see Ref.[8].

Proof. Due to the condition of the theorem, the charge $Q = \int d^3x \jmath_0(x)$ annihilates the vacuum:

$$Q|0\rangle = 0 \,. \tag{3.43}$$

Let $|n\rangle$ be an arbitrary state with vanishing 3-momentum. Then,

$$\langle n|Q|0\rangle = \langle n| \int d^3x \jmath_0(x)|0\rangle = 0 \,. \tag{3.44}$$

But because of the choice of the momentum $\mathbf{p}$ of the state $|n\rangle$ being equal zero, we have that $\langle n|\jmath_0(x)|0\rangle = e^{ip_0t}\langle n|\jmath_0(0)|0\rangle$ is independent of $\mathbf{x}$, and therefore Eq.(3.44) implies

*) Some symmetries connected with transformations of spin variables can be spontaneously broken even in quantum mechanical systems. The important examples are given by supersymmetric quantum mechanical systems with a spontaneously broken supersymmetry [6].

that

$$\langle n|j_0(x)|0\rangle = 0 \,. \tag{3.45}$$

This relation implies in turn that

$$\langle n|\partial^\mu j_\mu(x)|0\rangle = 0 \,. \tag{3.46}$$

But Lorentz invariance guarantees that, if this is valid in any one Lorentz frame, it is also valid in all Lorentz frames. Therefore Eq.(3.46) is valid for any momentum eigenstate $\langle n|$ with $p_0^2 - \mathbf{p}^2 > 0$. The only exception could be eigenstates with $p^2 = 0$. However, since $\varphi(x) \equiv \partial^\mu j_\mu(x)$ is a scalar field, we find that in this case $\langle n|\partial^\mu j_\mu(x)|0\rangle \sim p^2 \xrightarrow[p^2=0]{} 0$.

Thus the states $\langle n|$ in Eq. (3.46) are a complete set, and therefore

$$\partial^\mu j_\mu(x)|0\rangle = 0 \,. \tag{3.47}$$

Then the Federbush-Johnson (FJ) theorem [9], which states that if a local operator acting on the vacuum gives zero, the operator itself is equal to zero, implies that $\partial^\mu j_\mu(x) = 0$, i.e. the current is conserved (for comments concerning the FJ theorem, see below).

Thus the invariance of the vacuum implies the invariance of the World.

Let us comment some points in this theorem.

The theorem is heavily based on the Lorentz invariance and the locality. There is no analogue of the Coleman theorem in nonrelativistic field theory. In particular, the physical meaning of the FJ theorem is that any local operator in relativistic field theory includes a term containing creation operators only.*) This point, evident in free theory, is intimately connected with the PCT theorem [10]. The fact that a nonzero local operator contains such a term implies that it cannot annihilate the vacuum.

One can show that the action of a charge operator $Q = \int d^3x j_0(x)$ on the vacuum $|0\rangle$ is properly defined only if it annihilates $|0\rangle$. Indeed, let us assume that $Q|0\rangle = |\varphi\rangle \neq 0$. Then, because Q commutes with the momentum operator, $|\varphi\rangle$ is a translationally

*) Here under the creation operators we understand the creation operators $a_i^\dagger$ from the interpolating fields $\varphi_i^{(\text{int})}$ in a system, i.e., the operators a_i annihilate the physical vacuum.

invariant state, and therefore $\langle 0|j_0(x)|\varphi\rangle$ is independent of **x**. As the result, the norm of $|\varphi\rangle$ goes to infinite with $V \longrightarrow \infty$:

$$\langle\varphi|\varphi\rangle = \langle 0| \int d^3x\, j_0(x)|\varphi\rangle \sim V\ , \tag{3.48}$$

i.e. the action of Q on $|0\rangle$ is not properly defined.

In particular, since in the case of spontaneous symmetry breaking, the charge operators Q^a connected with the symmetry do not annihilate the vacuum, their action on $|0\rangle$ is not properly defined. This mathematical statement corresponds to less rigorous one that the action of the operators $U = \exp(iQ^a\theta^a)$ connects the different vacua $|\{\theta^a\}\rangle$ corresponding to the different Fock spaces $F_{\{\theta^a\}}$ (compare with the discussion in Secs.2.3 and 2.5). Although the operator U is not, strictly speaking, a well defined unitary operator, it is often visual and useful to use it in such a heuristic way.

We also emphasize that the local currents $j^a_\mu(x)$ are well defined objects even in the case they are connected with a spontaneously broken symmetry.

In the Coleman proof, momentum eigenstates are used. One can show that, although the action of Q is properly defined on a dense set of states in the Fock space, the set does not include the momentum eigenstates. This is reason why the above proof is not rigorous. However the rigorous proof (based on the same idea) of the theorem also exists [8].

The Coleman theorem tells us what happens if the vacuum is invariant under some group transformations. The Goldstone theorem establishes a general characteristic of relativistic field systems with spontaneous breakdown of continuous symmetries.

First of all, let us give a more precise definition of the spontaneous breakdown of a continuous symmetry in field theories.

Let us assume that a lagrangian density is invariant under some Lie group G. Then, because of the Noether theorem (see Sec.4.1), there exist the conserved currents $j^a_\mu(x)$, $\partial^\mu j^a_\mu(x) = 0$, with $a = 1, 2, ..., n$, where n is the number of the group generators. The conserved charge operators

$$Q^a = \int d^3x\, j^a_0(x) \tag{3.49}$$

are the generators of the group. The infinitesimal transformations of the fields $\varphi_i(x)$ in the lagrangian are

$$\delta\varphi_i(x) = -i\left[\theta^a Q^a, \varphi_i(x)\right] , \tag{3.50}$$

where θ^a are infinitesimal parameters.

Spontaneous breaking of the symmetry implies that

$$Q^{\bar{a}}|0\rangle \neq 0 . \tag{3.51}$$

at some $\bar{a}$. As we showed above, Eq. (3.51) implies that the action of $Q^{\bar{a}}$ on the vacuum is not properly defined. However, because of the locality, the commutators of $Q^{\bar{a}}$ with local fields are well defined and expressed through some linear combinations of the local fields. Spontaneous breaking of the symmetry can be characterized by the condition that there is some local fields $\phi_k(x)$ with

$$-\langle 0|\left[iQ^{\bar{a}}, \phi_k(x)\right]|0\rangle = \langle 0|\delta\phi_k(x)|0\rangle \neq 0 . \tag{3.52}$$

at some $\bar{a}$. Eq. (3.52) implies Eq. (3.51).

We emphasize that the fields $\phi_k(x)$ can be elementary (some of the fields $\varphi_i(x)$ in the lagrangian) or composite (like in superconductivity). In the latter case, we shall say that there is dynamic breakdown of the symmetry.

Let us formulate now the Goldstone theorem suggested by Goldstone [4] and proved by Goldstone, Salam and Weinberg [11].

Theorem. Let us consider a Lorentz and translational invariant local field theory with conserved currents j^a_μ relating to a Lie group G. Let us assume that this symmetry is spontaneously broken so that the equation (3.52) holds. Then there are massless spinless particles (Nambu-Goldstone (NG) bosons) which couple to both the currents $j^{\bar{a}}_\mu$ and fields ϕ_k.

Proof. Let us consider the Green's functions

$$G^{\bar{a}}_{\mu k}(q) = \int d^4x \exp(iqx)\langle 0|Tj^{\bar{a}}_\mu(x)\phi_k(0)|0\rangle . \tag{3.53}$$

The Lorentz and translational invariance implies that

$$G^{\bar{a}}_{\mu k}(q) = q_\mu f^{\bar{a}}_k(q^2) . \tag{3.54}$$

Since $\partial^\mu \jmath^{\bar{a}}_\mu = 0$, we find that

$$\partial^\mu \langle 0|T\jmath^{\bar{a}}_\mu(x)\phi_k(0)|0\rangle = \delta(x^0)\langle 0|\left[\jmath^{\bar{a}}_0(x), \phi_k(0)\right]|0\rangle \; . \tag{3.55}$$

Therefore this equation and Eq.(3.52) yield

$$\lim_{q\to 0} q^\mu G^{\bar{a}}_{\mu k}(q) = \lim_{q\to 0}\int d^4x \exp(iqx) i\partial^\mu \langle 0|T\jmath^{\bar{a}}_\mu(x)\phi_k(0)|0\rangle = -\langle 0|\delta\phi_k|0\rangle \; . \tag{3.56}$$

Because of the relation (3.54), we find from Eq. (3.56) that

$$G^{\bar{a}}_{\mu k}(q) \underset{q\to 0}{\simeq} \frac{-q_\mu}{q^2+i\epsilon}\langle 0|\delta\phi_k|0\rangle \; . \tag{3.57}$$

This relation implies that there is a NG boson state $|q\rangle$ which couples both to the $\jmath^{\bar{a}}_\mu$ and ϕ_k:

$$\begin{aligned} \langle 0|\jmath^{\bar{a}}_\mu(x)|q\rangle &= if^{\bar{a}} q_\mu e^{-iqx} \\ \langle q|\phi_k(x)|0\rangle &= Z_k^{1/2} e^{iqx} \end{aligned} \tag{3.58}$$

with $f^{\bar{a}}$ and $Z_k^{1/2}$ satisfying the relation

$$f^{\bar{a}} Z_k^{1/2} = \langle 0|\delta\phi_k|q\rangle \; , \tag{3.59}$$

where the state $|q\rangle$ is normalized as

$$\langle q|q'\rangle = (2\pi)^3 2q_0 \delta^3(\mathbf{q}-\mathbf{q}') \; . \tag{3.60}$$

We emphasize that this proof is heavily based on the Lorentz and translational invariance of a theory. There is an analogue of the Goldstone theorem, known as the Bogolyubov "$1/q^2$ - theorem", in condensed matter physics [12]. This theorem asserts that if there is spontaneous symmetry breaking in a system in which the interaction falls off sufficiently rapidly with the distance, then in the spectrum of the hamiltonian there necessarily exists a branch of elementary excitations whose energy $\epsilon(q)$ goes to zero as the momentum q goes to zero.

In particular the theorem implies that there must be a gapless mode in a superfluid Bose gas but that such a mode is absent in superconductors due to long range electromagnetic interactions.

On the other hand, there is no such a restriction for the validity of the Goldstone theorem. So it might seem that it is impossible to overcome it in relativistic systems without destroying such fundamental properties as the Lorentz and translational invariance or locality. However, as will be shown in the next section, there is a possibility to overcome it in the *physical* space of gauge field theories. In that case, the theorem is still valid, however, the corresponding NG bosons become unphysical particles like the longitudinal and time photon modes in QED.

3.5. Higgs Mechanism

To consider the fate of the Goldstone theorem in gauge theories, we begin with the simplest case of the Abelian model.

The lagrangian density of the model is

$$\mathcal{L} = -\frac{1}{4}F_{\mu\nu}F^{\mu\nu} + \mathcal{L}_s\left(\varphi, \varphi^\dagger, D_\mu\varphi, D_\mu\varphi^\dagger\right) , \tag{3.61}$$

where

$$F_{\mu\nu} = \partial_\mu A_\nu - \partial_\nu A_\mu \ , \ \mathcal{L}_s = D_\mu\varphi^\dagger D^\mu\varphi - U\left(\varphi^\dagger, \varphi\right) \tag{3.62}$$

with

$$\begin{aligned} D_\mu\varphi = \partial_\mu\varphi + igA_\mu\varphi, D_\mu\varphi^\dagger \ &= \ \partial_\mu\varphi^\dagger - igA_\mu\varphi^\dagger, \\ U\left(\varphi^\dagger, \varphi\right) = \mu^2\varphi^\dagger\varphi + \frac{\lambda}{3!}\left(\varphi^\dagger\varphi\right)^2 &; \lambda \geq 0 \ . \end{aligned} \tag{3.63}$$

Comparing Eq.(3.61) with Eq.(3.14), we see that this model is just the gauged Goldstone model.

The lagrangian density is invariant under the $U(1)$ gauge transformations:

$$\begin{aligned} &\varphi \longrightarrow \exp(i\omega(x))\varphi, \ \varphi^\dagger \longrightarrow \exp(-i\omega(x))\varphi^\dagger, \\ &D_\mu\varphi \longrightarrow \exp(i\omega(x))D_\mu\varphi, \ D_\mu\varphi^\dagger \longrightarrow \exp(-i\omega(x))D_\mu\varphi^\dagger, \\ &A_\mu \longrightarrow A_\mu - g^{-1}\partial_\mu\omega(x) \ . \end{aligned} \tag{3.64}$$

These transformations include in particular the transformations (3.16) with ω being independent of x (global $U(1)$ transformations).

Let us consider the model with $\mu^2 < 0$. The minima of the potential (3.62) are given by (see Eq.(3.17)):

$$\varphi^\dagger \varphi = v^2/2 = -3\mu^2/\lambda \, . \tag{3.65}$$

Introducing the angular field variables

$$\varphi = 2^{-1/2} \rho \exp(i\theta), \varphi^\dagger = 2^{-1/2} \rho \exp(-i\theta) \, , \tag{3.66}$$

we can rewrite the lagrangian density in the form

$$\begin{aligned} \mathcal{L} = & -\frac{1}{4} (\partial_\mu A_\nu - \partial_\nu A_\mu)(\partial^\mu A^\nu - \partial^\nu A^\mu) \\ & + \frac{1}{2} (\partial_\mu \rho')(\partial^\mu \rho') + \frac{1}{2} (\rho' + v)^2 (\partial_\mu \theta + g A_\mu)(\partial^\mu \theta + g A^\mu) - U(\rho' + v) \end{aligned} \tag{3.67}$$

with $\rho' = \rho - v$.

We note that under the transformations (3.64),

$$\theta(x) \longrightarrow \theta(x) + \omega(x) \, . \tag{3.68}$$

Therefore one can always choose the gauge in which

$$\theta(x) \equiv 0 \tag{3.69}$$

(the unitary gauge). As can be seen from Eq.(3.67), in this gauge, there is a massive scalar boson with $M_{\rho'} = \left(\frac{\lambda v^2}{3}\right)^{1/2}$, and also there is a massive vector boson with $M_A = gv$. The NG massless boson disappeared from the spectrum !

This phenomenon, discovered by Englert and Brout, Higgs, and Guralnik, Hagen and Kibble [13] (and which possibility was first pointed out by Schwinger and Anderson [14]) is known in the literature as the Higgs phenomenon.

A few comments are now in order. The lagrangian density (3.60) is nothing but a relativistic analogue of the Ginzburg-Landau (GL) model (see Sec.2.10). In particular the Higgs phenomenon is the relativistic analogue of the Meissner effect in superconductors. As in the case of the Meissner effect, it can be interpreted as a transformation of the NG boson to the longitudinal component of the massive vector field.

Why does the Goldstone theorem unfulfil in the unitary gauge with $\theta(x) = 0$? The answer is simple: in this gauge, the lagrangian (3.67) with $\theta = 0$ is no longer invariant under the $U(1)$ transformations (3.16). Since any trace of the $U(1)$ symmetry disappears in this gauge, the Goldstone theorem is no longer applicable either.

How does this phenomenon look in other gauges ?

Let us add a gauge term $-\frac{1}{2}[F(x)]^2$ to the lagrangian density (3.61), where (ξ is a gauge parameter)

$$F = -\xi^{1/2}\partial^\mu A_\mu(x) + \xi^{-1/2} gv\varphi_2 \,. \tag{3.70}$$

Here we use the fields φ_1 and φ_2 connected with φ through $\varphi = 2^{-1/2}(\varphi_1 + i\varphi_2)$; $\varphi_1^2 + \varphi_2^2 = v^2$ (see Eq.(3.17)). The gauge (3.70) is an example of the so called R_ξ gauges for gauge theories with spontaneous symmetry breaking. In these gauges, there is no term mixing A_μ with φ_2 in the quadratic part of lagrangians (see Eq. (3.71) below).

The quadratic part of the lagrangian density is in this gauge:

$$\begin{aligned}
\mathcal{L}_2 = & -\frac{1}{4}F_{\mu\nu}F^{\mu\nu} + \frac{1}{2}(gv)^2 A_\mu A^\mu + \frac{1}{2}\partial_\mu\varphi_1'\partial^\mu\varphi_1' \\
& + \frac{1}{2}\partial_\mu\varphi_2\partial^\mu\varphi_2 - \frac{1}{2}\left(\frac{\lambda v^2}{3}\right)\varphi_1'\varphi_1' + gv\partial^\mu\varphi_2 A_\mu \\
& - \frac{1}{2}[F(x)]^2 = -\frac{1}{4}F_{\mu\nu}F^{\mu\nu} + \frac{1}{2}(gv)^2 A_\mu A^\mu + \frac{1}{2}(\partial_\mu\varphi_1')(\partial^\mu\varphi_1') \\
& - \frac{1}{2}\left(\frac{\lambda v^2}{3}\right)\varphi_1'\varphi_1' + \frac{1}{2}(\partial_\mu\varphi_2)(\partial^\mu\varphi_2) - \frac{1}{2}\xi^{-1}g^2v^2\varphi_2^2 \\
& - \frac{1}{2}\xi\left(\partial^\mu A_\mu\right)^2 + gv\partial^\mu(\varphi_2 A_\mu) \,,
\end{aligned} \tag{3.71}$$

where $\varphi_1' = \varphi_1 - v$. Since the last term in Eq.(3.71) is a total derivative, it can be omitted.

Eq.(3.71) implies the following propagators for the fields in the tree approximation:

$$\begin{aligned}
D_{\mu\nu}^A &= \frac{1}{k^2 - g^2v^2}\left(g_{\mu\nu} - \frac{(\xi-1)k_\mu k_\nu}{\xi k^2 - g^2v^2}\right) \,, \\
D^{\varphi_1'} &= \frac{-1}{k^2 - \frac{\lambda v^2}{3}} \,, \\
D^{\varphi_2} &= \frac{-1}{k^2 - \xi^{-1}g^2v^2} \,.
\end{aligned} \tag{3.72}$$

Where is the NG boson in this gauge ? To answer this question, we note that the lagrangian density (3.71) preserves the $U(1)$ global symmetry only if $\xi \longrightarrow \infty$. As one can see in Eq.(3.72), in this limit, the propagator D^{φ_2} becomes the propagator of the massless NG boson. However this particle is unphysical like the longitudinal and time components of photon in QED. Indeed, its mass depends on the gauge parameter ξ; in particular $M_{\varphi_2} \longrightarrow \infty$ with $\xi \longrightarrow 0$.

One can see that the gauge with $\xi = 0$ is in fact unitary one in which there retain only physical degrees of freedom: the field φ_2, with $M_{\varphi_2} \longrightarrow \infty$, decouples, and the propagator of the vector field takes the canonical form

$$D^A_{\mu\nu} \underset{\xi \to 0}{\longrightarrow} \frac{1}{k^2 - g^2 v^2} \left(g_{\mu\nu} - \frac{k_\mu k_\nu}{g^2 v^2} \right) . \tag{3.73}$$

This example clearly shows the essence of the Higgs phenomenon. In the next section, we consider it in non-Abelian gauge field theories.

3.6. Higgs Phenomenon in Non-Abelian Gauge Field Theories

Let us consider some compact Lie group

$$G = G_1 \otimes G_2 \otimes \cdots \otimes G_n , \tag{3.74}$$

where every G_i is a simple Lie group or $U(1)$. An element g of G can be represented as

$$g = \exp [iQ^\alpha \omega_\alpha] ; \alpha = 1, 2, ..., N, \tag{3.75}$$

where Q^α are generators of the corresponding Lie algebra, and ω_α are continuous parameters.

An element g_x of the gauged group G can be represented as

$$g_x = \exp [iQ^\alpha \omega_\alpha(x)] , \tag{3.76}$$

where $\omega_\alpha(x)$ are functions of the space-time coordinate x^μ.

The generators Q^α satisfy the commutation relations

$$\left[Q^\alpha, Q^\beta \right] = i f_{\alpha\beta\gamma} Q^\gamma , \tag{3.77}$$

where $f_{\alpha\beta\gamma}$ are the structure constants of the algebra.

Let us consider the following fields:

a) N massless vector fields A_μ^α assigned to the adjoint representation of G;

b) K scalar fields φ_a assigned to a $D^{(K)}$ representation (possibly reducible) of G;

c) F spinor fields ψ_l assigned to some $D^{(F)}$ representation of G.

The infinitesimal gauge transformations of these fields take the form:

$$\begin{aligned}A_\mu^\alpha(x) &\to A_\mu^\alpha(x) - f_{\alpha\beta\gamma}A_\mu^\gamma(x)\delta\omega_\beta(x) - g_{\alpha\beta}^{-1}\partial_\mu\delta\omega_\beta(x) ,\\ \varphi_a(x) &\to \varphi_a(x) + i\sigma_{ab}^\alpha\varphi_b(x)\delta\omega_\alpha(x) ,\\ \psi_l(x) &\to \psi_l(x) + i\tau_{lm}^\alpha\psi_m(x)\delta\omega_\alpha(x) ,\end{aligned} \tag{3.78}$$

where σ^α and τ^α are matrices of the generators Q^α in the representations $D^{(K)}$ and $D^{(F)}$, respectively; $g_{\alpha\beta} = g_\alpha\delta_{\alpha\beta}$ and $g_\alpha = g_\beta$ when the indices α and β relate to the same subgroup G_i.

One can always choose the A_μ^α and φ_a fields to be real; then $f_{\alpha\beta\gamma}$ are real, and the matrices σ^α are imaginary and antisymmetric.

The most general renormalizable lagrangian density invariant under the transformations (3.78) takes the form:

$$\begin{aligned}\mathcal{L} = &-\frac{1}{4}F_{\mu\nu}^\alpha F^{\mu\nu,\alpha} + \frac{1}{2}(D_\mu\varphi)_a(D^\mu\varphi)_a\\ &+ i\bar{\psi}_l\gamma^\mu(D_\mu\psi)_l - \bar{\psi}m^{(0)}\psi - U(\varphi) - \bar{\psi}_l\Gamma_a^{lm}\varphi_a\psi_m ,\end{aligned} \tag{3.79}$$

where the covariant derivatives are

$$\begin{aligned}(D_\mu\varphi)_a &= \partial_\mu\varphi_a + ig_\alpha\sigma_{ab}^\alpha\varphi_b A_\mu^\alpha ,\\ (D_\mu\psi)_l &= \partial_\mu\psi_l + ig_\alpha\tau_{lm}^\alpha\psi_m A_\mu^\alpha ,\end{aligned} \tag{3.80}$$

and

$$F_{\mu\nu}^\alpha = \partial_\mu A_\nu^\alpha - \partial_\nu A_\mu^\alpha - g_\alpha f_{\alpha\beta\gamma}A_\mu^\beta A_\nu^\gamma . \tag{3.81}$$

The derivatives $D_\mu\varphi, D_\mu\psi$ are assigned to the representations $D^{(K)}$ and $D^{(F)}$, respectively, and $F^{\alpha}_{\mu\nu}$ is assigned to the adjoint representation:

$$\delta F^{\alpha}_{\mu\nu}(x) = -f_{\alpha\beta\gamma} F^{\gamma}_{\mu\nu} \delta\omega_{\beta}(x) \ . \tag{3.82}$$

Also, the invariance of $\mathcal{L}$ with respect to G implies that

$$\left[\tau^{\alpha}, \gamma_0 m^{(0)}\right] = 0 \ , \tag{3.83a}$$

$$[\tau^{\alpha}, \gamma_0 \Gamma_a] = -\sigma^{\alpha}_{ab} \gamma_0 \Gamma_b \ , \tag{3.83b}$$

$$\frac{\partial U(\varphi)}{\partial \varphi_a} \sigma^{\alpha}_{ab} \varphi_b = 0 \ . \tag{3.83c}$$

To get a renormalizable theory, the power of the potential $U(\varphi)$ has to be not larger than four.

Spontaneous breakdown of the G symmetry takes place if the equation

$$\frac{\partial U(\varphi)}{\partial \varphi_a} = 0 \tag{3.84}$$

has a nontrivial solution

$$\varphi_a = v_a \neq 0 \ , \tag{3.85}$$

and the mass matrix

$$M^2_{ab} = \frac{\partial^2 U}{\partial \varphi_a \partial \varphi_b} |_{\varphi_a = v_a} \tag{3.86}$$

has nonnegative eigenvalues.

The quadratic part of the lagrangian density (3.79) takes the form:

$$\begin{aligned} \mathcal{L}_{\text{free}} = &- \frac{1}{4} \left(\partial_{\mu} A^{\alpha}_{\nu} - \partial_{\nu} A^{\alpha}_{\mu}\right) \left(\partial^{\mu} A^{\nu,\alpha} - \partial^{\nu} A^{\mu,\alpha}\right) \\ &+ \frac{1}{2} (\mathcal{M}^2)_{\alpha\beta} A^{\alpha}_{\mu} A^{\mu,\beta} + \frac{1}{2} \partial_{\mu} \varphi'_a \partial^{\mu} \varphi'_a - \frac{1}{2} (M^2)_{ab} \varphi'_a \varphi'_b \\ &+ i \bar{\psi} \gamma_{\mu} \partial^{\mu} \psi - \bar{\psi} m \psi + i g_{\alpha} \partial^{\mu} \varphi'_a \sigma^{\alpha}_{ab} v_b A^{\alpha}_{\mu} \quad , \end{aligned} \tag{3.87}$$

where $\varphi'_a = \varphi_a - v_a$, and

$$\mathcal{M}^2_{\alpha\beta} = -g_{\alpha} g_{\beta} \sigma^{\alpha}_{ab} \sigma^{\beta}_{ac} v_b v_c \ , \tag{3.88}$$

$$m = m^{(0)} + \Gamma^a v_a \ . \tag{3.89}$$

The form of $\mathcal{L}_{\text{free}}$ implies that the matrices $\mathcal{M}$ and M are the mass matrices (in the tree approximation) of vector and scalar fields, respectively.

We note, however, that there is a term mixing A_μ^α and φ_a' in the lagrangian density (3.87); as will be shown below, it can be removed in some gauges.

Let us introduce the eigenstates for the matrices $\mathcal{M}^2$ and M^2:

$$\mathcal{M}^2_{\alpha\beta}\bar{E}^{(r)}_\beta = 0 \qquad (r = 1, ..., \bar{N}; 0 \leq \bar{N} \leq N) , \tag{3.90}$$

$$\mathcal{M}^2_{\alpha\beta}E^{(s)}_\beta = \mathcal{M}^2_s E^{(s)}_\alpha \neq 0 \qquad (s = 1, ..., N - \bar{N}) , \tag{3.91}$$

$$M^2_{ab}\bar{e}^{(p)}_b = 0 \qquad (p = 1, ..., \bar{K}; 0 \leq \bar{K} \leq K) , \tag{3.92}$$

$$M^2_{ab}e^{(q)}_b = M^2_q e^{(q)}_a \neq 0 \qquad (q = 1, ..., K - \bar{K}) . \tag{3.93}$$

One can show (for the proof, see, for example, Ref.[15]) that

$$\bar{K} = N - \bar{N} , \tag{3.94}$$

that is, the number of would be NG bosons coincides with the number of massive vector bosons. This of course reflects the essence of the Higgs phenomenon: NG bosons are transformed into longitudinal components of massive vector bosons. This point will become clearer below when we introduce concrete gauges.

Let us also indicate two useful relations:

$$M^2_{ab}\sigma^\alpha_{bc}v_c = 0 , \tag{3.95}$$

$$\sigma^\alpha_{bc}v_c = ig_\alpha^{-1}\sum_{p=1}^{\bar{K}} E^{(p)}_\alpha \mathcal{M}_p \bar{e}^{(p)}_b . \tag{3.96}$$

The relation (3.95) follows from Eq.(3.83c) and Eqs.(3.84), (3.86); the relation (3.96) follows from Eqs.(3.91) (3.92) and Eq.(3.95).

3.7. Higgs Phenomenon In Non-Abelian Gauge Field Theories. Concrete Gauges

As is well known, a consistent perturbative theory in gauge field theories requires fixing a gauge [16]. The most common way to do this is to add a gauge term, $-\frac{1}{2}[F^\alpha(x)]^2$, to the gauge invariant lagrangian (3.79):

$$\mathcal{L} \longrightarrow \tilde{\mathcal{L}} = \mathcal{L} - \frac{1}{2}[F^\alpha(x)]^2 \ , \tag{3.97}$$

where F^α is a functional of gauge and scalar fields.

The appearance of the gauge term in $\tilde{\mathcal{L}}$ is accompanied, in general case, by the appearance of additional (ghost) fields (the Faddeev-Popov (FP) fields [16]) in the lagrangian. These fields provide the unitarity of the S matrix in the subspace of physical states.

In this section we consider the Higgs phenomenon in non-Abelian gauge theories in some commonly used gauges. We limit ourselves here to considering the tree approximation; in this approximation, the FP fields can be neglected. We shall find the propagators for vector and scalar fields; the form of the propagator for spinor fields is the same in all gauges: $G(p) = (p_\mu\gamma^\mu - m)^{-1}$.

a) *Lorentz gauge*:

$$F^\alpha = -\xi_\alpha^{1/2}\partial^\mu A_\mu^\alpha \tag{3.98}$$

with $\xi_\alpha = \xi_\beta$ if α, β relate to the same subgroup G_i (there is no summation in α in (3.98)).

It is convenient to rewrite the propagators $D_{\mu\alpha,\nu\beta}$ and D_{ab} of vector and scalar fields, respectively, by using the eigenstates of the mass matrices $\mathcal{M}^2$ and M^2 (see Eqs. (3.90)–(3.93)):

$$\begin{aligned} D_{\mu\alpha,\nu\beta} = & \sum_s \frac{E_\alpha^{(s)} E_\beta^{(s)}}{k^2 - \mathcal{M}_s^2}\left(g_{\mu\nu} - \frac{k_\mu k_\nu}{k^2}\right) \\ & + \sum_r \frac{\bar{E}_\alpha^{(r)} \bar{E}_\beta^{(r)}}{k^2}\left(g_{\mu\nu} - \frac{k_\mu k_\nu}{k^2}\right) + \xi_\alpha^{-1}\delta_{\alpha\beta}\left(\frac{k_\mu k_\nu}{(k^2)\cdots}\right. \end{aligned}$$

$$D_{ab} = -\sum_q e_a^{(q)} e_b^{(q)} \frac{1}{k^2 - M_q^2} - \sum_p \bar{e}_a^{(p)} \bar{e}_b^{(p)} \frac{1}{k^2} . \tag{3.100}$$

We also note that there is the vertex

$$ig_\alpha \partial^\mu \varphi'_a \sigma^\alpha_{ab} v_b A^\alpha_\mu \tag{3.101}$$

in the lagrangian density (3.87). It mixes the vector and scalar fields.

This gauge preserves the invariance of the lagrangian under global G transformations. Therefore, in accordance with the Goldstone theorem, there are $\bar{K}$ massless NG bosons in this gauge. However, as it will be seen below, they are unphysical particles.

b) R_ξ *gauge* (compare with Eq.(3.70)):

$$F^\alpha = -\xi^{1/2} \partial^\mu A^\alpha_\mu + \xi^{-1/2} i g_\alpha \left(\varphi'_a \sigma^\alpha_{ab} v_b\right) \tag{3.102}$$

with $\varphi'_a = \varphi_a - v_a$.

In this case, the mixing A^α_μ-φ_b term disappears in the lagrangian.

We find

$$[D]^{-1}_{ab} = \left(M^2 - k^2\right)_{ab} - \xi^{-1} g^2_\alpha \sigma^\alpha_{ac} v_c \sigma^\alpha_{bd} v_d . \tag{3.103}$$

From here and Eq.(3.96), we get

$$D_{ab} = -\sum_q e_a^{(q)} e_b^{(q)} \frac{1}{k^2 - M_q^2} - \sum_p \bar{e}_a^{(p)} \bar{e}_b^{(p)} \frac{\xi}{\xi k^2 - \mathcal{M}_p^2} . \tag{3.104}$$

The propagator of vector fields is (compare with Eq.(3.72))

$$\begin{aligned} D_{\mu\alpha,\nu\beta} = & \sum_s E_\alpha^{(s)} E_\beta^{(s)} \frac{1}{k^2 - \mathcal{M}_s^2} \left(g_{\mu\nu} - \frac{(\xi - 1) k_\mu k_\nu}{\xi k^2 - \mathcal{M}_s^2} \right) \\ & + \sum_r \bar{E}_\alpha^{(r)} \bar{E}_\beta^{(r)} \frac{1}{k^2} \left(g_{\mu\nu} - \frac{(\xi - 1)}{\xi} \frac{k_\mu k_\nu}{k^2} \right) . \end{aligned} \tag{3.105}$$

As $\xi \to 0$, the propagator D_{ab} is reduced to the canonical form for $(K - \bar{K})$ massive scalar fields; the NG bosons disappear in this gauge. Also, as $\xi \to 0$, the vector propagator

$D_{\mu\alpha,\nu\beta}$ consists of two parts; the first part,

$$D'_{\mu\alpha,\nu\beta} = \sum_s E^{(s)}_\alpha E^{(s)}_\beta \frac{1}{k^2 - \mathcal{M}_s^2} \left(g_{\mu\nu} - \frac{k_\mu k_\nu}{\mathcal{M}_s^2} \right) , \tag{3.106}$$

is the canonical propagator for $(N - \bar{N})$ massive vector fields; the second one,

$$D''_{\mu\alpha,\nu\beta} = \lim_{\xi \to 0} \sum_r \bar{E}^{(r)}_\alpha \bar{E}^{(r)}_\beta \frac{1}{k^2} \left(g_{\mu\nu} - \frac{\xi - 1}{\xi} \frac{k_\mu k_\nu}{k^2} \right) , \tag{3.107}$$

coincides with the propagator of massless vector fields in the Lorentz gauge with $\xi = 0$.

A gauge in which only physical degrees of freedom are retained is called the unitary gauge. The propagators of all fields take the canonical form in this gauge. Because of the noncanonical form (3.107) for the propagator of massless vector fields, the R_ξ gauge with $\xi = 0$ is not unitary one. However one can generalize this gauge in order to get the canonical form (coincident with that in the Coulomb gauge) also for the propagator of massless vector fields. This is the so called generalized R_ξ gauge [17].

c) *Generalized R_ξ gauge.*

This is an example of a strictly unitary gauge.

Let us introduce the projection operators in the subspaces of massive and massless vector particles, respectively:

$$P_{\alpha\beta} = \sum_s E^{(s)}_\alpha E^{(s)}_\beta , \tag{3.108}$$

$$\bar{P}_{\alpha\beta} = \sum_r \bar{E}^{(r)}_\alpha \bar{E}^{(r)}_\beta . \tag{3.109}$$

The gauge function is now

$$\begin{aligned} F^\alpha = &- \xi^{1/2} P_{\alpha\beta} \partial^\mu A^\beta_\mu + \xi^{-1/2} i g_\alpha \left(\varphi'_a \sigma^\alpha_{ab} v_b \right) \\ &- \bar{P}_{\alpha\beta} \left[\eta^{1/2} \hat{r}^\mu A^\beta_\mu - \eta^{-1/2} \left(n^\nu \partial_\nu \right) n^\mu A^\beta_\mu \right] , \end{aligned} \tag{3.110}$$

where n^μ is some time-like vector, and $\hat{r}_\mu = \partial_\mu - (n^\nu \partial_\nu) n_\mu$.

This choice of F^α does not change the propagators D_{ab} and $D'_{\mu\alpha,\nu\beta}$ in Eqs.(3.104) and (3.106) but it changes the propagator $D''_{\mu\alpha,\nu\beta}$. It becomes now

$$\begin{aligned} D''_{\mu\alpha,\nu\beta} = & \sum_r \bar{E}_\alpha^{(r)} \bar{E}_\beta^{(r)} \frac{1}{k^2} \left(g_{\mu\nu} - n_\mu n_\nu - \frac{r_\mu r_\nu}{r^2} \right) \\ & + \sum_r \bar{E}_\alpha^{(r)} \bar{E}_\beta^{(r)} \frac{1}{\eta r^2 - (nk)^2} \left(\frac{r_\mu r_\nu}{r^2} - \eta n_\mu n_\nu \right) , \end{aligned} \tag{3.111}$$

where $r_\mu = k_\mu - (n \cdot k) n_\mu$.

With $n_\mu = (1, \mathbf{0})$ and $\eta \to \infty$, the propagator $D''_{\mu\alpha,\nu\beta}$ takes the Coulomb gauge form:

$$\begin{aligned} D''_{0\alpha,0\beta} = & -\sum_r \bar{E}_\alpha^{(r)} \bar{E}_\beta^{(r)} \frac{1}{\mathbf{k}^2} , \\ D''_{i\alpha,j\beta} = & -\sum_r \bar{E}_\alpha^{(r)} \bar{E}_\beta^{(r)} \left(\delta_{ij} - \frac{k_i k_j}{\mathbf{k}^2} \right) \frac{1}{k^2} , \\ D''_{0\alpha,j\beta} = & D_{j\alpha,0\beta} = 0 . \end{aligned} \tag{3.112}$$

Thus the gauge becomes unitary in this limit.

The present consideration clearly demonstrates the essence of the Higgs phenomenon. The main point is that in gauge theories, the Goldstone theorem can be overcome by choosing either the unitary gauge where the initial symmetry is destroyed or a covariant gauge in which the Goldstone theorem is valid but NG bosons become unphysical particles.

Our discussion is based on the tree approximation. A more rigorous treatment of the Higgs phenomenon can be found in Ref. [18].

Chapter 4

Introduction to Chiral Dynamics

4.1. Introduction

In this chapter, we consider the basic properties of chiral symmetry and chiral dynamics intimately related to the problem of generating fermion masses.

The notion of chiral symmetry is directly connected with the fact that there exist two inequivalent representations of the Lorentz group for spinors : $(\frac{1}{2}, 0)$ and $(0, \frac{1}{2})$ (see Sec.1.3 and Appendix A). While the kinetic term and gauge interactions do not mix fermions from these two different representations, the mass term mixes them. Therefore in the absence of the mass term, a wide class of field theories possesses an additional symmetry, the chiral symmetry, under which fermions from the representations $(\frac{1}{2}, 0)$ and $(0, \frac{1}{2})$ transform independently.

The main question which we will discuss in this chapter is how can one spontaneously break chiral symmetry ? We might expect that when this symmetry is broken, fermions will acquire dynamical masses.

4.2. Symmetries and Currents in the Lagrangian Formalism

Let us consider a field theory described by a lagrangian density $\mathcal{L}$ including a set of independent fields φ and their derivatives $\partial^\mu \varphi \equiv \varphi^\mu, \partial^\mu \partial^\nu \varphi \equiv \varphi^{\mu\nu}$, etc. In the usual case, $\mathcal{L}$ depends only on φ and their first derivatives φ^μ, but we will be also interested in effective theories with higher derivatives.

The description of continuous symmetries in quantum field theory is based on the analysis of currents connected with the symmetries. The existence of these currents is guaranteed by the Noether theorem which says that to every continuous symmetry of the lagrangian density there correspond conserved currents.

Following Gell-Mann and Levy [1] (see also the books [2, 3, 4]), we consider currents in a field theory with a lagrangian density $\mathcal{L}$ by using *local* infinitesimal transformations of the fields $\varphi_k(x)$:

$$\varphi_k(x) \longrightarrow \varphi_k'(x) = \varphi_k(x) + i\omega_\alpha(x) F_k^\alpha(\{\varphi\}) , \tag{4.1}$$

where $\omega_\alpha(x)$ are infinitesimal, linearly independent, functions, and the functions F_k^α are not specified at the moment.

The variation of the lagrangian density is

$$\begin{aligned} \delta\mathcal{L} =& \frac{\partial\mathcal{L}}{\partial\varphi_k}\delta\varphi_k + \frac{\partial\mathcal{L}}{\partial\left(\varphi_k^\mu\right)}\delta\varphi_{k;\mu} + \frac{\partial\mathcal{L}}{\partial\left(\varphi_k^{\mu_1\mu_2}\right)}\delta\varphi_{k;\mu_1\mu_2} + \cdots \\ =& \sum_{n=1} \frac{\partial\mathcal{L}}{\partial\left(\varphi_k^{\mu_1\cdots\mu_n}\right)}\delta\varphi_{k;\mu_1\cdots\mu_n} + \frac{\partial\mathcal{L}}{\partial\varphi_k}\delta\varphi_k , \end{aligned} \tag{4.2}$$

where $\varphi^{\mu_1\cdots\mu_n} \equiv (\partial^{\mu_1}\cdots\partial^{\mu_n})\varphi$.

Using Eq.(4.1), we can rewrite $\delta\mathcal{L}$ in the following form:

$$\begin{aligned} \delta\mathcal{L} =& \omega_\alpha K_\alpha(\varphi) + (\partial_\mu\omega_\alpha) K_\alpha^\mu(\varphi) + (\partial_{\mu_1}\partial_{\mu_2}\omega_\alpha) K_\alpha^{\mu_1\mu_2}(\varphi) + \cdots \\ =& \sum_{n=1} (\partial_{\mu_1}\cdots\partial_{\mu_n}\omega_\alpha) K_\alpha^{\mu_1\cdots\mu_n} + \omega_\alpha K_\alpha , \end{aligned} \tag{4.3}$$

where the functions $K^{\mu_1\cdots\mu_n}(\varphi)$ are determined from Eqs.(4.1), (4.2).

Let us consider now the variation (4.3) on solutions ($\bar\varphi_k(x)$) of the Euler-Lagrange (EL) equations of our system. The solutions $\bar\varphi_k(x)$ correspond to stationary points (in the functional sense) of the action $S = \int \mathcal{L}(\varphi)\, d^4x$ under general variations vanishing on the boundary. Therefore, considering the functions $\omega_\alpha(x)$ decreasing rather rapidly at infinity, we find

$$\begin{aligned} \delta S =& \int d^4x \left[\sum_{n=1} (\partial_{\mu_1}\cdots\partial_{\mu_n}\omega_\alpha) K_\alpha^{\mu_1\cdots\mu_n} + \omega_\alpha K_\alpha\right] \\ =& \int d^4x\,\omega_\alpha(x) \left[\sum_{n=1}(-1)^n (\partial_{\mu_1}\cdots\partial_{\mu_n}) K_\alpha^{\mu_1\cdots\mu_n} + K_\alpha\right] = 0 . \end{aligned} \tag{4.4}$$

This equation implies that

$$\sum_{n=1}(-1)^n (\partial_{\mu_1}\cdots\partial_{\mu_n}) K_\alpha^{\mu_1\cdots\mu_n} + K_\alpha = 0 , \tag{4.5}$$

which can be rewritten as

$$K_\alpha + \partial_\mu \jmath_\alpha^\mu = 0 \tag{4.6}$$

with the current

$$\jmath_\alpha^\mu \equiv \sum_{n=1} (-1)^{n+1} \left(\partial_{\mu_1} \cdots \partial_{\mu_n}\right) K_\alpha^{\mu\mu_1\cdots\mu_n} - K_\alpha^\mu \; . \tag{4.7}$$

Let us consider ω_α to be constants. In this case, the EL equations do not guarantee that the corresponding $\delta\mathcal{L} = 0$. In fact, Eqs.(4.4), (4.6) imply that in this case

$$\omega_\alpha \partial_\mu \jmath_\alpha^\mu = -\omega_\alpha K_\alpha = -\delta\mathcal{L} \; . \tag{4.8}$$

Therefore we find that if the transformations (4.1) with constant ω_α correspond to a symmetry of our system ($\delta\mathcal{L} = 0$), then the currents $\jmath_\alpha^\mu$ (4.7) are conserved (Noether's theorem).

The structure of the currents essentially simplifies for lagrangians involving only fields and their first derivatives. In this case, which is very important for applications, the expression (4.7) takes the form

$$\jmath_\mu^\alpha = -\frac{\partial \delta\mathcal{L}}{\partial\left(\partial^\mu \omega_\alpha\right)} = -i\frac{\partial\mathcal{L}}{\partial\left(\partial^\mu \varphi_k\right)} F_k^\alpha\left(\{\varphi\}\right) \; . \tag{4.9}$$

In the derivation of Eq.(4.9) we used Eqs.(4.1) and (4.2).

Let us consider now linear transformations with $F_k^\alpha(\{\varphi\}) = T_{ks}^\alpha \varphi_s$:

$$\varphi_k(x) \longrightarrow \varphi_k'(x) = \varphi_k(x) + i\omega_\alpha T_{ks}^\alpha \varphi_s(x) \; . \tag{4.10}$$

The currents (4.9) are now

$$\jmath_\mu^\alpha = -i\frac{\partial\mathcal{L}}{\partial\left(\partial^\mu \varphi_k\right)} T_{ks}^\alpha \varphi_s \; . \tag{4.11}$$

In a second quantized theory, the currents are operators. Their equal time commutators are determined by the canonical commutation relations:

$$\begin{aligned}
&[\varphi_k(t,\mathbf{x}), \varphi_s(t,\mathbf{y})] = 0 , \\
&[\pi_k(t,\mathbf{x}), \pi_s(t,\mathbf{y})] = 0 , \\
&[\varphi_k(t,\mathbf{x}), \pi_s(t,\mathbf{y})] = i\delta_{ks}\delta^3(\mathbf{x}-\mathbf{y})
\end{aligned} \tag{4.12}$$

with

$$\pi_k = \frac{\partial \mathcal{L}}{\partial(\partial_0\varphi_k)} \tag{4.13}$$

being the canonical conjugate moment (we specifically consider boson fields here although this consideration can be equally applied to the general case with substituting canonical commutators by canonical anticommutators for fermion fields).

The important point is that because of the commutation relations (4.12), the equal time commutators for the charge densities j_0^α (4.11) are model-independent, regardless whether or not the currents are conserved:

$$\begin{aligned}
\left[j_0^\alpha(t,\mathbf{x}), j_0^\beta(t,\mathbf{y})\right] &= -\left[\pi(t,\mathbf{x})T^\alpha\varphi(t,\mathbf{x}), \pi(t,\mathbf{y})T^\beta\varphi(t,\mathbf{y})\right] \\
&= -i\pi(t,\mathbf{x})\left[T^\alpha, T^\beta\right]\varphi(t,\mathbf{y})\delta^3(\mathbf{x}-\mathbf{y}) \\
&= if_{\alpha\beta\gamma}j_0^\gamma(t,\mathbf{x})\delta^3(\mathbf{x}-\mathbf{y}) ,
\end{aligned} \tag{4.14}$$

where the commutator

$$\left[T^\alpha, T^\beta\right] = if_{\alpha\beta\gamma}T^\gamma \tag{4.15}$$

determines the corresponding Lie algebra.

The commutators for the charges $Q^\alpha(t) = \int d^3x j_0^\alpha(t,\mathbf{x})$ are

$$\left[Q^\alpha(t), Q^\beta(t)\right] = if_{\alpha\beta\gamma}Q^\gamma . \tag{4.16}$$

The infinitesimal transformations (4.10) for the fields $\varphi_k(x)$ can be expressed through the commutator $[Q^\alpha(t), \varphi_k(t,\mathbf{x})]$:

$$\delta\varphi_k(x) = i\omega_\alpha(x)T_{ks}^\alpha\varphi_s(x) = -i\omega_\alpha(x)\left[Q^\alpha(t), \varphi_k(t,\mathbf{x})\right] . \tag{4.17}$$

In the next sections, we will apply this formalism to describe the chiral symmetry.

4.3. Chiral Symmetry in Theory of Free Fermions

The Dirac equation for free fermions is considered in Appendix A and Sec.1.3.

Let us begin by considering the lagrangian density for N free Dirac's fermions with the same mass m:

$$\mathcal{L} = i\bar{\psi}_k \gamma^\mu \partial_\mu \psi_k - m\bar{\psi}_k \psi_k, k = 1, 2, \cdots, N . \tag{4.18}$$

It is clear that $\mathcal{L}$ is invariant under $U(N)$ transformations:

$$\psi'_j = [\exp(i\omega_\alpha T^\alpha)]_{jk} \psi_k, \quad \alpha = 0, 1, \cdots, N^2 - 1 , \tag{4.19}$$

where $T^\alpha = \frac{\lambda^\alpha}{2}$ make up a basis of the $U(N)$ Lie algebra $(tr\lambda^\alpha\lambda^\beta = 2\delta_{\alpha\beta})$. At $N = 2$, the Pauli matrices $\tau^0, \boldsymbol{\tau}$ can be chosen as λ^α; at $N = 3$, a convenient choice for λ^α is the Gell-Mann matrices (see Sec. B.1 in Appendix B).

However, as it had been pointed out a long ago [5], the lagrangian density (4.18) possesses a larger symmetry (at least at $N > 1$). It is $O(2N)$, when $m \neq 0$, and it is $U(2N)$ for massless fermions.

To show this, we will use the Majorana representation for Dirac's matrices (see Eq.(A.32) in Appendix A). In this representation, all γ^μ matrices are imaginary, and therefore it is convenient to express the fields through real (Majorana) fermion fields (see Eq.(A.33) in Appendix A):

$$\psi_k = \psi_{k,1} + i\psi_{k,2} , \tag{4.20}$$

where $\psi_{k,1}$, $\psi_{k,2}$ are real. The lagrangian density (4.18) can be rewritten now as

$$\mathcal{L} = i\psi^\top_{k,a}\gamma^0\gamma^\mu\partial_\mu\psi_{k,a} - m\psi^\top_{k,a}\gamma^0\psi_{k,a} \tag{4.21}$$

with $k = 1, 2, \cdots, N$ and $a = 1, 2$. Since γ^μ are imaginary, it is clear that $\mathcal{L}$ (4.21) possesses $O(2N)$ symmetry.

Let us consider now the case with $m = 0$. The fields $\psi_{k,a}$ can be rewritten as (see Eq.(A.18))

$$\psi_{k,a} = \frac{1+\gamma_5}{2}\psi_{k,a} + \frac{1-\gamma_5}{2}\psi_{k,a} = \psi_{Lk,a} + \psi_{Rk,a}, \quad \gamma_5 = -i\gamma^0\gamma^1\gamma^2\gamma^3 \ . \tag{4.22}$$

Since γ_5 is imaginary, $\psi^*_{Lk,a}$ is $\psi^*_{Lk,a} = \psi_{Rk,a}$. Therefore $\mathcal{L}$ with $m = 0$ can be put in the form

$$\mathcal{L} = 2i\psi^\dagger_{Lk,a}\gamma^0\gamma^\mu\partial_\mu\psi_{Lk,a} = 2i\bar{\psi}_{Lk,a}\gamma^\mu\partial_\mu\psi_{Lk,a} \ , \tag{4.23}$$

which explicitly manifests the $U(2N)$ symmetry of $\mathcal{L}$ with massless fermions.

We note that amongst the $U(2N)$ and $O(2N)$ transformations there are transformations which mix fermion and antifermion fields. However, adding gauge interactions usually destroys the invariance under these transformations. Thus, for example, electromagnetic interactions take the form

$$eA_\mu\bar{\psi}_k\gamma^\mu\psi_k = 2ieA_\mu\bar{\psi}_{k,1}\gamma^\mu\psi_{k,2} \ , \tag{4.24}$$

thus mixing fields with $a = 1$ and $a = 2$.

The most important subgroups of the $O(2N)$ and $U(2N)$ are $U(N)$ and $U_L(N) \times U_R(N)$, respectively. The subgroup $U(N)$ is just the group of the transformations (4.19); under the chiral group $U_L(N) \times U_R(N)$, left-handed $\left(\psi_{L,k} = \frac{1+\gamma_5}{2}\psi_k\right)$ and right-handed $\left(\psi_{R,k} = \frac{1-\gamma_5}{2}\psi_k\right)$ fields transform independently.

It is convenient to discuss $U_L(N) \times U_R(N)$ in the spinor representation for γ^μ. In this representation, ψ_k is (see Eq.(1.23))

$$\psi_k = \begin{pmatrix} \psi_{Rk} \\ \psi_{Lk} \end{pmatrix} \equiv \begin{pmatrix} \xi_k \\ \eta_k \end{pmatrix} \ , \tag{4.25}$$

where $\psi_{Lk} = \frac{1+\gamma_5}{2}\psi_k$ and $\psi_{Rk} = \frac{1-\gamma_5}{2}\psi_k$ are assigned to the $\left(0, \frac{1}{2}\right)$ and $\left(\frac{1}{2}, 0\right)$ representations of the Lorentz group, respectively. The lagrangian density (4.18) can be rewritten as

$$\mathcal{L} = i\bar{\psi}_{Lk}\gamma^\mu\partial_\mu\psi_{Lk} + i\bar{\psi}_{Rk}\gamma^\mu\partial_\mu\psi_{Rk} - m\bar{\psi}_{Lk}\psi_{Rk} - m\bar{\psi}_{Rk}\psi_{LK} \ . \tag{4.26}$$

Without mass term, $\mathcal{L}$ is invariant under $U_L(N) \times U_R(N)$ group; $\psi_{Lk}\epsilon(N, 0)$ representation, and $\psi_{Rk}\epsilon(0, N)$. The mass term reduces this symmetry to $U(N)$ which is the diagonal subgroup of $U_L(N) \times U_R(N)$.

Since the current $\jmath^{\mu} = \bar{\psi}_k \gamma^{\mu} \psi_k = \bar{\psi}_{Lk} \gamma^{\mu} \psi_{Lk} + \bar{\psi}_{Rk} \gamma^{\mu} \psi_{Rk}$, the electromagnetic interactions preserve the chiral symmetry.

As is shown in Appendix A, at $m = 0$, right-handed fermions have helicity $s = +1$, and left-handed fermions have $s = -1$.

Charge conjugation transforms right-handed fields into left-handed ones (see Eq.(A.30):

$$\psi^{\alpha}_{cL} = -i\epsilon^{\alpha\beta} \psi^{*}_{R\beta} \,. \tag{4.27}$$

Therefore one can use only left-handed fields, ψ_L and ψ_{cL}, in a theory. The fields ψ_{cL} are assigned to the representation $(0, \bar{N}) \equiv (0, N^{*})$ of $U_L(N) \times U_R(N)$.

By using the general formalism of Sec.4.1, we finds N^2 chiral currents connected with $U_L(N) \times U_R(N)$:

$$\jmath^{\alpha}_{L\mu} = \bar{\psi}\gamma_{\mu} P_L \frac{\lambda^{\alpha}}{2} \psi \,, \tag{4.28}$$

$$\jmath^{\alpha}_{R\mu} = \bar{\psi}\gamma_{\mu} P_R \frac{\lambda^{\alpha}}{2} \psi \tag{4.29}$$

with $P_{L(R)} = \frac{1+(-)\gamma_5}{2}$. The charge densities satisfy the following equal time commutation relations (see Eq.(4.14)):

$$\left[\jmath^{\alpha}_{L0}(t,\mathbf{x}), \jmath^{\beta}_{L0}(t,\mathbf{y})\right] = i f_{\alpha\beta\gamma} \jmath^{\gamma}_{L0}(t,\mathbf{x})\, \delta^3(\mathbf{x}-\mathbf{y}) \,, \tag{4.30}$$

$$\left[\jmath^{\alpha}_{R0}(t,\mathbf{x}), \jmath^{\beta}_{R0}(t,\mathbf{y})\right] = i f_{\alpha\beta\gamma} \jmath^{\gamma}_{R0}(t,\mathbf{x})\, \delta^3(\mathbf{x}-\mathbf{y}) \,, \tag{4.31}$$

$$\left[\jmath^{\alpha}_{L0}(t,\mathbf{x}), \jmath^{\beta}_{R0}(t,\mathbf{y})\right] = 0 \,, \tag{4.32}$$

where $f_{\alpha\beta\gamma}$ are the structure constants of $U(N)$ algebra.

One can easily check that

$$\partial^{\mu} \jmath^{\alpha}_{\mu}(x) = 0 \,, \tag{4.33}$$

$$\partial^{\mu} \jmath^{\alpha}_{5\mu}(x) = i m \bar{\psi}(x) \gamma_5 \lambda^{\alpha} \psi(x) \,, \tag{4.34}$$

where

$$\jmath^{\alpha}_{\mu} = \bar{\psi}\gamma_{\mu} \frac{\lambda^{\alpha}}{2} \psi \,, \quad \jmath^{\alpha}_{5\mu} = \bar{\psi}\gamma_{\mu}\gamma_5 \frac{\lambda^{\alpha}}{2} \psi \,. \tag{4.35}$$

Thus, as it has to be, the mass term explicitly breaks the chiral $U_L(N) \times U_R(N)$ to its diagonal subgroup $U_{L+R}(N)$.

Using Eq.(4.17), we find the explicit transformation law for ψ_{Lk}, ψ_{Rk} fermions under the chiral transformations:

$$\delta_L \psi_{Lk} = i\omega_{L\alpha} T^{\alpha}_{ks} \psi_{Ls}, \delta_L \psi_{Rk} = 0 \; , \tag{4.36}$$

$$\delta_R \psi_{Lk} = 0 \; , \delta_R \psi_{Rk} = i\omega_{R\alpha} T^{\alpha}_{ks} \psi_{Rs} \; . \tag{4.37}$$

The main topic of the present book is how can one arrange spontaneous chiral symmetry breaking, that is how can one generate fermion masses, without destroying the conservation of the chiral currents? In the next section, we will consider one of the simplest realization of such a dynamics.

4.4. Linear σ-Model

Historically, the first models, in which fermions get masses despite the invariance of the action under chiral transformations, were invented by Nishijima and Gürsey [6, 7]. Nishijima and Gürsey built the models with a nonlinear realization of the $U_L(1) \times U_R(1)$ [6] and $SU_L(2) \times SU_R(2)$ [7] chiral symmetries, respectively, in which fermions were massive.*) We will consider a general scheme of the nonlinear realization of spontaneous chiral symmetry breaking in Chapter 5. Here we will discuss a simpler dynamical realization of this phenomenon (the linear σ-model) given by Gell-Mann and Levy [1].

Let us begin by considering the second Goldstone's model with the lagrangian density (see Eq.(3.14)):

$$\begin{aligned} \mathcal{L} =& \partial_\mu \varphi^\dagger \partial^\mu \varphi - U\left(\varphi^\dagger \varphi\right) \\ =& \frac{1}{2}\left(\partial_\mu \sigma \partial^\mu \sigma + \partial_\mu \eta \partial^\mu \eta\right) - U\left(\sigma^2 + \eta^2\right) \; , \end{aligned} \tag{4.38}$$

where $\varphi = \frac{1}{\sqrt{2}}(\sigma + i\eta)$,

$$U\left(\sigma^2 + \eta^2\right) = \frac{\mu^2}{2}\left(\sigma^2 + \eta^2\right) + \frac{\lambda}{4!}\left(\sigma^2 + \eta^2\right)^2 \; . \tag{4.39}$$

*) In hindsight, it is clear that in their pioneering papers, Nishijima and Gürsey gave the first dynamical realizations of spontaneous chiral symmetry breaking by using massless pseudoscalar fields in order to construct chiral invariant models with massive fermions. However, at that time, they did not recognize this conception.

Let us assign the following transformation law for σ and η under space inversion P:

$$P^{-1}\sigma P = \sigma, \; P^{-1}\eta P = -\eta \, , \tag{4.40}$$

i.e., σ and η are scalar and pseudoscalar, respectively. Then we can add to $\mathcal{L}$ the following P-invariant term for a Dirac fermion (for transformations of fermions under P see Eq.(A.34) in Appendix A):

$$\mathcal{L}_f = i\bar{\psi}\gamma^\mu \partial_\mu \psi - \sqrt{2}g\left(\sigma\bar{\psi}\psi - \eta\bar{\psi}i\gamma_5\psi\right) \, . \tag{4.41}$$

Taking into account the transformations (4.36), (4.37) of fermions under $U_L(1) \times U_R(1)$, we find that the lagrangian density $\tilde{\mathcal{L}} = \mathcal{L} + \mathcal{L}_f$ is $U_L(1) \times U_R(1)$ symmetric if

$$\begin{aligned} \delta_V\sigma &= \delta_V\eta = 0 \, , \\ \delta_A\sigma &= -\sqrt{2}\omega_A\eta \, , \;\; \delta_A\eta = \sqrt{2}\omega_A\sigma \, , \end{aligned} \tag{4.42}$$

where $Q_V = Q_L + Q_R \, , \;\; Q_A = Q_L - Q_R$ (the reason for introducing $\sqrt{2}$ in Eqs.(4.41), (4.42) is connected with our choice of the normalization, $tr\left(T^\alpha T^\beta\right) = \frac{1}{2}\delta^{\alpha\beta}$, for the chiral generators of $U_L(N) \times U_R(N)$).

We note that σ and η transform under $U_L(1) \times U_R(1)$ as

$$\sigma \sim -\bar{\psi}\psi \, , \; \eta \sim \bar{\psi}i\gamma_5\psi \, . \tag{4.43}$$

When μ^2 in the potential is negative, the $U_L(1) \times U_R(1)$ symmetry breaks down to $U_V(1) \equiv U_{L+R}(1)$, and in the tree approximation

$$\langle 0|\sigma|0\rangle = v = \left(-6\mu^2/\lambda\right)^{1/2} \, . \tag{4.44}$$

As a result, the fermion acquires a dynamical mass (see Eq.(4.41)):

$$m = \sqrt{2}gv \, . \tag{4.45}$$

The η-particle is a massless NG boson, and the mass of σ, in the tree approximation, is (see Eq.(3.20) in Chapter 3):

$$\mu_\sigma^2 = \lambda v^2/_3 \ . \tag{4.46}$$

By using the approach of Sec.4.1, we find the chiral currents in this model:

$$\begin{aligned} \jmath_{L\mu} &= \frac{1}{\sqrt{2}}\left[\bar{\psi}\gamma_\mu \frac{1+\gamma_5}{2}\psi - (\sigma\partial_\mu\eta - \eta\partial_\mu\sigma)\right] \ , \\ \jmath_{R\mu} &= \frac{1}{\sqrt{2}}\left[\bar{\psi}\gamma_\mu \frac{1-\gamma_5}{2}\psi + (\sigma\partial_\mu\eta - \eta\partial_\mu\sigma)\right] \ . \end{aligned} \tag{4.47}$$

Let us define the so called decay constant of the NG boson F as

$$iP_\mu F = \langle 0|\jmath_{5\mu}(0)|\eta; P\rangle = \langle 0|\left(\jmath_{L\mu}(0) - \jmath_{R\mu}(0)\right)|\eta; P\rangle \ . \tag{4.48}$$

Eqs.(4.44) and (4.46), (4.47) imply that in the tree approximation, F is

$$F = \sqrt{2}v \ . \tag{4.49}$$

Form here and Eq.(4.45) we find:

$$m = gF \ . \tag{4.50}$$

This relation is an example of the Goldberger-Treiman (GT) relation [8] between decay constant F, dynamical fermion mass and Yukawa coupling constant (see Sec.4.4 for a more detailed discussion of the GT relation).

Thus in this model, the fermion mass appears as a result of spontaneous chiral symmetry breaking.

Let us consider the linear σ-model with the chiral group $U_L(N) \times U_R(N)$ (or $SU_L(N) \times SU_R(N) \times U_V(1)$) [1, 9, 10]. We introduce boson fields as an $N \times N$ complex matrix M:

$$M = \frac{1}{\sqrt{2}}\left[(\sigma + i\eta)\lambda^0 + \left(\sigma^{\alpha'} + i\pi^{\alpha'}\right)\lambda^{\alpha'}\right] \tag{4.51}$$

with $\lambda^0 = \sqrt{\frac{2}{N}}\, I$ and $\alpha' = 1, 2, \cdots, N^2 - 1$. The field M is assigned to the $(N, \bar{N})$

representation of the $SU_L(N) \times SU_R(N)$ chiral group:

$$M \longrightarrow LMR^{\dagger} . \tag{4.52}$$

The general $U_L(N) \times U_R(N)$ and P invariant renormalizable lagrangian density for M takes the form:

$$\begin{aligned}\mathcal{L}_M =& \frac{1}{2} tr\left[\partial_\mu M^\dagger \partial^\mu M\right] - \frac{\lambda N}{4!}\left[tr\left(M^\dagger M M^\dagger M\right)\right.\\ &\left.+\frac{\delta}{N}\left(tr\left(M^\dagger M\right)\right)^2\right] - \frac{\mu^2}{2} tr\left(M^\dagger M\right) .\end{aligned} \tag{4.53}$$

If the symmetry is reduced from $U_L(N) \times U_R(N)$ to $SU_L(N) \times SU_R(N) \times U_V(1)$, one can add the following term to $\mathcal{L}_M$:

$$\gamma\left(\det M^\dagger + \det M\right) \tag{4.54}$$

which is renormalizable at $N \leq 4$.

The Yukawa interaction between M and fermions takes the form:

$$\mathcal{L}_Y = -\sqrt{2} g\left[\bar{\psi}_L M \psi_R + \bar{\psi}_R M^\dagger \psi_L\right] . \tag{4.55}$$

By using the approach of Sec.4.1, we find that the chiral currents are:

$$j^\alpha_{L\mu} = \bar{\psi}\gamma_\mu P_L \frac{\lambda^\alpha}{2}\psi + \frac{i}{2}\left[tr\left(M^\dagger \frac{\lambda^\alpha}{2}\partial_\mu M\right) - tr\left(\partial_\mu M^\dagger \frac{\lambda^\alpha}{2} M\right)\right] , \tag{4.56}$$

$$\begin{aligned}j^\alpha_{R\mu} &= \bar{\psi}\gamma_\mu P_R \frac{\lambda^\alpha}{2}\psi + \frac{i}{2}\left[tr\left(M \frac{\lambda^\alpha}{2}\partial_\mu M^\dagger\right) - tr\left(\partial_\mu M \frac{\lambda^\alpha}{2} M^\dagger\right)\right] ,\\ \alpha &= 0, 1, \cdots, N^2 - 1.\end{aligned} \tag{4.57}$$

If $\mu^2 < 0$, the minimum of the potential in Eq.(4.53) is at $M = v\frac{\lambda^0}{\sqrt{2}}$ with

$$v = \langle 0|\sigma|0\rangle = \left(\frac{-6\mu^2}{(1+\delta)\lambda}\right)^{1/2} , \tag{4.58}$$

and chiral $U_L(N) \times U_R(N)$ symmetry spontaneously breaks down to $U_V(N) \equiv U_{L+R}(N)$.

Eqs.(4.56), (4.57) imply that the decay constant F of the NG bosons π^α,

$$i\delta_{\alpha\beta}FP_\mu = \langle 0|\jmath^\alpha_{5\mu}(0)|\pi^\beta;P\rangle = \langle 0|\left(\jmath^\alpha_{L\mu}(0) - \jmath^\alpha_{R\mu}(0)\right)|\pi^\beta;P\rangle\ , \tag{4.59}$$

$(\pi^\alpha|_{\alpha=0} \equiv \eta)$ is

$$F = \left(\frac{2}{N}\right)^{1/2} v\ . \tag{4.60}$$

From Eq.(4.55) we find that the dynamical fermion mass m is

$$m = gF\ . \tag{4.61}$$

Let us consider the case with $N = 2$ in more detail.

The currents (4.56), (4.57) are in this case:

$$\jmath^0_{L\mu} = \frac{1}{2}\bar\psi\gamma_\mu P_L\psi - \frac{1}{2}\left(\sigma\partial_\mu\eta - \eta\partial_\mu\sigma + \sigma^{\alpha'}\partial_\mu\pi^{\alpha'} - \pi^{\alpha'}\partial_\mu\sigma^{\alpha'}\right)\ , \tag{4.62}$$

$$\begin{aligned}\jmath^{\alpha'}_{L\mu} =& \bar\psi\gamma_\mu P_L\frac{\tau^{\alpha'}}{2}\psi - \frac{1}{2}\left(\sigma\partial_\mu\pi^{\alpha'} - \pi^{\alpha'}\partial_\mu\sigma + \sigma^{\alpha'}\partial_\mu\eta - \eta\partial_\mu\sigma^{\alpha'}\right)\\ &+ \frac{1}{2}\epsilon_{\alpha'\beta'\gamma'}\left(\pi^{\beta'}\partial_\mu\pi^{\gamma'} + \sigma^{\beta'}\partial_\mu\sigma^{\gamma'}\right)\end{aligned} \tag{4.63}$$

with $\alpha' = 1,2,3$; $\jmath^0_{R\mu}, \jmath^{\alpha'}_{R\mu}$ are obtained from $\jmath^0_{L\mu}, \jmath^{\alpha'}_{L\mu}$ by substituting $P_L \longrightarrow P_R$ and $\eta \longrightarrow -\eta$, $\pi^{\alpha'} \longrightarrow -\pi^{\alpha'}$.

The interesting point at $N = 2$ is that in the model with $SU_L(2)\times SU_R(2)\times U_V(1)$ symmetry, we can retain only the fields σ and $\pi^{\alpha'}$, or η and $\sigma^{\alpha'}$, in the lagrangian. This follows from the fact that the representations $\{2\}$ and $\{\bar 2\}$ of $SU(2)$ are isomorphic: if $\varphi\epsilon\{2\}$, then $\tau_2\varphi^* \sim \varphi$. Therefore in this case

$$\tau_2 M^*\tau_2 \sim M\ , \tag{4.64}$$

and one can consider a theory with M satisfying the condition

$$\tau_2 M^*_\pm\tau_2 = \pm M_\pm\ . \tag{4.65}$$

Taking into account Eq.(4.51), we find that $M_\pm$ are

$$M_+ = \frac{1}{\sqrt{2}}\left(\sigma + i\pi^{\alpha'}\tau^{\alpha'}\right) , \tag{4.66}$$

$$M_- = \frac{i}{\sqrt{2}}\left(\eta - i\sigma^{\alpha'}\tau^{\alpha'}\right) . \tag{4.67}$$

The Gell-Mann-Levy lagrangian density for M_+ and fermions is

$$\begin{aligned}\mathcal{L} =& \frac{1}{2}\left(\partial_\mu\sigma\partial^\mu\sigma + \partial_\mu\boldsymbol{\pi}\partial^\mu\boldsymbol{\pi}\right) + i\bar{\psi}\gamma^\mu\partial_\mu\psi - \frac{\lambda}{4!}\left(\sigma^2+\boldsymbol{\pi}^2\right)^2 \\ &- \frac{\mu^2}{2}\left(\sigma^2+\boldsymbol{\pi}^2\right) - g\left(\sigma\bar{\psi}\psi - \boldsymbol{\pi}\bar{\psi}i\gamma_5\boldsymbol{\tau}\psi\right) .\end{aligned} \tag{4.68}$$

The chiral currents in this model are obtained from the currents (4.63) just omitting terms with η and $\sigma^{\alpha'}$:

$$\begin{aligned} j^{\alpha'}_{L\mu} =& \bar{\psi}\gamma_\mu P_L \frac{\tau^{\alpha'}}{2}\psi - \frac{1}{2}\left(\sigma\partial_\mu\pi^{\alpha'} - \pi^{\alpha'}\partial_\mu\sigma\right) + \frac{1}{2}\epsilon_{\alpha\beta\gamma}\pi^{\beta'}\partial_\mu\pi^{\gamma'} , \\ j^{\alpha'}_{R\mu} =& \bar{\psi}\gamma_\mu P_R \frac{\tau^{\alpha'}}{2}\psi + \frac{1}{2}\left(\sigma\partial_\mu\pi^{\alpha'} - \pi^{\alpha'}\partial_\mu\sigma\right) + \frac{1}{2}\epsilon_{\alpha'\beta'\gamma'}\pi^{\beta'}\partial_\mu\pi^{\gamma'} .\end{aligned} \tag{4.69}$$

As in the case of the $U_L(1)\times U_R(1)$ model, the mass of σ is equal to

$$\mu_\sigma^2 = \lambda v^2/3 \tag{4.70}$$

in the tree approximation.

From Eqs.(4.52), (4.66), we find the transformations of the σ and π^α fields:

$$\begin{aligned}\delta_L\sigma =& -\frac{1}{2}\omega_{L\alpha'}\pi^{\alpha'} ,\ \delta_L\pi^{\alpha'} = \frac{1}{2}\omega_{L\alpha'}\sigma - \frac{1}{2}\epsilon_{\alpha'\beta'\gamma'}\omega_{L\beta'}\pi^{\gamma'} , \\ \delta_R\sigma =& \frac{1}{2}\omega_{R\alpha'}\pi^{\alpha'} ,\ \delta_R\pi^{\alpha'} = -\frac{1}{2}\omega_{R\alpha'}\sigma - \frac{1}{2}\epsilon_{\alpha'\beta'\gamma'}\omega_{R\alpha'}\pi^{\gamma'} ,\end{aligned} \tag{4.71}$$

or (compare with Eq.(4.42))

$$\begin{aligned}\delta_V\sigma =& 0,\ \delta_V\pi^{\alpha'} = -\epsilon_{\alpha'\beta'\gamma'}\omega_{V\beta'}\pi^{\gamma'} , \\ \delta_A\sigma =& -\omega_{A\alpha'}\pi^{\alpha'} ,\ \delta_A\pi^{\alpha'} = \omega_{A\alpha'}\sigma .\end{aligned} \tag{4.72}$$

As is known from experiment, the Yukawa coupling constant for nucleons is large: $g^2_{\pi N}/4\pi \approx 14.5$. Therefore the tree approximation in the σ model (as well as the perturbative expansion in $g_{\pi N}$ itself) should not be reliable, and one might think that the model is not very useful for describing spontaneous chiral symmetry breaking in hadron dynamics. Fortunately, this is not so. The remarkable point is that the model perfectly works in the low energy, $q \leq 1\mathrm{GeV}$, region [2, 3]. We shall consider the explanation of this very important fact in Chapter 5.

4.5. The PCAC Hypothesis and the Goldberger-Treiman Relation

Although there are many interesting phenomenological applications of spontaneous chiral symmetry breaking in hadron dynamics, these are not our primary goal in this book. A comprehensive discussions of these applications can be found in the books [2, 3]. In this section, we consider just few points in that phenomenology intimately connected with the conception of dynamic chiral symmetry breaking.

Spontaneous breaking of the $SU_L(2) \times SU_R(2)$ chiral symmetry implies that there should be three pseudoscalar NG bosons, π. However there are no massless hadrons in nature. The lightest hadrons are pions with the mass $m_\pi \simeq 140\mathrm{MeV}$. This implies that the axial currents $j^\alpha_{5\mu}$ are not conserved.*) Indeed, let us consider the equation (4.59),

$$i\delta_{\alpha\beta} F P_\mu = \langle 0 | j^\alpha_{5\mu}(0) | \pi^\beta ; P \rangle \ . \tag{4.73}$$

From this equation we find that

$$\langle 0 | \partial^\mu j^\alpha_{5\mu}(0) | \pi^\beta ; P \rangle = \delta_{\alpha\beta} m_\pi^2 F \ , \tag{4.74}$$

thus yielding $\partial^\mu j^\alpha_{5\mu} \neq 0$ at $m_\pi^2 \neq 0$ and $F \neq 0$ (the decay constant $F = F_\pi$ for pions is $F_\pi \simeq 93\mathrm{MeV}$).

However, since pions are much lighter than other hadrons, one can think about them as " almost " NG bosons. This means that there is both spontaneous and explicit chiral symmetry breaking in nature, and for hadrons composed from two light quarks, u and d, spontaneous symmetry breaking dominates. This conception known as the partial conservation of the axial currents (PCAC) hypothesis proved to be extremely fruitful [2, 3].

---

*) Henceforth we understand that $\alpha = 1, 2, 3$.

Let us begin by considering the idealized case of massless pions. The axial currents $j^{\alpha}_{5\mu}$ are conserved. Let us show, following Nambu [11], that in this chiral limit there is an exact relation between F_{π}, the nucleon mass m_N, and the Yukawa coupling $g_{\pi N}$ (the Goldberger-Treiman (GT) relation [8]).*)

Let us consider the matrix element

$$\langle N(p_2)|j^{\alpha}_{5\mu}(0)|N(p_1)\rangle , \tag{4.75}$$

where $|N(p_2)\rangle, |N(p_1)\rangle$ are nucleon states. Because of the Lorentz and translational invariance and also because of the invariance of the hadronic interactions with respect to the isospin symmetry, space inversion P and charge conjugation C, the general structure of this matrix element is [2, 3]

$$\langle N(p_2)|j^{\alpha}_{5\mu}(0)|N(p_1)\rangle = \bar{u}(p_2)\left[\gamma_\mu G_1(q^2) + q_\mu G_2(q^2)\right]\gamma_5 \frac{\tau^\alpha}{2} u(p_1) , \tag{4.76}$$

where $q_\mu = (p_2 - p_1)_\mu$, and $G_1(q^2), G_2(q^2)$ are some functions of q^2. Since in the chiral limit $\partial^\mu j^{\alpha}_{5\mu} = 0$, we find

$$\begin{aligned} & i\bar{u}(p_2)\left[q^\mu \gamma_\mu G_1(q^2) + q^2 G_2(q^2)\right]\gamma_5 \frac{\tau^\alpha}{2} u(p_1) \\ =& i\bar{u}(p_2)\gamma_5 \frac{\tau^\alpha}{2} u(p_1)\left[2m_N G_1(q^2) + q^2 G_2(q^2)\right] = 0 . \end{aligned} \tag{4.77}$$

In this limit, the mass m_N is purely dynamical.

Eq.(4.77) implies that

$$G_2(q^2) = -\frac{2m_N G_1(q^2)}{q^2} . \tag{4.78}$$

The quantity $g_A \equiv G_1(0)$ is called the axial vector constant of nucleons. It is measured in the β -decay of neutron; $g_A \simeq 1.2$. Thus we conclude that there is a massless pole in $G_2(q^2)$. Of course it corresponds to massless pions. Indeed, Eq.(4.73) implies that the pole contribution in $\langle N(p_2)|j^{\alpha}_{5\mu}|N(p_1)\rangle$ is (see. Fig.4.1):

$$-q_\mu F_\pi g_{\pi N} \bar{u}(p_2)\gamma_5 \tau^\alpha u(p_1)\frac{1}{q^2} \tag{4.79}$$

(our definition of $g_{\pi N}$ agrees with g in the lagrangian density of the σ-model (4.68)).

*) The original derivation [8] of the GT relation was based on using dispersion relations.

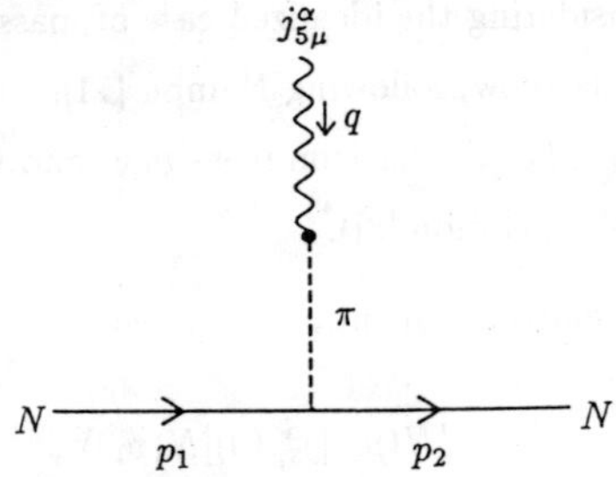

Fig. 4.1

From Eqs.(4.78) and (4.79) we find the GT relation

$$F_\pi = g_A m_N / g_{\pi N} \tag{4.80}$$

which is *exact* in the chiral limit. In the real world it fulfills at the level of 10% thus confirming that the dynamics of light hadrons is not so far from this limit.

Let us return to the real situation with $m_\pi^2 \neq 0$. Now the pion contribution to the matrix element (4.76) is somewhat different:

$$-q_\mu F_\pi \bar{g}_{\pi N} \bar{u}(p_2)\gamma_5 \tau^\alpha u(p_1) \frac{1}{q^2 - m_\pi^2} \tag{4.81}$$

(compare with Eq.(4.79)), where the parameters $F_\pi, \bar{g}_{\pi N}$ are defined at the pion mass shell $q^2 = m_\pi^2$ (the coupling $g_{\pi N}$ is determined at $q^2 = 0$).

Let us consider the derivation of the GT relation in this case. The combination

$$G(q^2) \equiv 2m_N G_1(q^2) + q^2 G_2(q^2) \tag{4.82}$$

is not zero now. The pole contribution in $G(q^2)$ is (see Eq.(4.81)):

$$\left[G(q^2)\right]_{\text{pole}} = \frac{2m_\pi^2 F_\pi \bar{g}_{\pi N}}{m_\pi^2 - q^2} \,. \tag{4.83}$$

Since now there is no pole at $q^2 = 0$ in $G_2(q^2)$, we find that

$$g_A \equiv G_1(0) = G(0)/2m_N \,. \tag{4.84}$$

The crucial step in estimating $G(0)$ is the assumption of an unsubtracted dispersion relation for $G(q^2)$:

$$G(q^2) = \frac{2m_\pi^2 F_\pi \bar{g}_{\pi N}}{m_\pi^2 - q^2} + \frac{1}{\pi} \int\limits_{9m_\pi^2}^{\infty} \frac{\rho(q'^2)}{q'^2 - q^2} dq'^2 \tag{4.85}$$

($9m_\pi^2$ corresponds to the three pion threshold).

The PCAC hypothesis assumes the pion pole dominance in the matrix elements of the axial currents $j_{5\mu}^\alpha$ and (at $m_\pi^2 \neq 0$) their divergences $\partial^\mu j_{5\mu}^\alpha$ in the low energy region. Specifically, in the present case, it means that $\rho(q^2)$ in Eq.(4.85) is $\rho(q^2) \sim m_\pi^2$. Therefore the integral in this relation is suppressed as $m_\pi^2 \longrightarrow 0$. Assuming that it can be neglected in the real world with $m_\pi \simeq 140\text{MeV}$, we find the GT relation:

$$g_A = G(0)/2m_N = F_\pi \bar{g}_{\pi N}/m_N \ . \tag{4.86}$$

We note, however, that in order to rewrite it in the form

$$g_A = F_\pi g_{\pi N}/m_N \tag{4.87}$$

with $g_{\pi N}^2/4\pi \simeq 14.5$, one has to assume that $\bar{g}_{\pi N}$ relating to $q^2 = m_\pi^2$ is close to $g_{\pi N}$ connected with $q^2 = 0$ (the smoothness assumption). This assumption in a necessary ingredient of the PCAC hypothesis.

Let us consider how the PCAC hypothesis is realized in the linear σ-model.

To break the $SU_L \times SU_R(2)$ symmetry in a P-invariant way, we add the term

$$\mathcal{L}_b = a\sigma \tag{4.88}$$

in the lagrangian density (4.68). Then, in the tree approximation, we find from the potential

$$V = \frac{\lambda}{4!}\left(\sigma^2 + \boldsymbol{\pi}^2\right)^2 + \frac{\mu^2}{2}\left(\sigma^2 + \boldsymbol{\pi}^2\right) - a\sigma \tag{4.89}$$

the following relations $\left(\mu^2 < 0\right)$:

$$\begin{aligned} a &= v\left(\mu^2 + \frac{\lambda}{6}v^2\right), \\ m_\pi^2 &= \mu^2 + \frac{\lambda}{6}v^2 , \\ m_\sigma^2 &= \mu^2 + \frac{\lambda}{2}v^2 , \end{aligned} \tag{4.90}$$

and

$$\partial^\mu j^\alpha_{5\mu} = a\pi^\alpha \, . \tag{4.91}$$

The relations (4.90) imply that $a = v m_\pi^2$, and therefore

$$\partial^\mu j^\alpha_{5\mu} = v m_\pi^2 \pi^\alpha = F_\pi m_\pi^2 \pi^\alpha \, . \tag{4.92}$$

Thus in the σ-model the divergence of the axial currents is proportional to the canonical pion field.

In the tree approximation, and in any order of perturbative expansion, the PCAC hypothesis fulfills in this model: the pion pole dominates in the matrix elements of $j^\alpha_{5\mu}$ and $\partial^\mu j^\alpha_{5\mu}$ in the low energy region.

We note that one can break the chiral symmetry by introducing in the lagrangian (instead the term (4.88)) a mass term for fermions:

$$\mathcal{L}_b = -m^{(0)}\bar{\psi}\psi \, . \tag{4.93}$$

That is what happens in quantum chromodynamics (QCD): quarks have nonzero bare masses (see Chapter 12). In this case,

$$\partial^\mu j^\alpha_{5\mu} = i m^{(0)} \bar{\psi}\gamma_5 \tau^\alpha \psi \, , \tag{4.94}$$

and because of $m^{(0)} \sim m_\pi^2$ (as will be shown in Sec. 9.3) and " smoothness" of the composite operator $\bar{\psi}\gamma_5\tau^\alpha\psi$ with the dimension $d = 3$, the PCAC hypothesis still fulfills there.

As we already know, in the tree approximation, the GT relation takes a special form in the linear σ-model (see Eq.(4.61)):

$$F_\pi = m_N / g_{\pi N} . \tag{4.95}$$

This is of course connected with the point that the axial-vector coupling $g_A = 1$ in the tree approximation there. Beyond this approximation, g_A is not more equal to one, however, other parameters change in such a way that the GT relation (4.80) is still valid.

The GT relation is an example of a low energy relation (theorem) in chiral dynamics: although parameters, like $g_A, g_{\pi N}$, etc., in these relations are model dependent, the low energy theorems themselves are universal [2, 3], that is, all models with spontaneous breakdown of the same chiral symmetry reproduce these relations.

This is the main reason why a toy model such as the linear σ-model is so useful. This is also the basic point in constructing the effective chiral lagrangians which we will discuss in Chapter 5.

In this chapter, we considered the basic characteristics of the chiral dynamics. The dynamical realization of the spontaneous chiral symmetry breaking was given in the linear σ-model. As we shall later see the basic points of the realization of chiral symmetry breaking in this toy model remain unchanged in such much more sophisticated field theories as QCD or models of electroweak symmetry breaking.

Chapter 5

Nonlinear Realization of Spontaneous Symmetry Breaking

5.1. Introduction

In this chapter we consider the nonlinear realization of spontaneous symmetry breaking and the theory of nonlinear effective lagrangians. This theory is extremely beautiful. This is in particular connected with its geometrical origin –in fact, the theory has some common features with gauge field theories.

But the importance of the nonlinear realization is connected not just with its beauty. Nonlinear chiral lagrangians give a powerful tool with which to study the low energy dynamics of hadrons. Also, nonlinear lagrangians give an important example of the effective lagrangian approach in quantum field theory. At present this approach is very commonly used to study both the nonperturbative dynamics of hadrons and the electroweak symmetry breaking problem (see Chapters 12 and 13).

5.2. Nonlinear Limit of the Linear σ-Model

In Sec.4.4, we considered the linear σ-models with the chiral symmetries $U_L(N) \times U_R(N)$ or $SU_L(N) \times SU_R(N) \times U_V(1)$. The models contain the boson fields

$$M = \frac{1}{\sqrt{2}} \left[(\sigma + i\eta) \lambda^0 + (\sigma^\alpha + i\pi^\alpha) \lambda^\alpha \right] \tag{5.1}$$

and fermion fields (here we understand that $\alpha = 1, 2, ..., N^2 - 1$). Besides massless NG bosons, there are also massive spinless particles σ and σ^α. If we are interested in the dynamics in low energy region with a characteristic momentum scale much smaller than masses of all these particles, one should expect that the massive particles will effectively decouple from the dynamics of NG bosons. In this case, it is desirable to have a low energy effective lagrangian not including these heavy bosons.

To derive such a lagrangian, it is convenient to consider the limit in which masses of the heavy bosons go to infinity [1]. Let us first consider this limit in the simplest linear σ-models with the chiral symmetries $U_L(1) \times U_R(1)$ and $SU_L(2) \times SU_R(2)$.

The lagrangian density of the first model is (see Eq. (4.38)):

$$\mathcal{L} = \frac{1}{2}\left(\partial_\mu \sigma \partial^\mu \sigma + \partial_\mu \eta \partial^\mu \eta\right) - \frac{\mu^2}{2}\left(\sigma^2 + \eta^2\right) - \frac{\lambda}{4!}\left(\sigma^2 + \eta^2\right)^2 . \tag{5.2}$$

When $\mu^2 < 0$, the mass of σ is

$$\mu_\sigma^2 = \lambda v^2/3 \tag{5.3}$$

with $v^2 = -6\mu^2/\lambda$; the decay constant $F = \sqrt{2}v$ (see Eqs.(4.44), (4.49)). We are interested in the limit $\mu_\sigma^2 \longrightarrow \infty$ with v being fixed. It is equivalent to that of $\lambda \longrightarrow \infty$, $\mu^2 \longrightarrow -\infty$ with μ^2/λ being fixed.

In this limit, the equation of motion of the σ field is

$$\partial^\mu \partial_\mu \sigma + \mu^2 \sigma + \frac{\lambda}{3!}\sigma\left(\sigma^2 + \eta^2\right) = 0 \longrightarrow \frac{\lambda}{6}\sigma\left[-v^2 + \left(\sigma^2 + \eta^2\right)\right] = 0 . \tag{5.4}$$

The stable vacuum corresponds to the nontrivial solution

$$\sigma^2 + \eta^2 = v^2 = F^2/2 . \tag{5.5}$$

Although the equation (5.5) looks exactly the same as that determining $\langle 0|\sigma|0\rangle$ in the tree approximation, its meaning is now essentially different. In this limit, it is an exact operator equation. Since the equation does not include time derivatives, it plays the role of a constraint.

The dynamical system with the lagrangian density (5.2) and the constraint (5.5) is equivalent to the system with the lagrangian density

$$\mathcal{L} = \frac{1}{2}\left(\partial_\mu \sigma \partial^\mu \sigma + \partial_\mu \eta \partial^\mu \eta\right) \tag{5.6}$$

and the same constraint. In fact, one can resolve this constraint by choosing $M = \sigma + i\eta$ as $M = \frac{F}{\sqrt{2}} \exp\left(\sqrt{2}i\tilde{\eta}/F\right)$. Then the lagrangian density (5.6) becomes:

$$\mathcal{L} = \frac{1}{2}\partial_\mu M \partial^\mu M^\dagger = \frac{F^2}{4}\partial_\mu \Sigma \partial^\mu \Sigma^\dagger = \frac{1}{2}\partial_\mu \tilde{\eta} \partial^\mu \tilde{\eta} \tag{5.7}$$

with $\Sigma \equiv \exp\left(\sqrt{2}i\tilde{\eta}/F\right)$. This lagrangian looks like the lagrangian for a free massless boson but this is not quite so. The point is that there is a constraint for the angular variable $\tilde{\eta} : 0 \le \tilde{\eta} < \sqrt{2}\pi F$. Therefore, this lagrangian corresponds to the free theory only perturbatively, when $\tilde{\eta} \ll \sqrt{2}\pi F$.

Let us write now the full lagrangian density including fermions (see Eqs. (4.41), (4.55)):

$$\mathcal{L} = \frac{F^2}{4}\partial_\mu \Sigma \partial^\mu \Sigma^\dagger + i\bar{\psi}\gamma^\mu \partial_\mu \psi - gF\left[\bar{\psi}_L \Sigma \psi_R + \bar{\psi}_R \Sigma^\dagger \psi_L\right] . \tag{5.8}$$

Introducing new fermion fields,

$$\chi_L = \xi^\dagger \psi_L, \ \chi_R = \xi \psi_R \tag{5.9}$$

with

$$\xi = \Sigma^{1/2} = \exp\left(i\tilde{\eta}/\sqrt{2}F\right) , \tag{5.10}$$

we can rewrite the lagrangian density as

$$\mathcal{L} = \frac{F^2}{4}\partial_\mu \Sigma \partial^\mu \Sigma^\dagger + \bar{\chi} i\gamma^\mu \partial_\mu \chi + \bar{\chi}\gamma^\mu \gamma_5 A_\mu \chi - m\bar{\chi}\chi \tag{5.11}$$

with $m = gF$ and

$$A_\mu = \frac{1}{2} i\left(\xi^\dagger \partial_\mu \xi - \xi \partial_\mu \xi^\dagger\right) . \tag{5.12}$$

The characteristic property of this lagrangian density is that the NG field appears only through its derivatives.

Taking into account the transformations properties of M and ψ_L, ψ_R under the chiral group (see Eqs.(4.36), (4.37), (4.42) and (4.52)), we find the transformations of χ_L, χ_R and $\tilde{\eta}$;

$$\delta_L \chi_L = \frac{i\omega_L}{2\sqrt{2}}\chi_L, \ \delta_L \chi_R = \frac{i\omega_L}{2\sqrt{2}}\chi_R , \tag{5.13}$$

$$\delta_R \chi_L = \frac{i\omega_R}{2\sqrt{2}}\chi_L, \ \delta_R \chi_R = \frac{i\omega_R}{2\sqrt{2}}\chi_R , \tag{5.14}$$

$$\delta_L \tilde{\eta} = \frac{1}{2}\omega_L F, \ \delta_R \tilde{\eta} = -\frac{1}{2}\omega_R F, \tag{5.15}$$

or

$$\delta_V \chi_L = \frac{i\omega_V}{\sqrt{2}}\chi_L, \ \delta_V \chi_R = \frac{i\omega_V}{\sqrt{2}}\chi_R , \tag{5.16}$$

$$\delta_A \chi_L = 0, \ \delta_A \chi_R = 0 , \tag{5.17}$$

$$\delta_V \tilde{\eta} = 0, \ \delta_A \tilde{\eta} = \omega_A F . \tag{5.18}$$

Thus we conclude that the new fermion fields χ_L and χ_R transform in the same way under the $U_L(1) \times U_R(1)$. This is the reason why the mass term $m\bar{\chi}\chi$ in Eq.(5.11) is chiral invariant.

Substituting $M = \frac{F}{\sqrt{2}} \exp\left(i\sqrt{2}\tilde{\eta}/F\right)$ into Eqs.(4.56), (4.57), we find the chiral currents (compare with Eq.(4.47)):

$$
\begin{aligned}
\jmath_{L\mu} =& \frac{1}{\sqrt{2}} \bar{\chi}\gamma_\mu \frac{1+\gamma_5}{2} \chi - \frac{F}{2} \partial_\mu \tilde{\eta} \ , \\
\jmath_{R\mu} =& \frac{1}{\sqrt{2}} \bar{\chi}\gamma_\mu \frac{1-\gamma_5}{2} \chi + \frac{F}{2} \partial_\mu \tilde{\eta} \ .
\end{aligned}
\tag{5.19}
$$

We also note that although the infinitesimal transformations (5.13)-(5.15) are linear, the transformations with "large" ω_L, ω_R are not. This is connected with the constraint $0 \leq \tilde{\eta} \leq \sqrt{2}\pi F$ which in turn leads to the constraint for the transformations:

$$
\tilde{\eta}' = \left(\tilde{\eta} + \frac{1}{2}\omega_L F\right) \quad (\text{mod } \sqrt{2}\pi F). \tag{5.20}
$$

Therefore they are nonlinear (that is, they depend on the value of $\tilde{\eta}$) and, moreover, they may have jumps equal to $\sqrt{2}F\pi n$.

The case of the Abelian chiral group $U_L(1) \times U_R(1)$ is in fact rather special. As we will see below, the nonlinear character of such transformations for non-Abelian chiral groups is much more explicit.

Let us consider now the limit $\mu_\sigma^2 \longrightarrow \infty$ in the $SU_L(2) \times SU_R(2)$ model with (see Eq.(4.68))

$$
\begin{aligned}
\mathcal{L} =& \frac{1}{2}\left(\partial_\mu\sigma\partial^\mu\sigma + \partial_\mu\boldsymbol{\pi}\partial^\mu\boldsymbol{\pi}\right) - \frac{\mu^2}{2}\left(\sigma^2 + \boldsymbol{\pi}^2\right) \\
& - \frac{\lambda}{4!}\left(\sigma^2 + \boldsymbol{\pi}^2\right)^2 + i\bar{\psi}\gamma^\mu\partial_\mu\psi - g\left(\sigma\bar{\psi}\psi - \boldsymbol{\pi}\bar{\psi}i\gamma_5\boldsymbol{\tau}\psi\right) \ .
\end{aligned}
\tag{5.21}
$$

As in the previous case, the limit corresponds to that with $\lambda \longrightarrow \infty$, $\mu^2 \longrightarrow -\infty$ and $F^2 = -6\mu^2/\lambda$ being fixed. Again we find that in this limit the dynamical system with the lagrangian density (5.21) is equivalent to that with

$$
\begin{aligned}
\mathcal{L} =& \frac{1}{2}\left(\partial_\mu\sigma\partial^\mu\sigma + \partial_\mu\boldsymbol{\pi}\partial^\mu\boldsymbol{\pi}\right) + i\bar{\psi}\gamma^\mu\partial_\mu\psi \\
& - g\left(\sigma\bar{\psi}\psi - \boldsymbol{\pi}\bar{\psi}i\gamma_5\boldsymbol{\tau}\psi\right)
\end{aligned}
\tag{5.22}
$$

and the constraint

$$\sigma^2 + \pi^2 = F^2 . \tag{5.23}$$

The constraint can be resolved by choosing $M = \frac{1}{\sqrt{2}}(\sigma + i\tau^\alpha\pi^\alpha)$ as

$$M = \frac{F}{\sqrt{2}} \exp(2i\pi/F) \equiv \frac{F}{\sqrt{2}}\Sigma \tag{5.24}$$

with

$$\pi \equiv \frac{1}{2}\tau^\alpha\pi^\alpha . \tag{5.25}$$

The lagrangian density becomes

$$\begin{aligned}\mathcal{L} =& \frac{1}{4}F^2 tr\partial_\mu\Sigma\partial^\mu\Sigma^\dagger + i\bar{\psi}\gamma^\mu\partial_\mu\psi \\ & - gF\left[\bar{\psi}_L\Sigma\psi_R + \bar{\psi}_R\Sigma^\dagger\psi_L\right] .\end{aligned} \tag{5.26}$$

By introducing the fields

$$\chi_L = \xi^\dagger\psi_L,\ \chi_R = \xi\psi_R \tag{5.27}$$

with

$$\xi = \Sigma^{1/2} = \exp(i\pi/F) , \tag{5.28}$$

we can rewrite this lagrangian density as (compare with Eq.(5.11));

$$\begin{aligned}\mathcal{L} =& \frac{1}{4}F^2 tr\partial_\mu\Sigma\partial^\mu\Sigma^\dagger + i\bar{\chi}\gamma^\mu(\partial_\mu - iV_\mu)\chi \\ & + \bar{\chi}\gamma^\mu\gamma_5 A_\mu\chi - m\bar{\chi}\chi ,\end{aligned} \tag{5.29}$$

where

$$V_\mu = \frac{1}{2}i\left(\xi^\dagger\partial_\mu\xi + \xi\partial_\mu\xi^\dagger\right) , \tag{5.30}$$

$$A_\mu = \frac{1}{2}i\left(\xi^\dagger\partial_\mu\xi - \xi\partial_\mu\xi^\dagger\right) . \tag{5.31}$$

This theory is nonrenormalizable.

The transformation law for Σ is of course the same as for M (see Eq.(4.52)):

$$\Sigma \longrightarrow L\Sigma R^{\dagger} . \tag{5.32}$$

However, the transformation law for $\xi = \Sigma^{1/2}$ is rather nontrivial. As we shall see in Sec.5.4, it is

$$\xi \longrightarrow L\xi U^{\dagger} = U\xi R^{\dagger} , \tag{5.33}$$

where the matrix U is determined from this equation as an implicit function of the chiral transformation (L, R) and the NG field $\pi = \frac{1}{2}\tau^{\alpha}\pi^{\alpha}$. The latter means that the transformation of ξ depends on the field π, and therefore it is nonlinear. We emphasize that since U depends on $\pi(x)$ it also depends on space-time coordinates x, i.e., U determines a kind of a local (gauge) transformation (see below and Sec. 5.3).

The simplest way to find the explicit form of the transformation law for π is to use the relation (5.32):

$$\Sigma' = \exp\left(2i\pi'/F\right) = L \exp\left(2i\pi/F\right) R^{\dagger} \tag{5.34}$$

with $L = \exp\left(\frac{1}{2}i\omega_{L\alpha}\tau^{\alpha}\right), R = \exp\left(\frac{1}{2}i\omega_{R\alpha}\tau^{\alpha}\right)$. When $L = R = V$, we have transformations from the vector subgroup $SU_V(2)$:

$$\Sigma' = V\Sigma V^{\dagger} . \tag{5.35}$$

These transformations are linear:

$$\pi' = V\pi V^{\dagger} . \tag{5.36}$$

However transformations with $L \neq R$, for example, axial transformations with $L = R^{\dagger}$, are nonlinear:

$$\begin{aligned}\left(1 + 2i\pi'/F - 2\pi'^2/F^2 + \cdots\right) = &\left(1 + i\omega - \frac{\omega^2}{2} + \cdots\right) \\ &\times \left(1 + 2i\pi/F - 2\pi^2/F^2 + \cdots\right) \\ &\times \left(1 + i\omega - \frac{\omega^2}{2} + \cdots\right)\end{aligned} \tag{5.37}$$

with $\omega \equiv \frac{1}{2}\omega_{\alpha}\tau^{\alpha}$. This relation determines π'_{α} as a function of π_{α} and ω_{α}:

$$\pi'_{\alpha} = \pi_{\alpha} + F\omega_{\alpha} + \cdots \tag{5.38}$$

From Eq.(5.33) we find the transformation law for the fermion fields χ_L and χ_R

(5.27):

$$\chi_L \longrightarrow U\chi_L \ , \quad \chi_R \longrightarrow U\chi_R \ . \tag{5.39}$$

Again, these transformations are linear for the $SU_V(2)$ subgroup and nonlinear for transformations with $L \neq R$. However, the remarkable point is that the fermion fields χ_L and χ_R transform now in the same way under the chiral transformations. Therefore the mass term $m\bar{\chi}\chi$ is chiral invariant.

Eq.(5.33) implies that V_μ and A_μ in Eqs.(5.30), (5.31) transform as

$$V_\mu \longrightarrow UV_\mu U^\dagger + iU\partial_\mu U^\dagger \ , \tag{5.40}$$

$$A_\mu \longrightarrow UA_\mu U^\dagger \ . \tag{5.41}$$

Therefore $D_\mu \equiv \partial_\mu - iV_\mu$ can be considered as the covariant derivative with the gauge field V_μ:

$$D_\mu\chi \longrightarrow UD_\mu\chi \tag{5.42}$$

(to check (5.42), it is convenient to use the identity $\partial_\mu\left(UU^\dagger\right) = \left(\partial_\mu U\right)U^\dagger + U\partial_\mu U^\dagger = 0$).

The chiral currents in the model can be obtained from Eqs.(4.56), (4.57) with $M = \frac{F}{\sqrt{2}}\Sigma = \frac{F}{\sqrt{2}}\exp\left(2i\pi/F\right)$:

$$\begin{aligned} J_{L\mu}^{\alpha} =& \bar{\chi}\gamma_\mu\frac{1+\gamma_5}{2}\frac{\tau^\alpha}{2}\chi + i\frac{F^2}{4}tr\left(\Sigma^\dagger\tau^\alpha\partial_\mu\Sigma\right) \ , \\ J_{R\mu} =& \bar{\chi}\gamma_\mu\frac{1-\gamma_5}{2}\frac{\tau^\alpha}{2}\chi + i\frac{F^2}{4}tr\left(\Sigma\tau^\alpha\partial_\mu\Sigma^\dagger\right) \ , \end{aligned} \tag{5.43}$$

where in the derivation of Eq.(5.43) we used the identity $\left(\partial_\mu\Sigma\right)\Sigma^\dagger = -\Sigma\left(\partial_\mu\Sigma^\dagger\right)$.

By using the formula [2]

$$\begin{aligned} \left(e^{\tilde{\pi}(x)}\partial_\mu e^{-\tilde{\pi}(x)}\right) =& \left[\frac{\left(1-e^{\Delta\tilde{\pi}}\right)}{\Delta\tilde{\pi}}\right]\partial_\mu\tilde{\pi} \\ =& -\left[\partial_\mu\tilde{\pi} + \frac{1}{2!}\left[\tilde{\pi}, \partial_\mu\tilde{\pi}\right] + \frac{1}{3!}\left[\tilde{\pi}, \left[\tilde{\pi}, \partial_\mu\tilde{\pi}\right]\right]\right] + \cdots \ , \end{aligned} \tag{5.44}$$

where $\tilde{\pi} \equiv 2i\pi/F$ and

$$\Delta_{\tilde{\pi}} \cdot R \equiv [\tilde{\pi}, R] \ , \tag{5.45}$$

we find that

$$\begin{aligned} \jmath^{\alpha}_{L\mu} =& \bar{\chi}\gamma_{\mu}\frac{1+\gamma_5}{2}\frac{\tau^{\alpha}}{2}\chi - \frac{F}{2}tr\left(\tau^{\alpha}\partial_{\mu}\pi\right) \\ & - \frac{i}{2}tr\left(\tau^{\alpha}\left[\pi, \partial_{\mu}\pi\right]\right) + \frac{1}{3F}tr\left(\tau^{\alpha}\left[\pi, \left[\pi, \partial_{\mu}\pi\right]\right]\right) + \cdots \ , \\ \jmath^{\alpha}_{R\mu} =& \bar{\chi}\gamma_{\mu}\frac{1+\gamma_5}{2}\frac{\tau^{\alpha}}{2}\chi + \frac{F}{2}tr\left(\tau^{\alpha}\partial_{\mu}\pi\right) \\ & - \frac{i}{2}tr\left(\tau^{\alpha}\left[\pi, \partial_{\mu}\pi\right]\right) - \frac{1}{3F}tr\left(\tau^{\alpha}\left[\pi, \left[\pi, \partial_{\mu}\pi\right]\right]\right) + \cdots \ . \end{aligned} \tag{5.46}$$

The consideration of the nonlinear limit in the general case of the $U_L(N) \times U_R(N)$ (or $SU_L(N) \times SU_R(N) \times U_V(1)$) theory with the lagrangian density (4.53), (4.54) can be done in the same way. The lagrangian density of the nonlinear theory with the $U_V(N)$ symmetry of the vacuum is (compare with Eq.(5.29))

$$\begin{aligned} \mathcal{L} =& \frac{1}{4}F^2 tr\partial_{\mu}\Sigma\partial^{\mu}\Sigma^{\dagger} + i\bar{\chi}\gamma^{\mu}\left(\partial_{\mu} - iV_{\mu}\right)\chi \\ & + \bar{\chi}\gamma^{\mu}\gamma_5 A_{\mu}\chi - m\bar{\chi}\chi \ , \end{aligned} \tag{5.47}$$

where $\Sigma = \exp(2i\pi/F)$ and $\xi = \Sigma^{1/2}$. The matrix π is

$$\pi = \frac{1}{2}\left(\eta\lambda^0 + \sum_{\alpha=1}^{N^2-1}\pi^{\alpha}\lambda^{\alpha}\right) \tag{5.48}$$

in the case of the $U_L(N) \times U_R(N)$ theory, and

$$\pi = \frac{1}{2}\sum_{\alpha=1}^{N^2-1}\pi^{\alpha}\lambda^{\alpha} \tag{5.49}$$

in the case of the $SU_L(N) \times SU_R(N) \times U_V(1)$ theory. The fields V_{μ} and A_{μ} are defined in Eqs.(5.30) and (5.31).

The chiral currents are given by Eqs.(5.43), (5.46) with $\tau^\alpha \longrightarrow \lambda^\alpha$:

$$\begin{aligned} j^\alpha_{L\mu} &= \bar\chi\gamma_\mu \frac{1+\gamma_5}{2}\frac{\lambda^\alpha}{2}\chi + i\frac{F^2}{4} tr\left(\Sigma^\dagger\lambda^\alpha\partial_\mu\Sigma\right) , \\ j^\alpha_{R\mu} &= \bar\chi\gamma_\mu \frac{1-\gamma_5}{2}\frac{\lambda^\alpha}{2}\chi + i\frac{F^2}{4} tr\left(\Sigma\lambda^\alpha\partial_\mu\Sigma^\dagger\right) , \end{aligned} \tag{5.50}$$

that is

$$\begin{aligned} j^\alpha_{L\mu} &= \bar\chi\gamma_\mu \frac{1+\gamma_5}{2}\frac{\lambda^\alpha}{2}\chi - \frac{F}{2} tr\left(\lambda^\alpha\partial_\mu\pi\right) , \\ &\quad - \frac{i}{2} tr\left(\lambda^\alpha\left[\pi,\partial_\mu\pi\right]\right) + \frac{1}{3F} tr\left(\lambda^\alpha\left[\pi\left[\pi,\partial_\mu\pi\right]\right]\right) + \cdots \\ j^\alpha_{R\mu} &= \bar\chi\gamma_\mu \frac{1-\gamma_5}{2}\frac{\lambda^\alpha}{2}\chi + \frac{F}{2} tr\left(\lambda^\alpha\partial_\mu\pi\right) , \\ &\quad - \frac{i}{2} tr\left(\lambda^\alpha\left[\pi,\partial_\mu\pi\right]\right) - \frac{1}{3F} tr\left(\lambda^\alpha\left[\pi\left[\pi,\partial_\mu\pi\right]\right]\right) + \cdots . \end{aligned} \tag{5.51}$$

In the next section, we will discuss the physical meaning of the nonlinear chiral lagrangians in more detail.

5.3. More on Nonlinear Chiral Theories

In the previous section, we derived the nonlinear chiral models as a special limit of the linear σ-models. Although such a derivation gives some insight in their dynamics, it is not necessary; one can study these models independently of the linear σ-models.

The general program of constructing nonlinear effective lagrangians was started in the $SU_L(2) \times SU_R(2)$ model by Weinberg [3] and was extended to the general case by Callan, Coleman, Wess and Zumino [2], and, independently, by Volkov [4]. This program reduces to the study of the general structure of lagrangians invariant under nonlinear transformations of fields (for a review see Ref.[5]).

The starting point is that these lagrangians are intimately connected with geometry. The transformation law (5.40) implies that the vector field

$$V_\mu = \frac{1}{2} i\left(\xi^\dagger\partial_\mu\xi + \xi\partial_\mu\xi^\dagger\right) \tag{5.52}$$

plays the role of a gauge field here. A fundamental object is the Maurer-Cartan 1-form

[5]:

$$\alpha_\mu(\pi) = i\xi^\dagger(\pi)\partial_\mu\xi(\pi) . \tag{5.53}$$

Since $\xi^\dagger(\pi) = \xi(-\pi)$ (see Eq. (5.28)), we see that

$$\begin{aligned} V_\mu &= \frac{1}{2}\left(\alpha_\mu(\pi) + \alpha_\mu(-\pi)\right), \\ A_\mu &= \frac{1}{2}\left(\alpha_\mu(\pi) - \alpha_\mu(-\pi)\right) \end{aligned} \tag{5.54}$$

(see Eqs.(5.30), (5.31)).

Because of the equations (5.40), (5.41), the transformation of $\alpha_\mu(\pi)$ under the chiral group is

$$\alpha_\mu(\pi) \longrightarrow U\alpha_\mu(\pi)U^\dagger + iU\partial_\mu U^\dagger . \tag{5.55}$$

In fact, as will be shown in Sec.5.4, the fields V_μ and A_μ coincide with the so called parallel ($\alpha_{\mu||}(\pi)$) and perpendicular ($\alpha_{\mu\perp}(\pi)$) components of $\alpha_\mu(\pi)$, respectively.

In the previous section, we introduced the fermion fields

$$\chi_L = \xi^\dagger\psi_L,\ \chi_R = \xi\psi_R \tag{5.56}$$

for the fields $\psi_L\epsilon(N,0)$ and $\psi_R\epsilon(0,N)$ transforming in the same way under the chiral group:

$$\chi_L \longrightarrow U\chi_L\ ,\chi_R \longrightarrow U\chi_R\ . \tag{5.57}$$

Therefore the matrix $U(\pi(x))$ is assigned to the vector subgroup $U_V(N)$. What has happened is that the field ξ in Eq.(5.56) changes the chiral charge density of the fermions so that the chiral charges become the same for the fermions χ_L and χ_R.

Let us show that this can be done for any matter field ψ assigned to some linear unitary representation D of $G = U_L(N) \times U_R(N)$ (or $SU_L(N) \times SU_R(N) \times U_V(1)$):

$$\psi \longrightarrow D(g)\psi,\ g \in G\ . \tag{5.58}$$

We note first that all unitary representations of G can be built on tensors

$$\psi^{\alpha_1\cdots\alpha_k\dot\alpha_1\cdots\dot\alpha_n}_{\beta_1\cdots\beta_m\dot\beta_1\cdots\dot\beta_s}\ , \tag{5.59}$$

where upper (lower) indices relate to $SU_L(N)(SU_R(N))$, and while $\alpha_i(\beta_j)$ are assigned to the $(N,0)((0,N))$ representation, $\dot\alpha_i\left(\dot\beta_j\right)$ are assigned to the $(\bar N,0)((0,\bar N))$ one.

Let us introduce a new field (compare with Eq.(5.56))

$$\begin{aligned}\chi^{\alpha_1\cdots\alpha_k\dot\alpha_1\cdots\dot\alpha_n}_{\beta_1\cdots\beta_m\dot\beta_1\cdots\dot\beta_s} =&\xi^\dagger_{\alpha_1\alpha'_1}\cdots\xi^\dagger_{\alpha_k\alpha'_k}\xi_{\beta_1\beta'_1}\cdots\xi_{\beta_m\beta'_m}\psi^{\alpha'_1\cdots\alpha'_k\dot\alpha'_1\cdots\dot\alpha'_n}_{\beta'_1\cdots\beta'_m\dot\beta'_1\cdots\dot\beta'_s}\\ &\times\xi_{\dot\alpha'_1\dot\alpha_1}\cdots\xi_{\dot\alpha'_n\alpha_n}\xi^\dagger_{\dot\beta'_1\dot\beta_1}\cdots\xi^\dagger_{\dot\beta'_s\dot\beta_s}.\end{aligned} \tag{5.60}$$

The tensor $\psi^{\alpha_1\cdots\alpha_k\dot\alpha_1\cdots\dot\alpha_n}_{\beta_1\cdots\beta_m\dot\beta_1\cdots\dot\beta_s}$ transforms under G as

$$\begin{aligned}\psi^{\alpha_1\cdots\alpha_k\dot\alpha_1\cdots\dot\alpha_n}_{\beta_1\cdots\beta_m\dot\beta_1\cdots\dot\beta_s}\longrightarrow& L_{\alpha_1\alpha'_1}\cdots L_{\alpha_k\alpha'_k}R_{\beta_1\beta'_1}\cdots R_{\beta_m\beta'_m}\psi^{\alpha'_1\cdots\alpha'_k\dot\alpha'_1\cdots\dot\alpha'_n}_{\beta'_1\cdots\beta'_m\dot\beta'_1\cdots\dot\beta'_s}\\ &\times L^\dagger_{\dot\alpha'_1\dot\alpha_1}\cdots L^\dagger_{\dot\alpha'_n\dot\alpha_n}R^\dagger_{\dot\beta'_1\dot\beta_1}\cdots R^\dagger_{\dot\beta'_s\dot\beta_s}\,.\end{aligned} \tag{5.61}$$

Taking into account the transformation law (5.33) for ξ, we find

$$\begin{aligned}\chi^{\alpha_1\cdots\alpha_k\dot\alpha_1\cdots\dot\alpha_n}_{\beta_1\cdots\beta_m\dot\beta_1\cdots\dot\beta_s}\longrightarrow& U_{\alpha_1\alpha'_1}\cdots U_{\alpha_k\alpha'_k}U_{\beta_1\beta'_1}\cdots U_{\beta_m\beta'_m}\chi^{\alpha'_1\cdots\alpha'_k\dot\alpha'_1\cdots\dot\alpha'_n}_{\beta'_1\cdots\beta'_m\dot\beta'_1\cdots\dot\beta'_s}\\ &\times U^\dagger_{\dot\alpha'_1\dot\alpha_1}\cdots U^\dagger_{\dot\alpha'_n\dot\alpha_n}U^\dagger_{\dot\beta'_1\dot\beta_1}\cdots U^\dagger_{\dot\beta'_s\dot\beta_s}\,.\end{aligned} \tag{5.62}$$

This completes the proof.

It is convenient to use more abstract notations to present the proof in a more transparent form. In these notations, the transformation of ψ (5.61) looks as

$$\psi\longrightarrow\psi' = D(g)\psi. \tag{5.63}$$

Let us introduce the element $\tilde\xi\epsilon G$ which is ξ and $\xi^\dagger$ in the representations $(N,0)$ and $(0,N)$, respectively, and the group element $\tilde U$ which is U in both these representations.

Eq.(5.33) implies that

$$D\left(\tilde\xi'^\dagger\right) = D\left(\tilde U\tilde\xi^\dagger g^\dagger\right)\,. \tag{5.64}$$

Therefore the transformation law for the field

$$\chi = D\left(\tilde\xi^\dagger\right)\psi \tag{5.65}$$

is

$$\chi\longrightarrow\chi' = D\left(\tilde\xi'^\dagger\right)\psi' = D\left(\tilde U\right)\chi\,. \tag{5.66}$$

Of course, $\tilde U$ depends on both the element g and the field $\pi(x)$.

Since $\tilde{U}$ depends on x, the transformation (5.66) can be considered as a local (gauge) transformation. Therefore the problem of the construction of G-invariant lagrangians is similar to that of constructing gauge (Yang-Mills) lagrangians.

For any matter field χ assigned to a representation D of G, we introduce the covariant derivative

$$D_\mu = \partial_\mu - iD\left(\tilde{V}_\mu\right) , \tag{5.67}$$

where

$$\tilde{V}_\mu \equiv \frac{1}{2}i\left(\tilde{\xi}^\dagger \partial_\mu \tilde{\xi} + \tilde{\xi}\partial_\mu \tilde{\xi}^\dagger\right) . \tag{5.68}$$

The kinetic term for the NG field π is (see Eq.(5.47))

$$\frac{1}{4}F^2 tr\partial_\mu \Sigma \partial^\mu \Sigma^\dagger = F^2 tr\left(A_\mu A^\mu\right) , \tag{5.69}$$

where A_μ is given in Eq.(5.31).

With the introducion of these quantities, the construction of G-invariants, and therefore G-invariant lagrangians, becomes straightforward. Let us discuss some subtle points connected with the present construction.

As it was shown above, any matter field ψ assigned to some linear representation of the chiral group G defines a field χ transformed nonlinearly under G:

$$\chi \longrightarrow \chi' = \rho(\tilde{U})\chi \tag{5.70}$$

with $\tilde{U}\epsilon U_V(N)$. Since $\rho(\tilde{U})$ is assigned now to the vector subgroup $U_V(N)$, fermion fields ψ_i assigned to different representations ρ_i of G which reduce to the same representation ρ on the subgroup $U_V(N)$ determine the same field χ.

Let us illustrate this on the example of the $SU(3)$ octet of spin-$\frac{1}{2}$ baryons χ. This field transforms as

$$\chi \longrightarrow \chi' = U\chi U^\dagger . \tag{5.71}$$

We can use the relation (5.65) to construct four different fields ψ_i from this field χ:

$$\begin{aligned} \psi_1 &= \xi\chi\xi^\dagger, \psi_2 = \xi^\dagger\chi\xi , \\ \psi_3 &= \xi\chi\xi, \psi_4 = \xi^\dagger\chi\xi^\dagger . \end{aligned} \tag{5.72}$$

with $\xi = \exp(i\pi/F), tr\pi = 0$.

Using Eqs.(5.33) and (5.71) we find that these fields are assigned to the following four different representations of $SU_L(3) \times SU_R(3)$:

$$\psi_1 \in (8,1), \psi_2 \in (1,8), \psi_3 \in (3,\bar{3}), \psi_4 \in (\bar{3},3). \tag{5.73}$$

Thus these four different fields lead to the same chiral lagrangian with the field χ.

This example clearly demonstrates that the real symmetry here is the $SU_V(3)$ symmetry of the vacuum. The initial spontaneously broken $SU_L(3) \times SU_R(3)$ is a dynamical one: it is reflected in the structure of the chiral lagrangian. In particular it leads to eight massless NG fields, and it determines the structure of the vertices which couple NG bosons and matter fields.

Let us indicate also the following point. Although one can choose any of the fields $\psi_i, i = 1,2,3,4$, or χ to work with (one can treat them as just equivalent variables), a certain choice might be practically preferable: the main reason is that we always work with some approximations for the chiral lagrangians and these approximations can destroy the equivalence between the different fields.

In our consideration above, we implicitly assumed that all NG bosons have the same decay constant F. However it is unnecessary in the case of the chiral $U_L(N) \times U_R(N)$ symmetry with $N \geq 2$. The point is that N^2 NG bosons are assigned to a reducible representation of the vacuum group $SU_V(N) \times U_V(1)$ in this case: $\left(N^2-1\right)$ bosons are assigned to the adjoint representation of $SU_V(N)$, and one boson is a singlet. The field ξ is $\xi = \xi_0 \xi_8$, where $\xi_0 = \exp\left(i\eta\lambda^0/F_0\right), \xi_8 = \exp\left(i\pi^\alpha\lambda^\alpha/F_8\right), tr\lambda^\alpha = 0$, and the decay constants F_0 and F_8 can be different. Therefore we can construct two different kinetic terms for NG bosons in this case:

$$F_0^2 tr\left(A_\mu^{(0)} A^{(0)\mu}\right) \text{ and } F_8^2 tr\left(A_\mu^{(8)} A^{(8)\mu}\right) \tag{5.74}$$

with

$$\begin{aligned} A_\mu^{(0)} &= \frac{1}{2} i\left(\xi_0^\dagger \partial_\mu \xi_0 - \xi_0 \partial_\mu \xi_0^\dagger\right) , \\ A_\mu^{(8)} &= \frac{1}{2} i\left(\xi_8^\dagger \partial_\mu \xi_8 - \xi_8 \partial_\mu \xi_8^\dagger\right) . \end{aligned} \tag{5.75}$$

This example again shows that the real symmetry is the vacuum symmetry $SU_V(N) \times U_V(1)$ in the case of chiral lagrangians.

In conclusion, let us consider a matter field connected with vector bosons assigned to the adjoint representation of $SU_V(N)$. The transformation law for this field is

$$\rho_\mu(x) \longrightarrow \rho'_\mu(x) = U\rho_\mu(x)U^\dagger . \tag{5.76}$$

As was noted by Weinberg [3], it is convenient to introduce another field

$$\rho_\mu^W = \rho_\mu(x) + V_\mu(x) \equiv \frac{\lambda^\alpha}{2}\rho_\mu^{\alpha W} \tag{5.77}$$

with $V_\mu = \frac{1}{2}i\left(\xi^\dagger\partial_\mu\xi + \xi\partial_\mu\xi^\dagger\right)$. Eqs.(5.40) and (5.76) imply that ρ_μ^W transformes under the chiral group as

$$\rho_\mu^W(x) \longrightarrow \rho_\mu'^W(x) = U\rho_\mu^W(x)U^\dagger + iU\partial_\mu U^\dagger , \tag{5.78}$$

i.e., it transforms as a gauge field. Therefore the covariant derivative of a matter field χ assigned to a representation D of $SU(N)$, $\partial_\mu - iD\left(\tilde{V}_\mu\right)$, can be replaced by the covariant derivative

$$\partial_\mu - iD\left(\tilde{\rho}_\mu^W\right) \equiv \partial_\mu - i\frac{1}{2}D\left(\lambda^\alpha\right)\rho_\mu^{\alpha W} . \tag{5.79}$$

The kinetic term for ρ_μ^W can be chosen as (compare with Eq.(3.79)):

$$-\frac{1}{4}F_{\mu\nu}^\alpha F^{\mu\nu,\alpha} \tag{5.80}$$

with

$$F_{\mu\nu}^\alpha = \partial_\mu\rho_\nu^{\alpha W} - \partial_\nu\rho_\mu^{\alpha W} + f_{\alpha\beta\gamma}\rho_\mu^{\beta W}\rho_\nu^{\gamma W} . \tag{5.81}$$

Even the mass term can be written in the gauge invariant form:

$$\frac{1}{2}M^2\left(\rho_\mu^{\alpha W} - V_\mu^\alpha\right)\left(\rho_\alpha^{\mu W} - V_\alpha^\mu\right) = \frac{1}{2}M^2\rho_\mu^\alpha\rho_\alpha^\mu . \tag{5.82}$$

We note, however, that, as for all matter fields, without additional information, there is arbitrariness in choosing a field to describe vector bosons. We could choose the field ρ_μ, for example, which transforms homogeneously under the chiral group, like all other matter fields. We shall return to this question in Chapter 12.

5.4. Nonlinear Realization of Spontaneous Symmetry Breaking. General Theory

In this section we consider the general theory of nonlinear realization of spontaneous symmetry breaking elaborated by Callan, Coleman, Wess and Zumino [2] and, independently, by Volkov [4] (for an excellent review, see Ref. [5]).

Let us consider a theory with a symmetry group G which is spontaneously broken to a subgroup H. We assume that G is compact.

The set of generators T^A of G can be divided into two parts: S^a are the generators of the unbroken subgroup H, and X^α are the remaining generators:

$$\left\{T^A\right\} = \{S^a \in \mathcal{H}, X^\alpha \in \mathcal{G} - \mathcal{H}\} \ , \tag{5.83}$$

where $\mathcal{G}$ and $\mathcal{H}$ are the corresponding Lie algebras.

The generators T^A are chosen in such a way that they satisfy the conditions:

$$tr\left(T^A T^B\right) = \frac{1}{2}\delta^{AB} \ , \tag{5.84}$$

$$tr\left(S^a X^\alpha\right) = 0 \ . \tag{5.85}$$

Eq.(5.85) in particular implies that $tr\left(S^a\left[S^b, X^\alpha\right]\right) = tr\left(\left[S^a, S^b\right]X^\alpha\right) = 0$, and therefore the element $[S^a, X^\alpha]$ is in $\mathcal{G} - \mathcal{H}$, i.e.

$$[\mathcal{H}, \mathcal{G} - \mathcal{H}] \subset \mathcal{G} - \mathcal{H} \ . \tag{5.86}$$

Let us consider the field (compare with Eq. (5.28))

$$\xi(\pi) = \exp\left(i\pi(x)/F\right), \pi(x) \equiv \sum_{\alpha \in \mathcal{G}-\mathcal{H}} \pi^\alpha(x) X^\alpha \tag{5.87}$$

with $\pi^\alpha(x)$ being the NG boson fields. $\xi(\pi)$ is an element of G relating to the coset space G/H (we choose the left coset; we recall that a left coset G/H identifies elements $g, g' \in G$ differing by right multiplication with an element of $H; g' = gh$).

Let us consider the product $g\xi(\pi)$, where g is some element of G. Then, due to the properties of the exponentials, the group element $g\xi(\pi)$ can be decomposed uniquely into the coset part $\xi' = \xi(\pi')$ and unbroken part $h \in H$:*)

$$g\xi(\pi) = \xi(\pi')h(\pi, g) . \tag{5.88}$$

This relation determines the nonlinear transformation of the field $\pi(x)$ under G:

$$\xi(\pi) \longrightarrow \xi' = \xi(\pi') = g\xi(\pi)h^\dagger(\pi, g) . \tag{5.89}$$

Let us verify that: a) this transformation determines a representation of the group and b) the transformation (5.89) determines a linear representation of the unbroken subgroup H.

Indeed, let us consider the product $g = g_2g_1$.

$$g_1\xi(\pi) = \xi(\pi')h_1(\pi, g_1) \tag{5.90}$$

and

$$g_2\xi(\pi') = \xi(\pi'')h_2(\pi', g_2) , \tag{5.91}$$

then

$$(g_2g_1)\xi(\pi) = \xi(\pi'')h_{21}(\pi'', g_2g_1) \tag{5.92}$$

with

$$h_{21}(\pi'', g_2g_1) = h_2(\pi', g_2)h_1(\pi, g_1) , \tag{5.93}$$

i.e., the transformation (5.89) satisfies the group product law.

---

*) Strictly speaking, the decomposition is unique in some neighborhood of the identity of G.

Let us consider the case with $g = h_0 \epsilon H$. Then

$$\xi(\pi')h(\pi, g) = h_0 \xi(\pi) h_0^{-1} h_0 \ . \tag{5.94}$$

Since the decomposition (5.88) is unique, and because of Eq. (5.86), we find that $h(\pi, g) = h_0$ (in particular it is independent of π) and

$$\xi(\pi') = h_0 \xi(\pi) h_0^{-1} \ , \tag{5.95}$$

$$\pi' = h_0 \pi h_0^{-1} \ , \tag{5.96}$$

i.e., we get the linear representation of the subgroup H.

An important special case is that where the coset G/H is a symmetric space, which means that

$$[\mathcal{G} - \mathcal{H}, \mathcal{G} - \mathcal{H}] \subset \mathcal{H} \ . \tag{5.97}$$

In particular, spontaneous chiral symmetry breaking with $G/H = [U_L(N) \times U_R(N)] / U_V(N)$or $[SU_L(N) \times SU_R(N) \times U_V(1)] / U_V(N)$ belongs to this class.

The condition (5.97) enables us to define the parity-like automorphism $\tau : g \longrightarrow \tau(g)$ such that

$$\tau(x) = x, \text{ for } x \in \mathcal{H}; \tau(x) = -x, \text{ for } x \in \mathcal{G} - \mathcal{H} \ . \tag{5.98}$$

Applying this automorphism to the relation (5.89)

$$\xi(\pi') = g\xi(\pi)h^{\dagger} \ , \tag{5.99}$$

we find

$$\xi(-\pi') = \xi^{\dagger}(\pi') = \tau(g)\xi^{\dagger}(\pi)h^{\dagger} \ . \tag{5.100}$$

Eqs. (5.99), (5.100) yield:

$$\Sigma(\pi') \equiv \xi^2(\pi') = g\xi^2(\pi)\tau(g^{-1}) = g\Sigma(\pi)\tau(g^{-1}) \ . \tag{5.101}$$

Let us return now to the spontaneous chiral symmetry breaking case. We will prove the relation (5.33).

Let us use the following representation for the chiral generators:

$$\frac{1}{2}\lambda_L^\alpha \equiv \frac{1}{2}(1+\gamma_5)\frac{\lambda^\alpha}{2}, \tag{5.102}$$

$$\frac{1}{2}\lambda_R^\alpha \equiv \frac{1}{2}(1-\gamma_5)\frac{\lambda^\alpha}{2}, \tag{5.103}$$

where γ_5 is Dirac's γ_5-matrix.

We note that in this representation, the parity automorphism τ is given by

$$\tau(x) = \gamma_0 x \gamma_0 . \tag{5.104}$$

The element $\xi(\pi) \in G/H$ can be written as

$$\begin{aligned} \xi(\pi) &= \exp\left(i\gamma_5\pi^\alpha\lambda^\alpha/2F\right) \\ &= \exp\left(i\pi^\alpha\lambda_L^\alpha/2F\right)\exp\left(-i\pi^\alpha\lambda_R^\alpha/2F\right) \\ &= \xi_L(\pi)\xi_R^\dagger(\pi) . \end{aligned} \tag{5.105}$$

The transformation law (5.89) takes now the form:

$$\xi_L(\pi') = g_L\xi_L(\pi)h_L^\dagger , \tag{5.106}$$

$$\xi_R^\dagger(\pi') = g_R\xi_R^\dagger(\pi)h_R^\dagger . \tag{5.107}$$

Here h in Eq.(5.89) is rewritten as

$$h \equiv \exp\left(v^\alpha\lambda^\alpha/2\right) = \exp\left(v^\alpha\lambda_L^\alpha/2\right)\exp\left(v^\alpha\lambda_R^\alpha/2\right) \equiv h_L h_R . \tag{5.108}$$

Applying the automorphism τ (5.104) to both sides of Eq.(5.107), we find:

$$\xi_L^\dagger(\pi') = \tau\left(g_R\right)\xi_L^\dagger(\pi)h_L^\dagger . \tag{5.109}$$

This equation and the equation (5.106) lead to the relation

$$g_L\xi_L(\pi)h_L^\dagger = h_L\xi_L(\pi)\tau\left(g_R^\dagger\right) \tag{5.110}$$

in this representation. Since all generators in this relation contain the projection operator $\frac{1+\gamma_5}{2}$, it is clear that it remains valid if we replace $\lambda_L^\alpha = \frac{1+\gamma_5}{2}\lambda^\alpha$ with λ^α. As a result, we get the relation (5.33).

Let us now consider the Lie-algebra-valued Maurer-Cartan 1-form:

$$\alpha_\mu(\pi) \equiv i\xi^\dagger(\pi)\partial_\mu\xi(\pi) \tag{5.111}$$

already briefly discussed in Sec.5.3 (see Eq.(5.53)). It is convenient to define the so called parallel $\left(\alpha_{\mu||}(\pi)\right)$ and perpendicular $\left(\alpha_{\mu\perp}(\pi)\right)$ components of $\alpha_\mu(\pi)$:

$$\alpha_{\mu||}(\pi) \equiv (2trS^a\alpha_\mu(\pi))\, S^a \epsilon \mathcal{H} \ , \tag{5.112}$$

$$\alpha_{\mu\perp}(\pi) \equiv (2trX^\alpha\alpha_\mu(\pi))\, X^\alpha \epsilon \mathcal{G} - \mathcal{H} \ . \tag{5.113}$$

The transformation law of α_μ following from Eq.(5.89) is

$$\alpha_\mu(\pi) \longrightarrow \alpha_\mu(\pi') = h(\pi,g)\alpha_\mu(\pi)h^\dagger(\pi,g) + ih(\pi,g)\partial_\mu h^\dagger(\pi,g) \ . \tag{5.114}$$

Since $h(\pi,g)\partial_\mu h^\dagger(\pi,g) \in \mathcal{H}$, this equation implies that

$$\alpha_{\mu||}(\pi) \longrightarrow \alpha_{\mu||}(\pi') = h(\pi,g)\alpha_{\mu||}(\pi)h^\dagger(\pi,g) + ih(\pi,g)\partial_\mu h^\dagger(\pi,g) \ , \tag{5.115}$$

$$\alpha_{\mu\perp}(\pi) \longrightarrow \alpha_{\mu\perp}(\pi') = h(\pi,g)\alpha_{\mu\perp}(\pi)h^\dagger(\pi,g) \ . \tag{5.116}$$

Comparing these equations with Eqs.(5.40), (5.41) (where $h = U$), we see that the quantities V_μ and A_μ in theories with spontaneous chiral symmetry breaking transform exactly as $\alpha_{\mu||}$ and $\alpha_{\mu\perp}$, respectively. Let us show that in those theories, $V_\mu = \alpha_{\mu||}$ and $A_\mu = \alpha_{\mu\perp}$. Indeed, since $\alpha_{\mu||} \in \mathcal{H}$ and $\alpha_{\mu\perp} \in \mathcal{G} - \mathcal{H}$, then $\tau\left(\alpha_{\mu||}(\pi)\right) = \alpha_{\mu||}(-\pi) = \alpha_{\mu||}(\pi)$ and $\tau\left(\alpha_{\mu\perp}(\pi)\right) = \alpha_{\mu\perp}(-\pi) = -\alpha_{\mu\perp}(\pi)$. This in turn implies that

$$\begin{aligned} \alpha_{\mu||} &= \frac{1}{2}\left(\alpha_\mu(\pi) + \alpha_\mu(-\pi)\right) = \frac{1}{2}i\left(\xi^\dagger\partial_\mu\xi + \xi\partial_\mu\xi^\dagger\right) \ , \\ \alpha_{\mu\perp} &= \frac{1}{2}\left(\alpha_\mu(\pi) - \alpha_\mu(-\pi)\right) = \frac{1}{2}i\left(\xi^\dagger\partial_\mu\xi - \xi\partial_\mu\xi^\dagger\right) \ . \end{aligned} \tag{5.117}$$

These two expressions exactly coincide with those for V_μ and A_μ, respectively (see Eqs.(5.30)(5.31)).

The construction of general G-invariant lagrangians can be done in the same way as the construction of chiral nonlinear lagrangians considered in the previous section. Let us discuss this in more detail.

The broken generators $\{X^\alpha\} \in \mathcal{G} - \mathcal{H}$ span some representation of the subgroup H which can be decomposed into some irreducible ones $\{X^{\alpha_i}\}, i = 1, 2, ..., n$. Then the quantity $\alpha_{\mu\perp}(\pi)$ can also be decomposed into H-irreducible pieces:

$$\alpha_{\mu\perp}(\pi) = \sum_i \alpha^{(i)}_{\mu\perp}(\pi) , \tag{5.118}$$

$$\alpha^{(i)}_{\mu\perp} = 2\left(tr X^{\alpha_i}\alpha_\mu(\pi)\right) X^{\alpha_i} . \tag{5.119}$$

Therefore the kinetic term is

$$\mathcal{L}^{(NG)}_k = \sum_{i=1}^n F_i^2 tr\left(\alpha^{(i)}_{\mu\perp}(\pi)\right)^2 , \tag{5.120}$$

where the decay constants F_i with different i are in general independent. This general form of $\mathcal{L}^{(NG)}_k$ was first pointed out by Boulware and Brown [6].

Of course, one can add terms with any power of $tr\left(\alpha^{(i)}_{\mu\perp}(\pi)\right)^2$ in the lagrangian.

For any matter field χ assigned to a representation D of G, the covariant derivative is (compare with Eq.(5.67)):

$$D_\mu\chi = \partial_\mu\chi - iD\left(\alpha_{\mu\parallel}(\pi)\right)\chi . \tag{5.121}$$

Since there is no renormalizability restriction for the form of lagrangians, one can again add invariant terms with any powers of $D_\mu\chi$ in the lagrangian.

Such phenomenological lagrangians describe low energy dynamics with $q^2 \leq C \min \{F_i^2\}$, where C is some characteristic dimensionless coefficient determined by an underlying fundamental dynamics.

We shall consider this point in more detail in the next section.

5.5. Where are Nonlinear Lagrangians Useful?

A general nonlinear lagrangian includes infinite number of terms organized in increasing powers of derivatives and numbers of fields. The lagrangian is of course nonrenormalizable.

Here we would like to add some comments about renormalizations in nonrenormalizable theories. Their nonrenormalizability does not prevent the realization of the *usual* renormalization procedure in perturbation theory. There are a *finite* number of counter terms in each order of the loop expansion in both renormalizable and nonrenormalizable theories. However, if the number of different types of counter terms is limited in renormalizable theories (four types of the counter terms in QED, for example), it is unlimited in nonrenormalizable theories and grows rapidly in higher orders of the expansion. Therefore, in perturbation theory, a nonrenormalizable theory looks as a renormalizable one but it contains a huge number of free parameters in higher orders of the loop expansion.

Where can such a complicated construction be useful? The answer to this question is that the nonlinear effective lagrangians give an excellent framework for the description of the low energy dynamics of such light particles as NG bosons.

Let us, following Weinberg [7], illustrate this point on the example of the nonlinear chiral lagrangian describing the spontaneous breakdown of the $SU_L(2) \times SU_R(2)$ chiral symmetry.

We consider the lagrangian including the NG (pion) field $\pi = \boldsymbol{\pi}\frac{\boldsymbol{\tau}}{2}$ only, i.e. we assume that all other fields (as ρ meson, nucleons, etc.) are integrated out.

The lagrangian density is an infinite series of operators of higher and higher dimensionality:

$$\begin{aligned}\mathcal{L} =& g_2 tr\left(D^\mu\pi \cdot D_\mu\pi\right) + g_4^{(1)}\left[tr\left(D^\mu\pi \cdot D_\mu\pi\right)\right]^2 \\ &+ g_4^{(2)} tr\left(D^\mu\pi \cdot D^\nu\pi\right) tr\left(D_\mu\pi \cdot D_\nu\pi\right) + \dots ,\end{aligned} \tag{5.122}$$

where the covariant derivative $D_\mu\pi$ is $D_\mu\pi \equiv \alpha_{\mu\perp}(\pi)$ (see Eqs.(5.117), (5.120)), $g_2 = F^2$, and $g_d^{(n)}$ are constants of dimensionality $[\text{mass}]^{4-d}$, d is the number of the covariant derivatives in the corresponding term.

We are interested in the low energy dynamics with a characteristic energy scale E, such that $E^2/F^2 \ll 1$ (in fact, $E^2/C^2F^2 \ll 1$ with a dimensionless constant C discussed below). Thus our interest is a series in powers of E or, what is the same, in powers of the derivatives of the field π.

Let us consider the matrix element M for a process involving N_e external pion lines:

$$S = i(2\pi)^4 \delta^4 \left(\sum_{i=1}^{N_e} p_i \right) M \,, \tag{5.123}$$

where S is the S-matrix element. The dimensionality of M is $[\text{mass}]^{D_1}$, where

$$D_1 = 4 - N_e \,. \tag{5.124}$$

Let us begin with the tree approximation. The leading term is the first one in Eq.(5.122) including two derivatives. Its contribution to M is proportional to the factor E^2. In particular, the matrix element for pion-pion scattering in the tree approximation is given by the Weinberg relation [8]:

$$M^{(1)}_{\alpha\beta\gamma\delta} = F^{-2} \left[\delta_{\alpha\beta}\delta_{\gamma\delta} s + \delta_{\alpha\gamma}\delta_{\beta\delta} t + \delta_{\alpha\delta}\delta_{\beta\gamma} u \right] \tag{5.125}$$

with Mandelstam variables $s = \left(p_\alpha + p_\beta\right)^2$, $t = \left(p_\alpha + p_\gamma\right)^2$, $u = \left(p_\alpha + p_\delta\right)^2$.

Let us consider now the contribution to the matrix elements of graphs with loops.

Since every pion field π^α appears with the factor F^{-1} in the lagrangian (we recall that $\pi = \pi^\alpha \tau^\alpha / 2F$) and since the dimensionality of $g_d^{(n)}$ is $[\text{mass}]^{4-d}$, we find that the dimensionality of the product of all coupling constants in a graph is $[\text{mass}]^{D_2}$, where

$$D_2 = \sum_d N_d (4 - d) - 2N_i - N_e \,, \tag{5.126}$$

where N_d is the number of vertices with d derivatives, and N_i is the number of internal pion lines in the graph.

Since pions are massless, the nonregularized loop integrals are formally homogeneous functions of external momenta $p_i \sim E$. If they were finite, the relations (5.124), (5.126) would imply that they are proportional to the factor

$$E^D \tag{5.127}$$

with

$$D = D_1 - D_2 = 4 + \sum_d N_d(d-4) + 2N_i \ . \tag{5.128}$$

However, since the loop integrals diverge, one has to use some regulazization introducing a dimensionful parameter σ in the integrals. The matrix element takes the form

$$M = E^D f\left(E/\sigma\right) \ , \tag{5.129}$$

and without knowing the function f, we can say nothing how M depends on E.

Fortunately this problem can be solved.

Since the lagrangian density (5.122) includes all terms consistent with $SU_L(2) \times SU_R(2)$ symmetry, one can remove the ultraviolet divergences by absorbing them into a renormalization of the infinite number of the coupling constants and the pion field.

Since the matrix elements themselves are renormalization scheme independent, one can use any renormalization scheme. For our purposes, the most convenient one is the minimal subtraction (MS) (or modified minimal subtraction ($\overline{MS}$)) scheme with the dimensional regulazization of the loop integrals (see Sec. C.2 in Appendix C) [9].

In the MS (or $\overline{MS}$) scheme, the dependence of the loop integrals on the renormalization scale σ is rather minor; σ appears there only through some logarithmic functions. Therefore the expressions for the matrix elements take the form (compare with Eq.(5.129)):

$$M = E^D f(E/\sigma) \ , \tag{5.130}$$

where f is some logarithmic function, and D is given in Eq.(5.128).

Using the well-known formula for the number of loops in a graph [9],

$$N_L = N_i - \sum_d N_d + 1 \ , \tag{5.131}$$

the power D can be rewritten as

$$D = 2 + \sum_d N_d(d-2) + 2N_L \ . \tag{5.132}$$

The relations (5.130) and (5.132) imply that the contribution of a graph to M is more and more suppressed with increasing the number of loops and the number of derivatives. In particular the dominant contribution corresponds to $N_L = 0$ (the tree approximation) and $d = 2$ (the first term in the lagrangian density (5.122)).

Moreover, since the powers D corresponding to the graphs with loops are not smaller than four, we conclude that there is no counter terms of the form of the first term in the lagrangian (5.122). This in turn implies that there are no renormalizations both of the coupling $g_2 = F^2$ and of the pion field. On the other hand, one-loop graphs with vertices from the first term (with two derivatives) can lead to the renormalization of the coupling constants $g_4^{(1)}$ and $g_4^{(2)}$, etc.

Let us consider the corrections of order $s^2 \sim E^4$ in the matrix element for pion-pion scattering. The relations (5.130), (5.132) imply that these corrections will arise from graphs in which there are any number of vertices with two derivatives ($d = 2$), and either one loop or one vertex with $d = 4$. The explicit expression for $M^{(2)}_{\alpha\beta\gamma\delta}$ is (we use the dimensional regularization) [7]:

$$\begin{aligned} M^{(2)}_{\alpha\beta\gamma\delta} =& \frac{\delta_{\alpha\beta}\delta_{\gamma\delta}}{16F^4}\Bigg[-\frac{s^2}{(2\pi)^2}\ln(-s) - \frac{1}{12\pi^2}\left(u^2 - s^2 + 3t^2\right)\ln(-t) \\ & - \frac{1}{12\pi^2}\left(t^2 - s^2 + 3u^2\right)\ln(-u) + \frac{1}{3\pi^2}\left(s^2 + t^2 + u^2\right) \\ & \cdot\left(\frac{1}{\epsilon} - \gamma + \ln 4\pi\right) + 8g_4^{(1)}s^2 + 4g_4^{(2)}\left(t^2 + u^2\right)\Bigg] + \text{crossed terms} \ , \end{aligned} \tag{5.133}$$

where $\epsilon = (4-n)/2$ with $n \longrightarrow 4, \gamma \simeq 0.58$ is the Euler constant, and the "crossed term" include the terms with the interchanges $\beta \leftrightarrow \gamma$ and $s \leftrightarrow t$ or $\beta \leftrightarrow \delta$ and $s \leftrightarrow u$. The divergences in the expression (5.133) can be removed by introducing renormalized coupling constants ($\overline{MS}$ scheme):

$$g_4^{(1)}(\sigma) \equiv g_4^{(1)} + \frac{1}{24\pi^2}\left[\frac{1}{\epsilon} - \gamma + \ln 4\pi - \ln\sigma^2\right] \ , \tag{5.134}$$

$$g_4^{(2)}(\sigma) \equiv g_4^{(2)} + \frac{1}{12\pi^2}\left[\frac{1}{\epsilon} - \gamma + \ln 4\pi - \ln\sigma^2\right] \ . \tag{5.135}$$

The expression (5.133) can be now rewritten as

$$M^{(2)}_{\alpha\beta\gamma\delta} = \frac{\delta_{\alpha\beta}\delta_{\gamma\delta}}{16F^4}\Bigg[-\frac{1}{2\pi^2}s^2\ln\left(\frac{-s}{\sigma^2}\right) - \frac{1}{12\pi^2}\left(u^2 - s^2 + 3t^2\right) \cdot \ln\left(\frac{-t}{\sigma^2}\right) - \frac{1}{12\pi^2}\left(t^2 - s^2 + 3u^2\right)\ln\left(\frac{-u}{\sigma^2}\right) + 8g_4^{(1)}(\sigma)s^2 + 4g_4^{(2)}(\sigma)\left(t^2+u^2\right)\Bigg] + \text{crossed terms}\ . \tag{5.136}$$

It is noticeable that the coefficients of the logarithmic terms can be calculated exactly. As was pointed by Lehman long ago [10], this fact follows from the perturbative unitarity of the scattering amplitude.

However the coefficients $g_4^{(1)}(\sigma)$, $g_4^{(2)}(\sigma)$ are unknown, and we should introduce some additional information to fix them. Let us discuss this point in more detail.

The nonlinear lagrangian (5.122) should be considered as an effective low energy lagrangian describing the dynamics of the light (massless) NG bosons. It is generated by some fundamental dynamics (like QCD in the case of hadronic dynamics) as a result of integrating out all "heavy" degrees of freedom. So we should consider the effective lagrangian as a model relevant below some scale Λ, which is a characteristic scale of the underlying fundamental dynamics. It is natural to assume that this scale is close to the mass of the lightest non-NG boson particle, as ρ meson in QCD. One might expect that a coupling constant $g_d^{(n)}$ of dimensionality $[\text{mass}]^{4-d}$ takes the order of

$$g_d^{(n)} \sim F^2(\Lambda)^{2-d}\ . \tag{5.137}$$

Then the derivative expansion is valid in the region with momentum $p \leq \Lambda$. One should understand that there is no real proof of the estimate (5.137). However, as we shall see in Sec. 12.14, the derivation of the low energy effective action in some toy models confirms it.

As to the scale Λ, there is an argument of Georgi and Manohar [11] leading to the following estimate for the upper bound of Λ:

$$\Lambda \leq 4\pi F\ . \tag{5.138}$$

The argument is based on the renormalization relations (5.134), (5.135) which we rewrite in the following way:

$$\bar{g}_4(\sigma) = \bar{g}_4 + \frac{1}{(4\pi)^2}\left[\frac{1}{\epsilon} - \gamma + \ln 4\pi - \ln \sigma^2\right] \tag{5.139}$$

with $\bar{g}_4 \equiv \frac{1}{2}\left(g_4^{(1)} + g_4^{(2)}\right), \bar{g}_4(\sigma) \equiv \frac{1}{2}\left(\bar{g}_4^{(1)}(\sigma) + \bar{g}_4^{(2)}(\sigma)\right)$. The conventional interpretation of the bare coupling constant $\bar{g}_4$ in $\overline{MS}$ scheme is as a running coupling at the cutoff scale Λ; this corresponds to replacing the factor $\left(\frac{1}{\epsilon} - \gamma + ln 4\pi\right)$ in Eq.(5.139) by $\ln \Lambda^2$:

$$\bar{g}_4(\sigma) = \bar{g}_4(\Lambda) + \frac{1}{(4\pi)^2} \ln \Lambda^2/\sigma^2 \ . \tag{5.140}$$

This relation tells us that at $\sigma < \Lambda$, a change in the renormalization scale of order one $\left(\ln \Lambda^2/\sigma^2 \sim 1\right)$ induces a change in $\bar{g}_4(\sigma)$ of order $\frac{1}{(4\pi)^2}$. Georgi and Manohar argue that this implies the restriction

$$\bar{g}_4(\Lambda) \geq \frac{1}{(4\pi)^2} \tag{5.141}$$

since the case with $\bar{g}_4(\Lambda) \ll 1/(4\pi)^2$ looks rather unnatural: a high energy dynamics with $p \geq \Lambda$ induces a coupling $\bar{g}_4(\Lambda)$ which is much smaller than that, say, $\bar{g}_4(\Lambda/2)$ induced by the low energy dynamics connected with the contribution of the NG bosons.

Relations (5.137) and (5.141) lead to the estimate (5.138).

Although this argument seems rather plausible, at present, we cannot rule out the situation with a high energy dynamics leading to $\bar{g}_4(\Lambda) \ll 1/(4\pi)^2$. The point is that the low energy dynamics leading to running of the coupling constant $\bar{g}_4(\sigma)$ at $\sigma < \Lambda$ does not determine uniquely a high energy dynamics fixing the value $\bar{g}_4(\Lambda)$. The estimate (5.138) is just a guess which should be checked in concrete models.

So far we considered nonlinear lagrangians in the strict chiral limit with NG bosons being massless. However, in the real world, pions are (although light) massive. As we already discussed in Sec.4.5, such a situation is described by the PCAC dynamics. In the PCAC dynamics, nonlinear chiral lagrangians include new terms breaking the chiral symmetry. We shall consider this problem in Sec. 12.13.

As we already indicated in Sec.4.5, the PCAC hypothesis assumes the pion pole dominance in the matrix elements of the axial currents and (at $m_\pi^2 \neq 0$) their divergences in low energy region. This hypothesis leads to numerous interesting relations (low energy theorems) [12]. The Goldberger-Treiman relation considered in Sec.4.5 is an example of them.

These low energy theorems can be obtained by using only *tree graphs* of the nonlinear chiral lagrangians. The discussion in this section explains why it is sufficient to use just the tree graphs to reproduce the theorems. Indeed, as was shown above, the tree approximation is the leading one in the momentum expansion for matrix elements. The pion pole dominance is realized in this approximation in the evident way.

Chapter 6

The Nambu-Jona-Lasinio Model. The First Acquaintance

6.1. Introduction

The Nambu-Jona-Lasinio (NJL) model [1] occupies a very special place in the subject of dynamic symmetry breaking in relativistic quantum field theory. This is in part because it was the first model in which this phenomenon was realized. But this is not the whole story: its dynamics contains many of the important ingredients of dynamic chiral symmetry breaking in quantum chromodynamics and, possibly, electroweak symmetry breaking (see Chapters 12, 14 and 15).

This short chapter is a brief introduction to the dynamics of the NJL model.

6.2. Dynamic Chiral Symmetry Breaking in the Nambu-Jona-Lasinio Model

The lagrangian density of the NJL model takes the form:

$$\mathcal{L} = i\bar{\psi}\gamma^\mu\partial_\mu\psi + G^{(0)}\sum_{\alpha=0}^{N^2-1}\left\{\left(\bar{\psi}\frac{\lambda^\alpha}{2}\psi\right)^2 + \left(\bar{\psi}\frac{\lambda^\alpha}{2}i\gamma_5\psi\right)^2\right\}, \tag{6.1}$$

where λ^α are the matrices from the fundamental representation of the flavor $U(N)$ group $(tr\lambda^\alpha\lambda^\beta = 2\delta_{\alpha\beta})$. The fermion field ψ carries the flavor index (a) and the color index (l), ψ_{al}. We assume that $l = 1, 2, \cdots, N_c$, i.e. ψ is assigned to the fundamental representation of the color group $SU(N_c)$.

By using the Fierz identity for λ^α matrices, $\sum_{\alpha=0}^{N^2-1}\frac{1}{2}\lambda^\alpha_{ab}\lambda^\alpha_{cd} = \delta_{ad}\delta_{cb}$ (considered in Sec. B.3 in Appendix B), the lagrangian density (6.1) can be rewritten in the following form:

$$\mathcal{L} = i\bar{\psi}\gamma^\mu\partial_\mu\psi + 2G^{(0)}\bar{\psi}^a_L\psi^b_R\bar{\psi}^b_R\psi^a_L\ . \tag{6.1a}$$

Because of the absence of the mass term, the lagrangian density (6.1), (6.1a) is invariant under the $U_L(N)\times U_R(N)$ chiral transformations.

Let us rewrite the lagrangian density as

$$\mathcal{L} = i\bar{\psi}\gamma^\mu\partial_\mu\psi - \bar{\psi}_L M\psi_R - \bar{\psi}_R M^\dagger\psi_L - \frac{1}{2G^{(0)}}tr(M^\dagger M) , \tag{6.2}$$

where M is (see Eq.(4.51)):

$$M = \frac{1}{\sqrt{2}}\left[(\sigma + i\eta)\lambda^0 + \left(\sigma^{\alpha'} + i\pi^{\alpha'}\right)\lambda^{\alpha'}\right] \tag{6.3}$$

with $\alpha' = 1, 2, \cdots, N^2 - 1$. Since there is no kinetic term for M, this field is auxiliary. The Euler-Lagrange equation for this field takes the form of a constraint:

$$M_{ab} = -2G^{(0)}\bar{\psi}_R^b\psi_L^a, M_{ab}^\dagger = -2G^{(0)}\bar{\psi}_L^b\psi_R^a . \tag{6.4}$$

The lagrangian density (6.2) reproduces Eq.(6.1a) upon application of the constraint (6.4).

The lagrangian density (6.2) coincides with that of the linear σ-model (see Eqs.(4.53) and (4.55)) in which the kinetic term for M is absent, and the coupling constants λ, g and μ^2 are $\lambda = 0, g = 1/\sqrt{2}$ and $\mu^2 = 1/G^{(0)}$.

We will consider the dynamics in the NJL model by using two different approaches. The first is rather heuristic, and can be considered as a relativistic analogue of the approach used in the BCS approximation in superconductivity (see Sec.2.8); the second, in fact, justifies the first one in the limit when the number of colors N_c goes to infinity.

Let us consider the equation of motion corresponding to the lagrangian density (6.1):

$$i\gamma^\mu\partial_\mu\psi + \frac{G^{(0)}}{2}\left(\bar{\psi}\lambda^\alpha\psi\right)\lambda^\alpha\psi + \frac{G^{(0)}}{2}\left(\bar{\psi}i\lambda^\alpha\gamma_5\psi\right)i\lambda^\alpha\gamma_5\psi = 0 \tag{6.5}$$

(the summation over α is implied). Let us approximate the composite operators $(\bar{\psi}\lambda^\alpha\psi)$ and $(\bar{\psi}i\lambda^\alpha\gamma_5\psi)$ in this equation as

$$\frac{G^{(0)}}{2}\left(\bar{\psi}\lambda^0\psi\right)\lambda^0 \longrightarrow \frac{G^{(0)}}{N}\langle 0|\bar{\psi}\psi|0\rangle \equiv -m_{dyn} , \tag{6.6}$$

$$\left(\bar{\psi}\lambda^\alpha\psi\right) = 0 \text{ if } \alpha \neq 0 , \tag{6.7}$$

$$\left(\bar{\psi}i\lambda^\alpha\gamma_5\psi\right) = 0 . \tag{6.8}$$

The similarity of this approximation with the BCS approximation (see Eq.(2.87)) is evident. The parameter m_{dyn} in equation (6.6) is the fermion dynamical mass (we recall that $\lambda^0 = \left(\frac{2}{N}\right)^{1/2} I$).

We will show that this approximation is justified when N_c is large.

With this substitution, the equation (6.5) reduces to a free Dirac equation:

$$i\gamma^\mu \partial_\mu \psi - m_{dyn}\psi = 0 \ , \tag{6.9}$$

where, however, m_{dyn} is not a free parameter but it is determined from the relation (6.6):

$$m_{dyn} = -\frac{G^{(0)}}{N}\langle 0|\bar{\psi}\psi|0\rangle = \frac{G^{(0)}}{N}\lim_{x\to 0} tr\langle 0|T\psi(x)\bar{\psi}(0)|0\rangle = i\frac{G^{(0)}}{N}\lim_{x\to 0} trG(x) \ , \tag{6.10}$$

where $G(x)$ is the fermion propagator:

$$G(x) = \frac{1}{(2\pi)^4}\int d^4p e^{-ipx}\frac{\hat{p}+m_{dyn}}{p^2 - m_{dyn}^2 + i\epsilon} \, , \hat{p} \equiv \gamma^\mu p_\mu \ . \tag{6.11}$$

Eqs.(6.9) and (6.11) lead to the following equation for m_{dyn}:

$$m_{dyn} = \frac{4iG^{(0)}N_c}{(2\pi)^4}\int d^4p \frac{m_{dyn}}{p^2 - m_{dyn}^2 + i\epsilon} \ . \tag{6.12}$$

Rotating the integral into Euclidean space ($p^0 \to ip_4$), we find the equation:

$$m_{dyn} = \frac{G^{(0)}N_c}{4\pi^2}\int_0^{\Lambda^2} dp^2 \frac{p^2 m_{dyn}}{p^2 + m_{dyn}^2} \ , \tag{6.13}$$

where Λ is the ultraviolet cutoff. This equation is equivalent to the following algebraic one:

$$m_{dyn}(\kappa - 1) = \kappa \frac{m_{dyn}^3}{\Lambda^2} \ln \frac{\Lambda^2 + m_{dyn}^2}{m_{dyn}^2} \tag{6.14}$$

with

$$\kappa \equiv \frac{G^{(0)}N_c\Lambda^2}{4\pi^2} \ . \tag{6.15}$$

There is always the trivial solution, $m_{dyn} = 0$, to this equation, and when $\kappa > 1$, there is also a nontrivial one with $m_{dyn} \neq 0$ corresponding to spontaneous breakdown of the chiral $U_L(N) \times U_R(N)$ symmetry to its vector (diagonal) subgroup: $U_L(N) \times U_R(N) \longrightarrow U_V(N)$.

We will show below that at $\kappa > 1$, the stable vacuum corresponds to the nontrivial solution. Like in the linear σ-model, there should be N^2 massless pseudoscalar NG bosons which now are fermion-antifermion bound states. We will consider the dynamics of these states in Sec.9.5.

We recall that because of the presence of the Fermi surface, dynamical $U(1)$ gauge symmetry breaking in the BCS model of superconductivity takes place at any positive value of the coupling constant g, i.e. the critical value g_c is $g_c = 0$ there (see Sec.2.8). Unlike the BCS model, in the NJL model, in order to generate a dynamical mass for fermions, the coupling constant κ must be larger than the critical value $\kappa_c = 1$. We will discuss the origin of the difference between these two dynamics in more detail in Chapter 10.

Let us now try to justify the approximation (6.6)-(6.8).

We begin by considering the path integral representation for the generating functional $Z(J)$ in the theory with the lagrangian density (6.2) (see Sec. C.3 in Appendix C):

$$Z(J) = K \int \mathcal{D}M\mathcal{D}M^\dagger \mathcal{D}\psi\mathcal{D}\bar{\psi} \exp\left\{i \int d^4x \,[\mathcal{L}(x) + F(J)]\right\} , \tag{6.16}$$

where $F(J) \equiv \sum_i J_i(x)\varphi_i(x)$ is a source term for some local fields $\varphi_i(x)$, and the normalization factor K ensures that $Z(0) = 1$. Let us rewrite this expression in the form

$$Z(J) = K \int \mathcal{D}M\mathcal{D}M^\dagger \exp\left\{i \int d^4x \left[-\frac{1}{2G^{(0)}} tr\left(M^\dagger M\right) + F(J)\right]\right\} Z_f\left(M, M^\dagger\right) , \tag{6.17}$$

where

$$Z_f\left(M, M^\dagger\right) = \int \mathcal{D}\psi\mathcal{D}\bar{\psi} \exp\left\{i \int d^4x \cdot \bar{\psi} i\mathfrak{D}\psi\right\} = C \left(\text{Det} i\mathfrak{D}\right)^{N_c} \tag{6.18}$$

with

$$i\mathfrak{D} = \left[i\gamma^\mu\partial_\mu - M\frac{1-\gamma_5}{2} - M^\dagger\frac{1+\gamma_5}{2}\right] \tag{6.19}$$

and C being some inessential normalization constant. In the derivation of Eq.(6.18) we used Eq.(C.32) in Appendix C; the determinant Det is taken in the functional sense.

Therefore the action of the effective theory for the field M is

$$S\left(M, M^{\dagger}\right) = -iN_c \ln \mathrm{Det} i\mathfrak{D} - \frac{1}{2G^{(0)}} \int d^4x tr(M^{\dagger} M) \ . \tag{6.20}$$

Eqs.(6.14) and (6.15) suggest that in the limit $N_c \longrightarrow \infty$, the effective coupling constant is

$$\tilde{G}^{(0)} \equiv G^{(0)} N_c \ . \tag{6.21}$$

Therefore

$$S\left(M, M^{\dagger}\right) = -N_c \left[i \ln \mathrm{Det} i\mathfrak{D} + \frac{1}{2\tilde{G}^{(0)}} \int d^4x tr\left(M^{\dagger} M\right)\right], \tag{6.22}$$

and the path integral (6.17) is dominated by stationary points of this action in the limit $N_c \longrightarrow \infty$,

$$\frac{\delta S}{\delta M} = 0 \ . \tag{6.23}$$

In other words, in this limit, the dynamics becomes effectively classical.

We are now interested in the vacuum solutions of the equation (6.23) with M being independent of x. As in the case of the linear σ-model, we are looking for the solution of the form

$$M = M^{\dagger} = v\frac{\lambda^0}{\sqrt{2}} = vN^{-\frac{1}{2}} \ ; \tag{6.24}$$

all other vacuum solutions can be obtained from this one by using chiral transformations.

At such a M, the action $S(M, M^{\dagger})$ can be easily calculated. Indeed, now the matrix element $\langle p'|i\mathfrak{D}|p\rangle$ is

$$\langle p'|i\mathfrak{D}|p\rangle = \delta^4(p - p') \times (\gamma^{\mu} p_{\mu} - vN^{-1/2}) \ . \tag{6.25}$$

This equation and the identity

$$\ln \mathrm{Det} i\mathfrak{D} = \mathrm{Tr} Ln i\mathfrak{D} \tag{6.26}$$

lead to the relation

$$\begin{aligned}\ln \mathrm{Det} i\mathfrak{D} &= \delta^4(p-p')\Big|_{p=p'} \times \int d^4p tr \ln(\gamma^\mu p_\mu - \upsilon N^{-1/2}) \\ &= \frac{1}{(2\pi)^4}\int d^4x \times 2N \int d^4p \ln(1-\frac{\upsilon^2 N^{-1}}{p^2}) + \text{constant}\,.\end{aligned} \tag{6.27}$$

Therefore the potential $V(\upsilon) \equiv -(d^4x)^{-1}S$ is

$$V = N_c\left[\frac{1}{2\tilde{G}^{(0)}}\upsilon^2 + \frac{2iN}{(2\pi)^4}\int d^4p \ln(1-\frac{\upsilon^2 N^{-1}}{p^2})\right]\,. \tag{6.28}$$

The equation (6.23) now takes the form

$$N^{-1/2}\upsilon = \frac{4iG^{(0)}N_c}{(2\pi)^4}\int d^4p \frac{N^{-1/2}\upsilon}{p^2-(N^{-1/2}\upsilon)^2+i\epsilon}\,. \tag{6.29}$$

It exactly coincides with Eq.(6.12) if we identify the dynamical mass m_{dyn} with $N^{-1/2}\upsilon$.

It is not difficult to check at $\kappa > 1$ that while the nontrivial solution $M = \upsilon\frac{\lambda^0}{\sqrt{2}}$ corresponds to a minimum of the potential V, the trivial one corresponds to its maximum. Therefore, as in the linear σ-model with $\mu^2 < 0$, the symmetric vacuum connected with the trivial solution is unstable.

Thus we have justified the approximation (6.6)-(6.8) in the limit $N_c \longrightarrow \infty$. We recall that the BCS approximation (2.87) in the theory of superconductivity (and the Bogolyubov approximation (2.29) in the theory of superfluidity) is valid in the limit of a large number of particles. In the NJL model, the color number N_c plays an analogous role. Since in this approximation all quantum fluctuations of the field M are neglected, it corresponds to the mean-field theory approximation.

6.3. The Structure of the Vacuum in the NJL Model

In the mean-field theory approximation, the equation of motion in the NJL model reduces to the free Dirac equation (6.9). Therefore to describe the structure of the vacuum in this model, we can use the results of Sec.1.4, in particular, Eq.(1.66). We find that the vacuum is

$$|0\rangle = \prod_{\mathbf{p},s=\pm 1} \left[\alpha_p - \beta_p b^\dagger_{1p}(s) d^\dagger_{1(-p)}(s)\right] |0_1\rangle \ , \tag{6.30}$$

where the coefficients α_p and β_p are defined in Eq.(1.64) and the operators $b^\dagger_{1p}(s), d^\dagger_{1p}(s)$ and the vacuum $|0_1\rangle$ relate to free massless theory (for simplicity, we consider the $U_L(1)\times U_R(1)$ chiral symmetry; all color indices are suppressed).

Since the operators $b^\dagger_{1p}(s)$ and $d^\dagger_{1(-p)}(s)$ have the same helicity (and therefore chirality), this vacuum is not eigenstate of the chiral operator $Q_5 = Q_L - Q_R$. In fact, since the left fermions have $s = -1$, and the right fermions have $s = +1$ (see Appendix A), the transformation properties of $b^\dagger_{1p}(s), d^\dagger_{1p}(s)$ under $U_5(\theta) = \exp(iQ_5\theta)$ are:

$$\begin{aligned} U_5^{-1}(\theta) b^\dagger_{1p}(\pm 1) U_5(\theta) &= e^{\mp i\theta} b^\dagger_{1p}(\pm 1) \\ U_5^{-1}(\theta) d^\dagger_{1p}(\pm 1) U_5(\theta) &= e^{\mp i\theta} d^\dagger_{1p}(\pm 1) \ . \end{aligned} \tag{6.31}$$

Therefore

$$|\theta\rangle \equiv U_5(\theta)|0\rangle = \prod_{p,\pm 1} \left\{\alpha_p - \beta_p e^{\pm 2i\theta} b^\dagger_{1p}(\pm 1) d^\dagger_{1(-p)}(\pm 1)\right\} |0_1\rangle \ . \tag{6.32}$$

Since the hamiltonian of the NJL model commutes with the chiral operator, all these vacua have the same energy, that is the vacuum degeneracy takes place in the model.

6.4. Dynamic Generation of a Majorana Mass

As we already noted in Sec.1.3, fermions may possess another type of a mass than Dirac's one. This is the Majorana mass leading to the violation of the conservation of the fermion number.

The lagrangian density for a free massive Majorana fermion takes the same form as for Dirac's one (up to the factor 1/2)

$$\mathcal{L} = \frac{i}{2}\bar\psi_M \gamma^\mu \partial_\mu \psi_M - \frac{m}{2}\bar\psi_M \psi_M \ , \tag{6.33}$$

however there is an additional restriction for ψ_M: it is invariant under the operation of charge conjugation (see Eq.(A.23)):

$$\psi_M = C\bar{\psi}_M^{\top} \ . \tag{6.34}$$

In the spinor representation of γ^μ, this relation is (see Eq.(A.29)):

$$\psi_M = \gamma^2 \psi_M^* \ . \tag{6.35}$$

The Majorana spinor ψ_M can be expressed through a left-handed spinor η ($\psi_L = \begin{pmatrix} 0 \\ \eta \end{pmatrix}$; see Eq.(A.16)) as

$$\psi_M = \begin{pmatrix} -\sigma_2 \eta^* \\ \eta \end{pmatrix} \ . \tag{6.36}$$

The lagrangian density (6.33) can be now rewritten (up to inessential surface terms) as

$$\mathcal{L} = i\eta^\dagger \partial_t \eta - i\eta^\dagger \sigma_i \partial_i \eta + \frac{m}{2}\eta^\dagger \sigma_2 \eta^* + \frac{m}{2}\eta^\top \sigma_2 \eta \ . \tag{6.37}$$

Let us consider a NJL-like model in which the Majorana mass is generated dynamically (such a model was first considered by Vaks and Larkin [2]); for simplicity, we will consider the case with one fermion flavor.

The lagrangian density of the model is (compare with Eq.(6.1)):

$$\mathcal{L} = \frac{i}{2}\bar{\psi}_M \gamma^\mu \partial_\mu \psi_M + \frac{G^{(0)}}{4}\left[\left(\bar{\psi}_M \psi_M\right)^2 + \left(\bar{\psi}_M i\gamma_5 \psi_M\right)^2\right] \ , \tag{6.38}$$

or, what is equivalent,

$$\mathcal{L} = i\eta^\dagger \partial_t \eta - i\eta^\dagger \sigma_i \partial_i \eta + \frac{G^{(0)}}{2}\left[\left(\eta^\dagger \sigma_2 \eta^*\right)\left(\eta^\top \sigma_2 \eta\right) + \left(\eta^\top \sigma_2 \eta\right)\left(\eta^\dagger \sigma_2 \eta^*\right)\right] \ . \tag{6.39}$$

It is clear that this $\mathcal{L}$ is invariant under the $U(1)$ transformations,

$$\eta \longrightarrow U^{-1}(\theta)\eta U(\theta) = e^{i\theta}\eta \tag{6.40}$$

connected with the fermion number conservation.

The equation of motion for ψ_M is (compare with Eq.(6.5)):

$$\begin{aligned} i\gamma^\mu \partial_\mu \psi_M + &\frac{G^{(0)}}{2}\left[\left(\bar{\psi}_M \psi_M\right)\psi_M + \psi_M\left(\bar{\psi}_M \psi_M\right)\right] \\ + &\frac{G^{(0)}}{2}\left[\left(\bar{\psi}_M i\gamma_5 \psi_M\right) i\gamma_5 \psi_M + i\gamma_5 \psi_M\left(\bar{\psi}_M i\gamma_5 \psi_M\right)\right] = 0 \,. \end{aligned} \tag{6.41}$$

In the mean-field approximation, the equation becomes (compare with Eq.(6.9))

$$i\gamma^\mu \partial_\mu \psi_M - m_{dyn}\psi_M = 0 \,, \tag{6.42}$$

where the dynamic Majorana mass is

$$m_{dyn} = -G^{(0)}\langle 0|\bar{\psi}_M \psi_M|0\rangle = iG^{(0)} \lim_{x\to 0} trG(x) \,, \tag{6.43}$$

and the propagator $G(x)$ is

$$G(x) = \frac{1}{(2\pi)^4}\int d^4p e^{-ipx}\frac{\hat{p} + m_{dyn}}{p^2 - m_{dyn}^2} \,. \tag{6.44}$$

Therefore the mass m_{dyn} is a solution of the same equation (6.12) which was derived for the dynamic Dirac mass. In particular we conclude that when the coupling constant $\kappa = G^{(0)} Nc\Lambda^2/4\pi^2$ is larger than 1, there is a nontrivial solution for m_{dyn} corresponding to the spontaneous breakdown of $U(1)$ symmetry.

The expression for the vacuum $|\theta\rangle$ takes the form (compare with Eq.(6.32))

$$|\theta\rangle = \prod_p{}' \left\{\alpha_p - \beta_p e^{-2i\theta} b^\dagger_{1p}(-1) b^\dagger_{1(-p)}(-1)\right\}\left\{\alpha_p - \beta_p e^{2i\theta} d^\dagger_{1p}(+1) d^\dagger_{1(-p)}(+1)\right\}|0_1\rangle \,, \tag{6.45}$$

where $b^\dagger_{1p}(-1)$ $\left(d^\dagger_{1p}(+1)\right)$ is a creation operator of a left-handed massless fermion (a right-handed massless antifermion), and α_p and β_p are defined in Eq.(1.64). The symbol $\prod'$ means that in order to avoid the double counting we consider here the vectors $\mathbf{n} = \mathbf{p}/|\mathbf{p}|$ lying in the upper semisphere ($n_z > 0$) only.

One can see that the vacuum $|\theta\rangle$ is not an eigenstate of the fermion number operator.

Chapter 7

The Effective Action Method in Quantum Field Theory

7.1. Introduction

The essence of the phenomenon of spontaneous symmetry breaking is connected with the inclusion of the vacuum degrees of freedom in the field dynamics. The important point is that these degrees of freedom are essentially classical in nature (as those described by the order parameter $\varphi_c(x) = \langle 0|\varphi(x)|0\rangle$ in the theory of superfluidity or in the Goldstone model). This is in turn connected with the point that the vacuum is a macroscopic state possessing a huge number of degrees of freedom (as the vacuum $|\theta\rangle$ in an almost ideal Bose gas (see Eq.(2.44)).

It is therefore important to get a regular method to deal with these vacuum degrees of freedom. Such a method introduced by Goldstone, Salam, and Weinberg [1], and by Jona-Lasinio [2] is referred as the method of the effective action.

The effective action $\Gamma(\varphi_c(x))$ is a functional of the classical field $\varphi_c(x) = \langle\psi|\varphi(x)|\psi\rangle$ (the order parameter) describing the dynamics of a state $|\psi\rangle$. The equation of motion is $\frac{\delta\Gamma}{\delta\varphi_c(x)} = 0$. Besides the ground state, the solutions of this equation describe different types of macroscopic excitations in the vacuum.

The method of the effective action had become a standard method to deal with relativistic theories with spontaneous symmetry breaking after the appearance of the work of Coleman and Weinberg [3]. Among many interesting results in that paper, one of the most important points was the demonstration of the fruitfulness of the renormalization scheme to deal with the effective action $\Gamma(\varphi_c(x))$ beyond the tree approximation in renormalizable theories.

In the present chapter, we will consider this method rather in detail.

7.2. Effective Action (Generating Functional for Proper Vertices)

In this section, we consider the basic properties of the effective action (generating functional for proper vertices) which is a main tool for studying spontaneous symmetry breaking in quantum field theory.

We begin by introducing the generating functional for Green's functions (for a thorough discussion of this formalism see the book [4]):

$$Z(J) = \langle 0|T \exp\left[i \int d^4x \varphi(x) J(x)\right] |0\rangle = \langle 0, \text{out}|0, \text{in}\rangle \ , \tag{7.1}$$

where the symbol $\varphi(x)$ represents all fields of a lagrangian, and $J(x)$ is a source for $\varphi(x)$. We emphasize that since the source $J(x)$ may depend on the time coordinate x^0, the in-vacuum and the out-vacuum (corresponding to $x^0 \longrightarrow -\infty$ and $x^0 \longrightarrow +\infty$, respectively) are in general different (the vacuum $|0\rangle$ is $|0\rangle = |0, \text{in}\rangle|_{J=0} = |0, \text{out}\rangle|_{J=0}$).

For simplicity, we will consider a theory with just one real scalar field $\varphi(x)$ with a mass m.

The functional $Z(J)$ can be expanded in a series:

$$Z(J) = 1 + \sum_{n=1}^{\infty} \frac{i^n}{n!} \int d^4x_1 \cdots d^4x_n G^{(n)}(x_1, \cdots, x_n) J(x_1) \cdots J(x_n) \tag{7.2}$$

with

$$G^{(n)}(x_1, \cdots, x_n) = (-i)^n \frac{\delta^n Z}{\delta J(x_1) \cdots \delta J(x_n)}|_{J=0} = \langle 0|T\varphi(x_1) \cdots \varphi(x_n)|0\rangle \tag{7.3}$$

being Green's functions.

Let us introduce the functional $W(J)$ which is

$$W(J) = \frac{1}{i} \ln Z(J) \ . \tag{7.4}$$

The corresponding series is:

$$W(J) = \sum_{n=1}^{\infty} \frac{i^{n-1}}{n!} \int d^4x_1 \cdots d^4x_n G_c^{(n)}(x_1, \cdots, x_n) J(x_1) \cdots J(x_n) \tag{7.5}$$

with $G_c^{(n)}$ being connected Green's functions, that, is the corresponding Feynman diagrams in perturbative theory for them are connected. To illustrate this, let us consider the few simplest Green's functions:

$$G_c^{(1)}(x_1) = \frac{\delta W}{\delta J(x_1)}|_{J=0} = \frac{1}{iZ(J)} \frac{\delta Z}{\delta J(x_1)}|_{J=0} = \langle 0|\varphi(0)|0\rangle \tag{7.6}$$

(since the vacuum $|0\rangle$ is translation invariant, $\langle 0|\varphi(x)|0\rangle = \langle 0|\varphi(0)|0\rangle$). We note that the definition of $Z(J)$ in Eq.(7.1) with $Z(0) = 1$ implies that all vacuum diagrams at $J = 0$ are removed. Therefore $G_c^{(1)}$ is really connected. Then, let us consider $G_c^{(2)}(x_1, x_2)$:

$$G_c^{(2)}(x_1, x_2) = \frac{1}{i} \frac{\delta^2 W}{\delta J(x_1) \delta J(x_2)} |_{J=0} = \langle 0|T\varphi(x_1)\varphi(x_2)|0\rangle - (\langle 0|\varphi(0)|0\rangle)^2 \ . \tag{7.7}$$

One can see that the contribution of disconnected diagrams in this expression is removed.

The same is valid for higher $G_c^{(n)}$.

The effective action is introduced as the Legendre transform of $W(J)$. Let us consider $\varphi_c(x, J)$ which is

$$\varphi_c(x, J) = \frac{\delta W}{\delta J(x)} = \frac{1}{Z(J)} \langle 0, \text{out}|\varphi(x)|0, \text{in}\rangle \ . \tag{7.8}$$

The effective action $\Gamma(\varphi_c)$ is defined as

$$\Gamma(\varphi_c) = \left[W(J) - \int d^4x \, J(x) \varphi_c(x) \right] |_{J(x)=J_c(x,\varphi_c)} \ . \tag{7.9}$$

Here we assumed that the relation (7.8) may be inverted.

First of all, let us show that

$$\frac{\delta \Gamma}{\delta \varphi_c(x)} = -J_c(x, \varphi_c) \ . \tag{7.10}$$

Indeed, we have from Eq.(7.9):

$$\frac{\delta \Gamma}{\delta \varphi_c(x)} = \int d^4y \left\{ \left[\frac{\delta W(J)}{\delta J(y)} - \varphi_c(y) \right] |_{J=J_c(\varphi_c)} \frac{\delta J_c(y, \varphi_c)}{\delta \varphi_c(x)} \right\} - J_c(x, \varphi_c) \ . \tag{7.11}$$

Because of the relation (7.8), this equation leads to the relation (7.10).

Let us define the notion of proper vertices $\Gamma^{(n)}(x_1, \cdots, x_n)$. They are amputated and one-particle irreducible Green's functions, i.e. in the perturbative expansion, they are defined in terms of amputated one-particle irreducible Feynman diagrams. This means the following. The amputated functions are defined as

$$G_{amp}^{(n)}(x_1, \cdots, x_n) \equiv (i^{n-1}) \prod_{k=1}^{n} \int dy_k \left[G^{(2)}(x_k, y_k) \right]^{-1} G^{(n)}(y_1, \cdots, y_n) \ , \tag{7.12}$$

i.e. they are determined through the convolution of Green's functions with the inverse two-point functions corresponding to each external coordinate x_k (the normalization factor $(i)^{n-1}$ is introduced here just for convenience).

One-particle irreducible Green's functions are those with corresponding diagrams remaining connected when an arbitrary internal line is cut. The proper vertices are both amputated and one-particle irreducible Green's functions. Let us demonstrate that the functional $\Gamma(\varphi_c)$ is the generating functional for proper vertices. For simplicity, we consider here the case when the only solution $\varphi_c(x, J)$ at $J = 0$ is $\varphi_c(x, 0) = v = 0$. The more interesting case with $\varphi_c(x, 0) \neq 0$ corresponding to spontaneous symmetry breaking will be discussed in Sec.7.4.

We will illustrate that the functions

$$\Gamma^{(n)}(x_1, \cdots, x_n) = \frac{\delta^n \Gamma}{\delta\varphi_c(x_1)\cdots\delta\varphi(x_n)}|_{\varphi_c=0} \tag{7.13}$$

in the series

$$\Gamma(\varphi_c) = \sum_{n=2}^{\infty} \frac{1}{n!} \int d^4x_1 \cdots d^4x_n \Gamma^{(n)}(x_1, \cdots, x_n)\varphi_c(x_1)\cdots\varphi_c(x_n) \tag{7.14}$$

are proper vertices by considering the two-point function $\Gamma^{(2)}(\Gamma^{(1)}(x) = \frac{\delta\Gamma}{\delta\varphi_c(x)}|_{\varphi_c=0}$ $= -J(x, 0) = 0$ due to our assumption).

Differentiating Eq.(7.8) and taking into account Eq.(7.10), we get:

$$\begin{aligned}
\delta^4(x-y) &= \frac{\delta}{\delta\varphi_c(y)}\left[\frac{\delta W(J)}{\delta J(x)}|_{J=J_c(\varphi_c)}\right] \\
&= \int d^4z \frac{\delta J_c(z,\varphi_c)}{\delta\varphi_c(y)} \frac{\delta^2 W(J)}{\delta J(z)\delta J(x)}|_{J=J_c(\varphi_c)} \\
&= -\int d^4z \frac{\delta^2\Gamma(\varphi_c)}{\delta\varphi_c(z)\delta\varphi_c(y)} \frac{\delta^2 W(J)}{\delta J(z)\delta J(x)}|_{J=J_c(\varphi_c)} \\
&\underset{J=0}{=} i^{-1}\int d^4z \Gamma^{(2)}(y,z) G_c^{(2)}(z,x)\ .
\end{aligned} \tag{7.15}$$

Because of this equation and the definition (7.12), we see that $\Gamma^{(2)}$ is an amputated Green's function. Let us introduce the self-energy operator $\Sigma(x, y)$:

$$\Gamma^{(2)}(x,y) = i\left[G_0^{(2)}(x,y)\right]^{-1} - \Sigma(x,y)\ , \tag{7.16}$$

where

$$iG_0^{(2)}(x,y) = D(x-y) = \frac{-1}{(2\pi)^8}\int d^4p d^4q e^{-i(px+qy)} \cdot (2\pi)^4\delta^4(p+q)(p^2-m^2)^{-1} \tag{7.17}$$

is the propagator of the free field with the mass m. The self-energy operator is determined by a set of one-particle irreducible diagrams. Indeed, in a theory with $J = 0$,

$\Sigma(x, y)$ depends only on the difference $x - y$, and the function $G_c^{(2)}(p, q)$ in momentum space is:

$$\begin{aligned} G_c^{(2)}(p,q) =& i(2\pi)^4\delta^4(p+q)\left(p^2 - m^2 - \Sigma(p)\right)^{-1} \\ =& (2\pi)^4\delta^4(p+q)\left[\frac{i}{p^2 - m^2} + \frac{i}{p^2 - m^2}\frac{1}{i}\Sigma(p)\frac{i}{p^2 - m^2} + \cdots\right] , \end{aligned} \tag{7.18}$$

where

$$\Sigma(p) = \int d^4x e^{ipx}\Sigma(x) \tag{7.19}$$

(see Fig. 7.1). It is clear that $\Sigma(p)$ is the one-particle irreducible contribution to the two-point function.

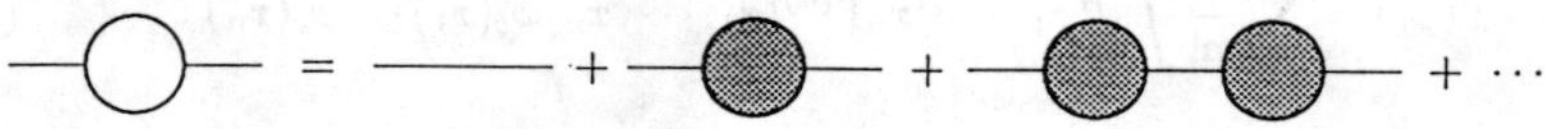

$$\frac{i}{p^2-m^2} \quad + \frac{i}{p^2-m^2}\frac{\Sigma}{i}\frac{i}{p^2-m^2} + \frac{i}{p^2-m^2}\frac{\Sigma}{i}\frac{i}{p^2-m^2}\frac{\Sigma}{i}\frac{i}{p^2-m^2} + \cdots$$

Fig. 7.1

One can check in the same way that the higher $\Gamma^{(n)}$ are also proper vertices (for a thorough discussion of this point see the book [4]).

Let us show that in the lowest order of perturbative expansion (the tree approximation) when the theory is reduced to the free one, the functional $\Gamma(\varphi_c)$ coincides with the classical action. Indeed, in this approximation, the only proper vertex is (see Eq.(7.16))

$$\begin{aligned} \Gamma^{(2)}(x_1, x_2) =& i\left[G_0^{(2)}(x_1, x_2)\right]^{-1} \\ =& \frac{1}{(2\pi)^8}\int d^4p_1 d^4p_2 e^{-i(p_1x_1+p_2x_2)}(2\pi)^4\delta^4(p_1+p_2)(p_1^2 - m^2) . \end{aligned} \tag{7.20}$$

Therefore the effective $\Gamma(\varphi_c)$ is:

$$\Gamma(\varphi_c) = \frac{1}{2}\int d^4x_1 d^4x_2 \Gamma^{(2)}(x_1, x_2)\varphi_c(x_1)\varphi_c(x_2) = \frac{1}{2}\int d^4x \left[\partial_\mu\varphi_c\partial^\mu\varphi_c - m^2\varphi_c^2\right] . \tag{7.21}$$

Beyond the tree approximation, the functional $\Gamma(\varphi_c)$ includes quantum corrections. This is the reason why $\Gamma(\varphi_c)$ is referred as the effective action.

We recall that in the discussion above, we assumed that $\varphi_c(x, J)$ in Eq.(7.8) is $\varphi_c = 0$ at $J = 0$. This assumption is invalid in the case of systems with spontaneous symmetry breaking when $v \equiv \varphi_c(x, 0) \neq 0$. In that case, instead of the expansion (7.14), we should consider the following one:

$$\Gamma(\varphi_c) = \sum_{n=1}^{\infty} \frac{1}{n!} \int d^4x_1 \cdots d^4x_n \Gamma_v^{(n)}(x_1, \cdots, x_n) \tilde{\varphi}_c(x_1) \cdots \tilde{\varphi}_c(x_n) \tag{7.22}$$

with $\tilde{\varphi}_c(x) = \varphi_c(x) - v$ and $\Gamma_v^{(n)}$ being proper vertices relating to the vacuum with $\frac{\delta W}{\delta J(x)}|_{J=0} = \langle 0|\varphi(x)|0\rangle = v$.

We will consider this case in detail in the following.

7.3. The Physical Meaning of the Effective Action

To understand deeper the physical meaning of the effective action, let us expand it in powers of derivatives of the field $\varphi_c(x)$ about an x-independent configuration $\varphi_c = \varphi_0$:

$$\varphi_c(x) = \varphi_0 + \tilde{\varphi}_c(x) \ . \tag{7.23}$$

We have:

$$\begin{aligned}\Gamma(\varphi_c) =& \Gamma(\varphi_0) + \int d^4x \frac{\delta\Gamma}{\delta\varphi_c(x)}|_{\varphi_c=\varphi_0} \tilde{\varphi}_c(x) \\ &+ \frac{1}{2} \int d^4x d^4y \frac{\delta^2\Gamma}{\delta\varphi_c(x)\delta\varphi_c(y)}|_{\varphi_c=\varphi_0} \tilde{\varphi}_c(x)\tilde{\varphi}_c(y) + \cdots \ ,\end{aligned} \tag{7.24}$$

which is (compare with Eq.(7.22)):

$$\Gamma(\varphi_c) = \Gamma(\varphi_0) + \int d^4x \Gamma_{\varphi_0}^{(1)}(0)\tilde{\varphi}_c(x) + \frac{1}{2}\int d^4x d^4y \Gamma_{\varphi_0}^{(2)}(x-y)\tilde{\varphi}_c(x)\tilde{\varphi}_c(y) + \cdots \ . \tag{7.25}$$

We use here the point that since the φ_0-configuration is x-independent, the proper vertices $\Gamma^{(n)}(x_1, \cdots, x_n)|_{\varphi_c=\varphi_0}$ depend only on defferences of coordinates. In particular, $\Gamma(\varphi_0)$ is

$$\Gamma(\varphi_0) = -\int d^4x V(\varphi_0) \tag{7.26}$$

with the effective potential

$$V(\varphi_0) = -\sum_{n=1}^{\infty} \frac{1}{n!} (\varphi_0)^n \tilde{\Gamma}^{(n)}(0, \cdots, 0) \ , \tag{7.27}$$

where $\tilde{\Gamma}^{(n)}(p_1, \cdots, p_n)$ are Fourier transforms of $\Gamma^{(n)}(x_1, \cdots, x_n)$ in Eq.(7.14):

$$\begin{aligned}\Gamma^{(n)}(x_1, \cdots, x_n) = \prod_{i=1}^{n} \left[\int \frac{d^4 p_i}{(2\pi)^4}\right] (2\pi)^4 \delta^4 \left(\sum_i p_i\right) \\ \cdot \tilde{\Gamma}^{(n)}(p_1, \cdots, p_n) \exp\left(-i \sum_i p_i x_i\right) .\end{aligned} \tag{7.28}$$

Using the Taylor expansion

$$\tilde{\varphi}_c(y) = \tilde{\varphi}_c(x) + \frac{\partial \tilde{\varphi}_c}{\partial x^\mu}(y-x)^\mu + \frac{1}{2}\frac{\partial^2 \tilde{\varphi}_c}{\partial x^\mu \partial x^\nu}(y-x)^\mu (y-x)^\nu + \cdots , \tag{7.29}$$

one can write

$$\begin{aligned}\Gamma(\varphi_c) = &-\int d^4 x V(\varphi_0) + \int d^4 x \tilde{\Gamma}^{(1)}_{\varphi_0}(0)\tilde{\varphi}_c(x) \\ &+\frac{1}{2}\int d^4 x \tilde{\Gamma}^{(2)}_{\varphi_0}(0,0)\tilde{\varphi}^2_c(x) + \frac{1}{4}\int d^4 x \frac{\partial^2 \tilde{\Gamma}^{(2)}_{\varphi_0}}{\partial p_\mu \partial p_\nu}|_{p=0} \frac{\partial \tilde{\varphi}_c}{\partial x^\mu}\frac{\partial \tilde{\varphi}_c}{\partial x^\nu} + \cdots\end{aligned} \tag{7.30}$$

All nonderivative terms sum up in the potential term

$$-\int d^4 x V\left(\varphi_c(x)\right) , \tag{7.31}$$

and the terms with two derivatives yield the kinetic term

$$\frac{1}{2}\int d^4 x Z\left(\varphi_c(x)\right) \partial_\mu \varphi_c \partial^\mu \varphi_c , \tag{7.32}$$

It is important that the functional form of $Z\left(\varphi_c(x)\right)$ coincides with that of $Z(\varphi_0)$, with φ_0 being x-independent, and $Z(\varphi_0)$ is expressed through the second derivative of the proper vertex $\tilde{\Gamma}^{(2)}_{\varphi_0}(p,-p)$ at $p=0$: $Z(\varphi_0) = \frac{1}{8} g_{\mu\nu} \frac{\partial^2 \tilde{\Gamma}^{(2)}_{\varphi_0}}{\partial p_\mu \partial p_\nu}|_{p=0}$.

Thus the derivative expansion

$$\Gamma(\varphi_c) = \int d^4 x \left[-V(\varphi_c) + \frac{1}{2} Z(\varphi_c)\partial_\mu \varphi_c \partial^\mu \varphi_c + \cdots\right] \tag{7.33}$$

reduces to calculating proper vertices around the point $p=0$.

We shall consider an approximation method for calculating the effective action in the next section. But before this, we shall discuss its physical interpretation [5, 6].

We begin by considering the path integral representation for the functional $Z(J)$:

$$Z(J) = \lim_{\substack{t_f \to +\infty \\ t_i \to -\infty}} N \int_{\varphi(t_i,\mathbf{x})=\varphi_i(\mathbf{x})}^{\varphi(t_f,\mathbf{x})=\varphi_f(\mathbf{x})} \mathcal{D}(\varphi) \exp\left[i \int d^4x \{\mathcal{L}(x) + J(x)\varphi(x)\}\right] , \tag{7.34}$$

where the φ integration is functional, $\mathcal{L}(x)$ is the lagrangian density of the theory, and the normalization factor N provides $Z(0) = 1$, i.e.

$$N^{-1} = \lim_{\substack{t_f \to +\infty \\ t_i \to -\infty}} \int_{\varphi(t_i,\mathbf{x})=\varphi_i(\mathbf{x})}^{\varphi(t_f,\mathbf{x})=\varphi_f(\mathbf{x})} \mathcal{D}(\varphi) \exp\left[i \int d^4x \mathcal{L}(x)\right] . \tag{7.35}$$

The symbol $\lim\limits_{\substack{t_f \to +\infty \\ t_i \to -\infty}} \int_{\varphi(t_i,\mathbf{x})=\varphi_i(\mathbf{x})}^{\varphi(t_f,\mathbf{x})=\varphi_f(\mathbf{x})} \mathcal{D}(\varphi)$ means that the path integral is evaluated along trajectories from $\varphi_i(\mathbf{x})$ to $\varphi_f(\mathbf{x})$ in the functional space; the initial time t_i and the final time t_f go to $-\infty$ and $+\infty$, respectively. A convenient choice is the periodic boundary conditions with $\varphi_f(\mathbf{x}) = \varphi_i(\mathbf{x})$.

We discuss path integrals in Sec.C.3 in Appendix C (for a thorough discussion of path integrals see the book [4]).

Let us consider a source J independent of $t \equiv x^0$. Then, as is discussed in Sec. C.3 in Appendix C, the functional $Z(J)$ coincides with the transition amplitude:

$$\tilde{N}\langle\varphi_f(\mathbf{x})|e^{-iH(J)(t_f-t_i)}|\varphi_i(\mathbf{x})\rangle = Z(J) , \tag{7.36}$$

where $H(J)$ is the hamiltonian corresponding to the lagrangian density

$$\mathcal{L}(J) = \mathcal{L} + J(x)\varphi(x) . \tag{7.37}$$

The normalization factor $\tilde{N}$ in Eq.(7.36) is chosen in such a way as to provide $Z(0) = 1$.

If we expand the transition amplitude in a complete set of energy eigenstates,

$$H(J)|n;J\rangle = E_n(J)|n;J\rangle \ , \tag{7.38}$$

then

$$Z(J) = \tilde{N} \sum_n \exp(-iE_n(J)T)\langle\varphi_f(\mathbf{x})|n;J\rangle\langle n;J|\varphi_i(\mathbf{x})\rangle, T = t_f - t_i \ . \tag{7.39}$$

With $T \longrightarrow \infty$, the strong oscillations damp the amplitude, and therefore the leading term in this expression corresponds to the lowest-lying energy eigenstate $|0;J\rangle$:

$$Z(J) \simeq \exp(-i(E_0(J) - E_0)T)\frac{\langle\varphi_f(\mathbf{x})|0;J\rangle\langle 0;J|\varphi_i(\mathbf{x})\rangle}{\langle\varphi_f(\mathbf{x})|0\rangle\langle 0|\varphi_i(\mathbf{x})\rangle} \ , \tag{7.40}$$

where

$$|0\rangle \equiv |0,J\rangle|_{J=0}, \ E_0 \equiv E_0(J)|_{J=0} \ . \tag{7.41}$$

Since $Z(J) = \exp(iW(J))$, Eq.(7.40) implies that, as $T \longrightarrow \infty$,

$$W(J) = (E_0 - E_0(J))T \ . \tag{7.42}$$

Henceforth we will use such a reference point for energy that $E_0 = 0$. Then,

$$W(J) = -E_0(J)T \ , \tag{7.43}$$

i.e. $-W(J)/T$ is the energy of the ground state of the hamiltonian $H(J)$ with the source term:

$$H(J) = H - \int d^3x J(\mathbf{x})\varphi(x) \ , \tag{7.44}$$

$$-W(J) = \langle 0;J|H|0;J\rangle T - \int d^4x J(\mathbf{x})\langle 0;J|\varphi(x)|0;J\rangle \ , \tag{7.45}$$

where $H = H(0)$ is the initial hamiltonian. Taking into account the definition of the effective action $\Gamma(\varphi_c)$ in Eq.(7.9), we find that

$$\Gamma(\varphi_c) = -\langle 0; J|H|0; J\rangle T \ , \tag{7.46}$$

where the source $J(\mathbf{x})$ leads to

$$\varphi_c(\mathbf{x}) = \langle 0; J|\varphi(x)|0; J\rangle \ . \tag{7.47}$$

Therefore it is convenient to use a slightly different notation for the ground state $|0; J\rangle$:

$$|0; J\rangle \longrightarrow |0; \varphi_c\rangle \ . \tag{7.48}$$

The relation (7.46) tells us that $-\Gamma(\varphi_c)/T$ is the average energy of the ground state $|0; \varphi_c\rangle$ satisfying the condition

$$\langle 0; \varphi_c|\varphi(x)|0; \varphi_c\rangle = \varphi_c(\mathbf{x}). \tag{7.49}$$

In particular, since (see Eq.(7.33))

$$\Gamma(\varphi_c) = \int d^4x \left[-V(\varphi_c) + \frac{1}{2}Z(\varphi_c)\partial_\mu\varphi_c\partial^\mu\varphi_c + \cdots\right] , \tag{7.50}$$

we find that for sources independent of x, when $\varphi_c = \text{const}$, the effective potential $V(\varphi_c)$ is

$$V(\varphi_c) = \langle 0; \varphi_c|\mathcal{H}|0; \varphi_c\rangle \tag{7.51}$$

($\mathcal{H}$ is the hamiltonian density), i.e. it is the energy density of the ground state $|0; \varphi_c\rangle$ with $\langle 0; \varphi_c|\varphi|0; \varphi_c\rangle = \varphi_c$.

This interpretation is connected with the following variational problem. As is known, the Schrödinger equation

$$(H - E)|\psi\rangle = 0$$

determines stationary points of the form $\langle\psi|H|\psi\rangle$ under the constraint that $|\psi\rangle$ is normalized, $\langle\psi|\psi\rangle = 1$. The energy E can be considered as a Lagrange multiplier in this variational problem, that is one can vary the form $\langle\psi|(H-E)|\psi\rangle$ without constraint. In the same way, the effective potential $V(\varphi_c) = \langle 0; \varphi_c|\mathcal{H}|0; \varphi_c\rangle$ can be treated as the vacuum energy density of the states $|0; \varphi_c\rangle$ which are stationary against arbitrary variation of $|\psi\rangle = |0; \varphi_c\rangle$, subject to the normalization constraint $\langle\psi|\psi\rangle = 1$ and the additional constraint $\langle\psi|\varphi(x)|\psi\rangle = \varphi_c$. The source J can be considered as an additional Lagrange multiplier in this variational problem.

7.4. The Loop Expansion

In this section, we consider the loop expansion [7,3] for calculating the effective action.

First of all, let us restore Planck's constant $\hbar$ in the functional $Z(J)$:

$$Z(J) = \exp\left(\frac{i}{\hbar}W(J)\right) = \lim_{\substack{t_f \to +\infty \\ t_i \to -\infty}} N \int_{\varphi(t_f,\mathbf{x})=\varphi_f(\mathbf{x})}^{\varphi(t_f,\mathbf{x})=\varphi_f(\mathbf{x})} \mathcal{D}(\varphi)\exp(iS(J)/\hbar) \tag{7.52}$$

with the classical action

$$S(J) = \int d^4x\,[\mathcal{L}(x) + J(x)\varphi(x)] \ . \tag{7.53}$$

As we know, the effective action is a generating functional of proper vertices. Let us show that the loop expansion (the $1/(N_L+1)$ expansion, where N_L is the number of loops in graphs corresponding to proper vertices) is equivalent to a power-series expansion in $\hbar$. Indeed, Eq.(7.52) implies that every vertex contains $\hbar^{-1}$, and every propagator contains $\hbar$. Therefore the power P of $\hbar$ associated with a graph is

$$P = N_i - N_v \ , \tag{7.54}$$

where N_i is the number of internal lines, and N_v is the number of vertices in the graph (recall that we consider amputated graphs and therefore external lines do not contribute to P). The number of loops, N_L, is, as one can rather easily check,

$$N_L = N_i - N_v + 1 \ . \tag{7.55}$$

Therefore the power P of $\hbar$ is

$$P = N_L - 1 \ , \tag{7.56}$$

i.e. the loop expansion yields a power-series expansion in $\hbar$ (the quasi-classical expansion).

Let us apply the loop expansion to calculating the effective acton. In our consideration we will closely follow the method introduced by Jackiw [8].

We begin by considering the effective potential. We will show that the potential $V(\varphi_c)$ can be calculated by using the following procedure. Let us introduce the modified action

$$\begin{aligned}\tilde{S}(\varphi_c,\varphi) &\equiv \int d^4x \tilde{\mathcal{L}}(\varphi_c,\varphi(x)) \\ &= S(\varphi+\varphi_c) - S(\varphi_c) - \int d^4x \varphi(x) \frac{\delta S(\varphi)}{\delta\varphi(x)}|_{\varphi=\varphi_c}\end{aligned} \tag{7.57}$$

(since we are interested now in calculating the effective potential, we consider φ_c independend of x). The new lagrangian density $\tilde{\mathcal{L}}(\varphi_c,\varphi)$ has quadratic and higher terms in $\varphi(x)$:

$$\tilde{\mathcal{L}}(\varphi_c,\varphi) = -\frac{1}{2}\int d^4y \varphi(x)\mathfrak{D}^{-1}\{\varphi_c;x,y\}\varphi(y) + \tilde{\mathcal{L}}_{\text{int}}\{\varphi_c;\varphi(x)\}\ . \tag{7.58}$$

The propagator $\mathfrak{D}\{\varphi_c;x,y\}$ satisfies

$$\mathfrak{D}^{-1}\{\varphi_c;x,y\} = \frac{-\delta^2 S(\varphi)}{\delta\varphi(x)\delta\varphi(y)}|_{\varphi=\varphi_c}\ . \tag{7.59}$$

Since φ_c is a constant, the propagator $\mathfrak{D}$ depends on relative coordinates $x-y$ only, and one can introduce the Fourier transform

$$\mathfrak{D}^{-1}\{\varphi_c,;p\} = \int d^4x e^{ipx}\mathfrak{D}^{-1}\{\varphi_c;x,0\}\ . \tag{7.60}$$

We will show that the expression for the effective potential is

$$\begin{aligned}V(\varphi_c) =& U(\varphi_c) - \frac{i}{2}\hbar\int\frac{d^4p}{(2\pi)^4}\ln\left(\mathfrak{D}^{-1}\{\varphi_c,p\}\right) \\ &+ i\hbar\left\langle T\exp\left(\frac{i}{\hbar}\int d^4x\mathcal{L}_{\text{int}}\{\varphi_c,\varphi(x)\}\right)\right\rangle_{1PI} + C\ .\end{aligned} \tag{7.61}$$

Let us explain the meaning of these terms (C is a constant).

The first term is just the classical potential. It corresponds to the tree approximation.

The second term corresponds to the one-loop approximation. To understand its meaning better, let us consider a theory with the lagrangian density

$$\mathcal{L} = \frac{1}{2}\partial^\mu\varphi\partial_\mu\varphi - U(\varphi) \tag{7.62}$$

with $U(\varphi) = \frac{1}{2}\mu^2\varphi^2 + U_0(\varphi)$. The inverse propagator $\mathcal{D}^{-1}(\varphi_c;p)$ is

$$\mathfrak{D}^{-1}(\varphi_c;p) = -p^2 + \mu^2 + U_0''(\varphi_c) , \tag{7.63}$$

where the primes indicate the derivatives in φ. The second term in Eq.(7.61) is

$$\begin{aligned} V^{(1)}(\varphi_c) &= -\frac{i}{2}\hbar\int\frac{d^4p}{(2\pi)^4}\ln\left(-p^2+\mu^2+U_0''(\varphi_c)\right) \\ &= -\frac{i}{2}\hbar\int\frac{d^4p}{(2\pi)^4}\ln(-p^2+\mu^2) + \frac{i}{2}\hbar\int\frac{d^4p}{(2\pi)^4}\sum_{n=1}^{\infty}\frac{1}{n}\left[\frac{U_0''(\varphi_c)}{p^2-\mu^2}\right]^n . \end{aligned} \tag{7.64}$$

The first term on the right hand side of this equation corresponds to a free scalar field with mass μ; we shall return to this term in a moment. The second term is the sum of the contributions of one-loop diagrams including n propagators $\frac{i}{p^2-\mu^2}$ and n vertices $-iU_0''(\varphi_c)$; all external lines carry zero external momentum. These diagrams are shown in Fig 7.2 in the case of $U_0(\varphi) = \frac{\lambda\varphi^4}{4!}$.

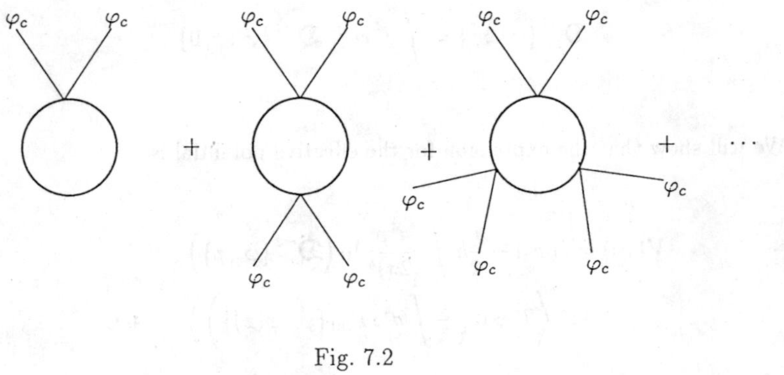

Fig. 7.2

As to the term

$$-\frac{i}{2}\hbar\int\frac{d^4p}{(2\pi)^4}\ln(-p^2+\mu^2) , \tag{7.65}$$

after the Wick rotation $p_0 \to ip_4$ into Euclidean space, it becomes

$$\frac{\hbar}{2}\int \frac{d^4p}{(2\pi)^4}\ln(p^2+\mu^2)\,. \tag{7.66}$$

The meaning of this (badly divergent) expression becomes transparent after the integration over p_0; up to an infinite constant, which can be removed by an appropriate vacuum energy density normalization, it is:

$$\frac{\hbar}{2}\int \frac{d^3p}{(2\pi)^3}(\mathbf{p}^2+\mu^2)^{1/2}\,, \tag{7.67}$$

i. e. it coincides with the zero-point energy density of a free scalar field with the mass μ.

In the same way, we find that $V^{(1)}(\varphi_c)$ in Eq.(7.64) is

$$V^{(1)}(\varphi_c) = \frac{\hbar}{2}\int \frac{d^3p}{(2\pi)^3}\left(\mathbf{p}^2+\mu^2+U_0''(\varphi_c)\right)^{1/2}. \tag{7.68}$$

Therefore we can interpret $V^{(1)}$ as the first quantum correction to the vacuum energy density of the dynamical system with the potential $\tilde{U}(\varphi(x))$ connected with the action $\tilde{S}(\varphi_c,\varphi)$ in Eq.(7.57). The correction corresponds to the approximation in which the potential $\tilde{U}$ is approximated by the harmonic oscillator potential

$$\tilde{U}_h(\varphi) = \frac{1}{2}\tilde{U}''(\varphi_c)(\varphi-\varphi_c)^2 = \frac{1}{2}(\mu^2+U_0''(\varphi_c))\cdot(\varphi-\varphi_c)^2 \tag{7.69}$$

near the extremum with $\varphi(\mathbf{x}) = \varphi_c$ (we recall that $\tilde{U}(\varphi_c) = 0$ and $\tilde{U}'(\varphi_c) = 0$; see Eq.(7.57)).

The third term in Eq.(7.61) corresponds to the contribution of n-loop one-particle irreducible graphs ($n \geq 2$). More precisely, it is yielded by the following operation. Compute the vacuum expectation value of

$$T\exp\left(\frac{i}{\hbar}\int d^4x\mathcal{L}_{\text{int}}\left\{\varphi_c,\varphi(x)\right\}\right) \tag{7.70}$$

with $\mathfrak{D}\{\varphi_c;p\}$ as the propagator; keep only one-particle irreducible graphs, and remove an over-all factor of space-time volume $\int d^4x$.

Examples of one-particle irreducible diagrams are shown in Fig 7.3. The contribution one-particle reducible diagrams, as that one shown in Fig.7.4, must be omitted.

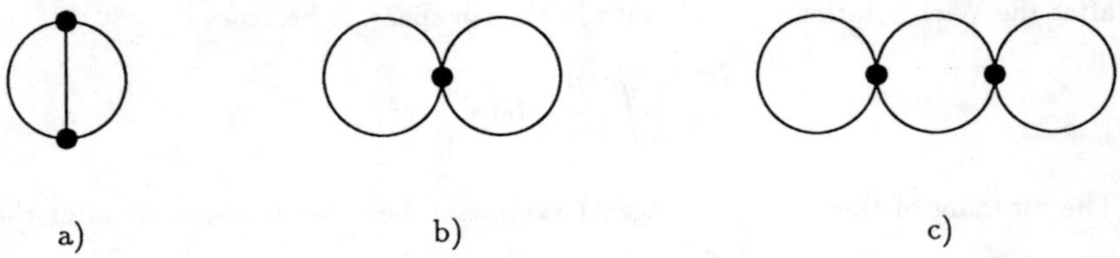

Fig. 7.3. Examples of one-particle irreducible diagrams

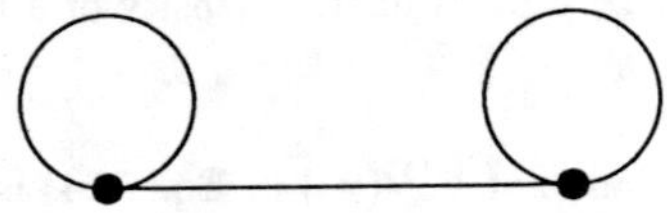

Fig. 7.4. An example of an one-particle reducible diagram

Let us now prove the relation (7.61).

We begin by considering the representation (7.34) for $Z(J)$:

$$Z(J) = N \int \mathcal{D}(\varphi) \exp\left(\frac{i}{\hbar}\left[S(\varphi) + \int d^4x J(x)\varphi(x)\right]\right) \tag{7.71}$$

(we omitted the boundary values $\varphi_i(\mathbf{x}), \varphi_f(\mathbf{x})$ in the path integral (7.71)). Shifting the integration variable φ as $\varphi \longrightarrow \varphi + \varphi_c$, we get:

$$Z(J) = N \exp\left(\frac{i}{\hbar}\left[S(\varphi_c) + \int d^4x J(x)\varphi_c\right]\right) Z_1(J) , \tag{7.72}$$

$$Z_1(J) = \int \mathcal{D}(\varphi) \exp\left(\frac{i}{\hbar}\left[S(\varphi + \varphi_c) - S(\varphi_c) + \int d^4x J(x)\varphi(x)\right]\right) . \tag{7.73}$$

Therefore,

$$W(J) = S(\varphi_c) + \int d^4x J(x)\varphi_c + W_1(J) - i\hbar \ln N , \tag{7.74}$$

$$W_1(J) = -i\hbar \ln Z_1(J) . \tag{7.75}$$

By definition, φ_c is

$$\varphi_c = \frac{\delta W(J)}{\delta J} \,. \tag{7.76}$$

This equation defines J as an implicit functional of φ_c : $J = J(\varphi_c)$. In particular, φ_c = const implies that $J(\varphi_c)$ is also independent of x. However, it is convenient to consider a general $\varphi_c(x)$ and put $\varphi_c(x) = \varphi_c$ = const only at the end of the calculations.

Then we obtain the identity:

$$\varphi_c(x) = \frac{\delta W(J)}{\delta J(x)} = \varphi_c(x) + \int d^4y \left[\frac{\delta S(\varphi_c)}{\delta\varphi_c(y)} + J(\varphi_c(y)) + \frac{\delta W_1(J)}{\delta\varphi_c(y)}\right]\frac{\delta\varphi_c(y)}{\delta J(x)} \,, \tag{7.77}$$

i.e. we get

$$\frac{\delta S(\varphi_c)}{\delta\varphi_c(x)} + \frac{\delta W_1}{\delta\varphi_c(x)} + J(\varphi_c(x)) = 0 \,. \tag{7.78}$$

The effective action is:

$$\Gamma(\varphi_c) = W(J) - \int d^4x J(x)\varphi_c(x) = S(\varphi_c) + W_1(J(\varphi_c)) - i\hbar \ln N \,. \tag{7.79}$$

Substituting $J(\varphi_c)$ from Eq.(7.78) into Eq.(7.75), we find the functional equation for $W_1(J(\varphi_c))$:

$$\begin{aligned} W_1 = -\, i\hbar \ln \int \mathcal{D}(\varphi) \exp\Bigg\{\frac{i}{\hbar}\Bigg[& S(\varphi+\varphi_c) - S(\varphi_c) \\ & - \int d^4x\varphi(x)\frac{\delta S(\varphi_c)}{\delta\varphi_c(x)} - \int d^4x\varphi(x)\frac{\delta W_1}{\delta\varphi_c(x)}\Bigg]\Bigg\} \,. \end{aligned} \tag{7.80}$$

Because of Eq.(7.57), it can be rewritten as

$$W_1 = -i\hbar \ln \int \mathcal{D}(\varphi) \exp\left\{\frac{i}{\hbar}\left[\tilde{S}(\varphi,\varphi_c) - \int d^4x\varphi(x)\frac{\delta W_1}{\delta\varphi_c(x)}\right]\right\} \,. \tag{7.81}$$

We will show that W_1 includes the contribution connected with one-particle irreducible graphs generated by the action $\tilde{S}(\varphi,\varphi_c)$. To do this, we use the following trick. Let us introduce the functional

$$W(\varphi_c, K) = -i\hbar \ln Z(\varphi_c, K) \ , \tag{7.82}$$

$$Z(\varphi_c, K) = \int \mathcal{D}(\varphi) \exp\left\{ \frac{i}{\hbar} \left[\tilde{S}(\varphi, \varphi_c) + \int d^4x \varphi(x) K(x) \right] \right\} \ , \tag{7.83}$$

where the source $K(x)$ is independent of φ_c now. It is clear that

$$W_1(\varphi_c) = W(\varphi_c, K)|_{K=-\frac{\delta W_1}{\delta \varphi_c}} \ . \tag{7.84}$$

When $K = \bar{K}$ with $\bar{K}$ satisfying the equation

$$\left. \frac{\delta W(\varphi_c, K)}{\delta K} \right|_{K=\bar{K}} = \varphi = 0 \ , \tag{7.85}$$

the functional $W(\varphi_c, \bar{K})$ coincides with the effective action $\Gamma(\varphi_c, \varphi)|_{\varphi=0}$, which is just the contribution of all one-particle irreducible vacuum graphs generated by the action $\tilde{S}(\varphi, \varphi_c)$ (see Eqs.(7.13), (7.14)).

Let us show that $\bar{K}$ is equal to $-\frac{\delta W_1}{\delta \varphi_c}$. This together with Eq.(7.84) will imply that $W_1(\varphi_c)$ yields the contribution of those one-particle irreducible vacuum graphs.

Thus we should show that

$$\left. \frac{\delta W(\varphi_c; K)}{\delta K} \right|_{K=-\frac{\delta W_1}{\delta \varphi_c}} = -i\hbar Z^{-1}(\varphi_c; K) \left. \frac{\delta Z(\varphi_c; K)}{\delta K} \right|_{K=-\frac{\delta W_1}{\delta \varphi_c}} = 0 \ . \tag{7.86}$$

The functional derivative of $Z(\varphi_c, K)$ in K is

$$\begin{aligned} \frac{\delta Z(\varphi_c, K)}{\delta K(x)} = & \frac{i}{\hbar} \int \mathcal{D}(\varphi) \varphi(x) \exp\left\{ \frac{i}{\hbar} \left[S(\varphi + \varphi_c) - S(\varphi_c) \right. \right. \\ & \left. \left. - \int d^4z \varphi(z) \frac{\delta S(\varphi_c)}{\delta \varphi_c(z)} + \int d^4z \varphi(z) K(z) \right] \right\} . \end{aligned} \tag{7.87}$$

Therefore we should show that

$$\begin{aligned} & \int \mathcal{D}(\varphi) \varphi(x) \exp\left\{ \frac{i}{\hbar} \left[S(\varphi + \varphi_c) - S(\varphi_c) \right. \right. \\ & \qquad \left. \left. - \int d^4z \varphi(z) \frac{\delta S(\varphi_c)}{\delta \varphi_c(z)} - \int d^4z \varphi(z) \frac{\delta W_1}{\delta \varphi_c(z)} \right] \right\} = 0 \ . \end{aligned} \tag{7.88}$$

To establish this relation, let us take the functional derivative in both sides of Eq.(7.81):

$$\begin{aligned}\frac{\delta W_1}{\delta\varphi_c(x)} =& Z_1^{-1}\int \mathcal{D}(\varphi)\bigg\{\frac{\delta S(\varphi+\varphi_c)}{\delta\varphi_c(x)} - \frac{\delta S(\varphi_c)}{\delta\varphi_c(x)} \\ &- \int d^4y\varphi(y)\frac{\delta^2}{\delta\varphi_c(x)\delta\varphi_c(y)}\Big[S(\varphi_c)+W_1(\varphi_c)\Big]\bigg\} \\ &\cdot \exp\bigg\{\frac{i}{\hbar}\Big[S(\varphi+\varphi_c) - S(\varphi_c) \\ &- \int d^4z\varphi(z)\frac{\delta S(\varphi_c)}{\delta\varphi_c(z)} - \int d^4z\varphi(z)\frac{\delta W_1}{\delta\varphi_c(z)}\Big]\bigg\}\,.\end{aligned} \tag{7.89}$$

The last trick in the derivation is to rewrite the first term in the first curly brackets as

$$\begin{aligned}&\int \mathcal{D}(\varphi)\frac{\delta S(\varphi+\varphi_c)}{\delta\varphi_c(x)}\exp\bigg\{\frac{i}{\hbar}\Big[S(\varphi+\varphi_c) - S(\varphi_c) \\ &\qquad - \int d^4z\varphi(z)\frac{\delta S(\varphi_c)}{\delta\varphi_c(z)} - \int d^4z\varphi(z)\frac{\delta W_1}{\delta\varphi_c(z)}\Big]\bigg\} \\ &= -i\hbar\int \mathcal{D}(\varphi)\frac{\delta}{\delta\varphi(x)}\exp\bigg\{\frac{i}{\hbar}\Big[S(\varphi+\varphi_c) - S(\varphi_c) \\ &\qquad - \int d^4z\varphi(z)\frac{\delta S(\varphi_c)}{\delta\varphi_c(z)} - \int d^4z\varphi(z)\frac{\delta W_1}{\delta\varphi_c(z)}\Big]\bigg\} \\ &+ \int \mathcal{D}(\varphi)\bigg\{\frac{\delta S(\varphi_c)}{\delta\varphi_c(x)} + \frac{\delta W_1}{\delta\varphi_c(x)}\bigg\}\exp\bigg\{\frac{i}{\hbar}\Big[S(\varphi+\varphi_c) \\ &\qquad - S(\varphi_c) - \int d^4z\varphi(z)\frac{\delta S(\varphi_c)}{\delta\varphi_c(z)} - \int d^4z\varphi(z)\frac{\delta W_1}{\delta\varphi_c(z)}\Big]\bigg\}\,.\end{aligned} \tag{7.90}$$

The first integral yields an inessential contribution independent of $\varphi_c(x)$. Reinserting the remainder into Eq.(7.89), we get

$$\begin{aligned}\frac{\delta W_1}{\delta\varphi_c(x)} =& Z_1^{-1}\int \mathcal{D}(\varphi)\bigg\{\frac{\delta W_1}{\delta\varphi_c(x)} - \int d^4y\varphi(y)\frac{\delta^2}{\delta\varphi_c(x)\delta\varphi_c(y)}\Big[S(\varphi_c)+W_1(\varphi_c)\Big]\bigg\} \\ &\cdot\exp\bigg\{\frac{i}{\hbar}\Big[\tilde{S}(\varphi,\varphi_c) - \int d^4z\varphi(z)\frac{\delta W_1}{\delta\varphi_c(z)}\Big]\bigg\}\,,\end{aligned} \tag{7.91}$$

which implies that

$$\begin{aligned}&\int d^4y\frac{\delta^2}{\delta\varphi_c(x)\delta\varphi_c(y)}\left[S(\varphi_c)+W_1(\varphi_c)\right]\int\mathcal{D}(\varphi)\varphi(y) \\ &\cdot\exp\left\{\frac{i}{\hbar}\left[\tilde{S}(\varphi,\varphi_c) - \int d^4z\varphi(z)\frac{\delta W_1}{\delta\varphi_c(z)}\right]\right\} = 0\,.\end{aligned} \tag{7.92}$$

Now let us return to $\varphi_c = \text{const}$. Then, taking into account Eq.(7.79), we find

$$\int d^4y\Gamma^{(2)}(x-y)\int \mathcal{D}(\varphi)\varphi(y)\exp\left\{\frac{i}{\hbar}\left[\tilde{S}(\varphi,\varphi_c)-\int d^4z\varphi(z)\frac{\delta W_1}{\delta\varphi_c(z)}\right]\right\}=0\,. \quad (7.93)$$

Since the factor $\Gamma^{(2)}(x-y)$ in this convolution is not zero, we get

$$\int \mathcal{D}(\varphi)\varphi(x)\exp\left\{\frac{i}{\hbar}\left[\tilde{S}(\varphi,\varphi_c)-\int d^4z\varphi(z)\frac{\delta W_1}{\delta\varphi_c(z)}\right]\right\}=0\,, \quad (7.94)$$

i.e. we get the wanted relation (7.88).

Thus we have proved that $W_1(\varphi_c)$ is given by the contribution of one-particle irreducible graphs connected with the action $\tilde{S}(\varphi,\varphi_c)$. Now, using Eq.(7.79), it is straightforward to prove the relation (7.61). Indeed, since $\varphi_c = \text{const}$, the first term on the right hand side of Eq.(7.79) is just $S(\varphi_c) = -U(\varphi_c)\int d^4x$, i.e. it leads to the classical contribution $U(\varphi_c)$ in the effective potential V. The one-loop contribution is

$$\begin{aligned} V^{(1)}(\varphi_c) &= -\left(\int d^4x\right)^{-1} W_1^{(1)}(\varphi_c)\\ &= \left(\int d^4x\right)^{-1} i\hbar\ln\left\{\int \mathcal{D}(\varphi)\exp\left[\frac{-i}{\hbar}\int d^4yd^4z\right.\right. \\ &\qquad \left.\left.\cdot\mathfrak{D}^{-1}\{\varphi_c;y-z\}\varphi(y)\varphi(z)\right]\right\}, \end{aligned} \quad (7.95)$$

where (see Eq.(7.59))

$$\mathfrak{D}^{-1}\{\varphi_c;x-y\} = -\frac{\delta^2 S}{\delta\varphi(x)\delta\varphi(y)}|_{\varphi=\varphi_c} = -\frac{\delta^2\tilde{S}}{\delta\varphi(x)\delta\varphi(y)}|_{\varphi=0}\,. \quad (7.96)$$

A Gaussian functional integral $\int \mathcal{D}(\varphi)\exp\left[-i\int dxdy\varphi(x)A(x,y)\varphi(y)\right]$ is equal (see Sec.C.3 in Appendix C) to $N_0(\text{Det}\hat{A})^{-1/2}$, where $\hat{A}$ is the operator defined by the kernel $A(x,y)$, and N_0 is a normalization constant; the determinant Det is taken in the functional sense. Therefore,

$$V^{(1)} = i\hbar(\int d^4x)^{-1}\ln(N_0\text{Det}\hat{\mathfrak{D}})^{-1/2} = i\hbar(\int d^4x)^{-1}\ln N_0 - \frac{i\hbar}{2}(\int d^4x)^{-1}\text{TrLn}\hat{\mathfrak{D}}^{-1} \quad (7.97)$$

(the trace Tr and the logarithm Ln are taken here in the functional sense; we use the identiry $\ln \mathrm{Det}\hat{A} = \mathrm{TrLn}\hat{A}$). Since the operator $\hat{\mathfrak{D}}^{-1}$ is diagonal in momentum space,

$$\langle p'|\hat{\mathfrak{D}}^{-1}|p\rangle = \delta^4(p'-p)\mathfrak{D}^{-1}\{\varphi_c;p\}\;, \tag{7.98}$$

we find

$$\begin{aligned}\mathrm{Tr}\delta(p'-p)\ln\mathfrak{D}^{-1}\{\varphi_c;p\} =& \delta^4(0)\int d^4p \ln\mathfrak{D}^{-1}\{\varphi_c;p\} \\ =& \int \frac{d^4p}{(2\pi)^4}\ln\mathfrak{D}^{-1}\{\varphi_c;p\}\int d^4x\end{aligned} \tag{7.99}$$

(here we use the identity $\delta^4(p)|_{p=0} = \frac{1}{(2\pi)^4}\int d^4x$). Therefore

$$V^{(1)} = i\hbar(d^4x)^{-1}\ln N_0 - \frac{i\hbar}{2}\int\frac{d^4p}{(2\pi)^4}\cdot\ln\mathfrak{D}^{-1}\{\varphi_c;p\}\;, \tag{7.100}$$

i.e. up to some inessential constant, we get the second term in Eq.(7.61).

The contribution of n-loop graphs with $n \geq 2$ is given by the third term on the right hand side in Eq.(7.61).

So far we have considered the effective potential. However, using the derivative expansion method, we also can calculate derivative terms in the effective action. For instance, the function $Z(\varphi_c)$ in the kinetic term (see Eq.(7.32)) can be determined if we consider $\varphi_c = \varphi_0 = \mathrm{const}$ and then we use Eq.(7.30):

$$g^{\mu\nu}Z(\varphi_0) = \frac{1}{2}\frac{\partial^2\tilde{\Gamma}^{(2)}_{\varphi_0}}{\partial p_\mu \partial p_\nu}\bigg|_{p=0}\;, \tag{7.101}$$

where $\tilde{\Gamma}^{(2)}_{\varphi_0}(p)$ is the two-point proper vertex defined by the action $\tilde{S}(\varphi,\varphi_c)|_{\varphi_c=\varphi_0}$. The whole procedure of the determination of the effective action reduces to calculating the proper vertices (connected with the action $\tilde{S}$) around the point $p=0$.

Of course, calculating the effective action, one should use a renormalization procedure to remove divergences. In particular the term $-\frac{i}{2}\hbar\int\frac{d^4p}{(2\pi)^4}\ln(-p^2+\mu^2)$ in $V^{(1)}$ (independent of φ_c) may be removed by an appropriate choice of the constant C in Eq.(7.61).

We will calculate the effective action in some models in the next section.

7.5. One-Loop Effective Potential. Concrete Examples

In this section, we will calculate the one-loop effective potential in some models. We will also derive a general structure of the one-loop effective potential in renormalizable models.

We begin by considering the simplest case of φ^4-theory:

$$\mathcal{L} = \frac{1}{2}(\partial_\mu \varphi^{(0)})(\partial^\mu \varphi^{(0)}) - \frac{1}{2}\mu^{(0)^2}\varphi^{(0)^2} - \frac{\lambda^{(0)}}{4!}\varphi^{(0)^4} - C \tag{7.102}$$

where the simbol (0) means that all these quantities are bare ones, and C is a constant. Of course we can rewrite $\mathcal{L}$ in terms of renormalizable quantities:

$$\mathcal{L} = \frac{1}{2}Z_2(\partial_\mu\varphi)(\partial^\mu\varphi) - \frac{1}{2}\mu^2 Z_2\varphi^2 - \frac{1}{2}(\mu^{(0)^2} - \mu^2)Z_2\varphi^2 - Z_1\frac{\lambda}{4!}\varphi^4 - C\ , \tag{7.103}$$

where

$$\begin{aligned}\varphi^{(0)} &= Z_2^{1/2}\varphi\ ,\\ \lambda^{(0)} &= Z_1 Z_2^{-2}\lambda\ ,\end{aligned} \tag{7.104}$$

and the renormalization constants Z_1, Z_2 and $\Delta\mu^2 = \mu^{(0)^2} - \mu^2$ are defined by some renormalization conditions (see below).

It is convenient to rewrite $\mathcal{L}$ (7.103) in the form

$$\begin{aligned}\mathcal{L} =& \frac{1}{2}(\partial_\mu\varphi)(\partial^\mu\varphi) - \frac{1}{2}\mu^2\varphi^2 - \frac{\lambda}{4!}\varphi^4\\ &+\frac{1}{2}A(\partial_\mu\varphi)(\partial^\mu\varphi) - \frac{1}{2}B\varphi^2 - \frac{1}{4!}D\varphi^4 - C\end{aligned} \tag{7.105}$$

with $A = Z_2 - 1$, $B = \mu^{(0)^2}Z_2 - \mu^2$, $D = (Z_1 - 1)\lambda$. The four last terms on the right hand side of Eq.(7.105) are counter terms removing divergences in the perturbative loop expansion.

Let us calculate the effective potential $V(\varphi_c)$ and the function $Z(\varphi_c)$ up to order $\hbar$.

To lowest order (tree approximation),

$$V^{(0)}(\varphi_c) = U(\varphi_c) = \frac{\mu^2}{2}\varphi_c^2 + \frac{\lambda}{4!}\varphi_c^4, \quad Z^{(0)}(\varphi_c) = 1\,. \tag{7.106}$$

Using the relation (7.61), we find that the one-loop contribution to V, $V^{(1)}$, is

$$\begin{aligned} V^{(1)}(\varphi_c) = & -\frac{i}{2}\hbar \int \frac{d^4p}{(2\pi)^4} \ln(-p^2 + \mu^2 + \frac{\lambda}{2}\varphi_c^2) \\ & + \frac{1}{2}B^{(1)}\varphi_c^2 + \frac{1}{4!}D^{(1)}\varphi_c^4 + C^{(1)}\,. \end{aligned} \tag{7.107}$$

Performing the Wick rotation $p_0 \to ip_0$, we find

$$\begin{aligned} V^{(1)}(\varphi_c) = & \frac{\hbar}{2} \int \frac{d^4p}{(2\pi)^4} \ln(p^2 + \mu^2 + \frac{\lambda}{2}\varphi_c^2) + \frac{1}{2}B^{(1)}\varphi_c^2 + \frac{1}{4!}D^{(1)}\varphi_c^4 + C^{(1)} \\ = & \hbar\left[\frac{\Lambda^2}{32\pi^2}(\mu^2 + \frac{\lambda}{2}\varphi_c^2) + \frac{1}{64\pi^2}(\mu^2 + \frac{\lambda}{2}\varphi_c^2)^2\left[\ln\left(\frac{\mu^2 + \frac{\lambda\varphi_c^2}{2}}{\Lambda^2}\right) - \frac{1}{2}\right]\right] \\ & + \frac{1}{2}B^{(1)}\varphi_c^2 + \frac{1}{4!}D^{(1)}\varphi_c^4 + C^{(1)}\,, \end{aligned} \tag{7.108}$$

where Λ is the ultraviolet cutoff (we omitted terms vanishing as Λ goes to infinity).

We can use different types of renormalization conditions to define $A^{(1)}, B^{(1)}, D^{(1)}$ and $C^{(1)}$. All physical results are independent of their choice; different renormalization conditions correspond to different parameterizations of the same theory.

Some conventional renormalization conditions are the following:

a) $\mu^2 < 0$; Ref. [9]:

$$\begin{aligned} & \frac{dV}{d\varphi_c}\Big|_{\varphi_c = v = \left(-\frac{6\mu^2}{\lambda}\right)^{1/2}} = 0\,, \\ & \frac{d^2V}{d\varphi_c^2}\Big|_{\varphi_c = v} = \frac{\lambda v^2}{3} = -2\mu^2\,, \\ & Z(\varphi_c)|_{\varphi_c = v} = 1\,. \end{aligned} \tag{7.109}$$

These normalization conditions are chosen to preserve the tree approximation relations for V', V'' and Z at the minimum $\varphi_c = v$.

Eqs.(7.108), (7.109) imply that the one-loop effective potential V_1 is (henceforth we will again omit the Planck constant $\hbar$):

$$\begin{aligned} V_1 = & V^{(0)} + V^{(1)} = \frac{\mu^2}{2}\varphi_c^2 + \frac{\lambda}{4!}\varphi_c^4 + \frac{(\mu^2 + \frac{\lambda\varphi_c^2}{2})^2}{64\pi^2} \ln\left(\frac{\mu^2 + \frac{\lambda\varphi_c^2}{2}}{-2\mu^2}\right) \\ & - \frac{7\lambda\mu^2}{128\pi^2}\varphi_c^2 - \frac{3\lambda^2}{512\pi^2}\varphi_c^4 + \tilde{C} \end{aligned} \tag{7.110}$$

with $\tilde{C} = -\frac{\mu^2 v^2}{2}\left(1 - \frac{7\lambda}{64\pi^2}\right) - \frac{\lambda}{4!}v^4\left(1 - \frac{9\lambda}{64\pi^2}\right)$ (our choice of $C^{(1)}$ ensures that $V = 0$ at the minimum $\varphi_c = v$);

b) $\overline{MS}$ renormalization scheme.

This renormalization scheme is considered in Sec.C.2 in Appendix C. In the scheme, renormalization of V is performed simply by omitting terms proportional to positive powers of $1/\bar{\epsilon} \equiv 2/(4-n) - \gamma + \ln 4\pi$ with $n \longrightarrow 4$ and $\gamma \simeq 0.58$ being the Euler constant.

The one-loop effective potential is

$$V_1 = V^{(0)} + V^{(1)} = \frac{\mu^2}{2}\varphi_c^2 + \frac{\lambda}{4!}\varphi_c^4 + \frac{(\mu^2 + \frac{\lambda\varphi_c^2}{2})^2}{64\pi^2}\left[\ln\left(\frac{\mu^2 + \frac{\lambda\varphi_c^2}{2}}{\sigma^2}\right) - \frac{3}{2}\right] + \tilde{C} \tag{7.111}$$

(here we do not specify $\tilde{C}$ defining a reference point of the vacuum energy density). The parameter σ is the renormalization point; it is arbitrary, and observables do not depend on it (see the next section).

c) $\mu^2 = 0$; Ref.[3]:

$$\begin{aligned} & V(0) = 0 , \\ & \frac{d^2V}{d\varphi_c^2}|_{\varphi_c=0} = \mu^2 = 0 , \\ & \frac{d^4V}{d\varphi_c^4}|_{\varphi_c=\sigma} = \lambda , \\ & Z(\varphi_c)|_{\varphi_c=\sigma} = 1 . \end{aligned} \tag{7.112}$$

The one-loop effective potential is

$$V_1 = V^{(0)} + V^{(1)} = \frac{\lambda}{4!}\varphi_c^4 + \frac{\lambda^2\varphi_c^4}{256\pi^2}\left(\ln\frac{\varphi_c^2}{\sigma^2} - \frac{25}{6}\right) . \tag{7.113}$$

Let us discuss the physical meaning of these expressions in more detail.

The loop expansion is a perturbative expansion in powers of λ and $\ln\frac{M^2(\varphi_c)}{\sigma^2}$ with $M^2(\varphi_c) = \frac{d^2V^{(0)}}{d\varphi_c^2} = \mu^2 + \frac{\lambda\varphi_c^2}{2}$; in general, the n-loop potential has terms of order $\left[\ln\frac{M^2(\varphi_c)}{\sigma^2}\right]^n$, since each momentum integration may lead to a logarithmic divergence, which turns into $\ln\frac{M^2(\varphi_c)}{\sigma^2}$ as a result of renormalization. We emphasize that the factor $M^2(\varphi_c)$ in the logarithm appears because of the relations (7.57), (7.61). The n-loop potential will contain terms of order $\lambda^{n+1}\left[\ln\frac{M^2(\varphi_c)}{\sigma^2}\right]^n$.

Therefore the expansion is reliable only if λ and $\lambda\ln\frac{M^2(\varphi_c)}{\sigma^2}$ are small. Under these conditions, higher-order contributions do not change qualitatively the physical picture of the tree approximation. However, as will be shown soon, they may be essential in theories with more than one coupling constant [3].

This discussion also implies that the loop expansion can be reliable only in a rather small region of values of the field φ_c, where $\ln\frac{M^2(\varphi_c)}{\sigma^2}$ is not very large. In the next section, we will see how the renormalization group improves the situation.

As we already noted above, the parameter σ in the expressions (7.111), (7.113) is arbitrary. To illustrate this, let us change σ and λ for σ' and $\lambda' = \lambda + \frac{3\lambda^2}{32\pi^2}\ln\sigma'^2/\sigma^2$ in the renormalization condition (7.112) :

$$\frac{d^4V}{d\varphi_c^4}|_{\varphi_c=\sigma'} = \lambda' = \lambda + \frac{3\lambda^2}{32\pi^2}\ln\sigma'^2/\sigma^2 \,. \tag{7.114}$$

Then one can check that the potential in Eq.(7.113) takes the form:

$$V_1 = \frac{\lambda'}{4!}\varphi_c^4 + \frac{\lambda'^2\varphi_c^4}{256\pi^2}\left(\ln\frac{\varphi_c^2}{\sigma'^2} - \frac{25}{6}\right) + 0(\lambda^3)\,. \tag{7.115}$$

Thus we see that, to order in which we are working, this is just a reparametrization of the potential. The transformation $\sigma, \lambda \longrightarrow \sigma', \lambda'$ is of course renormalization group transformation discussed in more detail in the next section.

We are now ready to describe the structure of the one-loop effective potential in a general renormalizable model. First of all, we note that the relations (7.61) and (7.63) can be easily generalized for theories including gauge fields and fermion fields as well as scalar fields. It is very convenient to work in the Landau gauge in this case. In particular, since in this gauge, there is no direct coupling of the Faddeev-Popov (FP) ghost fields (present in non-Abelian gauge theories) [4] to the spinless fields, there is no need to consider them in the one-loop effective potential calculations (for comments concerning the calculations in other gauges, see the end of this section).*)

The inverse propagator $\mathfrak{D}^{-1}_{\mu\alpha,\nu\beta}\{\varphi_c;p\}$ of gauge fields is (see Eq.(3.99)):

$$\mathfrak{D}^{-1}_{\mu\alpha,\nu\beta}\{\varphi_c;p\} = \sum_s E^{(s)}_\alpha E^{(s)}_\beta \left(k^2 - \mathcal{M}^2_s(\varphi_c)\right)\left(g_{\mu\nu} - \frac{k_\mu k_\nu}{k^2}\right) . \tag{7.116}$$

Their contribution to the one-loop effective potential is

$$V^{(1)}_g = -3\frac{i}{2}\sum_s \int \frac{d^4p}{(2\pi)^4} \ln\left(-p^2 + \mathcal{M}^2_s(\varphi_c)\right) , \tag{7.117}$$

where the factor of 3 appears here due to the polarization sum.

We note that in the general case, φ_c is a vector assigned to some representation of the gauge group (for simplicity of notations, we consider the case with only one spinless field having a nontrivial classical part $\varphi_c \neq 0$).

The contribution $V^{(1)}_g$ is

$$V^{(1)}_g = \frac{3}{64\pi^2}\sum_s \mathcal{M}^4_s(\varphi_c) \ln \mathcal{M}^2_s(\varphi_c) \tag{7.118}$$

plus quartic and quadratic terms, which will be absorbed in counter terms through renormalizations.

The contribution of fermions may be computed in a similar way. The Yukawa interactions lead to the mass term

$$\mathcal{L}_m = -\bar{\psi}_l m_{lk}(\varphi_c)\psi_k \tag{7.119}$$

with

$$m_{lk}(\varphi_c) = S_{lk}(\varphi_c) + iP_{lk}(\varphi_c)\gamma_5 , \tag{7.120}$$

where S and P are linear functions of φ_c. The one-loop contribution of fermions is

$$V^{(1)}_f = i\int \frac{d^4p}{(2\pi)^4} tr \ln\left(S^{-1}\{\varphi_c, p\}\right) \tag{7.121}$$

---

*) In general covariant gauges, the ghost fields may couple to scalar fields because of the mixing term $\sim \partial^\mu \varphi A_\mu$ in the lagrangian density (see Eq.(3.87)).

with

$$S^{-1}\{\varphi_c, p\} = p_\mu \gamma^\mu - m(\varphi_c) \ . \tag{7.122}$$

Of course, there is the standard minus sign for fermion loops (compare Eq.(7.121) with Eq.(7.117)).*) $V_f^{(1)}$ is

$$V_f^{(1)} = -\frac{1}{64\pi^2} tr \left[(mm^\dagger(\varphi_c))^2 \ln(mm^\dagger(\varphi_c)) \right] \tag{7.123}$$

plus quartic and quadratic terms, which will be absorbed in counter terms.

The trace runs over both Dirac and internal indices in the equation (7.123). Therefore if, for example, $m_{lk}(\varphi_c) = m_l(\varphi_c)\delta_{lk}$, we find that

$$V_f^{(1)} = -\frac{1}{16\pi^2} \sum_l (m_l m_l^*(\varphi_c))^2 \ln(m_l m_l^*(\varphi_c)) \ . \tag{7.124}$$

The contribution of spinless fields has the form

$$V_s^{(1)} = \frac{1}{64\pi^2} tr \left[(MM^\dagger(\varphi_c))^2 \ln(MM^\dagger(\varphi_c)) \right] \tag{7.125}$$

plus of course quartic and quadratic terms absorbed in counter terms. Note that in theories with NG bosons, their contribution must also be included: their mass is zero only at the minimum with $\varphi_c = v$.

We can summarize this consideration with the following general expression for the one-loop effective potential:

$$V^{(1)} = \sum_i (-1)^F \eta_i \frac{(M_i M_i^*(\varphi_c))^2}{64\pi^2} \ln(M_i M_i^*(\varphi_c))^2 \tag{7.126}$$

(up to terms which can be absorbed in counter terms). Here the sum is over all fields which are eigenstates of the mass matrix in the vacuum with $\langle 0|\varphi|0\rangle = \varphi_c$; $F = 1$ and $F = 0$ for fermions and bosons, respectively, and η_i is the number of degrees of freedom of the i-th field.

---

*) Since ψ is a complex field, there is no factor $\frac{1}{2}$ in the relation (7.121).

So far we considered the effective potential. The function $Z(\varphi_c)$ in the effective action can be determined from Eq.(7.101). In particular it is not difficult to check that in the one-loop approximation it is

$$Z^{(1)}(\varphi_c) = \frac{\lambda^2}{96\pi^2} \frac{\varphi_c^2}{(2\mu^2 + \lambda\varphi_c^2)} \tag{7.127}$$

in the model (7.103). If we choose the renormalization conditions (7.109), we find that the counter term $A^{(1)}$ is

$$A^{(1)} = -\frac{\lambda^2}{96\pi^2} \frac{v^2}{(2\mu^2 + \lambda v^2)} = -\frac{\lambda}{64\pi^2} \,. \tag{7.128}$$

Therefore $Z_1 = 1 + Z^{(1)}(\varphi_c) + A^{(1)}$ is

$$Z_1(\varphi_c) = 1 + \frac{\lambda^2}{96\pi^2} \frac{\varphi_c^2}{(2\mu^2 + \lambda\varphi_c^2)} - \frac{\lambda}{64\pi^2} \,. \tag{7.129}$$

Let us apply these results to some models.

I. The Coleman-Weinberg (CW) model [3].

The CW model is the Abelian gauge model (considered in Sec. 3.5) with the additional restriction

$$\frac{d^2V}{d\varphi_c^2}|_{\varphi_c=0} = \mu^2 = 0 \,, \tag{7.130}$$

i.e. in the classical (tree) approximation, it is just scalar electrodynamics with massless charge bosons.

The lagrangian density of this model is (see Eq.(3.61):

$$\begin{aligned}\mathcal{L} = -\frac{1}{4}F_{\mu\nu}F^{\mu\nu} &+ \frac{1}{2}\left(\partial_\mu\varphi_1 - gA_\mu\varphi_2\right)\left(\partial^\mu\varphi_1 - gA^\mu\varphi_2\right) \\ &+ \frac{1}{2}\left(\partial_\mu\varphi_2 + gA_\mu\varphi_1\right)\left(\partial^\mu\varphi_2 + gA^\mu\varphi_1\right) - \frac{\lambda}{4!}\left(\varphi_1^2 + \varphi_2^2\right)^2\end{aligned} \tag{7.131}$$

(for simplicity, we omitted the counter terms). Using Eqs.(7.118) and (7.125), we find that the one-loop potential V_1 takes the form:

$$\begin{aligned} V_1(\varphi_c) =& \frac{\lambda}{4!}\varphi_c^4 + \frac{5\lambda^2}{1152\pi^2}\varphi_c^4 \ln \varphi_c^2 \\ &+ \frac{3g^4}{64\pi^2}\varphi_c^4 \ln \varphi_c^2 + \frac{1}{2}B\varphi_c^2 + \frac{1}{4!}D\varphi_c^4 + C\,, \end{aligned} \tag{7.132}$$

where $\varphi_c^2 \equiv \varphi_{c1}^2 + \varphi_{c2}^2$. The counter terms B, D and C are determined from the renormalization conditions (7.112). As the result, we find:

$$V_1(\varphi_c) = \frac{\lambda}{4!}\varphi_c^4 + \left(\frac{5\lambda^2}{1152\pi^2} + \frac{3g^4}{64\pi^2}\right)\varphi_c^4\left(\ln\frac{\varphi_c^2}{\sigma^2} - \frac{25}{6}\right)\,. \tag{7.133}$$

The model has two coupling constants, λ and g. Let us assume that $\lambda^2 \ll g^4$ and show that under this condition, there is a nontrivial minimum with $\varphi_c = v \neq 0$ (we of course assume that $g^2 \ll 1$).

At $\lambda^2 \ll g^4$, the potential takes the form:

$$V_1(\varphi) \simeq \frac{\lambda}{4!}\varphi_c^4 + \frac{3g^4}{64\pi^2}\varphi_c^4\left(\ln\frac{\varphi_c^2}{\sigma^2} - \frac{25}{6}\right)\,. \tag{7.134}$$

The minimum v is determined from the equation:

$$V_1'(\varphi_c)|_{\varphi_c=v} = 0 = \frac{\lambda}{6}v^3 + \frac{3g^4}{16}v^3\left(\ln\frac{v^2}{\sigma^2} - \frac{25}{6}\right) + \frac{3g^4}{32\pi^2}v^3\,. \tag{7.135}$$

Since the renormalization parameter σ is arbitrary, we can choose $\sigma = v$. Then we find

$$\left(\frac{\lambda}{6} - \frac{11g^4}{16\pi^2}\right)v = 0\,, \tag{7.136}$$

or

$$\lambda = \frac{33}{8\pi^2}g^4\,. \tag{7.137}$$

The relation (7.137) justifies our assumption that $\lambda^2 \ll g^4$. Moreover, it fixes the value of λ!

What has happened is the following. The initial lagrangian contained two free parameters, λ and g. Renormalization led to the appearance of the additional parameter σ in the theory. At the particular value of this parameter, $\sigma = v$, the relation (7.137) takes place. Therefore we can consider this as a reparameterization of the theory in the term of the parameters g and v. Since the coupling constant λ is dimensionless, and v is dimensional, the transition $(g, \lambda) \longrightarrow (g, v)$ is called the dimensional transmutation.

Let us show that in the vacuum with $\varphi_c = v$, the following relation takes place between the mass of scalar (M) and the mass of vector boson ($\mathcal{M}$):

$$\frac{M^2}{\mathcal{M}^2} = \frac{3g^2}{8\pi^2} \,. \tag{7.138}$$

Indeed, in this approximation,

$$\mathcal{M}^2 = g^2 v^2 \,, \tag{7.139}$$

and

$$M^2 = V_1''(\varphi_c)|_{\varphi_c = v} = \frac{3g^4}{8\pi^2} v^2 \,. \tag{7.140}$$

Combining these two equations, we find the relation (7.138).

The possibility of calculating the ratio $M^2/\mathcal{M}^2$ in the model may appear miraculous. But one should understand that the miracle was created by our hands, in fact, by the additional constraint $\mu^2 = 0$ in the renormalization condition $V''(\varphi_c)|_{\varphi_c = 0} = \mu^2$.

It is useful to consider the same model with slightly different renormalization conditions:

$$V(\varphi_c)|_{\varphi_c = \sigma} = \frac{\lambda}{4!}\sigma^4 \tag{7.141}$$

instead of $V''''|_{\varphi_c = \sigma} = \lambda$.

One can easily check that the potential V_1 becomes now:

$$V_1(\varphi_c) = \frac{\lambda}{4!}\varphi_c^4 + \left(\frac{5\lambda^2}{1152\pi^2} + \frac{3g^4}{64\pi^2}\right)\varphi_c^4 \ln\frac{\varphi_c^2}{\sigma^2} \,. \tag{7.142}$$

Then we get

$$V_1'(\varphi_c)|_{\varphi_c = \sigma = v} = 0 = \left(\frac{\lambda}{6} + \frac{3g^4}{32\pi^2}\right) v^3 \,, \tag{7.143}$$

i.e.

$$\lambda = -\frac{9g^4}{16\pi^2}\,. \tag{7.144}$$

This relation is different from that in Eq.(7.137). However, one can check that the ratio $M^2/\mathcal{M}^2$ is still the same as in Eq.(7.138). Also, one can easily check that the value $V_1(\varphi_c)$ at the minimum $\varphi_c = v$ is $V_1(v) = -\frac{3g^4}{128\pi^2}v^4$ in both of these renormalization schemes. This example gives a good illustration of the independence of the physical observables of renormalization schemes.

The present analysis can be directly extended to the non-Abelian case [3].

II. The Linde-Weinberg bound [10, 11].

Let us add the mass term $-\frac{\mu^2}{2}\varphi^2$ with $\mu^2 < 0$ in the lagrangian density (7.131). Using the renormalization conditions (7.109), we find the following one-loop potential at $\lambda^2 \ll g^4$:

$$V_1(\varphi_c) = \frac{\mu^2}{2}\varphi_c^2\left(1 - \frac{9g^4}{8\pi^2\lambda}\right) + \frac{\lambda}{4!}\varphi_c^4\left(1 - \frac{27g^4}{16\pi^2\lambda}\right) + \frac{3g^4}{64\pi^2}\varphi_c^4 \ln\left(\frac{-\lambda\varphi_c^2}{6\mu^2}\right) \tag{7.145}$$

(we choose the constant C such that $V(0) = 0$).

This expression shows that when $\lambda < \frac{9g^4}{8\pi^2}$, the potential acquires an additional minimum at $\varphi_c = 0$ (see Fig. 7.5). When $\lambda < \frac{9g^4}{16\pi^2}$, this minimum becomes even deeper than the minimum with $\varphi_c = v = \left(-\frac{6\mu^2}{\lambda}\right)^{1/2}$. In other words, spontaneous symmetry breaking disappears at such a small λ that $\lambda < \frac{9g^4}{16\pi^2}$ (see Fig. 7.5).

Since the mass of scalar is

$$M^2 = V_1''|_{\varphi_c = v} = \frac{\lambda}{3}v^2\,, \tag{7.146}$$

and the mass of vector boson is $\mathcal{M}^2 = g^2v^2$, we find that spontaneous symmetry breaking takes place if

$$M^2 > \frac{3g^4}{16\pi^2}\mathcal{M}^2\,, \tag{7.147}$$

i.e. in the nonsymmetric vacuum, scalar cannot be very light. The bound (7.147) was first derived by Linde and Weinberg [10,11].

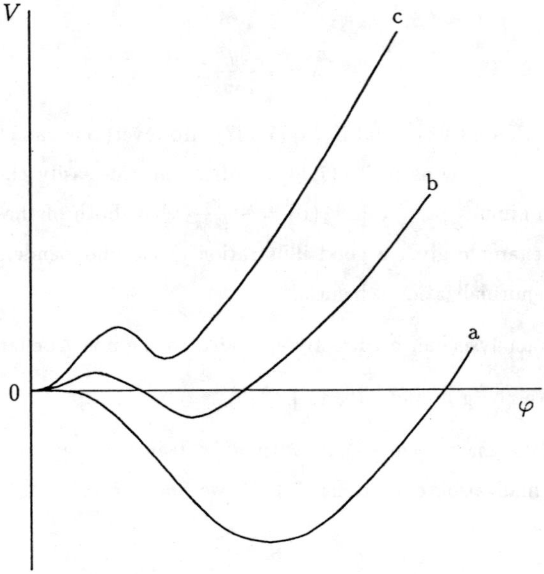

Fig. 7.5. The form of the effective potential at different values of λ: a) $\lambda > \frac{9g^4}{8\pi^2}$; b) $\frac{9g^4}{8\pi^2} > \lambda > \frac{9g^4}{16\pi^2}$; c) $\lambda < \frac{9g^4}{16\pi^2}$

So far we have neglected a contribution of fermions assuming implicitly that their Yukawa couplings are small (i.e. that they are very light). However, as we shall see, heavy fermions may change the form of the potential drastically.

III. The σ model with heavy fermions [12].

In the previous example, we saw how quantum fluctuations of vector fields lead to the restoration of the symmetry of the vacuum. We will show that quantum fluctuations of fermions enhance symmetry breaking. This effect was first pointed out by Krive and Linde [12].

Let us consider the $U_L(1) \times U_R(1)$ linear σ-model (see Eq.(4.38)):

$$\begin{aligned}\mathcal{L} =& \frac{1}{2}\left(\partial_\mu\sigma\partial^\mu\sigma + \partial_\mu\eta\partial^\mu\eta\right) - \frac{\mu^2}{2}(\sigma^2+\eta^2) - \frac{\lambda}{4!}(\sigma^2+\eta^2)^2 \\ &+ i\bar{\psi}\gamma^\mu\partial_\mu\psi - i\sqrt{2}g\left(\sigma\bar{\psi}\psi - \eta\bar{\psi}i\gamma_5\psi\right)\end{aligned} \tag{7.148}$$

(we again omitted all counter terms).

Let us consider the case with $\lambda \ll g^2 \ll 1$ ("heavy" fermions). Then, using the renormalization conditions (7.109), we get the following one-loop potential:

$$V_1(\varphi_c) = \frac{\mu^2}{2}\varphi_c^2 + \frac{\lambda}{4!}\varphi_c^4 - \frac{g^4}{4\pi^2}\varphi_c^4 \ln\frac{\varphi_c^2}{v^2} + \frac{3g^4}{8\pi^2}\varphi_c^4 - \frac{g^4}{2\pi^2}\varphi_c^2 v^2 + \frac{\lambda v^4}{4!} + \frac{g^4 v^4}{8\pi^2} + V_1(v)\,, \tag{7.149}$$

where $v^2 = -\frac{6\mu^2}{\lambda}$. One can see from this expression that fermions give a negative contribution at large φ_c, and the expression (7.149) be unbounded from below (see Fig. 7.6).

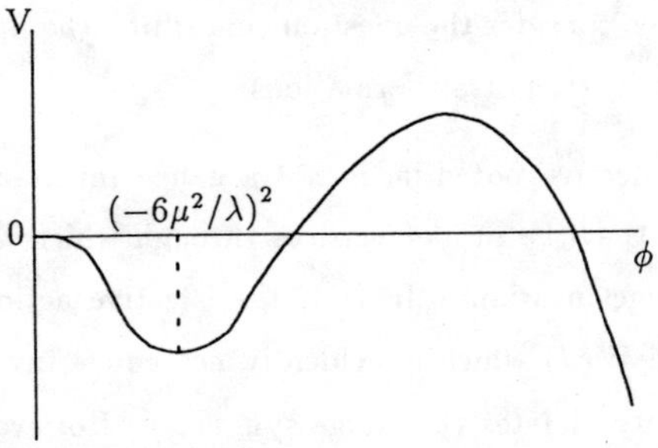

Fig. 7.6.

Since the loop expansion is reliable only if $\lambda, g^2 \ll 1$ and $g^2 \ln\frac{\varphi_c^2}{v^2} \ll 1$, we cannot conclude that the exact potential $V(\varphi_c)$ is also unbounded from below. Let us however show that in the region of applicability of the expression (7.149), there are values of $\varphi_c > v$ at which potential V_1 becomes lower than the value of $V_1(v)$ at the perturbative vacuum with $\varphi_c = v$.

Indeed, let us take $\varphi_c = \tilde{v} = v(1+\epsilon)\exp\left(\frac{\lambda}{g^4}\right)$ with $1 \gg \epsilon \gg \frac{\lambda}{g^4}$. At such a $\tilde{v}$, the expression (7.149) is still reliable. One can check that $V_1(\tilde{v}) < V_1(v)$ (see Fig. 7.6).

Thus we conclude that in the theory with heavy fermions, the minimum of $V(\varphi_c)$ at $\varphi_c = v$ is not global minimum. The global minimum (if it exists at all) may exist only in the region with large φ_c, where nonperturbative dynamics is essential.

As we will see in Chapter 15, the dynamics of heavy fermions (in particular, of the t quark) may play the important role in triggering electroweak symmetry breaking.

The most interesting case for the application of the stability criterion of the vacuum with spontaneous symmetry breaking is the standard electroweak theory with the gauge group $SU_L(2) \times U_Y(1)$ (see Chapter 13). The constraint that the vacuum of this theory corresponds to the spontaneous breakdown of the $SU_L(2) \times U_Y(1)$ to the electromagnetic group $U_Q(1)$ leads to the following bound for the mass of the Higgs boson M_h [13]:

$$M_h > 1.85(m_t - 85\text{GeV}) , \tag{7.150}$$

where m_t is the mass of the (heaviest) t quark.

In conclusion, let us consider the question concerning the properties of the effective potential with respect to gauge transformations.

First of all, the effective potential is *not* a gauge invariant object. This follows already from the fact that the proper vertices through which the effective potential is expressed are not gauge invariant. In fact, the effective action is a Legendre transform of the functional $W(J)$ which is evidently not gauge invariant: the source term $\int d^4x J(x)\varphi(x)$ explicity violates the gauge symmetry. However the values of the potential at its extremuma corresponding to $J = 0$ are gauge invariant quantities. This implies that there is no real problem with this issue. As always, physical observables (such as, for example, values of V at extremuma corresponding to the vacuum energy density) are gauge invariant; different gauges yield different descriptions of the dynamics, and some gauges can be more useful than others in the consideration of certain problems, but the physics is the same if the transformation from one gauge to another is realized in an appropriate way, just as the gauge transformations for Green's functions are realized [14,15].

7.6. Renormalization Group Improvement of the Effective Action

As is known, the renormalization group (RG) method allows to improve the perturbative expansion for Green'n functions [4,16]. For example in gauge theories, the perturbative expansion is an expansion in powers of the gauge coupling g and $g^2 \ln \frac{p^2}{\sigma^2}$; it is reliable only if both g and $g^2 |\ln \frac{p^2}{\sigma^2}|$ are small. RG replaces this expansion with that in powers of the running coupling which takes the following form in the one-loop approximation:

$$\bar{g}^2(p^2) = \frac{g^2}{1 + C \ln \frac{p^2}{\sigma^2}} , \tag{7.151}$$

where a constant C is $C > 0$ and $C < 0$ in asymptotically free gauge theories and non-asymptotically free theories, like QED, respectively. One can see that if g is small, $\bar{g}^2(p^2)$ remains small in a rather large range of $\ln \frac{p^2}{\sigma^2}$, including all positive $\ln \frac{p^2}{\sigma^2}$ when $C > 0$ and all negative $\ln \frac{p^2}{\sigma^2}$ when $C < 0$. Thus the new expansion in $\bar{g}^2(p^2)$ essentially enlarges the range of validity of the ordinary perturbative expansion.

As was first pointed out by Coleman and Weinberg [3], RG can also be helpful in improving the loop expansion for the effective action. Let us first consider how RG works in the massless $\lambda\varphi^4$ model.

The starting point is that the effective action is independent of the choice of the renormalization-group parameter σ:

$$\sigma \frac{d}{d\sigma} \Gamma(\varphi_c, \lambda; \sigma) = 0 . \tag{7.152}$$

Since the action depends on σ both explicitly and implicity through φ_c and λ, this equation is equivalent to the following one:

$$\left[\sigma \frac{\partial}{\partial \sigma} + \beta \frac{\partial}{\partial \lambda} - \gamma \int d^4x \varphi_c(x) \frac{\delta}{\delta \varphi_c(x)} \right] \Gamma = 0 , \tag{7.153}$$

where

$$\begin{aligned} \beta =& \sigma \frac{d\lambda}{d\sigma} , \\ \gamma =& \frac{1}{2} \sigma \frac{d \ln Z_2}{d\sigma} \end{aligned} \tag{7.154}$$

with Z_2 being renormalization constant of the field $\varphi(x)$ ($\varphi = Z_2^{-1/2} \varphi^{(0)}$, where $\varphi^{(0)}$ is the bare field).

Henceforth we will use the $\overline{MS}$ renormalization scheme (see Sec.C.2 in Appendix C). In this scheme, β and γ depend only on the coupling constant λ both in massless and massive theories.

Applying the equation (7.153) to the derivative expansion of Γ (7.33), we get:

$$\left(\sigma\frac{\partial}{\partial\sigma}+\beta\frac{\partial}{\partial\lambda}-\gamma\varphi_c\frac{\partial}{\partial\varphi_c}\right)V=0\ , \tag{7.155}$$

$$\left(\sigma\frac{\partial}{\partial\sigma}+\beta\frac{\partial}{\partial\lambda}-\gamma\varphi_c\frac{\partial}{\partial\varphi_c}-2\gamma\right)Z=0\ . \tag{7.156}$$

The solutions of these equations can be found in the following way. Let us introduce the parameter $\sigma(t)=e^t\sigma$. Then let us introduce the running coupling constant $\bar{\lambda}(t)$ satisfying the equation

$$\frac{d\bar{\lambda}(t)}{dt}=\beta(\bar{\lambda}(t));\ \bar{\lambda}(0)=\lambda\ , \tag{7.157}$$

and the running field $\bar{\varphi}_c(t)$ satisfying the equation

$$\frac{d\bar{\varphi}_c(t)}{dt}=-\gamma(\bar{\lambda}(t))\bar{\varphi}_c(t);\ \bar{\varphi}_c(0)=\varphi_c\ . \tag{7.158}$$

Then we find from Eqs.(7.155), (7.156) that

$$\begin{aligned}&\left(\sigma\frac{\partial}{\partial\sigma}+\beta(\bar{\lambda})\frac{\partial}{\partial\bar{\lambda}}-\gamma(\bar{\lambda})\bar{\varphi}_c\frac{\partial}{\partial\bar{\varphi}_c}\right)V(\bar{\varphi}_c(t),\bar{\lambda}(t);e^t\sigma)\\&=\left(\frac{\partial}{\partial t}+\frac{d\bar{\lambda}}{dt}\frac{\partial}{\partial\bar{\lambda}}+\frac{d\ln\bar{\varphi}_c}{dt}\frac{\partial}{\partial\ln\bar{\varphi}_c}\right)V\left(\bar{\varphi}_c(t),\bar{\lambda}(t);e^t\sigma\right)\\&=\frac{d}{dt}V\left(\bar{\varphi}_c(t),\bar{\lambda}(t);e^t\sigma\right)=0\ ,\end{aligned} \tag{7.159}$$

and

$$\left(\frac{d}{dt}-2\bar{\gamma}(\bar{\lambda}(t))\right)Z\left(\bar{\varphi}_c(t),\bar{\lambda}(t);e^t\mu\right)=0\ . \tag{7.160}$$

The solutions of these equations are:

$$V\left(\bar{\varphi}_c(t),\bar{\lambda}(t);e^t\sigma\right)=V\left(\varphi_c,\lambda;\sigma\right)\ , \tag{7.161}$$

$$Z\left(\bar{\varphi}_c(t),\bar{\lambda}(t);e^t\sigma\right)=Z\left(\varphi_c,\lambda;\sigma\right)\exp\left[2\int_0^t dt'\gamma(\bar{\lambda}(t'))\right]\ . \tag{7.162}$$

The solutions (7.161), (7.162) contain full information of the renormalization group equations: once V and Z are known at a certain value of $\sigma(t)|_{t=t_0}=e^{t_0}\sigma$, they are determined at all values of $\sigma(t)$.

In the $\overline{MS}$ scheme, the one-loop V and Z are

$$V_1 = \frac{\lambda}{4!}\varphi_c^4 + \frac{\lambda^2\varphi_c^4}{256\pi^2}\left[\ln\frac{\lambda\varphi_c^2}{2\sigma^2} - \frac{3}{2}\right] , \tag{7.163}$$

$$Z_1 = 1 + \frac{\lambda}{96\pi^2} \tag{7.164}$$

at $\mu^2 = 0$ (see Eqs.(7.111), (7.127)). The β-function $\beta(\lambda)$ is

$$\beta = \frac{3}{16\pi^2}\lambda^2 \tag{7.165}$$

and $\gamma = 0$. Let us show how the relations (7.161), (7.162) can improve the one-loop approximation.

To avoid a potentially large logarithm, we choose $\sigma = \sigma_0$, where $\ln\frac{\lambda\varphi_c^2}{2\sigma_0^2} = 0$. Then the relation (7.161) tells us that the effective potential $V(\varphi_c, \lambda; \sigma)$ at any other value of σ is

$$V(\varphi_c, \lambda; \sigma) = V\left(\bar{\varphi}_c(t_0), \bar{\lambda}(t_0); \sigma_0\right) , \tag{7.166}$$

where $t_0 = \frac{1}{2}\ln\frac{\sigma_0^2}{\sigma^2} = \frac{1}{2}\ln\frac{\lambda\varphi_c^2}{2\sigma^2}$. Therefore

$$V(\varphi_c, \lambda; \sigma) = \frac{\bar{\lambda}(t_0)}{4!}\bar{\varphi}_c^4(t_0) - \frac{3\bar{\lambda}^2(t_0)}{512\pi^2}\bar{\varphi}_c^4(t_0) . \tag{7.167}$$

The β-function (7.165) yields

$$\bar{\lambda}(t_0) = \lambda\left(1 - \frac{3\lambda}{32\pi^2}\ln\frac{\lambda\varphi_c^2}{2\sigma^2}\right)^{-1} , \tag{7.168}$$

and $\gamma = 0$ yields $\bar{\varphi}_c(t_0) = \varphi_c$. Therefore the RG improved potential is

$$V = \frac{\bar{\lambda}}{4!}\varphi_c^4 - \frac{3\bar{\lambda}^2}{512\pi^2}\varphi_c^4 \tag{7.169}$$

with

$$\bar{\lambda} = \lambda\left(1 - \frac{3\lambda}{32\pi^2}\ln\frac{\lambda\varphi_c^2}{2\sigma^2}\right)^{-1} . \tag{7.170}$$

The RG improved function Z is (see Eq.(7.164))

$$Z = 1 + \frac{\bar{\lambda}}{96\pi^2} . \tag{7.171}$$

Of course, at $|\ln\frac{\lambda\varphi_c^2}{\sigma^2}| \ll 1$, we get the one-loop expressions (7.163), (7.164) from Eqs.(7.169), (7.171).

Let us consider now the extension of this procedure to the massive ($\mu^2 \neq 0$) case. In fact this extension has been elaborated only recently in Refs.[17,18,19]. In our discussion, we will closely follow Ref.[18].

To clarify some subtle points in this problem, let us consider the expression (7.111) for V_1 in more detail:

$$V_1 = V^{(0)} + V^{(1)} , \tag{7.172}$$

where

$$V^{(0)} = U = \frac{1}{2}\mu^2\varphi_c^2 + \frac{1}{4!}\lambda\varphi_c^4 + \tilde{C} , \tag{7.173}$$

$$V^{(1)} = \frac{1}{64\pi^2}M_{\varphi_c}^4\left(\ln\frac{M_{\varphi_c}^2}{\sigma^2} - \frac{3}{2}\right) \tag{7.174}$$

with

$$M_{\varphi_c}^2 = \frac{1}{2}\lambda\varphi_c^2 + \mu^2 . \tag{7.175}$$

If in the massless theory the only scale is set by φ_c and therefore all logarithms have the form $\ln\frac{\lambda\varphi_c^2}{\sigma^2}$, now we have an additional dimensional parameter, the mass μ. However, as we can see, the logarithm in $V^{(1)}$ (7.174) appears only in the form $\ln\frac{M_{\varphi_c}^2}{\sigma^2}$. This is not accident: such a logarithmic structure of the potential follows from the relations (7.61), (7.63). There is however the following subtle point connected with the vacuum-energy term corresponding to the value $V(\varphi_c = 0)$. This term is

$$V_1(\varphi_c = 0) = \tilde{C} + \frac{1}{64\pi^2}\mu^4\left(\ln\frac{\mu^2}{\sigma^2} - \frac{3}{2}\right) , \tag{7.176}$$

i.e. it depends on the renormalization parameter σ. To remove this dependence, one should accept that the parameter $\tilde{C}$ itself depends on σ. Since the dependence of the second term on the right hand side of Eq.(7.176) on σ disappears at $\mu = 0$, we write $\tilde{C}$ as

$$\tilde{C} = h\mu^4 = (\sigma^2)^{-\epsilon}Z_h^{-1}h_0\mu^4 , \tag{7.177}$$

where $\epsilon = \frac{4-n}{2}$, and h_0 is the bare parameter in the vacuum-energy term. In other words, there is an additional counter term connected with the renormalization of h_0 in the massive theory [17,18,19].

One might think that a way to avoid introducing the new counter term is to use the following normalization condition for the potential.

$$V(\varphi_c)|_{\varphi_c=0} = 0\ . \tag{7.178}$$

In the one-loop approximation, it implies that

$$V_1(\varphi_c) \longrightarrow V_1(\varphi_c) - V_1(0) = V_1(\varphi_c) - \tilde{C} - \frac{\mu^4}{64\pi^2}\left(\ln\frac{\mu^2}{\sigma^2} - \frac{3}{2}\right)\ . \tag{7.179}$$

Thus this subtraction introduces a new kind of the logarithmic term, $\ln\frac{\mu^2}{\sigma^2}$, in the potential, and this prevents us from using the RG improvement technique discussed before.

Now we can write the equations for $V(\varphi_c, \lambda, \mu, h; \sigma)$ and $Z(\varphi_c, \lambda, \mu, h; \sigma)$. They are (compare with Eqs.(7.155), (7.156):

$$\left(\sigma\frac{\partial}{\partial\sigma} + \beta\frac{\partial}{\partial\lambda} - \gamma_m\mu\frac{\partial}{\partial\mu} + \beta_h\frac{\partial}{\partial h} - \gamma\varphi_c\frac{\partial}{\partial\varphi_c}\right)V = 0\ , \tag{7.180}$$

$$\left(\sigma\frac{\partial}{\partial\sigma} + \beta\frac{\partial}{\partial\lambda} - \gamma_m\mu\frac{\partial}{\partial\mu} + \beta_h\frac{\partial}{\partial h} - \gamma\varphi_c\frac{\partial}{\partial\varphi_c} - 2\gamma\right)Z = 0\ , \tag{7.181}$$

where

$$\gamma_m = -\frac{d\ln\mu}{d\ln\sigma}\ , \tag{7.182}$$

$$\beta_h = \frac{dh}{d\ln\sigma}\ . \tag{7.183}$$

In the same way as before, we find the analogues of the relations (7.161), (7.162):

$$V\left(\bar{\varphi}_c(t), \bar{\lambda}(t), \bar{\mu}(t), \bar{h}(t); e^t\sigma\right) = V\left(\varphi_c, \lambda, \mu, h; \sigma\right)\ , \tag{7.184}$$

$$Z\left(\bar{\varphi}_c(t), \bar{\lambda}(t), \bar{\mu}(t), \bar{h}(t); e^t\sigma\right) = Z\left(\varphi_c, \lambda, \mu, h; \sigma\right)\exp\left[2\int_0^t dt'\gamma(\bar{\lambda}(t'))\right]\ , \tag{7.185}$$

where $\bar{\lambda}(t)$ and $\bar{\varphi}_c(t)$ are determined from Eqs.(7.157), (7.158), and the new running parameters $\bar{\mu}(t)$ and $\bar{h}(t)$ satisfy the equations:

$$\frac{d\ln\bar{\mu}(t)}{dt} = -\gamma_m(\bar{\lambda}(t)) \; ; \; \bar{\mu}(0) = \mu \; , \tag{7.186}$$

$$\frac{d\bar{h}(t)}{dt} = \beta_h\left(\bar{\lambda}(t), \bar{h}(t)\right) \; ; \; \bar{h}(0) = h \; . \tag{7.187}$$

Note that the vacuum-energy parameter h affects only the evolution of itself.

As in the massless case, the relations (7.184), (7.185) imply that the RG-improved V and Z are:

$$V(\varphi_c, \lambda, \mu, h; \sigma) = V\left(\bar{\varphi}_c(t_0), \bar{\lambda}(t_0), \bar{\mu}(t_0), \bar{h}(t_0); \sigma_0\right) \; , \tag{7.188}$$

$$Z(\varphi_c, \lambda, \mu, h; \sigma) = Z\left(\bar{\varphi}_c(t_0), \bar{\lambda}(t_0), \bar{\mu}(t_0), \bar{h}(t_0); \sigma_0\right) \exp\left[-2\int_0^{t_0} dt' \gamma(\bar{\lambda}(t'))\right] \; , \tag{7.189}$$

where $t_0 = \frac{1}{2}\ln\frac{\sigma_0^2}{\sigma^2} = \frac{1}{2}\ln\frac{\lambda\varphi_c^2 + 2\mu^2}{2\sigma^2}$.

In the one loop approximation, $\bar{\lambda}(t)$ and $\bar{\varphi}_c(t)$ are the same as in the massless case, and γ_m and β_h are:

$$\begin{aligned} \gamma_m &= -\frac{1}{32\pi^2}\lambda \; , \\ \beta_h &= -\frac{1}{8\pi^2}\lambda h + \frac{1}{32\pi^2} \; . \end{aligned} \tag{7.190}$$

Using the relations (7.188) and Eqs.(7.172), (7.177), we find the following expression for the RG improved effective potential:

$$V = \frac{1}{2}\bar{\mu}^2\bar{\varphi}_c^2 + \frac{1}{4!}\bar{\lambda}\bar{\varphi}_c^4 + \bar{h}\bar{\mu}^4 - \frac{3}{128\pi^2}\left(\frac{1}{2}\bar{\lambda}\bar{\varphi}_c^2 + \bar{\mu}^2\right)^2 \tag{7.191}$$

with

$$
\begin{aligned}
\bar{\varphi}_c =& \varphi_c \,, \\
\bar{\lambda} =& \lambda \left(1 - \frac{3\lambda}{32\pi^2} \ln \frac{\frac{1}{2}\lambda\varphi_c^2 + \mu^2}{\sigma^2}\right)^{-1} \,, \\
\bar{\mu}^2 =& \mu^2 \left(1 - \frac{3\lambda}{32\pi^2} \ln \frac{\frac{1}{2}\lambda\varphi_c^2 + \mu^2}{\sigma^2}\right)^{-\frac{1}{3}} \,, \\
\bar{h}\bar{\mu}^4 =& h\mu^4 + \frac{1}{2}\frac{\mu^4}{\lambda}\left[1 - \left(1 - \frac{3\lambda}{32\pi^2} \ln \frac{\frac{\lambda\varphi_c^2}{2} + \mu^2}{\sigma^2}\right)^{\frac{1}{3}}\right] \,.
\end{aligned}
\tag{7.192}
$$

The parameter h is arbitrary; one can for example choose $h = 0$.

Taking into account Eq.(7.127), we find the RG improved expression for Z:

$$
Z = 1 + \frac{\bar{\lambda}^2}{96\pi^2} \frac{\varphi_c^2}{(2\bar{\mu}^2 + \bar{\lambda}\varphi_c^2)} \,. \tag{7.193}
$$

The method, with some minor modifications, may be extended to theories with few coupling constants and to the multi-mass-scale case [18].

7.7. More About the Physical Meaning of the Effective Potential.

As we already showed in Sec.7.3, the effective potential $V(\varphi_c)$ is the expectation value of the energy density in the state which minimizes $\langle\psi|\mathcal{H}|\psi\rangle$ subject to the condition that $\langle\psi|\varphi(x)|\psi\rangle = \varphi_c$. Since $\mathcal{H}$ is an hermitian operator, this implies that $V(\varphi_c)$ must be real. It can also be shown that the potential must be everywhere convex ($V'' \geq 0$) [5].

However a look at the expression (7.111) for the one-loop potential in φ^4 theory,

$$
V_1 = \frac{\mu^2}{2}\varphi_c^2 + \frac{\lambda}{4!}\varphi_c^4 + \frac{\left(\mu^2 + \frac{\lambda\varphi_c^2}{2}\right)^2}{64\pi^2}\left[\ln\left(\frac{\mu^2 + \frac{\lambda\varphi_c^2}{2}}{\sigma^2} - i\epsilon\right) - \frac{3}{2}\right] + \tilde{C} \,, \tag{7.194}
$$

reveals that when $\mu^2 < 0$, these two restrictions fulfill not for all values φ_c: at $\left(\mu^2 + \frac{\lambda\varphi_c^2}{2}\right) < 0$, the potential becomes complex and nonconvex.*) Of course, since this expression

*) The $i\epsilon$ in Eq.(7.194) insures that the imaginary part is negative; $i\epsilon$ is a trace of the definition of the propagator of φ in the Minkowski space (see Eq.(1.16)).

for V is approximate, one might think that this means just a breakdown of the loop expansion. We will argue, however, following Weinberg and Wu [20], that the situation is in fact more interesting .

First of all let us recall a similar situation taking place for quasi-stationary states in quantum mechanics. Although the hamiltonian is a hermitian (self-adjoint) operator, energy of quasi-stationaly states is complex, $E = E_0 - \frac{i}{2}\Gamma$. The resolution of this puzzle is well known [21]. At large distances the wave function of quasi-stationary states contains the factor

$$\exp\left[ir2m\left(E_0 - \frac{i}{2}\Gamma\right)^{1/2}\right] , \tag{7.195}$$

which increases exponentially as $r \longrightarrow \infty$. Therefore the normalization integral $\int |\psi|^2 d^3x$ for this wave function badly diverges, and it does not lie in the Hilbert space of square integrable functions L_2, where the self-adjointness of the hamiltonian is proved. This of course does not prevent quasi-stationary states from being interesting physical objects.

Let us return to the consideration of the expression (7.194) for the effective potential. When $\left(\mu^2 + \frac{\lambda\varphi_c^2}{2}\right) < 0$, the imaginary part of V_1 is

$$\mathrm{Im} V_1 = \frac{-\left(\mu^2 + \frac{\lambda\varphi_c^2}{2}\right)^2}{64\pi} . \tag{7.196}$$

It is instructive to obtain this result in somewhat different way by using the relation (7.68). Then

$$\mathrm{Im} V_1 = \mathrm{Im}\frac{1}{2(2\pi)^3}\int d^3p\left(p^2 + \mu^2 + \frac{\lambda\varphi_c^2}{2} - i\epsilon\right)^{1/2} = \frac{-\left(\mu^2 + \frac{\lambda\varphi_c^2}{2}\right)^2}{64\pi} . \tag{7.197}$$

Thus the imaginary part appears as the result of zero-point fluctuations of the tachyonic field with $M^2_{\varphi_c} = \mu^2 + \frac{\lambda\varphi_c^2}{2} < 0$.

Since the potential yields the energy density of a state, its imaginary part should be a signature of an instability of this state. In the case of quasi-stationary states, Γ is equal to the probability of decay per unit time; in the present case, $-2\mathrm{Im}V(\varphi_c)$ should be equal to the decay probability per unit time per unit volume of a system of infinite spatial extent with $\langle\psi|\varphi|\psi\rangle = \varphi_c$.

We recall that the wave function of quasi-stationary states is a solution of Schrödinger equation with the asymptotics at $r \longrightarrow \infty$ being an outgoing wave. The crucial point is to find a proper definition for the states corresponding to a complex effective potential.

Let us begin, following Weinberg and Wu [20], by considering the classical approximation for a system with the classical potential $U(\varphi) = \frac{\mu^2}{2}\varphi^2 + \frac{\lambda}{4!}\varphi^4, \mu^2 < 0$, having the form shown in Fig.7.7.

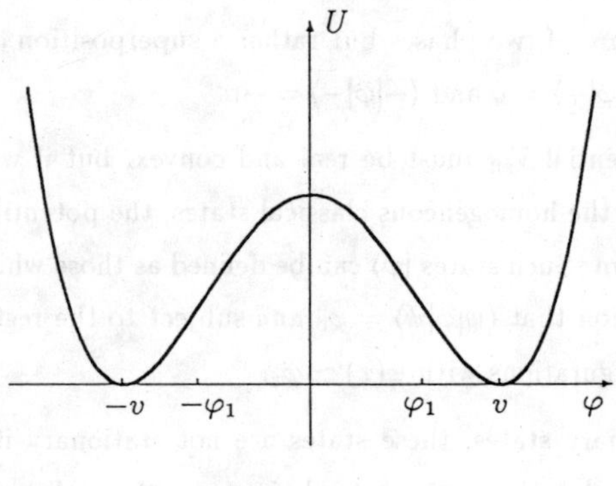

Fig. 7.7. A potential with $U''(\phi) < 0$ in the region $-\varphi_1 < \varphi < \varphi_1$.

As the classical analogue of $\langle\psi|\varphi(x)|\psi\rangle$, we take the spatial average of $\varphi, \bar{\varphi}$. A classical *effective* potential $U_{\text{eff}}(\bar{\varphi})$ is then defined as the minimum value of the energy density among all states with a given value of $\bar{\varphi}$. Let us consider the connection between $U(\varphi)$ and $U_{\text{eff}}(\varphi)$.

When $|\bar{\varphi}| \geq v$, the energy density is minimized by a homogeneous configuration with $\varphi(x) = \bar{\varphi}$ everywhere, and $U(\varphi)$ coincides with $U_{\text{eff}}(\varphi)$. However, when $|\bar{\varphi}| < v$, the configuration which minimizes the energy is inhomogeneous. In this configuration $\varphi(x) = v$ in a fraction $f = (\bar{\varphi}+v)/(2v)$ of space and $\varphi(x) = v$ elsewhere. Of course there is a surface energy connected with the boundary between the two regions, but, in the infinite-volume limit, this can be neglected. Therefore $U_{\text{eff}}(\bar{\varphi}) = 0$ in the whole region with $|\bar{\varphi}| \leq v$. What we have gotten now is nothing but the famous Maxwell construction originally introduced to describe the coexistence of the gas and liquid phases.

The classical effective potential U_{eff} and the potential U are essentially different in the region with $|\bar{\varphi}| < v$. We also note that $U''_{\text{eff}} \geq 0$ everywhere, i.e. it is everywhere convex, although $U'' < 0$ at $|\bar{\varphi}| < v$. It is however clear that the initial potential U is a very useful object. In particular, if we are interested in the description of only homogeneous states (which are classically unstable at $|\bar{\varphi}| < v$), we should use this potential and not U_{eff}.

In the quantum theory, one can introduce an analogue of U_{eff}, the effective potential V_{eff} (we retain the notation V for the perturbative potential). The main difference from the classical case is that now the state minimizing the energy density is not an inhomogeneous mixture of two phases but rather a superposition of the vacuum states $|+\rangle$ and $|-\rangle$ with $\langle+|\varphi|+\rangle = v$ and $\langle-|\varphi|-\rangle = -v$.

The effective potential V_{eff} must be real and convex, but if we are interested in a quantum analogue of the homogeneous classical states, the potential V becomes appropriate to describe them. Such states $|\psi\rangle$ can be defined as those which minimize $\langle\psi H|\psi\rangle$ subject to the condition that $\langle\psi|\varphi|\psi\rangle = \varphi_c$ *and* subject to the restriction that they be concentrated on configurations with $\varphi(x) \simeq \varphi_c$.

Like quasi-stationary states, these states are not stationary if $\text{Im}V \neq 0$. On the other hand, they can play the important role in some physical processes in which such an unstable state is realized as the initial one. The description of the evalution of these states is a very interesting problem which can find interesting application for example in cosmology [22].

We only note here that since in the quasi-classical approximation, the potential $\tilde{U}$ connected with the action $\tilde{S}$ (7.57) is

$$\tilde{U}(\varphi) = \frac{1}{2}\tilde{U}''(\varphi_c)(\varphi - \varphi_c)^2 \ , \tag{7.198}$$

the dynamics with $\tilde{U}'' < 0$ (yielding a complex potential V) is intimately connected with that of the upside-down harmonic oscillator [20] with the hamiltonian

$$H = \frac{\hat{p}^2}{2m} + \frac{m\omega^2\hat{q}^2}{2}; \omega^2 < 0 \ . \tag{7.199}$$

The detailed discussion of this issue can be found in Refs.[20,22].

Thus we conclude this chapter by saying that the effective potential is a very convenient tool in studying theories with spontaneous symmetry breaking. The potential admits a clear physical interpretation, and it can be useful in describing not only the ground state of a system but also unstable excited states playing here the same role as quasi-stationary states in quantum mechanics.

Chapter 8

Effective Action for Composite Operators and Equations for Green's Functions on the Nonperturbative Vacuum

8.1. Introduction

As we saw in Sec.2.8, the fundamental object in describing the superconductivity phenomenon is the electron propagator G (2.89), and in particular its anomalous part F (2.90) describing the electron-electron condensate. In fact, $F_{\alpha\beta}(x) = -i\langle 0|T\psi_\alpha(x)\psi_\beta(0)|0\rangle$ can be considered as a bilocal order parameter characterizing the dynamical breakdown of the gauge $U(1)$ symmetry in a superconductor.

The same approach can be used for describing dynamic symmetry breaking in relativistic quantum field theory. Therefore it is very important to have a general formalism for the derivation of the equations for propagators (and Green's functions in general) on the nonperturbative vacuum.

Such a formalism, known as the effective action for composite operators, was first elaborated in condensed matter physics [1, 2, 3] and was then extended to relativistic field theories [4, 5, 6]. A very convenient approach to deal with the effective action for composite operators was introduced by Cornwall, Jackiw and Tomboulis [6]. The approach is based on functional methods, in particular, or the path integral representation for Green's functions. We will use it in this chapter.

The effective action for composite operators $\Gamma(\varphi_c, G)$ is a functional both of the vacuum average $\varphi_c(x) = \langle 0|\varphi(x)|0\rangle$ and of the propagator $G(x,y) = i\langle 0|T\varphi(x)\varphi(y)|0\rangle$ (the symbol $\varphi(x)$ represents all fields of the initial lagrangian). The variational equations

$$\frac{\delta\Gamma}{\delta\varphi_c(x)} = 0\,, \tag{8.1}$$

$$\frac{\delta\Gamma}{\delta G(x,y)} = 0 \tag{8.2}$$

determine φ_c and G on the vacuum. As we will see, the equation (8.2) coincides with the Schwinger-Dyson equation for the propagator G. We will also show that the second variational derivative of Γ in G leads to the Bethe-Salpeter equation describing bound states. The advantage of the present derivation of these dynamical equations as compared to the original derivations [7, 8, 9] is that it gives these equations on the nonperturbative vacuum.

The effective action $\Gamma(\varphi_c, G)$ generalizes the effective action $\Gamma(\varphi_c)$ considered in the previous chapter. However there are some subtle points in the interpretation of $\Gamma(\varphi_c, G)$ which were fully clarified only recently [10, 11, 12]. This in particular relates to the stability criterion for the vacuum described by φ_c and G. We will discuss these issues in detail in Sec.8.7.

8.2. Effective Action for Composite Operators. The General Formalism

To simplify the notations, we will consider the derivation of the effective action for composite operators in a theory including spinless fields only; the generalization to theories with gauge fields and fermion fields is straightforward, and will be indicated at the end of this section.

Let us consider the path integral representation for the generating functional $Z(J, I)$ with the local source $J(x)$ and the bilocal source $I(x, y)$ (compare with Eq.(7.34)):

$$Z(J,I) = \exp(iW(J,I)) = N\int \mathcal{D}(\varphi)\exp\left\{i\Big[S(\varphi) + \int d^4x\varphi(x)J(x) + \int d^4xd^4y\varphi(x)\varphi(y)I(x,y)\Big]\right\} , \tag{8.3}$$

where N is a normalization factor, the symbol $\varphi(x)$ represents all fields of the initial lagrangian (all group indices of $\varphi(x)$ are omitted), and the action $S(\varphi) = \int d^4x\mathcal{L}(x)$ may be written as

$$S(\varphi) = -\int d^4xd^4y\varphi(x)D^{-1}(x-y)\varphi(x) + S_{\text{int}}(\varphi) , \tag{8.4}$$

where $D(x-y)$ is the free propagator

$$\begin{aligned} D(x-y) &= \frac{1}{(2\pi)^4}\int d^4k\frac{1}{m^2-k^2-i\epsilon}e^{-ik(x-y)} , \\ D^{-1}(x-y) &= (\Box + m^2)\delta^4(x-y) , \end{aligned} \tag{8.5}$$

and S_{int} is the interaction part of the action.

The functional $Z(J,I)$ can be expanded in a series (integration over repeated variables is understood here):

$$Z(J,I) = \sum_{n=0}^{\infty} \frac{i^n}{n!} G^{(n)}(J; x_1y_1, x_2y_2, \cdots, x_ny_n) I(x_1,y_1) I(x_2,y_2) \cdots I(x_n,y_n) \ , \tag{8.6}$$

where

$$G^{(n)}(J; x_1y_1, \cdots, x_ny_n) = \langle 0,_{\text{out}} | T\varphi(x_1)\varphi(y_1), \cdots, \varphi(x_n)\varphi(y_n) | 0,_{\text{in}} \rangle_J \ . \tag{8.7}$$

These Green's functions relate to the theory with the local source $J(x)$ (since the source $J(x)$ may be time dependent, the in-vacuum and the out-vacuum are in general different).

We expand $W(J,I)$ as

$$W(J,I) = i^{-1} \ln Z(J,I) = \sum_{n=0}^{\infty} \frac{i^{n-1}}{n!} G_c^{(n)}(J; x_1y_1, \cdots x_ny_n) I(x_1,y_1) \cdots I(x_n,y_n) \ . \tag{8.8}$$

Comparing this equation with equation (8.6), we find the relations between $G^{(n)}$ and $G_c^{(n)}$, for example,

$$G_c^{(1)}(J; xy) = \frac{1}{Z(J)} G^{(1)}(J; xy) \ , \tag{8.9}$$

$$\begin{aligned} G_c^{(2)}(J; x_1y_1, x_2y_2) =& \frac{1}{Z(J)} G^{(2)}(J; x_1y_1, x_2y_2) \\ &- \frac{1}{Z^2(J)} G^{(1)}(J; x_1y_1) G^{(1)}(J; x_2y_2) \ , \end{aligned} \tag{8.10}$$

$$\begin{aligned} G_c^{(3)}(J; x_1y_1, x_2y_2, x_3y_3) =& \frac{1}{Z(J)} G^{(3)}(J; x_1y_1, x_2y_2, x_3y_3) \\ &- \frac{1}{Z^2(J)} \sum_{\{i_1<i_2\neq i_3\}=1}^{3} G_c^{(2)}(J; x_{i_1}y_{i_1}, x_{i_2}y_{i_2}) G(J; x_{i_3}y_{i_3}) \\ &- \frac{1}{Z^3(J)} G^{(1)}(J; x_1y_1) G^{(1)}(J; x_2y_2) G^{(1)}(J; x_3y_3) \ , \end{aligned} \tag{8.11}$$

where

$$Z(J) \equiv Z(J,I)|_{I=0} \ . \tag{8.12}$$

It is clear from these relations that Green's functions $G_c^{(n)}$ for bilocal operators are analogues of connected Green's functions for local operators considered in Sec.7.2.

Indeed, considering each pair $(x_i y_i)$ as one point, one can see that all contributions which factorize into pieces are subtracted in $G_c^{(n)}$; the factor $Z^{-1}(J)$ removes the disconnected vacuum contributions from $G_c^{(n)}$. In other words, the generalization from the local case to the bilocal one is that the points (x_i) are replaced by pairs of points $(x_i y_i)$.

The effective action $\Gamma(\varphi_c, G)$ is introduced as a double Legendre transform of $W(J, I)$. We define

$$\begin{aligned} \frac{\delta W(J,I)}{\delta J(x)} =& \varphi_c(x) , \qquad (8.13) \\ \frac{\delta W(J,I)}{\delta I(x,y)} =& \varphi_c(x)\varphi_c(y) - iG(x,y) , \qquad (8.14) \end{aligned}$$

and set

$$\begin{aligned} \Gamma(\varphi_c, G) =& W(J,I) - \int d^4x \varphi_c(x) J(x) - \int d^4x d^4y \varphi_c(x)\varphi_c(y) I(x,y) \\ &+ i \int d^4x d^4y G(x,y) I(x,y) . \end{aligned} \qquad (8.15)$$

Eqs.(8.13), (8.14) and (8.15) imply that

$$\begin{aligned} \frac{\delta \Gamma}{\delta \varphi_c(x)} =& - J(x) - 2 \int d^4y \varphi_c(y) I(x,y) , \qquad (8.16) \\ \frac{\delta \Gamma}{\delta G(x,y)} =& iI(x,y) . \qquad (8.17) \end{aligned}$$

In the physical situation corresponding to vanishing sources J and I, Eqs.(8.16) and (8.17) reduce to the stationarity equations (8.1) and (8.2).

Let us recall that the effective action $\Gamma(\varphi_c)$ considered in Chapter 7 is the generating functional for the proper vertices $\Gamma^{(n)}(x_1, \cdots, x_n)$. One should expect that $\Gamma(\varphi_c, G)$ plays a similar role for Green's functions which are bilocal analogues of the proper vertices $\Gamma^{(n)}$.

Let us find the relationship between $\Gamma(\varphi_c, G)$ and $W(J, I)$ for the case when $J = 0$ and $\varphi_c = 0$. This case may correspond to dynamical symmetry breakdown, which is characterized by the point that Eq.(8.1) has only the trivial solution $\varphi_c(x) = 0$ but Eq.(8.2) admits symmetry-breaking solutions.

Then we find from Eqs.(8.14) and (8.17):

$$\begin{aligned}&\int d^4x d^4y \frac{\delta^2\Gamma}{\delta G(x_1,y_1)\delta G(x,y)} \frac{\delta^2 W}{\delta I(x,y)\delta I(x_2,y_2)} \\ &= \frac{1}{2}\left[\delta_{x_1x_2}\delta_{y_1y_2}\delta^4(x_1-x_2)\delta^4(y_1-y_2)+\delta_{x_1y_2}\delta_{y_1x_2}\delta^4(x_1-y_2)\delta^4(y_1-x_2)\right] ,\end{aligned} \tag{8.18}$$

where $\delta_{x_1x_2}, \delta_{x_iy_j}, \delta_{y_iy_j}$ represent symbolically Kronecker's δ-functions connected with internal group indices. Therefore $i\frac{\delta^2\Gamma}{\delta G(x_1y_1)\delta G(x_2y_2)}$ is the inverse propagator of the bilocal field, ${G_c^{(2)}}^{-1}(x_1y_1, x_2y_2)$.

At $J = 0$ and $I = 0$, the translation invariance implies that the solutions to the equation (8.2) are functions only of $x-y$, $G(x,y)|_{\substack{J=0\\I=0}} = G(x-y)$, where $G(x-y)$ is the full propagator of the field φ. Introducing now the bilocal field

$$\bar{G}(xy) \equiv G(x,y) - G(x-y) , \tag{8.19}$$

one can expand the action Γ in this bilocal field as:

$$\Gamma(G) = \Gamma(G(x-y)) + \sum_{n=2}^{\infty} \frac{(-i)^n}{n!}\Gamma^{(n)}(x_1y_1,\cdots,x_ny_n)\cdot \bar{G}(x_1y_1)\cdots\bar{G}(x_ny_n) . \tag{8.20}$$

Eq.(8.18) implies that

$$\Gamma^{(2)}(x_1y_1,x_2y_2) = i\left[G_c^{(2)}(x_1y_1,x_2y_2)\right]^{-1} . \tag{8.21}$$

In the same way as the relation (8.18) was derived, we find the relations between $\Gamma^{(3)}, \Gamma^{(4)}$ and $G_c^{(3)}, G_c^{(4)}$:

$$\begin{aligned}&G_c^{(3)}(x_1y_1,x_2y_2,x_3y_3) = iG_c^{(2)}(x_1y_1,x_1'y_1')G_c^{(2)}(x_2y_2,x_2'y_2') \\ &\times G_c^{(2)}(x_3y_3,x_3'y_3')\Gamma^{(3)}(x_1'y_1',x_2'y_2',x_3'y_3') ,\end{aligned} \tag{8.22}$$

$$\begin{aligned}&G_c^{(4)}(x_1y_1,x_2y_2,x_3y_3,x_4y_4) = iG_c^{(2)}(x_1y_1,x_1'y_1')G_c^{(2)}(x_2y_2,x_2'y_2') \\ &\times G_c^{(2)}(x_3y_3,x_3'y_3')G_c^{(2)}(x_4y_4,x_4'y_4')\Gamma^{(4)}(x_1'y_1',x_2'y_2',x_3'y_3',x_4'y_4') \\ &- 3G_c^{(2)}(x_1y_1;x_1'y_1')G_c^{(2)}(x_2y_2,x_2'y_2')\Gamma^{(3)}(x_1'y_1',x_2'y_2',xy) \\ &\times G_c^{(2)}(xy,x'y')\Gamma^{(3)}(x'y',x_3'y_3',x_4'y_4')G_c^{(2)}(x_3'y_3',x_3y_3)G_c^{(2)}(x_4'y_4',x_4y_4) .\end{aligned} \tag{8.23}$$

The structure of these relations suggests that $\Gamma^{(n)}$ can be treated as bilocal proper vertices: they are amputated bilocal Green's functions, and one cannot decompose $\Gamma^{(n)}$ into two disconnected pieces by removing one $G_c^{(2)}$ factor.

8.3. THE LOOP EXPANSION FOR $\Gamma(\varphi_c, G)$

In this section, following Cornwall, Jackiw and Tomboulis [6], we will derive the loop expansion for the effective action $\Gamma(\varphi_c, G)$. We will show that the required series is

$$\Gamma(\varphi_c, G) = S(\varphi_c) + \frac{i}{2}\mathrm{TrLn}G^{-1} + \frac{i}{2}\mathrm{Tr}\left[\mathfrak{D}^{-1}(\varphi_c)G\right] + \Gamma_2(\varphi_c, G) + C\ , \tag{8.24}$$

where $S(\varphi_c)$ is the classical action and

$$\mathfrak{D}^{-1}(\varphi_c) = -\frac{\delta^2 S(\varphi_c)}{\delta\varphi_c(x)\delta\varphi_c(y)} = D^{-1}(x-y) - \frac{\delta^2 S_{\mathrm{int}}(\varphi_c)}{\delta\varphi_c(x)\delta\varphi_c(y)}\ , \tag{8.25}$$

and the trace, the logarithm and the product $\mathfrak{D}^{-1}G$ are taken in the functional sense; C is a constant. The term $\Gamma_2(\varphi_c, G)$ is computed as follows. In the classical action $S(\varphi)$ shift the field φ by $\varphi_c(x)$. An interaction part $S_{\mathrm{int}}(\varphi, \varphi_c)$ in the new action $S(\varphi + \varphi_c)$ is formed by terms cubic and higher in φ. $\Gamma_2(\varphi_c, G)$ is given by all the two-loop and higher two-particle irreducible vacuum graphs in a theory with vertices determined by $S_{\mathrm{int}}(\varphi, \varphi_c)$ and propagators set equal to $G(x, y)$ (see Fig.8.1).

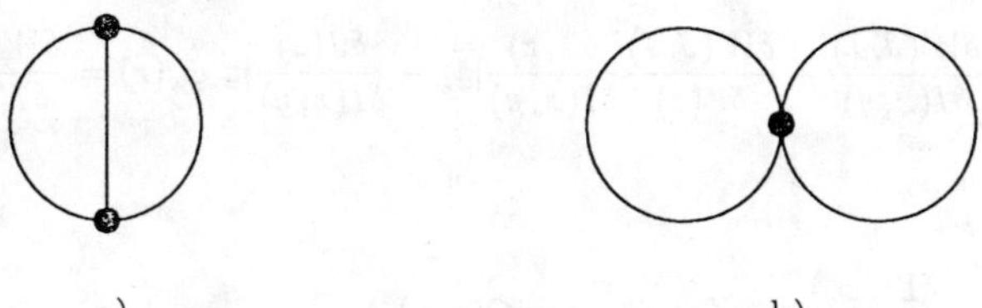

Fig. 8.1. Examples of two-particle irreducible graphs: a) ϕ^3 theory; b) ϕ^4 theory.

The crucial point for the derivation of this relation will be the following observation: the functional derivatives of the action $\Gamma(\varphi_c, G)$ in $\varphi_c(x)$ generate two-particle irreducible Green's functions expressed through diagrams with the propagator G. Indeed, if a diagram is not two-particle irreducible, it can be decomposed into two disconnected pieces when two particle lines are cut (see Fig. 8.2). This in turn implies that there is a propagator correction in this diagram. However, since G is the full propagator, such a diagram does not contribute to $\Gamma(\varphi_c, G)$.

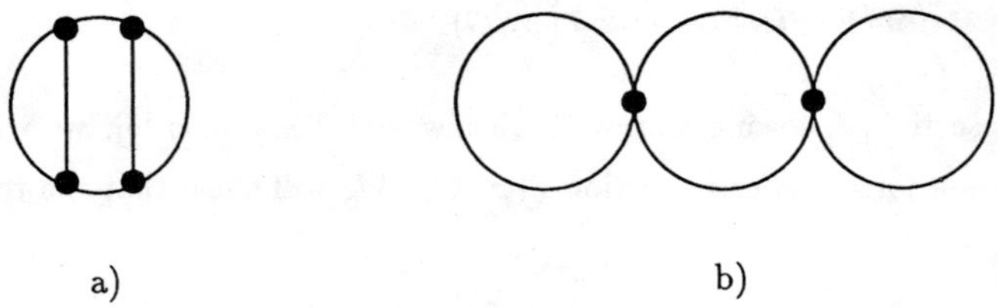

Fig. 8.2. Examples of two-particle reducible graphs: a) ϕ^3 theory; b) ϕ^4 theory.

This observation in particular implies that $\Gamma(0, G)$ is the sum of all two-particle irreducible vacuum graphs of the theory.

We begin the derivation with the introduction of the following Legendre transform of $W(J, I)$ (at fixed I):

$$\Gamma^I(\varphi_c) = W(J, I) - \varphi(x)J(x) \tag{8.26}$$

(we recall that integration over repeated variables is understood).

Since

$$\frac{\delta\Gamma^I(\varphi_c)}{\delta I(x,y)} = \frac{\delta W(J,I)}{\delta I(x,y)} + \frac{\delta W(J,I)}{\delta J(z)}\frac{\delta J(z)}{\delta I(x,y)}|_{\varphi_c} - \frac{\delta J(z)}{\delta I(x,y)}|_{\varphi_c}\varphi_c(z) = \frac{\delta W(J,I)}{\delta I(x,y)}\ , \tag{8.27}$$

we find that

$$\frac{\delta\Gamma^I(\varphi_c)}{\delta I(x,y)} = \varphi_c(x)\varphi_c(y) - iG(x,y)\ , \tag{8.28}$$

$$\Gamma(\varphi_c, G) = \Gamma^I(\varphi_c) - \varphi_c(x)\varphi_c(y)I(x,y) + iG(x,y)I(x,y) \tag{8.29}$$

(compare with Eqs.(8.14), (8.15)).

According to Eqs.(8.3) and (8.26), $\Gamma^I(\varphi_c)$ is the effective action (introduced in Chapter 7) for a theory with the classical action

$$S^I(\varphi) = S(\varphi) + \varphi(x)\varphi(y)I(x,y)\ . \tag{8.30}$$

Hence according to the analysis in Sec.7.4, the functional Γ^I is

$$\Gamma^I(\varphi_c) = S^I(\varphi_c) + W_1^I(\varphi_c)\ , \tag{8.31}$$

where $W_1^I(\varphi_c)$ satisfies the functional relation (see Eq.(7.80)):

$$\begin{aligned}W_1^I(\varphi_c) =& -i\ln\int \mathcal{D}(\varphi)\exp\left\{i\left[S^I(\varphi+\varphi_c)-S^I(\varphi_c)-\varphi(x)\frac{\delta S^I(\varphi_c)}{\delta\varphi_c(x)}\right.\right.\\ &\left.\left.-\varphi(x)\frac{\delta W_1^I(\varphi_c)}{\delta\varphi_c(x)}\right]\right\}\\ =& -i\ln\int \mathcal{D}(\varphi)\exp\left\{i\left[\frac{-1}{2}\varphi(x)\left(\mathfrak{D}^{-1}\{\varphi_c;x,y\}-2I(x,y)\right)\varphi(y)\right.\right.\\ &\left.\left.+S_{\text{int}}(\varphi_c,\varphi)-\varphi(x)\frac{\delta W_1^I(\varphi_c)}{\delta\varphi_c(x)}\right]\right\}.\end{aligned} \tag{8.32}$$

We will now treat Eq.(8.24) as a definition of $\Gamma_2(\varphi, G)$. Then equating (8.24) with (8.29) and using Eqs.(8.30), (8.31), we find

$$\Gamma_2(\varphi_c, G) + \text{const} = \frac{-i}{2}\text{Tr}\left[\mathfrak{D}^{-1}(\varphi_c) - 2I^T\right]G - \frac{i}{2}\text{TrLn}G^{-1} + W_1^I(\varphi_c)\,. \tag{8.33}$$

The relation (8.24) will be proved if it can be shown that $\Gamma_2(\varphi_c, G)$, as given by Eq.(8.33), is the sum of two and higher loop two-particle irreducible vacuum graphs in the theory with vertices determined by $S_{\text{int}}(\varphi, \varphi_c)$ and propagators set equal to $G(x, y)$.

We first express I in Eq.(8.33) through φ_c and G by using Eqs.(8.17) and (8.24):

$$I^T = -\frac{1}{2}G^{-1} + \frac{1}{2}\mathfrak{D}^{-1} - i\frac{\delta\Gamma_2}{\delta G}\,. \tag{8.34}$$

(all quantities here have the operator meaning). Then, taking into account Eq.(8.32), we find that the relation (8.33) is equivalent to the following one:

$$\begin{aligned}\Gamma_2(\varphi_c, G) + \text{const} =& \text{Tr}G\frac{\delta\Gamma_2}{\delta G} - i\ln\int \mathcal{D}(\varphi)\exp\left\{i\left[\frac{1}{2}\varphi(-G)^{-1}\varphi + S_{\text{int}}(\varphi_c,\varphi)\right.\right.\\ &\left.\left.-\varphi\frac{\delta W_1^I(\varphi_c)}{\delta\varphi_c} - i\varphi\frac{\delta\Gamma_2(\varphi_c,G)}{\delta G}\varphi\right]\right\} + i\ln\int \mathcal{D}(\varphi)\exp\left\{i\left[\frac{1}{2}\varphi(-G)^{-1}\varphi\right]\right\}.\end{aligned} \tag{8.35}$$

Let us recall that $\Gamma(0, G)$ is the sum of all two-particle irreducible vacuum graphs. According to Eqs.(8.15) and (8.17), $\Gamma(0, G)$ is

$$\begin{aligned}\Gamma(0,G) + \text{const} =& \text{Tr}G\frac{\delta\Gamma(0,G)}{\delta G} - i\ln\int \mathcal{D}(\varphi)\cdot\exp\left\{i\left[-\frac{1}{2}\varphi D^{-1}\varphi\right.\right.\\ &\left.\left.+S_{\text{int}}(\varphi) + \varphi J_0 - i\varphi\frac{\delta\Gamma(0,G)}{\delta G}\varphi\right]\right\},\end{aligned} \tag{8.36}$$

where J_0 is that value of J which makes $\frac{\delta W}{\delta J} = \varphi_c$ vanish, that is, all tadpole graphs are removed.

Let us compare now Eqs.(8.35) and (8.36). First, using Eq.(8.34), we rewrite Eq.(8.32) in the form:

$$W_1^I(\varphi_c) = -i \ln \int \mathcal{D}(\varphi) \exp\{i[\frac{1}{2}\varphi(-G)^{-1}\varphi + S_{\rm int}(\varphi_c, \varphi) - i\varphi \frac{\delta \Gamma_2(\varphi_c, G)}{\delta G}\varphi - \varphi \frac{\delta W_1^I(\varphi_c)}{\delta \varphi_c}]\} .$$

As it follows from the consideration in Sec.7.4, $\frac{\delta W_1^I(\varphi_c)}{\delta \varphi_c}$ satisfying this equation is precisely that value of the external current which makes tadpole vanish in the theory with the action $\frac{1}{2}\varphi(-G)^{-1}\varphi + S_{\rm int}(\varphi_c, \varphi) - i\varphi \frac{\delta \Gamma_2(\varphi_c G)}{\delta G}\varphi$. Therefore, comparing the form of Eq.(8.35) with the form of Eq.(8.36), we conclude that

$$\Gamma_2(\varphi_c, G) - i \ln \int \mathcal{D}(\varphi) \exp\left\{ i \left[\frac{1}{2}\varphi(-G)^{-1}\varphi \right] \right\} = \Gamma_2(\varphi_c, G) + \frac{i}{2} \mathrm{Tr}\mathrm{Ln} G^{-1} + \mathrm{const} \quad (8.37)$$

is the sum of all two-particle irreducible vacuum graphs in the theory governed by the action $\frac{1}{2}\varphi(-G)^{-1}\varphi + S_{\rm int}(\varphi_c, \varphi)$. The term $\frac{i}{2}\mathrm{TnLn}G^{-1}$ includes the contribution of the one-loop vacuum diagrams. Therefore $\Gamma_2(\varphi_c, G)$ is indeed the sum of two and higher loop two-particle irreducible vacuum diagrams.

This completes the proof of the relation (8.24).

Let us now introduce the notion of the effective potential for composite operators.

If, as it happens frequently, one is interested only in translation-invariant solutions with $\varphi_c(x)$ being constant, and $G(x, y)$ being a function only of $(x - y)$, the effective action takes the form:

$$\Gamma(\varphi_c, G) = -V(\varphi_c, G) \int d^4x \, , \quad (8.38)$$

where V is called the effective potential for composite operators.

The loop expansion series for $V(\varphi_c, G)$ follows from the relation (8.24):

$$\begin{aligned} V(\varphi_c, G) = U(\varphi_c) &- \frac{i}{2} \int \frac{d^4p}{(2\pi)^4} \, \mathrm{tr} \ln(G^{-1}(p)) \\ &- \frac{i}{2} \int \frac{d^4p}{(2\pi)^4} \, \mathrm{tr} \left[\mathfrak{D}^{-1}\{\varphi_c; p\} G(p) \right] + V_2(\varphi_c, G) + \mathrm{const} \, , \end{aligned} \quad (8.39)$$

where $U(\varphi_c)$ is the classical potential,

$$\begin{aligned} G(p) &= \int d^4x e^{ipx} G(x) \, , \\ \mathfrak{D}\{\varphi_c, p\} &= \int d^4x e^{ipx} \mathfrak{D}\{\varphi_c; x\} \, , \end{aligned} \quad (8.40)$$

and $-V_2(\varphi_c, G)$ is the sum of all two and higher loop two-particle irreducible vacuum graphs of the theory with vertices given by $S_{\text{int}}(\varphi_c, \varphi)$ and propagator $G(p)$ (the trace tr is not functional here; it applies to the component degrees of freedom).

The effective potential satisfies the equations.

$$\frac{\partial V(\varphi_c, G)}{\partial \varphi_c} = J + 2\varphi_c \int d^4x I(x) \,, \tag{8.41}$$

$$\frac{\delta V(\varphi_c, G)}{\delta G(p)} = -i \int d^4x e^{ipx} I(x) \,. \tag{8.42}$$

(in this case, the source $I(x, y)$ depends only on $(x - y)$). The stationarity requirements are

$$\frac{\partial V(\varphi_c, G)}{\partial \varphi_c} = 0 \,, \tag{8.43}$$

$$\frac{\delta V(\varphi_c, G)}{\delta G(p)} = 0 \,. \tag{8.44}$$

We will discus the physical meaning of $V(\varphi_c, G)$ in Sec.8.7.

In conclusion, let us indicate the expressions for $\Gamma(\varphi_c, G)$ and $V(\varphi_c, G)$ in theories including also fermions and gauge bosons. The contribution of gauge bosons is given by the same expressions if the Lorentz indices μ and ν in their propagators are treated as component degrees of freedom. The case of fermions is slightly more subtle. Because of Fermi statistics, all factors $\frac{1}{2}$ in the expressions (8.24), (8.39) should be replaced by -1 (we consider complex Dirac fermion fields), that is, for example,

$$\begin{aligned} &-\frac{i}{2}\int \frac{d^4p}{(2\pi)^4} \operatorname{tr}\ln(G^{-1}(p)) - \frac{i}{2}\int \frac{d^4p}{(2\pi)^4} \operatorname{tr}\left[\mathfrak{D}^{-1}\{\varphi_c; p\}G(p)\right] \\ &\longrightarrow i\int \frac{d^4p}{(2\pi)^4} \operatorname{tr}\ln(G^{-1}(p)) + i\int \frac{d^4p}{(2\pi)^4} \operatorname{tr}\left[S^{-1}\{\varphi_c; p\}G(p)\right] \end{aligned} \tag{8.45}$$

in the effective potential.

8.4. The Two-Loop Effective Potential in the Nambu-Jona-Lasinio Model and QED

In this section, we calculate the two-loop effective potential $V(G)$ in the Nambu-Jona-Lasinio (NJL) model and QED.

The lagrangian density of the NJL model is given in Eq.(6.1):

$$\mathcal{L} = i\bar{\psi}\gamma^\mu\partial_\mu\psi - \bar{\psi}m^{(0)}\psi + G^{(0)}\sum_{\alpha=0}^{N^2-1}\left\{\left(\bar{\psi}\frac{\lambda^\alpha}{2}\psi\right)^2 + \left(\bar{\psi}\frac{\lambda^\alpha}{2}i\gamma_5\psi\right)^2\right\}, \tag{8.46}$$

where, for simplicity, we consider the same bare mass for all fermion flavors. The two-loop effective action $\Gamma(G)$ is (see Eqs.(8.24) and (8.45)):

$$\Gamma(G) = -i\mathrm{Tr}\left\{\mathrm{Ln}G^{-1} + S^{-1}G\right\} + \Gamma_2(G) + \mathrm{const}\ , \tag{8.47}$$

where $\psi_c = \langle 0|\psi|0\rangle = 0$ and $\Gamma_2(G)$ is connected with the two-loop diagrams shown in Fig. 8.3:

$$\begin{aligned}\Gamma_2(G) = -\,G^{(0)}\sum_{\alpha=0}^{N^2-1}\int d^4x\Bigg\{&\left[\mathrm{tr}\,\frac{\lambda^\alpha}{2}G(x,x)\right]^2 + \left[\mathrm{tr}\,i\gamma_5\frac{\lambda^\alpha}{2}G(x,x)\right]^2\\ &-\mathrm{tr}\left[\frac{\lambda^\alpha}{2}G(x,x)\frac{\lambda^\alpha}{2}G(x,x)\right] - \mathrm{tr}\left[i\gamma_5\frac{\lambda^\alpha}{2}G(x,x)i\gamma_5\frac{\lambda^\alpha}{2}G(x,x)\right]\Bigg\}\,.\end{aligned} \tag{8.48}$$

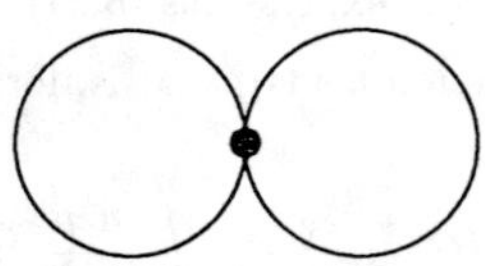

Fig. 8.3. The general type of diagrams contributing to the effective action Γ_2 (8.48) in the Nambu-Jona-Lasinio model

We are interested in studying a translation-invariant propagator

$$G(p) = \frac{1}{\hat{p}A(p^2) - B(p^2)}\,, \tag{8.49}$$

where A and B are some functions of p^2 (the color and flavor indices are omitted here).

The corresponding effective potential is:

$$
\begin{aligned}
V(G) =& i\int \frac{d^4p}{(2\pi)^4}\,\mathrm{tr}\left[\ln(\hat{p}A(p^2)-B(p^2)) + (\hat{p}-m^{(0)})(\hat{p}A(p^2)-B(p^2))^{-1}\right] \\
&-\frac{8G^{(0)}N_c^2N}{(2\pi)^8}\left[\int d^4p\frac{B(p^2)}{p^2A^2(p^2)-B^2(p^2)}\right]^2 + C\,,
\end{aligned}
\tag{8.50}
$$

where N_c is the number of the fermion colors. This expression takes the following form in Euclidean region:

$$
\begin{aligned}
V(G) = -\frac{N_cN}{4\pi^2}\Bigg\{\int_0^{\Lambda^2} dp^2p^2\left[\frac{1}{2}\ln\left(A^2(p^2)+\frac{B^2(p^2)}{p^2}\right)+\frac{p^2A(p^2)+m^{(0)}B(p^2)}{p^2A^2(p)+B^2(p)}-1\right] \\
+\frac{G^{(0)}N_c}{8\pi^2}\left[\int_0^{\Lambda^2} dp^2\frac{p^2B(p^2)}{p^2A^2(p^2)+B^2(p^2)}\right]^2\Bigg\}\,,
\end{aligned}
\tag{8.51}
$$

where the constant C in Eq.(8.50) was fixed from the normalization condition $V(G)|_{G=S=(\hat{p}-m^{(0)})^{-1}} = 0$.

The Schwinger-Dyson equation for the propagator G, that is for the functions $A(p^2)$ and $B(p^2)$, can be derived either from the stationarity condition (8.2) or (8.44):

$$
\frac{\delta\Gamma}{\delta G(y,x)} = iG^{-1}(x,y) - iS^{-1}(x-y) - \delta^4(x-y)\frac{G^{(0)}}{N}\,\mathrm{tr}\,G(x,x) = 0\,. \tag{8.52}
$$

Solutions of this equations are translation invariant functions, $G(x,y) = G(x-y)$. Therefore we get (see Fig.8.4)

$$
G^{-1}(x,y) = S^{-1}(x-y) - \delta^4(x-y)\frac{iG^{(0)}}{N}\,\mathrm{tr}\,G(0)\,. \tag{8.53}
$$

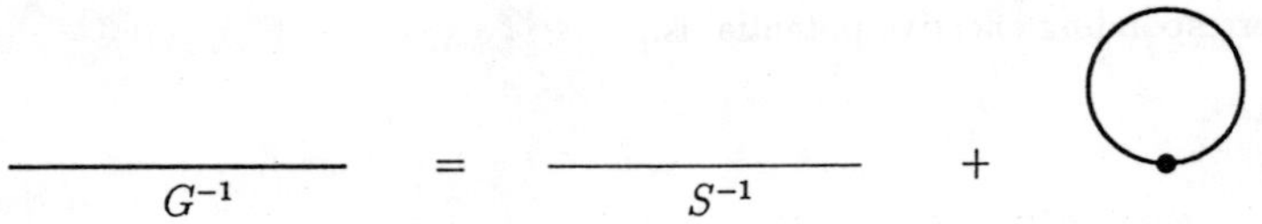

Fig. 8.4. Graphical representation of the Schwinger-Dyson equation in the Nambu-Jona-Lasinio model. A thin solid line corresponds to the free fermion propagator; a bold line corresponds to the full fermion propagator.

From here, we find the following equations for the functions $A(p^2)$ and $B(p^2)$ in Euclidean region:

$$A(p^2) = 1 , \tag{8.54}$$

$$B(q^2) = m^{(0)} + \frac{G^{(0)} N_c}{4\pi^2} \int_0^{\Lambda^2} dp^2 \frac{p^2 B(p^2)}{p^2 + B^2(p^2)} . \tag{8.55}$$

The solution of the equation (8.55) is a constant independent of $q^2, B(q^2) = m$. One can see that at $m^{(0)} = 0$ this equation coincides with equation (6.13) for the dynamical fermion mass obtained in the mean-field approximation.

Let us consider now the effective action in QED. The lagrangian density is:

$$\mathcal{L} = -\frac{1}{4} F_{\mu\nu} F^{\mu\nu} + i\bar{\psi}\gamma^\mu \partial_\mu \psi - m^{(0)} \bar{\psi}\psi - g\bar{\psi}\gamma^\mu \psi A_\mu - \frac{1}{2}\xi \left(\partial_\mu A^\mu\right)^2 . \tag{8.56}$$

The effective action takes the form:

$$\begin{aligned} \Gamma(G, \Delta) = -\, i\mathrm{Tr}\Big[\ln G^{-1} + S^{-1}(\psi_c, A_c) G - \frac{1}{2} \mathrm{Ln} \Delta^{-1} \\ - \frac{1}{2} \mathfrak{D}^{-1}(\psi_c, A_c) \Delta \Big] + \Gamma_2(G, \Delta) + C , \end{aligned} \tag{8.57}$$

where $\Delta_{\mu\nu}$ is the full photon propagator. The free photon propagator $D_{\mu\nu} = \mathfrak{D}_{\mu\nu}(\psi_c, A_c)|_{\substack{A_c=0 \\ \psi_c=0}}$ is

$$D_{\mu\nu}(k) = \frac{1}{k^2} \left(g_{\mu\nu} - \frac{\xi - 1}{\xi} \frac{k_\mu k_\nu}{k^2} \right) . \tag{8.58}$$

Let us consider the approximation in which $\Delta_{\mu\nu} = D_{\mu\nu}$, the source $J = 0$, and there is no source $I(x,y)$ for the photon field in the generating functional. Therefore in this case, Γ_2 includes the contribution of diagrams which are two-particle irreducible with respect to fermion lines only.

In the two-loop approximation, the Γ_2-contribution is connected with the graphs shown in Fig. 8.5. It is

$$\begin{aligned}\Gamma_2 =& \frac{g^2}{2}\int d^4x d^4y \,\mathrm{tr}\,[G(x,y)\gamma^\mu G(y,x)\gamma^\nu D_{\mu\nu}(x-y)] \\ &-\frac{g^2}{2}\int d^4x d^4y \,\mathrm{tr}(\gamma^\mu G(x,x)) D_{\mu\nu}(x-y)\,\mathrm{tr}(\gamma^\nu G(y,y))\,.\end{aligned} \tag{8.59}$$

The first term and the second one on the right hand side of this equation are connected with the graphs in Fig.8.5a and in Fig.8.5b, respectively.

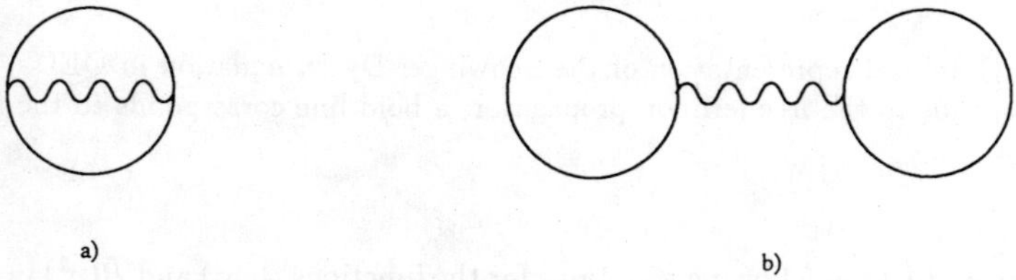

Fig. 8.5. Two-loop diagrams contributing to Γ_2 (8.59) in QED. Solid lines correspond to fermions; wavy lines correspond to photons.

In this approximation, the effective potential takes the following form in Euclidean region:

$$\begin{aligned}V(G) =& -\frac{N}{4\pi^2}\left\{\int_0^{\Lambda^2} dp^2 p^2\left[\frac{1}{2}\ln\left(A^2(p^2)+\frac{B^2(p^2)}{p^2}\right)+\frac{p^2A(p^2)+m^{(0)}B(p^2)}{p^2A^2(p^2)+B^2(p^2)}-1\right]\right. \\ &+\frac{g^2}{32\pi^2}\int_0^{\Lambda^2} dp^2\int_0^{\Lambda^2} dk^2\frac{p^2k^2}{(p^2A^2(p^2)+B^2(p^2))(k^2A^2(k^2)+B^2(k^2))} \\ &\cdot\left[\left(\frac{k^2}{p^2}\theta(p^2-k^2)+\frac{p^2}{k^2}\theta(k^2-p^2)\right)\xi^{-1}A(p^2)A(k^2)\right. \\ &\left.\left.+\left(\frac{1}{p^2}\theta(p^2-k^2)+\frac{1}{k^2}\theta(k^2-p^2)\right)(3+\xi^{-1})B(p^2)B(k^2)\right]\right\}\,.\end{aligned} \tag{8.60}$$

In the derivation of this expression we used some formulas in Sec. B.4 in Appendix B to integrate the angular variables. We note that the second term on the right-hand side of Eq.(8.59) does not give contribution both to $V(G)$ and to the Schwinger-Dyson equation for G. The Schwinger-Dyson equation takes the following form in this approximation (see Fig.8.6):

$$G^{-1}(x-y) = S^{-1}(x-y) + g^2 i\gamma^{\mu} G(x-y)\gamma^{\nu} D_{\mu\nu}(x-y) . \tag{8.61}$$

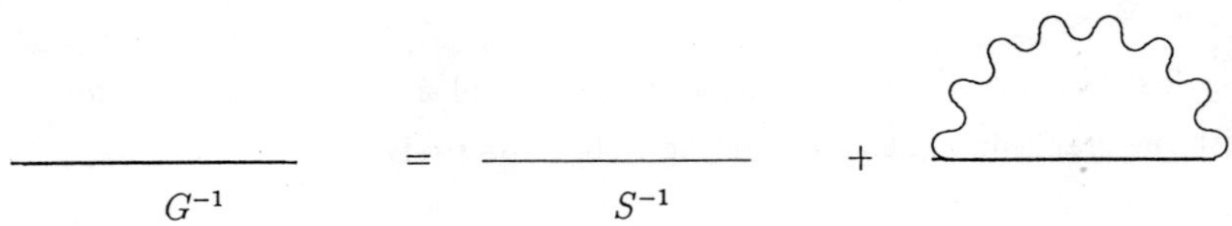

Fig. 8.6. Graphical representation of the Schwinger-Dyson equation in QED. A thin solid line corresponds to the free fermion propagator; a bold line corresponds to the full fermion propagator.

From here we find the following equations for the functions $A(p^2)$ and $B(p^2)$ in Euclidean region:

$$\begin{aligned} A(p^2) =& 1 + \xi^{-1}\frac{g^2}{16\pi^2 (p^2)^2}\int_0^{p^2} dk^2 \frac{(k^2)^2 A(k^2)}{k^2 A^2(k^2) + B^2(k^2)} \\ &+ \xi^{-1}\frac{g^2}{16\pi^2}\int_{p^2}^{\Lambda^2} dk^2 \frac{A(k^2)}{k^2 A^2(k^2) + B^2(k^2)} . \end{aligned} \tag{8.62}$$

$$\begin{aligned} B(p^2) =& m^{(0)} + \frac{g^2(3+\xi^{-1})}{16\pi^2 p^2}\int_0^{p^2} dk^2 \frac{k^2 B(k^2)}{k^2 A^2(k^2) + B^2(k^2)} \\ &+ \frac{g^2(3+\xi^{-1})}{16\pi^2}\int_{p^2}^{\Lambda^2} dk^2 \frac{B(k^2)}{k^2 A^2(k^2) + B^2(k^2)} . \end{aligned} \tag{8.63}$$

One can see that these equations drastically simplify in the Landau gauge with $\xi \longrightarrow \infty$. In this gauge, the function $A(p^2)$ is $A(p^2) = 1$, and the equation for $B(p^2)$ takes the form:

$$B(p^2) = m^{(0)} + \frac{3g^2}{16\pi^2 p^2} \int_0^{p^2} dk^2 \frac{k^2 B(k^2)}{k^2 + B^2(k^2)} + \frac{3g^2}{16\pi^2} \int_{p^2}^{\Lambda^2} dk^2 \frac{B(k^2)}{k^2 + B^2(k^2)} \,. \tag{8.64}$$

We will analyze these equations in Chapter 10.

8.5. The Schwinger-Dyson Equation and the Bethe-Salpeter Equation. General Consideration

The Schwinger-Dyson (SD) equation for the propagator G coincides with the stationarity equation (8.2):

$$\frac{\delta\Gamma}{\delta G(x,y)} = 0 \,. \tag{8.65}$$

Taking into account the expression (8.24) for Γ, we find the following equation

$$G^{-1} = \mathfrak{D}^{-1}(\varphi_c) - \Sigma(\varphi_c, G) \,, \tag{8.66}$$

where

$$\Sigma(\varphi_c, G) = 2i\frac{\delta\Gamma_2}{\delta G} \tag{8.67}$$

is called the self-energy operator.

Let us now derive the Bethe-Salpeter (BS) equation describing two-particle bound states.

We begin by considering the Green's function $G_c^{(2)}(J; x_1y_1, x_2y_2)$ which is (see Eq.(8.10))

$$\begin{aligned} G_c^{(2)}(J; x_1y_1, x_2y_2) =& \frac{1}{Z(J)} G^{(2)}(J; x_1y_1, x_2y_2) - \frac{1}{Z^2(J)} \\ &\times G^{(1)}(J; x_1y_1) G^{(1)}(J; x_2y_2) \end{aligned} \tag{8.68}$$

with $G^{(n)}$ defined in Eq.(8.6). The BS equation for $G_c^{(2)}$ takes the form (see Eqs.(8.18) and (8.21)):

$$\begin{aligned}&\frac{1}{i}\Gamma^{(2)}(J;x_1y_1,xy)G_c^{(2)}(J;xy,x_2y_2)\\ &=\frac{1}{2}\Big[\delta_{x_1x_2}\delta_{y_1y_2}\delta^4(x_1-x_2)\delta^4(y_1-y_2)\\ &\qquad+\delta_{x_1y_2}\delta_{y_1x_2}\delta^4(x_1-y_2)\delta^4(y_1-x_2)\Big] ,\end{aligned} \tag{8.69}$$

where the BS operator $\frac{1}{i}\Gamma^{(2)}(J;x_1y_1,xy)$ is

$$\begin{aligned}\frac{1}{i}\Gamma^{(2)}(J;x_1y_1,x_2y_2) =& i\frac{\delta^2\Gamma}{\delta G(x_1,y_1)\delta G(x_2,y_2)}\\ =& -\frac{1}{4}(G^{-1}(y_2,x_1)G^{-1}(y_1,x_2)+G^{-1}(y_2,y_1)G^{-1}(x_1,x_2))\\ &+i\frac{\delta^2\Gamma_2}{\delta G(x_1,y_1)\delta G(x_2,y_2)} .\end{aligned} \tag{8.70}$$

The equation (8.69) can be rewritten in a more conventional form:

$$\begin{aligned}G_c^{(2)}(J;x_1y_1,x_2y_2) =& -[G(x_1,y_2)G(x_2,y_1)+G(x_1,x_2)G(y_2,y_1)]\\ &+i[G(x_1,y)G(x,y_1)+G(x_1,x)G(y,y_1)]\frac{\delta^2\Gamma_2}{\delta G(x,y)\delta G(x',y')}\\ &\cdot G_c^{(2)}(J;x'y',x_2y_2) ,\end{aligned} \tag{8.71}$$

or

$$\begin{aligned}G_c^{(2)}(J;x_1y_1,x_2y_2) =& G_0^{(2)}(J;x_1y_1,x_2y_2)+G_0^{(2)}(J;x_1y_1,xy)\\ &\cdot K(J;xy,x'y')G_c^{(2)}(J;x'y',x_2y_2) ,\end{aligned} \tag{8.72}$$

where

$$G_0^{(2)}(J;x_1y_1,x_2y_2) = -[G(x_1,y_2)G(x_2,y_1)+G(x_1,x_2)G(y_2,y_1)] , \tag{8.73}$$

$$K(J;x_1y_1,x_2y_2) = -i\frac{\delta^2\Gamma_2}{\delta G(x_1,y_1)\delta G(x_2,y_2)} . \tag{8.74}$$

Let us now consider the BS equation in a theory without sources ($J(x)=I(x,y)=0$). We will also consider the case in which the fields $\varphi(x_1),\varphi(x_2)$ and $\varphi(y_1),\varphi(y_2)$ in Green's function $G_c^{(2)}(x_1y_1,x_2y_2)$ describe particles with a mass m_x and a mass m_y, respectively. For the sake of simplicity, we will consider the case with $G(x_1, y_2)G(x_2, y_1) = 0$ when (see Eq.(8.73))

$$G_0^{(2)}(J; x_1y_1, x_2y_2) = -G(x_1, x_2)G(y_2, y_1) \tag{8.75}$$

(it may happen either because of $m_x \neq m_y$ or because of selection rules following from some symmetry).

Because of the translational invariance of the theory, $G_c^{(2)}(x_1y_1, x_2y_2)$ and $K(x_1y_1, x_2y_2)$ are functions of three variables only:

$$\begin{aligned} z_1 =& x_1 - y_1 \ , \\ z_2 =& y_2 - x_2 \ , \\ z_3 =& \eta_x(x_1 - x_2) + \eta_y(y_1 - y_2) \end{aligned} \tag{8.76}$$

with $\eta_x = \frac{m_x}{m_x+m_y}, \eta_y = \frac{m_y}{m_x+m_y}$. Introducing the Fourier transforms,

$$\begin{aligned} G^{(2)}_{c;n_1m_1,n_2m_2}(x_1y_1, x_2y_2) =& \frac{1}{(2\pi)^{12}} \int dpdqdP \\ & \cdot e^{-i(pz_1+qz_2+Pz_3)} G^{(2)}_{c;n_1m_1,n_2m_2}(p, q; P) \ , \end{aligned} \tag{8.77}$$

$$\begin{aligned} K_{n_1m_1,n_2m_2}(x_1y_1, x_2y_2) =& \frac{1}{(2\pi)^{12}} \int dpdqdPe^{-i(pz_1+qz_2+Pz_3)} \ . \\ & \cdot K_{n_1m_1,n_2m_2}(p, q; P) \ , \end{aligned} \tag{8.78}$$

where we now explicitly indicate the component indices n_i, m_i, the BS equation can be written in the following form:

$$\begin{aligned} G^{(2)}_{c;n_1m_1,n_2m_2}(p, q; P) =& G^{(2)}_{0;n_1m_1,n_2m_2}(p, q; P) + \frac{1}{(2\pi)^8} \int d^4p'd^4q' G^{(2)}_{0;n_1m_1,nm}(p, p'; P) \\ & K_{nm,n'm'}(p', q'; P) G^{(2)}_{c;n'm',n_2m_2}(q', q; P) , \end{aligned} \tag{8.79}$$

where

$$G^{(2)}_{0;n_1m_1,n_2m_2}(p, q; P) = -(2\pi)^4\delta^4(p - q)G_{n_1n_2}(p + \eta_xP)G_{m_2m_1}(p - \eta_yP) \ . \tag{8.80}$$

The graphical illustration of Eq.(8.79) is given in Fig.8.7.

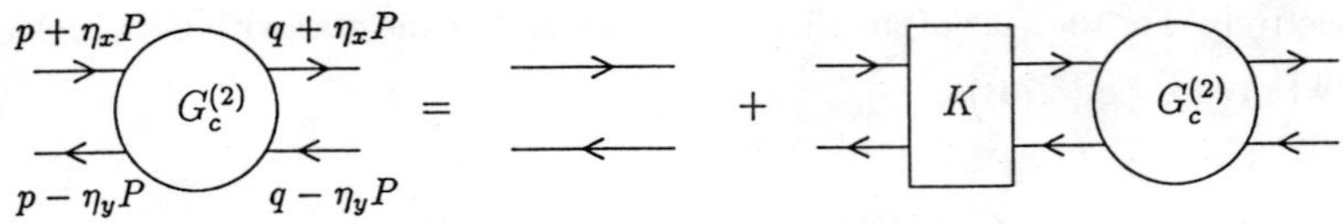

Fig. 8.7.

Let us consider now the homogeneous BS equation. Let $|P_b,1\rangle, |P_b,2\rangle, \cdots |P_b,k\rangle$ be a degenerate set of bound states with a momentum P_b. The BS wave function for $|P_b,r\rangle$ and its conjugate are defined as

$$\chi_{r;nm}(x,y;P_b) = \langle 0|T\varphi_n(x)\varphi_m(y)|P_b,r\rangle\ , \tag{8.81}$$

$$\bar{\chi}_{r;mn}(y,x;P_b) = \langle P_b,r|T\varphi_n(x)\varphi_m(y)|0\rangle = \left(\langle 0|\bar{T}\varphi_n(x)\varphi_m(y)|P_b,r\rangle\right)^*\ , \tag{8.82}$$

where $\bar{T}$ orders the operators in an antichronological order (we recall that $\varphi_n(x)$ are real fields), and the states $|P_b,r\rangle$ are normalized as

$$\langle P_b',r'|P_b,r\rangle = (2\pi)^3 2P_{b0}\delta^3\left(\mathbf{P}_b' - \mathbf{P}_b\right)\ . \tag{8.83}$$

Let us show that because of the translation invariance, the BS wave functions can be written in the form:

$$\begin{aligned}\chi_{r;nm}(x,y;P_b) &= e^{-iP_bX}\chi_{r;nm}(x-y;P_b)\\ \bar{\chi}_{r;nm}(x,y;P_b) &= e^{iP_bX}\bar{\chi}_{r;nm}(x-y;P_b)\ ,\end{aligned} \tag{8.84}$$

where $X = \eta_x x + \eta_y y = \frac{m_x x + m_y y}{m_x + m_y}$ is the center mass coordinate. Indeed, since under a translation through a

$$U(a)|P_b,r\rangle = e^{iP_ba}|P_b,r\rangle\ , \tag{8.85}$$

we have

$$\begin{aligned}\chi_{r;nm}(x,y;P_b) &= \langle 0|U^{-1}(a)U(a)T\varphi_n(x)\varphi_m(y)U^{-1}(a)U(a)|P_b,r\rangle\\ &= \langle 0|T\varphi_n(x+a)\varphi_m(y+a)|P_b,r\rangle e^{iP_ba}\\ &= e^{iP_ba}\cdot\chi_{r;nm}(x+a,y+a)\ .\end{aligned} \tag{8.86}$$

Taking in particular $a = -X = -(\eta_x x + \eta_y y)$, we are led to the form (8.84).

Using the standard formalism (see for example Refs.[13,14]), one can show that Green's function $G_c^{(2)}(p,q;P)$ has a pole term at $P^2 = P_b^2$:

$$\left[G^{(2)}_{c;n_1m_1,n_2m_2}(p,q;P)\right]_{\text{pole}} = i\,\frac{\sum\limits_{r=1}^{k}\chi_{r;n_1m_1}(p,P_b)\bar{\chi}_{r;n_2m_2}(q,P_b)}{P^2 - P_b^2 + i\epsilon}\,, \tag{8.87}$$

where $\chi_{r;nm}(p,P_b)$ ($\bar{\chi}_{r;nm}(p,P_b)$) is the Fourier transform of $\chi_{r;nm}(z;P_b)$ ($\bar{\chi}_{r;nm}(z;P_b)$):

$$\begin{aligned} \chi_{r;nm}(p,P_b) &= \int d^4z e^{ipz}\chi_{r;nm}(z;P_b)\,, \\ \bar{\chi}_{r;nm}(p,P_b) &= \int d^4z e^{ipz}\bar{\chi}_{r;nm}(z;P_b)\,. \end{aligned} \tag{8.88}$$

Taking into account the linear independence of $\chi_1,\chi_2,\cdots,\chi_k$, we find the homogeneous BS equation from Eqs.(8.79), (8.87):

$$\begin{aligned} &-G^{-1}_{nn_1}(p+\eta_xP_b)\chi_{r;n_1m_1}(p,P_b)G^{-1}_{m_1m}(p-\eta_yP_b) \\ &= \frac{1}{(2\pi)^4}\int d^4qK_{nm,n_2m_2}(p,q;P_b)\chi_{r;n_2m_2}(q,P_b)\,. \end{aligned} \tag{8.89}$$

The graphical illustration of this equation is given in Fig.8.8.

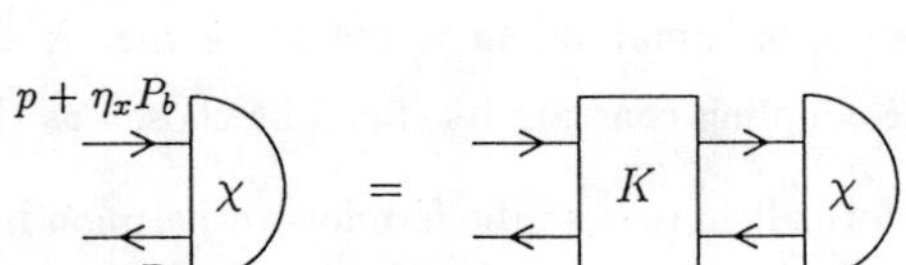

Fig. 8.8.

Sometimes it is convenient to rewrite the BS equations (8.79) and (8.89) in the following operator form:

$$G_c^{(2)} = G_0^{(2)} + G_0^{(2)}KG_c^{(2)} \tag{8.90}$$

or

$$G_c^{(2)} = \left(G_0^{(2)^{-1}} - K\right)^{-1}\,, \tag{8.91}$$

and

$$G_0^{(2)^{-1}} \chi = K\chi \,. \tag{8.92}$$

Let us derive the normalization condition for the BS wave function from this equations [13, 14].

We begin with the introduction of the auxiliary parameter η in the kernel K:

$$G_c^{(2)}(\eta) = \left(G_0^{(2)^{-1}} - \eta K \right)^{-1} \,, \tag{8.93}$$

$$G_0^{(2)^{-1}} \chi = \eta K \chi \,. \tag{8.94}$$

Differentiating both sides of Eq.(8.93) in η, we find

$$\frac{\partial G_c^{(2)}}{\partial \eta} = G_c^{(2)} K G_c^{(2)} \,. \tag{8.95}$$

By comparing the residues of the double pole at $P^2 = P_b^2$ in both sides of this equation and taking into account Eq.(8.94), we obtain the normalization condition:

$$\eta \frac{\partial P_b^2}{\partial \eta} = i\bar{\chi} G_0^{(2)^{-1}} \chi \,. \tag{8.96}$$

We note that in those approximations in which the kernel K linearly depends on a coupling constant, the coupling constant itself can be chosen as the parameter η.

Let us adapt this formalism to describe fermion-antifermion bound states which are very important for describing dynamical chiral symmetry breaking.

In this case, Green's function $G_c^{(2)}(x_1y_1, y_2x_2)$ is (see Eq.(8.8)):

$$G_c^{(2)}(x_1y_1, y_2x_2) = \frac{1}{Z(0)} G^{(2)}(x_1y_1, y_2x_2) - \frac{1}{Z^2(0)} G^{(1)}(x_1y_1) G^{(1)}(y_2x_2) \,, \tag{8.97}$$

where

$$G^{(2)}(x_1y_1, y_2x_2) = \langle 0|T\psi(x_1)\bar{\psi}(y_1)\psi(y_2)\bar{\psi}(x_2)|0\rangle \tag{8.98}$$

(such a notation for the arguments of $G^{(2)}$ ($G^{(2)}(x_1y_1, y_2x_2)$ and not $G^{(2)}(x_1y_1, x_2y_2)$) will be convenient for the description of bound states composed from a fermion and an antifermion with different masses, m_x and m_y).

The BS equation for $G_c^{(2)}$ is (compare with Eq.(8.79)):

$$G^{(2)}_{c;n_1m_1,m_2n_2}(p,q;P) = G^{(2)}_{0;n_1m_1,m_2n_2}(p,q;P) + \frac{1}{(2\pi)^8}\int d^4p'd^4q'G^{(2)}_{0;n_1m_1,mn}(p,p';P)$$
$$K_{mn,n'm'}(p',q';P)G^{(2)}_{c;n'm',m_2n_2}(q',q;P)\ , \tag{8.99}$$

where

$$G^{(2)}_{0;n_1m_1,m_2n_2}(p,q;P) = (2\pi)^4\delta^4(p-q)G_{n_1n_2}(p+\eta_xP)G_{m_2m_1}(p-\eta_yP)\ , \tag{8.100}$$

and the BS kernel $K_{m_1n_1,n_2m_2}(p,q;P)$ is the Fourier transform (8.78) of the function

$$K_{m_1n_1,n_2m_2}(y_1x_1,x_2y_2) = -i\frac{\delta^2\Gamma_2}{\delta G_{m_1n_1}(y_1,x_1)\delta G_{n_2m_2}(x_2,y_2)}\ . \tag{8.101}$$

The BS wave functions χ_r and $\bar{\chi}_r$ are

$$\begin{aligned}\chi_{r;nm}(x,y;P_b) =& \langle 0|T\psi_n(x)\bar{\psi}_m(y)|P_b,r\rangle\ ,\\ \bar{\chi}_{r;nm}(x,y;P_b) =& \langle P_b,r|T\psi_m(y)\bar{\psi}_n(x)|0\rangle\\ =& \gamma_0\left(\langle 0|\bar{T}\psi_n(x)\bar{\psi}_m(y)|P_b,r\rangle\right)^*\gamma_0\ .\end{aligned} \tag{8.102}$$

They can be written as

$$\begin{aligned}\chi_{r;nm}(x,y;P_b) = e^{-iP_bX}\chi_{r;nm}(x-y;P_b) =& e^{-iP_bX}\frac{1}{(2\pi)^4}\int d^4p\exp[-ip(x-y)]\\ & \chi_{r;nm}(p;P_b)\ ,\\ \bar{\chi}_{r;nm}(x,y;P_b) =& e^{iP_bX}\bar{\chi}_{r;nm}(x-y;P_b)\\ =& e^{iP_bX}\frac{1}{(2\pi)^4}\int d^4p\exp[ip(x-y)]\\ & \bar{\chi}_{r;nm}(p;P_b)\ ,\end{aligned} \tag{8.103}$$

where $X = \frac{m_xx+m_yy}{m_x+m_y}$. The pole term at $P^2 = P_b^2$ in $G_c^{(2)}$ has the form:

$$\left[G^{(2)}_{c;n_1m_1,m_2n_2}(p,q;P)\right]_{\text{pole}} = i\frac{\sum_{r=1}^{k}\chi_{r;n_1m_1}(p,P_b)\bar{\chi}_{r;n_2m_2}(q,P_b)}{P^2-P_b^2+i\epsilon}\ . \tag{8.104}$$

The homogeneous BS equation is:

$$\begin{aligned} G^{-1}_{nn_1}(p+\eta_x P_b)\chi_{r;n_1m_1}(p,P_b)G^{-1}_{m_1m}(p-\eta_y P_b) \\ =\frac{1}{(2\pi)^4}\int d^4q K_{mn,n_2m_2}(p,q;P_b)\chi_{r;n_2m_2}(q,P_b)\ . \end{aligned} \tag{8.105}$$

The normalization condition for the BS wave function coincides with that in Eq.(8.96). We would like to indicate (without a derivation) another useful form of the normalization condition for the practically important case when the BS kernel is independent of the momentum P_b [13, 14]:

$$\begin{aligned} itr\int \bar{\chi}(q,P_b)[(\frac{\partial}{\partial P_b^\mu}G^{-1}(q+\frac{1}{2}P_b))\chi(q,P_b)G^{-1}(q-\frac{1}{2}P_b) \\ +G^{-1}(q+\frac{1}{2}P_b)\chi(q,P_b)\frac{\partial}{\partial P_b^\mu}G^{-1}(q-\frac{1}{2}P_b)]\frac{d^4q}{(2\pi)^4}=-2P_b^\mu\ . \end{aligned} \tag{8.105a}$$

In conclusion of this section, we would like to make the following remarks.

The definition of the BS wave function (see Eqs.(8.81), (8.102)) implies that it plays the role of a form factor for a certain bound state $|P_b,r\rangle$. As it was shown by Mandelstam and Nishijima, the S-matrix elements of bound states can be expressed through their BS wave functions [14, 15]:

$$\begin{aligned} (S-1)_{r_1r_2\cdots r_k}=(i)^{1-k}\chi_{r_1;n_1m_1}(x_1y_1)\chi_{r_2;n_2m_2}(x_2y_2)\cdots\chi_{r_k;n_km_k}(x_ky_k) \\ G^{(k)}_{\mathrm{amp};n_1m_1,\cdots,n_km_k}(x_1y_1,\cdots x_ky_k)\ , \end{aligned} \tag{8.106}$$

where $G^{(k)}_{\mathrm{amp}}$ is the amputated, connected, k-bilocal Green's function (compare with Eq. (7.12)):

$$\begin{aligned} G^{(k)}_{c;n_1m_1,\cdots,n_km_k}(x_1y_1,\cdots,x_ky_k) \\ \equiv(i)^{1-k}G^{(2)}_{c;n_1m_1,n_1'm_1'}(x_1y_1,x_1'y_1')\cdots G^{(2)}_{c;n_km_k,n_k'm_k'}(x_ky_k,x_k'y_k') \\ \cdot G^{(k)}_{\mathrm{amp};n_1'm_1',\cdots n_k'm_k'}(x_1'y_1',\cdots,x_k'y_k')\ . \end{aligned} \tag{8.107}$$

Thus, the BS wave functions contain all necessary information for the description of the dynamics of bound states.

The present derivation of the SD and BS equations has certain advantages as compared to the original ones [7, 8, 9]: it allows to derive these equations not only on the perturbative vacuum but also on the nonperturbative one. This point is primary important for the description of the dynamical symmetry breaking phenomenon.

8.6. The Bethe-Salpeter equation in the Nambu-Jona-Lasinio model and QED

In this section we will derive the BS equation in the NJL model and QED. We will consider the approximations connected with the expressions (8.48) and (8.59) for $\Gamma_2(G)$ in the NJL model and QED, respectively.

In the NJL model, the BS kernel is

$$\begin{aligned}K_{m_1n_1,n_2m_2}(y_1x_1,x_2y_2) =& -i\frac{\delta^2\Gamma_2}{\delta G_{m_1n_1}(y_1,x_1)\delta G_{n_2m_2}(x_2,y_2)} \\ =& i\frac{G^{(0)}}{2}\Big[(\lambda^\alpha)_{n_1m_1}(\lambda^\alpha)_{m_2n_2} + (i\gamma_5\lambda^\alpha)_{n_1m_1}(i\gamma_5\lambda^\alpha)_{m_2n_2} \\ & - (\lambda^\alpha)_{n_1n_2}(\lambda^\alpha)_{m_2m_1} - (i\gamma_5\lambda^\alpha)_{n_1n_2}(i\gamma_5\lambda^\alpha)_{m_2m_1}\Big] \\ & \times \delta^4(x_1-y_1)\delta^4(x_1-x_2)\delta^4(x_1-y_2)\ .\end{aligned} \tag{8.108}$$

Therefore in this approximation, the homogeneous BS equation in the NJL model has the form:

$$\begin{aligned}& G^{-1}_{nn_1}(p+\eta_x P_b)\chi_{r;n_1m_1}(p,P_b)G^{-1}_{m_1m}(p-\eta_y P_b) \\ =& \frac{1}{(2\pi)^4}\int d^4q K_{mn,n_2m_2}(p,q;P_b)\chi_{r;n_2m_2}(q,P_b)\ ,\end{aligned} \tag{8.109}$$

where G satisfies the SD equation (8.52), and the BS kernel is

$$\begin{aligned}K_{mn,n_2m_2}(p,q;P) =& i\frac{G^{(0)}}{2}\Big[(\lambda^\alpha)_{nm}(\lambda^\alpha)_{m_2n_2} + (i\gamma_5\lambda^\alpha)_{nm}(i\gamma_5\lambda^\alpha)_{m_2n_2} \\ & - (\lambda^\alpha)_{nn_2}(\lambda^\alpha)_{m_2n} - (i\gamma_5\lambda^\alpha)_{nn_2}(i\gamma_5\lambda^\alpha)_{m_2m}\Big]\ .\end{aligned} \tag{8.110}$$

We note that it is independent of momenta.

In QED, we find the following BS kernel from Eq.(8.59):

$$\begin{aligned}K_{m_1n_1,n_2m_2}(y_1x_1,x_2y_2) =& -i\frac{\delta^2\Gamma_2}{\delta G_{m_1n_1}(y_1,x_1)\delta G_{n_2m_2}(x_2,y_2)} \\ =& -ig^2\delta^4(x_1-x_2)\delta^4(y_1-y_2)\gamma^\mu_{n_1n_2}\gamma^\nu_{m_2m_1}D_{\mu\nu}(y_1-x_1) \\ & + ig^2\delta^4(x_1-y_1)\delta^4(y_2-x_2)\gamma^\mu_{n_1m_1}\gamma^\nu_{m_2n_2}D_{\mu\nu}(x_1-x_2)\ .\end{aligned} \tag{8.111}$$

Therefore in this approximation, the homogeneous BS equation in QED has the form:

$$\begin{aligned} &G^{-1}_{nn_1}(p+\eta_x P_b)\chi_{r;n_1m_1}(p,P_b)G^{-1}_{m_1m}(p-\eta_y P_b) \\ =&\frac{1}{(2\pi)^4}\int d^4q K_{mn,n_2m_2}(p,q;P_b)\chi_{r;n_2m_2}(q,P_b)\ , \end{aligned} \tag{8.112}$$

where G satisfies the SD equation (see Eqs.(8.62), (8.63)), and the BS kernel is (see Fig. 8.9)

$$K_{mn,n_2m_2}(p,q;P) = -ig^2\gamma^\mu_{nn_2}\gamma^\nu_{m_2m}D_{\mu\nu}(p-q) + ig^2\gamma^\mu_{nm}\gamma^\nu_{m_2n_2}D_{\mu\nu}(P)\ . \tag{8.113}$$

The BS wave function χ_r is a matrix carrying both Dirac indices and indices connected with an inner group symmetry.

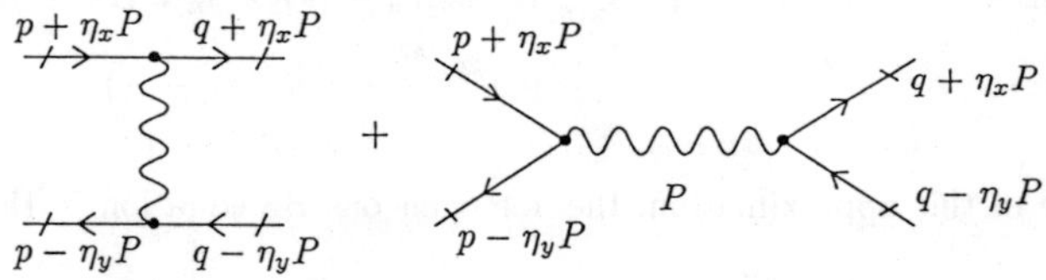

Fig. 8.9. Graphical representation of the Bethe-Salpeter kernel (8.113) in QED. The slashes on the external fermion lines indicate that these lines are amputated.

The transformation properties of the bound state wave function follow from those of the constituent fields. For example, taking into account the space inversion transformation for $\psi(x)$ (see Eqs.(A.34), (A.39)),

$$\psi(x) \longrightarrow \psi_p(x) = U_p^{-1}\psi(x)U_p = i\gamma^0\psi(i_s x)\ ,$$

with $i_s x = (x^0, -\mathbf{x})$, we find for a spinless bound state $|P_b\rangle$ that

$$\begin{aligned} \chi(x,y;P_b) = \langle 0|T\psi(x)\bar\psi(y)|P_b\rangle =&\langle 0|U_p^{-1}T\psi(x)\bar\psi(y)U_pU_p^{-1}|P_b\rangle \\ =&\eta_p^{-1}\gamma_0\langle 0|T\psi(i_s x)\bar\psi(i_s y)|i_s P_b\rangle\gamma_0 \\ =&\eta_p^{-1}\gamma_0\chi(i_s x, i_s y; i_s P_b)\gamma_0\ , \end{aligned} \tag{8.114}$$

where η_p is the parity of the bound state $|P_b\rangle$, $U_p|P_b\rangle = \eta_p|i_s P_b\rangle$. This relation implies

that

$$\gamma^0 \chi(i_s p, i_s P_b)\gamma^0 = \eta_P \chi(p, P) \ . \tag{8.115}$$

In the same way, one can find restrictions for χ following from other symmetry transformations.

We will analyze the BS equation in the NJL model in Chapter 9. The BS equation in QED will be considered in Chapter 10.

8.7. Physical Interpretation of the Effective Action for Composite Operators. The Stability Criterion

As was shown in Sec.7.3, the effective potential $V(\varphi_c)$ for elementary fields φ admits a clear physical interpretation. It yields the minimum value of the energy density among all normalized states $|\psi\rangle$ with a given value of the average $\varphi_c = \langle\psi|\varphi|\psi\rangle$. One should expect that there exists an analogous interpretation of the effective potential for composite operators $V(\varphi_c, G)$. However, as we will see below, the situation with the interpretation of $V(\varphi_c, G)$ is rather subtle.

First of all, following the method of Sec.7.3, it is easy to show that when the sources $J(x)$ and $I(x, y)$ are time-translation invariant functions of the form

$$J(x) = J(\mathbf{x}), I(x, y) = \delta(x^0 - y^0) J(\mathbf{x}, \mathbf{y}) \ , \tag{8.116}$$

the action $\Gamma(\varphi_c, G)$ indeed admits a simple interpretation: $\left(\int dx^0\right)^{-1} \Gamma(\varphi_c, G)$ yields the expectation value of the energy in that state $|\psi\rangle$ which minimizes $\langle\psi|H|\psi\rangle$ subject to the constraint that $\langle\psi|\psi\rangle = 1$ and

$$\langle\psi|\varphi(x)|\psi\rangle = \varphi_c(\mathbf{x}) \ , \tag{8.117}$$

$$\langle\psi|\varphi(x)\varphi(y)|\psi\rangle|_{x^0=y^0} = \varphi_c(\mathbf{x})\varphi_c(\mathbf{y}) - iG(\mathbf{x}, \mathbf{y}) \ . \tag{8.118}$$

In particular, when φ_c is a constant and $G(\mathbf{x}, \mathbf{y}) = G(\mathbf{x} - \mathbf{y})$, the effective potential $V(\varphi_c, G) = \left(\int d^4x\right)^{-1} \Gamma(\varphi_c, G)$ yields the minimum expectation value of the energy density among all states satisfying the constraints (8.117), (8.118).

However, if φ_c is a constant, the effective potential $V(\varphi_c, G)$ can be defined for any translation invariant source $I(x, y)$ depending only on the difference $x - y$: indeed, $I(x, y) = I(x-y)$ implies that $G(x, y) = G(x-y)$. However, if $I(x-y)$ is not proportional to $\delta(x^0 - y^0)$, the source term $I(x - y)\varphi(x)\varphi(y)$ is nonlocal in time, and therefore it is unclear how one can define properly the hamiltonian operator in a theory with such a source.*) What is the interpretation of $V(\varphi_c, G)$ in this general case?

This question was intensively discussed in the literature [16, 17, 18, 19] (for a thorough discussion see the review [20]). In particular, as was shown by Haymaker and Matsuki [18], the physical vacuum defined by the stationary conditions (8.1) and (8.2) corresponds to a saddle point (and not a minimum) of the effective potential $V(\varphi_c, G)$. This leads to the question what is the stability criterion in systems described by the effective potential for composite operators?

One way to resolve this puzzle is to introduce a modified effective action $\tilde{\Gamma}(\varphi_c, G)$ leading to the same equations (8.1), (8.2) for φ_c and G but admitting the usual stability criterion for the physical vacuum; in particular the saddle point corresponding to the vacuum in $V(\varphi_c, G)$ should be transformed into the minimum of the modified potential $\tilde{V}(\varphi_c, G)$. Such a modified effective action was constructed by Casalbuoni, De Curtis, Dominici and Gatto [21]. However, as was pointed out by Fukuda [10], Fukuda, Komachiya and Ukita [11], the ambiguity-free stability criterion can be in fact formulated for the original effective action $\Gamma(\varphi_c, G)$. Let us discuss their approach in more detail.

The main idea is that the stability conditions should be based on the considerations of only those fluctuations around a solution of the equations of motions which are consistent with the corresponding dynamics. Let us illustrate this point on the example of a classical system with the action

$$S = \int dt\, L(t) \;, \tag{8.119}$$

where

$$L(t) = \sum_{i=1}^{N} \frac{m\dot{q}_i^2}{2} - U(q_i) \;. \tag{8.120}$$

---

*) Because of the translation invariance, there are corresponding conserved currents in the classical theory. In particular, there is a conserved current connected with the time-translation invariance (at least for the sources $I(x-y)$ decreasing rapidly enough with $|x^0 - y^0| \longrightarrow \infty$). However there is no known way to introduce the hamiltonian operator in a *quantum* theory which is nonlocal in time.

The equation of motion is

$$-\frac{\delta S}{\delta q_i(t)} = \frac{d}{dt}\frac{\partial L}{\partial \dot{q}_i(t)} - \frac{\partial L}{\partial q_i(t)} = 0\ . \tag{8.121}$$

Let us consider a static solution $q_i^{(0)}$ to this equation (this case is just what we are interested in since the vacuum corresponds to a static solution). Let us now consider the variations of this solution, $q_i(t) = q_i^{(0)} + \Delta q_i(t)$, which are admitted by the equation of motion (8.121). Then up to the first order in $\Delta q_i(t)$, we get

$$\int dt' \frac{\delta^2 S}{\delta q_i(t)\delta q_j(t')}|_{q_i=q_i^{(0)}} \Delta q_j(t') = 0\ . \tag{8.122}$$

Since the solution $q_i^{(0)}$ is static, the function $S_{ij}^{(2)}(t,t') \equiv \frac{\delta^2 S}{\delta q_i(t)\delta q_j(t')}|_{q_i=q_i^{(0)}}$ is a function of $t-t'$. Therefore realizing the Fourier transformation

$$\Delta q_i(\omega) = \int dt e^{i\omega t} \Delta q_i(t)\ , \tag{8.123}$$

$$S_{ij}^{(2)}(\omega,-\omega') = \delta(\omega-\omega')S_{ij}^{(2)}(\omega) = \int dt dt' e^{(i\omega t - i\omega' t')} S_{ij}^{(2)}(t-t')\ , \tag{8.124}$$

we get the equation

$$S_{ij}^{(2)}(\omega)\Delta q_j(\omega) = 0\ . \tag{8.125}$$

In the case of lagrangian (8.120), the function $S_{ij}^{(2)}(\omega)$ is

$$S_{ij}^{(2)}(\omega) = \left(-m\omega^2\delta_{ij} + U_{ij}\right)\ , \tag{8.126}$$

where $U_{ij} = \frac{\partial^2 U}{\partial q_i \partial q_j}|_{q_i=q_i^{(0)}}$. Let us introduce $v^{(k)}$ which are the eigenvalues of the matrix U_{ij}. The solution of the equation (8.125) takes the form

$$\Delta q_j(\omega) = \sum_k C_{jk}\delta(\omega^2 - \omega^{(k)^2})\ , \tag{8.127}$$

where $\omega^{(k)^2} = v^{(k)}/m$ and C_{ij} are some coefficients. The functions $\Delta q_j(t)$ are

$$\Delta q_j(t) = \frac{1}{2\pi}\sum_k \left(c_{jk}e^{i\omega^{(k)}t} + c_{jk}^* e^{-i\omega^{(k)}t}\right)\ . \tag{8.128}$$

The stability criterion in classical mechanics is simple: $\omega^{(k)} = (v^{(k)}/m)^{1/2}$ must be real, i.e. the eigenvalues $v^{(k)}$ must be positive. The physical meaning of this criterion is clear: if some of $\omega^{(k)}$ are complex, $\Delta q_i(t)$ blows up as $t \longrightarrow \pm\infty$.

Fukuda has suggested that this criterion be extended to quantum field theories. Let us first discuss how the criterion works in theories described by the effective action $\Gamma(\varphi_c)$ for elementary fields considered in Chapter 7.

The effective potential $\Gamma(\varphi_c)$ is (see Eq.(7.9))

$$\Gamma(\varphi_c) = W(J) - \int d^4x J(x)\varphi_c(x) \, . \tag{8.129}$$

The vacuum is described by a solution $\varphi_c = \bar{\varphi}_c$ of the equation

$$\frac{\delta\Gamma(\varphi_c)}{\delta\varphi_c(x)} = 0 \, . \tag{8.130}$$

The vacuum solution $\bar{\varphi}_c = \langle 0|\varphi(x)|0\rangle$ is a constant. Let us now consider the variations of the vacuum solution, $\varphi_c(x) = \bar{\varphi}_c + \Delta\varphi_c(x)$, which are admitted by the equation of motion (8.130). Then the analogue of the equation (8.125) is

$$\int d^4y \left(\frac{\delta^2\Gamma}{\delta\varphi_c(x)\delta\varphi_c(y)}\right)|_{\varphi_c=\bar{\varphi}_c}\Delta\varphi_c(y) = 0 \, . \tag{8.131}$$

As we showed in Sec.7.2, $-\Gamma^{(2)}(x-y) = -\frac{\delta^2\Gamma}{\delta\varphi_c(x)\delta\varphi_c(y)}|_{\varphi_c=\bar{\varphi}_c}$ is the inverse propagator of the field $\varphi(x)$. Therefore $\Delta\varphi_c(y)$ describes the excitations which couple to $\varphi(x)$. The stability criterion is clear: masses of all these excitations must be real, in particular there must not be tachyons. This criterion coincides with the stability conditions for the vacuum which we discussed in Chapter 7.

The generalization of this criterion to theories described by the effective action for composite operators $\Gamma(\varphi_c, G)$ is straightforward. The vacuum is described by a translation invariant solution, $\varphi_c = \bar{\varphi}_c, G(x,y) = \bar{G}(x-y)$, to the equations:

$$\frac{\delta\Gamma}{\delta\varphi_c} = 0 \, , \tag{8.132}$$

$$\frac{\delta\Gamma}{\delta G} = 0 \, . \tag{8.133}$$

The equation (8.133) is the Schwinger-Dyson equation for the propagator. For the sake of simplicity, let us ignore the variations of φ_c, that is $\bar{\varphi}_c \longrightarrow \bar{\varphi}_c, \bar{G}(x-y) \longrightarrow G(x,y) =$

$\bar{G}(x-y)+\Delta G(x,y)$. Then the analogue of the equation (8.131) is the following equation:

$$\int d^4x' d^4y' \left[\frac{\delta^2\Gamma}{\delta G(x,y)\delta G(x',y')}\right]|_{G=\bar{G}}\ \Delta G(x',y') = 0\ . \tag{8.134}$$

But this is nothing else but the homogeneous BS equation considered in Sec. 8.5. The stability criterion is clear: there must not be solutions to this equation corresponding to tachyonic bound states.

Thus we conclude that the effective action for composite operators is a very convenient tool for studying the dynamics of bound states and, in particular, for studying theories with dynamical symmetry breaking.

Chapter 9

Dynamical Chiral Symmetry Breaking. General Consideration

9.1. Introduction

In Section 4.4, we considered spontaneous chiral symmetry breaking in the linear σ-model with the $U_L(N) \times U_R(N)$ (or $SU_L(N) \times SU_R(N) \times U_V(1)$) chiral group. Let us recall that at $\mu^2 < 0$ in this model:

a) in the symmetric (unstable) phase, with $\langle 0|M|0\rangle = \frac{v}{\sqrt{2}}\lambda^0 = 0$, there are $2N^2$ tachyons (N^2 scalar and N^2 pseudoscalar ones) assigned to the $(N, \bar{N}) \oplus (\bar{N}, N)$ representation of the $SU_L(N) \times SU_R(N)$ group;

b) in the asymmetric (stable) phase with $\langle 0|M|0\rangle = \frac{v}{\sqrt{2}}\lambda^0 \neq 0$, the $U_L(N) \times U_R(N)$ ($SU_L(N) \times SU_R(N) \times U_V(1)$) symmetry is broken to $U_V(N)$, and N^2 ($N^2 - 1$ in the case of the $SU_L(N) \times SU_R(N) \times U_V(1)$) pseudoscalar tachyons become massless NG bosons; the rest of tachyons acquire a real mass ($M^2 > 0$);

c) in the case when there are both spontaneous and explicit chiral symmetry breakings, the dynamics of the partial conservation of the axial currents (PCAC) is realized.

In field theories with no fundamental spinless fields, this scenario can be realized only through the formation of the $2N^2$ fermion-antifermion bound states assigned to the $(N, \bar{N}) \oplus (\bar{N}, N)$ representation of the $SU_L(N) \times SU_R(N)$. Therefore in this case the problem of spontaneous (dynamic) chiral symmetry breaking becomes in many respects the problem of the determination of the dynamics which gives rise to such bound states as tachyons and NG bosons in the symmetric and asymmetric phases, respectively.

We will be interested in two kinds of models with dynamic symmetry breaking: the Nambu-Jona-Lasinio model already described briefly in Chapter 6, and gauge models without fundamental scalar fields. The latter are in turn divided into two classes: vector-like gauge theories and chiral gauge theories. Their definitions are given below.

As was indicated in Sec. 4.3, one can use only left-handed fields ψ_L to describe fermions in a theory. Vector-like gauge theories are those ones in which the fields ψ_L make up a real representation of the gauge group. Otherwise, gauge theories are called chiral gauge theories.

The most important example of a vector-like theory is Quantum Chromodynamics (QCD) with the color gauge group $SU_c(3)$. Left quarks and left antiquark are assigned to the real representation $(3) \oplus (\bar{3})$ of the $SU_c(3)$ (see Chapter 12).

The $SU_L(2) \times U_Y(1)$ electroweak theory is the most important example of a chiral gauge theory (see Chapter 13).

In this chapter, we will consider only vector-like gauge theories.

Dynamic chiral symmetry breaking is a rather complicated phenomenon. At present it can be understood only in rather simple model approximations. Therefore it is important to get the restrictions on the mechanism of this phenomenon following from the general principles of quantum field theory.

In this chapter, we will discuss such restrictions. We will also analyze the Bethe-Salpeter (BS) equations for the bound states in the NJL model. This analysis will provide a basis for studying the BS equations in the more complicated case of gauge theories.

9.2. Chiral Ward-Takahashi Identities

We will discuss here the chiral Ward-Takahashi (WT) identities in a field theory without fundamental spinless fileds. It can be the NJL model with the lagrangian density (6.1) or a vector-like gauge theory with the lagrangian density (compare with Eq.(3.79)):

$$\mathcal{L} = -\frac{1}{4}F^K_{\mu\nu}F^{\mu\nu K} + i\bar{\psi}\gamma^\mu D_\mu\psi - \bar{\psi}\hat{m}^{(0)}\psi \,, \tag{9.1}$$

where

$$F^K_{\mu\nu} = \partial_\mu A^K_\nu - \partial_\nu A^K_\mu - g f_{KST} A^S_\mu A^T_\nu \,, \tag{9.2}$$

$$(D_\mu\psi)_{al} = \partial_\mu\psi_{al} + \frac{i}{2} g t^K_{lm}\psi_{am} A^K_\mu \,, \tag{9.3}$$

$$(\hat{m}^{(0)}\psi)_{al} = m^{(0)}_{ab}\psi_{bl} + i\gamma_5\tilde{m}^{(0)}_{ab}\psi_{bl} \,. \tag{9.4}$$

Here $K, S, T = 1, 2, \cdots, N_c^2 - 1$, the matrices t^K define the representation of the gauge group G_c to which the fermion fields are assigned, and a, b are the flavor fermion indices (for simplicity, we consider a simple gauge group, and we use here the notations for color indices of the gauge fields different from those in Chapter 3).*)

If the mass matrix $\hat{m}^{(0)}$ is zero, the lagrangian density (9.1) is formally invariant under the $U_L(N) \times U_R(N)$ chiral transformation (see Eqs.(4.36), (4.37)). We wrote "formally" since, as is known, in quantum gauge theories, the singlet axial current $j^0_{5\mu} = \bar{\psi}\frac{\lambda^0}{2}\gamma_\mu\gamma_5\psi$ is not conserved even in the chiral limit:

$$\partial^\mu j^0_{5\mu} = \frac{-\left(\frac{N}{2}\right)^{1/2} g^2}{16\pi^2}\epsilon^{\mu\nu\delta\sigma} tr F_{\delta\sigma} F_{\mu\nu} , \tag{9.5}$$

where $F_{\mu\nu} \equiv \frac{1}{2}t^K F^K_{\mu\nu}$ (the Adler-Bell-Jackiw (ABJ) anomaly [1, 2]). A thorough discussion of the ABJ anomaly is given in the book [3]. We will return to this issue in Chapter 12.

Therefore, unlike the NJL model, the chiral group of the gauge theory (9.1) with $\hat{m}^{(0)} = 0$ is $SU_L(N) \times SU_R(N) \times U_V(1)$.

When $\hat{m}^{(0)} \neq 0$, we get (compare with Eqs.(4.33), (4.34)):

$$\partial^\mu j^\alpha_\mu = \frac{i}{2}\bar{\psi}\left[\hat{m}^{(0)}, \lambda^\alpha\right]\psi; \quad {\scriptstyle \alpha=0,1,\cdots,N^2-1} , \tag{9.6}$$

$$\partial^\mu j^\alpha_{5\mu} = \frac{i}{2}\bar{\psi}\left\{\hat{m}^{(0)}, \gamma_5\lambda^\alpha\right\}\psi; \quad {\scriptstyle \alpha=1,\cdots,N^2-1} . \tag{9.7}$$

In the NJL model $\partial^\mu j^0_{5\mu}$ is

$$\partial^\mu j^0_{5\mu} = \frac{i}{2}\bar{\psi}\left\{\hat{m}^{(0)}, \gamma_5\lambda^0\right\}\psi , \tag{9.8}$$

but in gauge theories,

$$\partial^\mu j^0_{5\mu} = -\left(\frac{N}{2}\right)^{1/2}\frac{g^2}{16\pi^2}\epsilon^{\mu\nu\delta\sigma} tr F_{\delta\sigma} F_{\mu\nu} + \frac{i}{2}\bar{\psi}\left\{\hat{m}^{(0)}, \gamma_5\lambda^0\right\}\psi . \tag{9.9}$$

The chiral limit with $\hat{m}^{(0)} = 0$ is very special: the invariance of the lagrangian density under the chiral transformations guarantees that the mass term in the fermion

*) We use the normalization $tr(t^K t^{K'}) = 2\delta^{KK'}$ in the fundamental representation of the gauge group G_c.

propagator will never appear in any order of perturbative theory. This implies that on the perturbative symmetric vacuum, the propagator takes the following simple form

$$G_{ab}(p) = \frac{1}{\hat{p}A(p^2)}\delta_{ab} \,. \tag{9.10}$$

Indeed, if

$$U_\alpha(\theta)|0\rangle = |0\rangle \,, \tag{9.11}$$

$$U_{5\alpha}(\theta)|0\rangle = |0\rangle \,, \tag{9.12}$$

where the operators $U_\alpha, U_{5\alpha}$ are defined as

$$\begin{cases} U_\alpha^{-1}(\theta)\psi U_\alpha(\theta) = \exp(i\theta\lambda^\alpha)\psi \,, \\ U_\alpha^{-1}(\theta)\bar{\psi} U_\alpha(\theta) = \bar{\psi}\exp(-i\theta\lambda^\alpha) \,, \end{cases}$$
$$\begin{cases} U_{5\alpha}^{-1}(\theta)\psi U_{5\alpha}(\theta) = \exp(i\theta\gamma_5\lambda^\alpha)\psi \,, \\ U_{5\alpha}^{-1}(\theta)\bar{\psi} U_{5\alpha}(\theta) = \bar{\psi}\exp(i\theta\gamma_5\lambda^\alpha) \,, \end{cases} \tag{9.13}$$

then

$$G(x) = -\,i\langle 0|T\psi(x)\bar{\psi}(0)|0\rangle = \exp(i\theta\lambda^\alpha)G(x)\exp(-i\theta\lambda^\alpha) \,, \tag{9.14}$$

$$G(x) = \exp(i\theta\gamma_5\lambda^\alpha)G(x)\exp(i\theta\gamma_5\lambda^\alpha) \,. \tag{9.15}$$

The relations (9.14), (9.15) imply that

$$[\lambda^\alpha, G] = 0, \quad \{\gamma_5\lambda^\alpha, G\} = 0 \,. \tag{9.16}$$

These relations lead to the form (9.10) for the fermion propagator.

If the chiral $SU_L(N) \times SU_R(N)$ symmetry is spontaneously broken, the relations (9.12) are violated, and a dynamical mass term in the fermion propagator can appear:

$$G^{-1}(p) = \hat{p}A(p^2) - B_{\text{dyn}}(p^2) \,. \tag{9.17}$$

Let us discuss how this phenomenon manifests itself in the structure of Green's functions including the chiral currents by using the WT identities. In particular, we will show how the BS wave function of NG bosons is expressed through the dynamical mass function $B_{\text{dyn}}(p^2)$.

In order to derive the WT identities, we will use the approach used in Sec.4.2 for the derivation of the Noether theorem, that is, we will use the local chiral infinitesimal transformations:

$$\delta_V \psi = i\omega_\alpha(x)\frac{\lambda^\alpha}{2}\psi, \delta_V \bar{\psi} = -i\omega_\alpha(x)\bar{\psi}\frac{\lambda^\alpha}{2}, \tag{9.18}$$

$$\delta_A \psi = i\omega_{5\alpha}(x)\gamma_5\frac{\lambda^\alpha}{2}\psi, \delta_A \bar{\psi} = i\omega_{5\alpha}(x)\bar{\psi}\gamma_5\frac{\lambda^\alpha}{2}. \tag{9.19}$$

Let us consider Green's functions of some local operators $\varphi^{(s)}, {}_{s=1,2,\cdots,n}$, which are some local multilinear functions of the fermion fields:

$$\langle 0|T\left(\varphi^{(1)}(x_1)\cdots\varphi^{(n)}(x_n)\right)|0\rangle . \tag{9.20}$$

We will use the path integral representation for them:

$$\langle 0|T\left(\varphi^{(1)}(x_1)\cdots\varphi^{(n)}(x_n)\right)|0\rangle = N\int \mathcal{D}(A_\mu)\mathcal{D}(\psi)\mathcal{D}(\bar{\psi})\exp(iS)\varphi^{(1)}(x_1)\cdots\varphi^{(n)}(x_n), \tag{9.21}$$

where N is a normalization factor (see Sec.C.3 in Appendix C).

Let us perform the change (9.18), (9.19) of variables in the integral (9.21). A change of variables does not change the value of the integral. Therefore, taking into account the relation (see Eq.(4.3))

$$\begin{aligned}\delta S = \int d^4y\Big\{&\omega_\alpha(y)\left[-\frac{i}{2}\bar{\psi}\left[\hat{m}^{(0)}, \lambda^\alpha\right]\psi + \partial^\mu j^\alpha_\mu\right] \\ &+ \omega_{5\alpha}(y)\left[-\frac{i}{2}\bar{\psi}\left\{\hat{m}^{(0)}, \gamma_5\lambda^\alpha\right\}\psi + \partial^\mu j^\alpha_{5\mu}\right]\Big\},\end{aligned} \tag{9.22}$$

we get the following WT identities:

$$\begin{aligned}&\partial^\mu_y\langle 0|Tj^\alpha_\mu(y)\varphi^{(1)}(x_1)\cdots\varphi^{(n)}(x_n)|0\rangle \\ &\quad= \frac{i}{2}\langle 0|T\bar{\psi}(y)\left[\hat{m}^{(0)}, \lambda^\alpha\right]\psi(y)\varphi^{(1)}(x_1)\cdots\varphi^{(n)}(x_n)|0\rangle \\ &\qquad+ i\delta^4(y-x_1)\langle 0|T\frac{\partial\varphi^{(1)}(x_1)}{\partial\omega_\alpha(x_1)}\cdots\varphi^{(n)}(x_n)|0\rangle \\ &\qquad+\cdots+ i\delta^4(y-x_n)\langle 0|T\varphi^{(1)}(x_1)\cdots\frac{\partial\varphi^{(n)}(x_n)}{\partial\omega_\alpha(x_n)}|0\rangle,\end{aligned} \tag{9.23}$$

$$
\begin{aligned}
&\partial_y^\mu \langle 0|T j_{5\mu}^\alpha(y)\varphi^{(1)}(x_1)\cdots\varphi^{(n)}(x_n)|0\rangle \\
&\qquad = \frac{i}{2}\langle 0|T\bar\psi(y)\left\{\hat m^{(0)},\gamma_5\lambda^\alpha\right\}\psi(y)\varphi^{(1)}(x_1)\cdots\varphi^{(n)}(x_n)|0\rangle \\
&\qquad + i\delta^4(y-x_1)\langle 0|T\frac{\partial\varphi^{(1)}(x_1)}{\partial\omega_{5\alpha}(x_1)}\cdots\varphi^{(n)}(x_n)|0\rangle \\
&\qquad + \cdots + i\delta^4(y-x_n)\langle 0|T\varphi^{(1)}(x_1)\cdots\frac{\partial\varphi^{(n)}(x_n)}{\partial\omega_{5\alpha}(x_n)}|0\rangle\ ,
\end{aligned}
\tag{9.24}
$$

where

$$
\begin{aligned}
\delta^4(y-x_i)\frac{\partial\varphi^{(i)}(x_i)}{\partial\omega_\alpha(x_i)} &\equiv \frac{\delta\varphi^{(i)}(x_i)}{\delta\omega_\alpha(y)}\ , \\
\delta^4(y-x_i)\frac{\partial\varphi^{(i)}(x_i)}{\partial\omega_{5\alpha}(x_i)} &\equiv \frac{\delta\varphi^{(i)}(x_i)}{\delta\omega_{5\alpha}(y)}\ .
\end{aligned}
\tag{9.25}
$$

In fact, in gauge theories, because of the ABJ anomaly (9.9), the WT identity (9.24) with $\alpha = 0$ has to be modified:

$$
\begin{aligned}
&\partial_y^\mu \langle 0|T j_{5\mu}^0(y)\varphi^{(1)}(x_1)\cdots\varphi^{(n)}(x_n)|0\rangle \\
&= -\left(\frac{N}{2}\right)^{1/2}\frac{g^2}{16\pi^2}\epsilon^{\mu\nu\delta\sigma}\langle 0|T tr\left(F_{\delta\sigma}(y)F_{\mu\nu}(y)\right)\varphi^{(1)}(x_1)\cdots\varphi^{(n)}(x_n)|0\rangle \\
&+ \frac{i}{2}\langle 0|\bar\psi(y)\left\{\hat m^{(0)},\gamma_5\lambda^0\right\}\psi(y)\varphi^{(1)}(x_1)\cdots\varphi^{(n)}(x_n)|0\rangle \\
&+ i\delta^4(y-x_1)\langle 0|T\frac{\partial\varphi^{(1)}(x_1)}{\partial\omega_{50}(x_1)}\cdots\varphi^{(n)}(x_n)|0\rangle \\
&+ \cdots + i\delta^4(y-x_n)\langle 0|T\varphi^{(1)}(x_1)\cdots\frac{\partial\varphi^{(n)}(x_n)}{\partial\omega_{50}(x_n)}|0\rangle\ .
\end{aligned}
\tag{9.26}
$$

The derivation of this identity in the path integral formalism was realized by Fujikawa [4].

We are interested in the WT identities for the following Green's functions:

$$\tilde\Gamma^\alpha(x,y) = \frac{1}{2}\langle 0|T\bar\psi(0)\lambda^\alpha\psi(0)\psi(x)\bar\psi(y)|0\rangle\ , \tag{9.27}$$

$$\tilde\Gamma_5^\alpha(x,y) = \frac{1}{2}\langle 0|T\bar\psi(0)\gamma_5\lambda^\alpha\psi(0)\psi(x)\bar\psi(y)|0\rangle\ , \tag{9.28}$$

$$\tilde\Gamma_\mu^\alpha(x,y) = \langle 0|T j_\mu^\alpha(0)\psi(x)\bar\psi(y)|0\rangle\ , \tag{9.29}$$

$$\tilde\Gamma_{5\mu}^\alpha(x,y) = \langle 0|T j_{5\mu}^\alpha(0)\psi(x)\bar\psi(y)|0\rangle\ . \tag{9.30}$$

We introduce their Fourier transforms

$$G(p)\Gamma_A^\alpha(p,k)G(k) = \int dx\,dy \exp[ipx - iky] \cdot \tilde{\Gamma}_A^\alpha(x,y) \ , \tag{9.31}$$

where $\tilde{\Gamma}_A^\alpha(x,y)$ is a generic notation for $\tilde{\Gamma}^\alpha(x,y), \tilde{\Gamma}_5^\alpha(x,y), \tilde{\Gamma}_\mu^\alpha(x,y), \tilde{\Gamma}_{5\mu}^\alpha(x,y)$, and $\Gamma_A^\alpha(p,k)$ is the corresponding amputated Green's function. Then the relations (9.23), (9.24) lead to the following non-anomalous WT identities:

$$P^\mu\Gamma_\mu^\alpha(p,k) = [\hat{m}^{(0)}, \Gamma^\alpha(p,k)] - G^{-1}(p)\frac{\lambda^\alpha}{2} + \frac{\lambda^\alpha}{2}G^{-1}(k) \ , \tag{9.32}$$

$$P^\mu\Gamma_{5\mu}^\alpha\Gamma_5^\alpha(p,k) = \{\hat{m}^{(0)}, \Gamma_5^\alpha(p,k)\} - G^{-1}(p)\frac{\lambda^\alpha}{2}\gamma_5 - \frac{\lambda^\alpha}{2}\gamma_5 G^{-1}(k) \ , \tag{9.33}$$

where $P = p - k$.

Let us use the WT identity (9.33) for studying the BS wave function of NG bosons. In the chiral limit with $\hat{m}^{(0)} = 0$, this identity takes the form

$$P^\mu\Gamma_{5\mu}^\alpha(p,k) = -G^{-1}(p)\frac{\lambda^\alpha}{2}\gamma_5 - \frac{\lambda^\alpha}{2}\gamma_5 G^{-1}(k) \ . \tag{9.34}$$

Eq.(9.30) implies that the pole contribution of pseudoscalar NG bosons in $\Gamma_{5\mu}^\alpha(p,k)$ is (compare with Eq.(4.79)):

$$\left[G(p)\Gamma_{5\mu}^\alpha(p,k)G(k)\right]_{\text{pole}} = \frac{FP_\mu\chi^{(p);\alpha}}{P^2}(q,P) \ , \tag{9.35}$$

where

$$\chi^{(p);\alpha}(q,P) = e^{iPX}\int d^4(x-y)\exp[iq(x-y)]\langle 0|T\psi(x)\bar{\psi}(y)|P,\alpha\rangle$$

is the BS wave function of NG bosons; here $P = p - k$, and, since $\eta_{x(y)} = \frac{m_x(y)}{m_x + m_y} = \frac{1}{2}$ at $m_x = m_y = m_{\text{dyn}}$, q is $q = \eta_y p + \eta_x k = \frac{1}{2}(p+k)$ and X is $X = \eta_x x + \eta_y y = \frac{1}{2}(x+y)$ (see Eqs.(8.103) - (8.105)). The decay constant F of NG bosons is:

$$\langle P,\alpha|j_{5\mu}^\alpha(0)|0\rangle = -iFP_\mu \ . \tag{9.36}$$

The BS wave function $\chi^{(p);\alpha}(q,P)$ can be expanded in the Dirac matrices:

$$\chi^{(p);\alpha}(q,p) = \left[\chi_1^{(p)} + \chi_2^{(p)}\hat{P} + \chi_3^{(p)}\hat{q} + \chi_4^{(p)}\sigma^{\mu\nu}(P_\mu q_\nu - P_\nu q_\mu)\right]\frac{\lambda^\alpha}{2}\gamma_5 \ , \tag{9.37}$$

where $\sigma^{\mu\nu} = \frac{i}{2}[\gamma^\mu, \gamma^\nu]$, and $\chi_i^{(p)}$ are some functions of the Lorentz invariants q^2 and Pq ($P^2 = 0$ for NG bosons).

Let us consider the limit $P^\mu = p^\mu - k^\mu \longrightarrow 0$ in Eq.(9.34). We get

$$G^{-1}(q)\chi^{(p),\alpha}(q)G^{-1}(q) = \frac{\lambda^\alpha}{2}\gamma_5(-q^2 + B^2_{\text{dyn}}(q^2))\chi_1^{(p)}(q^2) = \lambda^\alpha\gamma_5\frac{B_{\text{dyn}}(q^2)}{F} . \quad (9.38)$$

This is a remarkable relation: it shows that the BS wave function of NG bosons at $P^\mu = 0$ is completely determined by the fermion propagator. We will frequently use this relation in the following.

In conclusion, let us derive the equations for Green's functions $\tilde{\Gamma}^\alpha_A(x,y)$.

From the definition (8.97) of Green's function $G_c^{(2)}$ and Eqs.(9.27) - (9.30), we find the following relation:

$$\tilde{\Gamma}^\alpha_{A;nm}(x,y) = -G^{(2)}_{c;nm,m'n'}(xy,00)\left(A\frac{\lambda^\alpha}{2}\right)_{n'm'} , \quad (9.39)$$

where $A = I, \gamma_5, \gamma_\mu$ or $\gamma_\mu\gamma_5$. Therefore from the BS equation (8.99) for $G_c^{(2)}$, we get the following equation for $\Gamma^\alpha_A(q,k)$:

$$\Gamma^\alpha_{A;nm}(p,k) = -\left(A\frac{\lambda^\alpha}{2}\right)_{nm} + \frac{1}{(2\pi)^4}\int d^4q K_{mn,n'm'}(\eta_y p + \eta_x k, q; p-k)$$
$$\times \left[G_{n'n''}(q+\eta_x(p-k))\Gamma^\alpha_{A;n''m''}(q+\eta_x(p-k), q-\eta_y(p-k))G_{m''m'}(q-\eta_y(p-k))\right] , \quad (9.40)$$

where $K_{mn,n'm'}$ is the BS kernel (see Fig. 9.1).

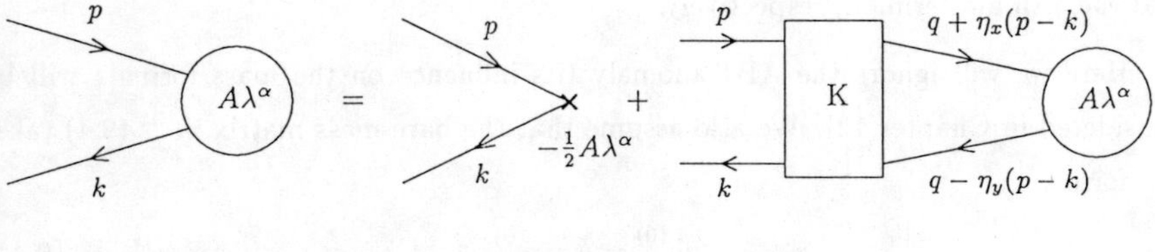

Fig. 9.1. Graphical representation of the BS equation for $\Gamma^\alpha_A(p,k)$.

9.3. The Exact Relation for the Mass of Pseudoscalar Bosons

In this section we will derive the exact relation for the mass of pseudoscalar bosons. We will see that in the first order of the expansion in $\hat{m}^{(0)}$, this relation reduces to the PCAC formula of Gell-Mann, Oakes and Renner [5].

The exact formula will be very useful in the analysis of the ultraviolet asymptotic behavior of the BS wave function of pseudoscalar bosons.

Let us consider a theory with dynamical chiral symmetry breaking in which the mass term $\bar{\psi}\hat{m}^{(0)}\psi$ explicitly violates the chiral symmetry.

We consider the chiral group $SU_L(N)\times SU_R(N)$ (or $U_L(N)\times U_R(N)$), and ψ_{La} (ψ_{Rb}) fermions are assigned to the $(N,0)$ ($(0,N)$) representation of the chiral group; $a,b = 1,\cdots,N$. The BS wave function $\chi^{(p)}_{ab}$ of a colorless pseudosalar boson composed of the a-th fermion and the b-th antifermion can be expanded in the Dirac matrices:

$$\chi^{(p)}_{ab}(q,P) = \frac{1}{\sqrt{2}}\left[\chi^{(p)}_{ab;1} + \chi^{(p)}_{ab;2}\hat{P} + \chi^{(p)}_{ab;3}\hat{q} + \chi^{(p)}_{ab;4}\sigma^{\mu\nu}(P_\mu q_\nu - P_\nu q_\mu)\right]\gamma_5 \,, \tag{9.41}$$

where $\sigma^{\mu\nu} = \frac{i}{2}[\gamma^\mu,\gamma^\nu]$, and $\chi^{(p)}_{ab,i}$ are some functions of the Lorentz invariants P^2, q^2 and Pq (the coefficient $\frac{1}{\sqrt{2}}$ in Eq. (9.41) ensures that $\chi_{ab;i}$ becomes equal $\chi^{(p)}_i$ from Eq. (9.37) in the chiral limit).

The BS wave function $\chi^{(p)}_{ab}(q,P)$ is the Fourier transform (8.103) of the function

$$\chi^{(p)}_{ab}(x,y) = \langle 0|T\psi_a(x)\bar{\psi}_b(y)|P\rangle \,, \tag{9.42}$$

that is,

$$\chi^{(p)}_{ab}(x,y) = e^{-iPX}(2\pi)^{-4}\int d^4q\exp[-iq(x-y)]\chi^{(p)}_{ab}(q,P) \,, \tag{9.43}$$

where $X = \frac{m_a x + m_b y}{m_a + m_b}$ with m_a, m_b are physical (and not bare) masses of the a-th fermion and the b-th antifermion, respectively.

Here we will ignore the ABJ anomaly (its influence on the mass formula will be considered in Chapter 12). We also assume that the bare mass matrix $\hat{m}^{(0)}$ (9.4) takes the form

$$\hat{m}^{(0)}_{ab} = \delta_{ab} m^{(0)}_a \,. \tag{9.44}$$

In fact, it is easy to show that in vector-like theories, the matrix $\hat{m}^{(0)}$ can be always reduced to the form (9.44) by an appropriate chiral transformation of fermion fields.

Let us consider the matrix element

$$i\left(m_a^{(0)}+m_b^{(0)}\right)\langle 0|\bar{\psi}_b(0)\gamma_5\psi_a(0)|P\rangle\,. \tag{9.45}$$

On the one hand, using the equation

$$\partial^\mu\left(\bar{\psi}_b\gamma_\mu\gamma_5\psi_a\right)=i\left(m_a^{(0)}+m_b^{(0)}\right)\bar{\psi}_b\gamma_5\psi_a \tag{9.46}$$

(compare with Eq.(9.7)), we find that it is

$$-iP^\mu\langle 0|\bar{\psi}_b(0)\gamma_\mu\gamma_5\psi_a(0)|P\rangle\,. \tag{9.47}$$

On the other hand, it can be expressed through the BS wave function:

$$\begin{aligned}
i\left(m_a^{(0)}+m_b^{(0)}\right)\langle 0|\bar{\psi}_b(0)\gamma_5\psi_a(0)|P\rangle &= -\,i\left(m_a^{(0)}+m_b^{(0)}\right)\lim_{x,y\to 0} tr\left(\gamma_5\chi_{ab}^{(p)}(x,y)\right)\\
&= -\,i\left(m_a^{(0)}+m_b^{(0)}\right)\frac{1}{(2\pi)^4}tr\left(\gamma_5\int d^4q\chi_{ab}^{(p)}(q,P)\right)\\
&= -\,\frac{4iN_c\left(m_a^{(0)}+m_b^{(0)}\right)}{\sqrt{2}(2\pi)^4}\int d^4q\chi_{ab;1}^{(p)}\,.
\end{aligned} \tag{9.48}$$

Let us introduce the decay constant F_{ab}:

$$\langle 0|\bar{\psi}_b(0)\gamma_\mu\gamma_5\psi_a(0)|P\rangle=i\sqrt{2}F_{ab}P_\mu\,. \tag{9.49}$$

Then from Eqs.(9.46) - (9.49), we get the following mass relation:

$$M_{ab}^2\equiv P^2=-\frac{2iN_c\left(m_a^{(0)}+m_b^{(0)}\right)}{F_{ab}(2\pi)^4}\int d^4q\chi_{ab;1}^{(p)}\,. \tag{9.50}$$

In first order in $m_a^{(0)},m_b^{(0)}$, this relation reduces to the PCAC relation [5]

$$M_{ab}^2=-\frac{<\bar{\psi}\psi>}{F^2}\left(m_a^{(0)}+m_b^{(0)}\right)\,, \tag{9.51}$$

where F is $F\equiv F_{ab}$ in the chiral limit with $\hat{m}^{(0)}=0$, and the chiral condensate $<\bar{\psi}\psi>$ is

$$\begin{aligned}
<\bar{\psi}\psi>&\equiv\lim_{x\to 0}\frac{1}{N}tr\langle 0|T\bar{\psi}(0)\psi(x)|0\rangle=-\lim_{x\to 0}\frac{i}{N}trG(x)\\
&=-\frac{4iN_c}{(2\pi)^4}\int d^4q\frac{B_{\rm dyn}(q^2)}{q^2A^2(q^2)-B_{\rm dyn}^2(q^2)}
\end{aligned} \tag{9.52}$$

(it is calculated in the chiral limit). Indeed, taking into account Eq.(9.38), we get the PCAC formula (9.51) from the relation (9.50).

The relations (9.50), (9.51), (9.52) yield a nontrivial constraint for the ultraviolet asymptotic behavior of the functions $\chi^{(p)}_{ac;1}$ and B_{dyn}. Let us discuss this point in more detail.

First of all, we would like to indicate that $m^{(0)}_a, <\bar{\psi}\psi>, A(q^2), B(q^2), \chi^{(p)}_{ab}(x,y)$ are bare quantities, that is, they depend on ultraviolet cutoff Λ regularizing the theory. In particular,

$$m^{(0)}_a = Z^{(\sigma)}_m(\Lambda) m^{(\sigma)}_{\text{cur};a} \ , \tag{9.53}$$

$$A(q^2) = Z^{(\sigma)^{-1}}_2(\Lambda) A^{(\sigma)}(q^2) \ , \tag{9.54}$$

$$B(q^2) = Z^{(\sigma)^{-1}}_2(\Lambda) B^{(\sigma)}(q^2) \ , \tag{9.55}$$

$$\chi^{(p)}_{ab}(x,y) = Z^{(\sigma)}_2(\Lambda) \chi^{(p)}_{(\sigma);ab}(x,y) \ , \tag{9.56}$$

where $m^{(\sigma)}_{\text{cur};a}$ is the renormalized (current) mass, and $A^{(\sigma)}, B^{(\sigma)}, \chi^{(p)}_{(\sigma);ab}$ are the renormalized functions relating to the subtraction point σ [6] (the mass function $B(q^2)$ is equal to $B_{\text{dyn}}(q^2)$ in the chiral limit). The renormalization constants $Z^{(\sigma)}_m, Z^{(\sigma)}_2$ can be calculated in perturbation theory or in renormalization group (RG) improved perturbation theory. $Z^{(\sigma)}_2$ is the renormalization constant of fermion fields,

$$\psi = \left(Z^{(\sigma)}_2\right)^{1/2} \psi^{(\sigma)} \ . \tag{9.57}$$

Therefore the relations (9.54) - (9.56) follow from Eq.(9.57) and from the definitions of the fermion propagator,

$$G(x) = -i\langle 0|T\psi(x)\bar{\psi}(0)|0\rangle = \frac{1}{(2\pi)^4}\int d^4q e^{-iqx}\left(\hat{q}A(q) - B(q^2)\right)^{-1} \ , \tag{9.58}$$

and the BS wave function (9.42).

On the other hand, such physical quantities as M^2_{ac} and F_{ac} should be finite in the continuum limit $\Lambda \longrightarrow \infty$ in local theories. Therefore the relations (9.50), (9.51) imply that

$$\lim_{\Lambda\longrightarrow\infty} Z^{(\sigma)}_m Z^{(\sigma)}_2 \int^{\Lambda} d^4q \chi^{(p)}_{(\sigma);ab;1} = \text{finite} \ , \tag{9.59}$$

$$\lim_{\Lambda \longrightarrow \infty} Z_m^{(\sigma)} <\bar{\psi}\psi> = - \lim_{\Lambda \longrightarrow \infty} Z_m^{(\sigma)} \frac{4iN_c Z_2^{(\sigma)}}{(2\pi)^4}$$

$$\times \int^{\Lambda} d^4 q \frac{B_{\mathrm{dyn}}^{(\sigma)}(q^2)}{q^2 A^{(\sigma)^2}(q^2) - B_{\mathrm{dyn}}^{(\sigma)^2}(q^2)} = \text{finite} , \tag{9.60}$$

where the symbol $\int^{\Lambda}$ implies the ultraviolet cutoff must be put in these expression after the Wick rotation $(q_0 \to iq_4)$ into Euclidean space.

As an example, let us consider these constraints in QCD with the gauge group $SU(N_c)$. We will use the Landau gauge with the gluon propagator

$$D_{\mu K,\nu N} = \delta_{KN} \frac{1}{k^2} \left(g_{\mu\nu} - \frac{k_\mu k_\nu}{k^2} \right) . \tag{9.61}$$

In the one-loop approximation, in this gauge, the renormalization constants $Z_m^{(\sigma)}, Z_2^{(\sigma)}$ are [6]:

$$Z_m^{(\sigma)} = \left(\frac{\ln\left(\sigma^2/\Lambda_{\mathrm{QCD}}^2\right)}{\ln\left(\Lambda^2/\Lambda_{\mathrm{QCD}}^2\right)} \right)^{c_m} , \tag{9.62}$$

$$Z_2^{(\sigma)} = 1 , \tag{9.63}$$

where Λ_{QCD} is the QCD scale parameter ($\Lambda_{\mathrm{QCD}} \sim 300\mathrm{MeV}$ in the real QCD with $N_c = 3$), $c_m = \frac{3(N_c^2-1)}{16\pi^2 b N_c}$, and $b = \frac{11N_c - 2N}{24\pi^2}$ is the first coefficient of the QCD β function (in order to preserve the property of asymptotic freedom, we consider the case with $N < \frac{11N_c}{2}$).

Let us make the following, reasonable, assumption: as $q^2 \to \infty$ in Euclidean space, the structure of the leading asymptotic term in $\chi_{(\sigma);ab;1}^{(p)}(q^2, Pq, P^2)$ has the form

$$\chi_{(\sigma);ab;1}^{(p)} \sim F(P^2) f(q^2) ,$$

where F and f are some functions. Then, we find that the constraint (9.59) implies that

$$\chi_{(\sigma);ab;1}^{(p)} \underset{q^2 \gg \Lambda_{\mathrm{QCD}}^2}{\sim} (q^2)^{-2} \left(\ln q^2/\Lambda_{\mathrm{QCD}}^2\right)^{c_m - 1} . \tag{9.64}$$

The constraint (9.60) leads to the following asymptotic relation:

$$\frac{B_{\rm dyn}^{(\sigma)}}{A^{(\sigma)2}} \underset{q^2 \gg \Lambda^2_{\rm QCD}}{\sim} \frac{1}{q^2} \left(\ln q^2/\Lambda^2_{QCD}\right)^{c_m - 1} . \tag{9.65}$$

In the derivation of this expression, we used the following formula for the angle integration in Euclidean region (see Sec. B.4 in Appendix B):

$$\int d^4q f(q^2) = \pi^2 \int dq^2 q^2 f(q^2) . \tag{9.66}$$

We will discuss the relations (9.64), (9.65) in more detail in Chapter 12.

9.4. The Exact Relation for the Decay Constant

In this section, we consider the exact relation for the decay constant [7, 8].

The decay constant F_{ab} is defined as

$$\langle 0|\bar{\psi}_a(0)\gamma^\mu\gamma_5\psi_b(0)|P\rangle = i\sqrt{2}F_{ab}P^\mu , \tag{9.67}$$

where $|P\rangle$ is a pseudoscalar state composed of the a-th fermion and the b-th antifermion. The left-hand side of this relation can be expressed through the BS wave function

$$\chi_{ab}^{(p)}(x,y) = \langle 0|T\psi_a(x)\bar{\psi}_b(y)|P\rangle \tag{9.68}$$

in the following way:

$$\begin{aligned}\langle 0|\bar{\psi}_a(0)\gamma^\mu\gamma_5\psi_b(0)|P\rangle &= -\lim_{\substack{x\to 0\\ y\to 0}} tr\left(\gamma^\mu\gamma_5\chi_{ab}^{(p)}(x,y)\right)\\ &= \frac{4}{\sqrt{2}}N_c\frac{1}{(2\pi)^4}\int d^4q\left[P^\mu\chi_{ab;2}^{(p)} + q^\mu\chi_{ab;3}^{(p)}\right] ,\end{aligned} \tag{9.69}$$

where the Lorentz invariant functions $\chi_{ab;2}^{(p)}, \chi_{ab;3}^{(p)}$ appears in the expansion

$$\chi_{ab}^{(p)}(q,P) = \frac{1}{\sqrt{2}}\left[\chi_{ab;1}^{(p)} + \chi_{ab;2}^{(p)}\hat{P} + \chi_{ab;3}^{(p)}\hat{q} + \chi_{ab;4}^{(p)}\sigma^{\mu\nu}\left(P_\mu q_\nu - P_\nu q_\mu\right)\right]\gamma_5 . \tag{9.70}$$

Therefore

$$iF_{ab}P^\mu = \frac{2N_c}{(2\pi)^4}\int d^4q\left[P^\mu\chi_{ab;2}^{(p)} + q^\mu\chi_{ab;3}^{(p)}\right] . \tag{9.71}$$

This relation can be simplified further in the chiral limit if the invariance of vector-like theories under charge conjugation and parity are taken into account (in the chiral limit, pseudoscalars are massless NG bosons).

In Sec.8.6 we showed that parity invariance leads to the relation (see Eq.(8.115))

$$\gamma^0 \chi(i_s q, i_s P)\, \gamma^0 = \eta_P \chi(q, P) \ , \tag{9.72}$$

where $i_s q = (q^0, -\mathbf{q}), i_s P = (P^0, -\mathbf{P})$ and η_p is the parity of the bound state. In the same way, one finds that charge conjugation invariance yields the relation

$$C\chi^T(-q, P)C^{-1} = \eta_C \chi(q, P) \ , \tag{9.73}$$

where η_c is the charge parity of the bound state.

Since $J^{PC} = 0^{-+}$ for the multiplet of NG bosons, the relation (9.73) implies that

$$\begin{aligned} \chi_2^{(p)}(q^2, Pq) &= \chi_2^{(p)}(q^2) + (Pq)^2 \chi_{2;1}^{(p)}(q^2) + \cdots \ , \\ \chi_3^{(p)}(q^2, Pq) &= (Pq)\chi_3^{(p)}(q^2) + (Pq)^3 \chi_{3;1}^{(p)}(q^2) + \cdots \ , \end{aligned} \tag{9.74}$$

where the functions $\chi_i^{(p)}$ are defined in Eq.(9.37) (we recall that $\chi_{ab;i}^{(p)} = \chi_i^{(p)}$ in the chiral limit). Substituting these expressions into Eq.(9.71) and performing the Wick rotation into Euclidean space, we get (see Eq.(9.66) and the formulae in Sec.B.4 in Appendix B):

$$F = \frac{N_c}{32\pi^2} \int duu \left(4\chi_2^{(p)}(u) - u\chi_3^{(p)}(u) \right) \ , \tag{9.75}$$

where $F \equiv F_{ab}$ in the chiral limit, and $u = q^2$ (note here that no higher power terms in the expansions (9.74) contribute to F because they yeild P^2 proportional terms after $d^4 q$ integration and $P^2 = 0$ for NG bosons).

We will frequently use the relation (9.75) in the following.

9.5. Bound states and Decay Constant in the Nambu-Jona-Lasinio Model

In this section we will consider the BS equations and the general relations obtained above in the Nambu-Jona-Lasinio model.

The lagrangian density of the model takes the form (see Eq.(6.1)):

$$\mathcal{L} = i\bar{\psi}\gamma^\mu \partial_\mu \psi - \bar{\psi}\hat{m}^{(0)}\psi + G^{(0)} \sum_{\alpha=0}^{N^2-1} \left\{ \left(\bar{\psi}\frac{\lambda^\alpha}{2}\psi \right)^2 + \left(\bar{\psi}\frac{\lambda^\alpha}{2} i\gamma_5 \psi \right)^2 \right\} . \tag{9.76}$$

As in Chapter 6, we will analyze the model in the limit in which the number of fermion colors N_c goes to infinity.

As was shown in Sec. 6.2 and Sec.8.4, in this model, in leading order of the $1/N_c$ expansion, the Schwinger-Dyson (SD) equation for the fermion propagator takes the following form in the chiral limit $\left(\hat{m}^{(0)} = 0\right)$:

$$A(q^2) = 1 \tag{9.77}$$

$$B_{\rm dyn}(q^2) = \frac{4iG^{(0)}N_c}{(2\pi)^4}\int d^4p \frac{B_{\rm dyn}(p^2)}{p^2 - B^2_{\rm dyn}(p^2) + i\epsilon}\ . \tag{9.78}$$

The solution of Eq.(9.78) is $B_{\rm dyn} = m_{\rm dyn}$ independent of q^2. Therefore this equation is equivalent to the following algebraic one (see Eq.(6.14)):

$$m_{\rm dyn}(\kappa - 1) = \kappa \frac{m^3_{\rm dyn}}{\Lambda^2} \ln \frac{\Lambda^2 + m^2_{\rm dyn}}{m^2_{\rm dyn}}\ , \tag{9.79}$$

where the dimensionless coupling constant κ is

$$\kappa = \frac{G^{(0)}N_c\Lambda^2}{4\pi^2}\ . \tag{9.80}$$

We consider the limit $N_c \longrightarrow \infty$ with κ being fixed.

The nontrivial solution $m_{\rm dyn} \neq 0$ to the equation (9.79) exists if $\kappa > 1$. The value $\kappa = \kappa_c = 1$ is the critical coupling constant separating two phases: at $\kappa < 1$, the model is in the symmetric phase ($m_{\rm dyn} = 0$) with the $U_L(N) \times U_R(N)$ being unbroken, and at $\kappa > 1$, it is in the phase with the spontaneously broken chiral symmetry, $U_L(N) \times U_R(N) \longrightarrow U_V(N)$. Since $m_{\rm dyn}$ goes continuously to zero as $\kappa \longrightarrow \kappa_c = 1$ from the side of the broken phase, the critical value $\kappa_c = 1$ corresponds to the continuous (second order) phase transition.*)

If the bare mass $\hat{m}^{(0)}_{ab} = m^{(0)}_a \delta_{ab}$ is not zero, the SD equation can be derived from the effective action for composite operators $\Gamma(G)$ (8.47), in which as a free fermion propagator, one must take

$$S^{-1}(q) = \hat{q} - \hat{m}^{(0)}\ . \tag{9.81}$$

Then we find the following SD equation for the propagator $G_{ab}(q) = \delta_{ab}(A_a(q^2)\hat{q} - B_a(q^2))^{-1}$ from Eq.(8.2):

$$A_a(q^2) = 1\ , \tag{9.82}$$

*) Phase transitions (critical phenomena) are discussed in more detail in Chapter 11.

$$B_a(q^2) = m_a^{(0)} + \frac{4iG^{(0)}N_c}{(2\pi)^4}\int d^4p \frac{B_a(p^2)}{p^2 - B_a^2(p^2) + i\epsilon} . \tag{9.83}$$

The equation (9.83) is equivalent to the following algebraic equation:

$$m_a(1-\kappa) = m_a^{(0)} - \kappa\frac{m_a^3}{\Lambda^2}\ln\frac{\Lambda^2 + m_a^2}{m_a^2} \tag{9.84}$$

with $m_a \equiv B_a$.

Let us consider now the BS equations for

$$\chi_{ab}^{(p)}(q,P) = \frac{1}{\sqrt{2}}\left[\chi_{ab;1}^{(p)} + \chi_{ab;2}^{(p)}\hat{P} + \chi_{ab;3}^{(p)}\hat{q} + \chi_{ab;4}^{(p)}\sigma^{\mu\nu}\left(P_\mu q_\nu - P_\nu q_\mu\right)\right]\gamma_5 , \tag{9.85}$$

$$\chi_{ab}^{(s)}(q,P) = \frac{1}{\sqrt{2}}\left[\chi_{ab;1}^{(s)} + \chi_{ab;2}^{(s)}\hat{P} + \chi_{ab;3}^{(s)}\hat{q} + \chi_{ab;4}^{(s)}\sigma^{\mu\nu}\left(P_\mu q_\nu - P_\nu q_\mu\right)\right] \tag{9.86}$$

describing colorless pseudoscalar and scalar bosons composed of the a-th fermion and the b-th antifermion.

Using the expression (8.108) for the BS kernel in the NJL model, we find the following equaitons in the leading order of the $1/N_c$ expansion:

a) pseudoscalar $\left(J^{PC} = 0^{-+}\right)$ states:

$$\begin{aligned}\left(\frac{P^2}{4} - q^2 + m_a m_b\right)\chi_{ab;1}^{(p)} - \left(\bar{m}_{ab}P^\mu + \Delta m_{ab}q^\mu\right)\chi_{ab;\mu}^{(p)} - 2i\left(P^2q^2 - (Pq)^2\right)\chi_{ab;4}^{(p)} \\ = -\frac{i\kappa}{\Lambda^2\pi^2}\int d^4k\,\chi_{ab;1}^{(p)} ,\end{aligned} \tag{9.87}$$

$$\begin{aligned}\left(\frac{P^2}{4} - q^2 - m_a m_b\right)\chi_{ab;\mu}^{(p)} + \left(2q_\mu q^\lambda - \frac{1}{2}P_\mu P^\lambda\right)\chi_{ab;\lambda}^{(p)} \\ + \left(\bar{m}_{ab}P_\mu + \Delta m_{ab}q_\mu\right)\chi_{ab;1}^{(p)} - i\left(4\bar{m}_{ab}P^\lambda + \Delta m_{ab}q^\lambda\right)\chi_{ab;\mu\lambda}^{(p)} = 0 ,\end{aligned} \tag{9.88}$$

$$\begin{aligned}\left(\frac{P^2}{4} - q^2 - m_a m_b\right)\chi_{ab;\lambda\mu}^{(p)} - \frac{i}{2}\left(P_\lambda q_\mu - P_\mu q_\lambda\right)\chi_{ab;1}^{(p)} - i\bar{m}_{ab}\left(q_\lambda\chi_{ab;\mu}^{(p)} - q_\mu\chi_{ab;\lambda}^{(p)}\right) \\ - i\frac{\Delta m_{ab}}{4}\left(P_\lambda\chi_{ab;\mu}^{(p)} - P_\mu\chi_{ab;\lambda}^{(p)}\right) = 0 ;\end{aligned} \tag{9.89}$$

b) scalar (0^{++}) states:

$$\begin{aligned}\left(\frac{P^2}{4} - q^2 - m_a m_b\right)\chi_{ab;1}^{(s)} + \left(2\bar{m}_{ab}q^\mu - \frac{\Delta m_{ab}}{2}P^\mu\right)\chi_{ab;\mu}^{(s)} + 2i\left((Pq)^2 - P^2q^2\right)\chi_{ab;4}^{(s)} \\ = -\frac{i\kappa}{\Lambda^2\pi^2}\int d^4k\,\chi_{ab;1}^{(s)} ,\end{aligned} \tag{9.90}$$

$$\begin{aligned}\left(\frac{P^2}{4}-q^2+m_am_b\right)\chi^{(s)}_{ab;\mu}+\left(2q_\mu q^\lambda-\frac{1}{2}P_\mu P^\lambda\right)\chi^{(s)}_{ab;\lambda}\\-\left(2\bar{m}_{ab}q_\mu-\frac{1}{2}\Delta m_{ab}P_\mu\right)\chi^{(s)}_{ab;1}-2i\left(\bar{m}_{ab}P^\lambda-\Delta m_{ab}q^\lambda\right)\chi^{(s)}_{\lambda\mu}=0\ ,\end{aligned}\tag{9.91}$$

$$\begin{aligned}\left(\frac{P^2}{4}-q^2+m_am_b\right)\chi^{(s)}_{ab;\lambda\mu}-\frac{i}{2}\left(P_\lambda q_\mu-P_\mu q_\lambda\right)\chi^{(s)}_{ab;1}\\+\frac{i}{2}\bar{m}_{ab}\left(P_\lambda\chi^{(s)}_{ab;\mu}-P_\mu\chi^{(s)}_{ab;\lambda}\right)-\frac{i}{2}\Delta m_{ab}\left(q_\lambda\chi^{(s)}_{ab;\mu}-q_\mu\chi^{(s)}_{ab;\lambda}\right)=0\ .\end{aligned}\tag{9.92}$$

Here we use the following notations:

$$\chi^{(i)}_{ab;\mu}=P_\mu\chi^{(i)}_{ab;2}+q_\mu\chi^{(i)}_{ab;3},\ \chi^{(i)}_{ab;\lambda\mu}=\left(P_\lambda q_\mu-P_\mu q_\lambda\right)\chi^{(i)}_{ab;4};\ i=p,s\ ,\tag{9.93}$$

$$\bar{m}_{ab}=\frac{m_a+m_b}{2},\ \Delta m_{ab}=m_b-m_a\ .\tag{9.94}$$

The graphic picture of the BS equation for $\chi^{(i)}_{ab}$ is shown in Fig.9.2.

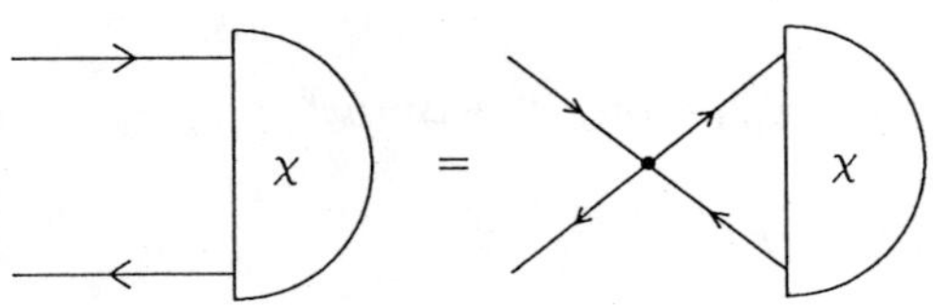

Fig. 9.2. Graphical representation of the homogeneous Bethe-Salpeter equation in the Nambu-Jona-Lasinio model.

Before starting the analysis of these equations, some observations should be made:

1) the right-hand sides in Eqs.(9.88), (9.89), (9.91) and (9.92) equal zero only in leading order in $1/N_c$. This point will be crucial in solving these equations;

2) the equations for scalar and pseudoscalar bosons coincide in the symmetric phase with $m_a=0$. This is because in the symmetric phase these bosons are assigned to the $(N,\bar{N})\oplus(\bar{N},N)$ representation of the $SU_L(N)\times SU_R(N)$ group. Since under parity $P,(N,\bar{N})\longrightarrow(\bar{N},N)$, this representation is irreducible with respect to the extended group including parity P.

In addition, the equations for the invariant functions $\chi^{(i)}_{ab;1}, \chi^{(i)}_{ab;4}$ decouple from the equations for the functions $\chi^{(i)}_{ab;2}, \chi^{(i)}_{ab;3}$ in the symmetric phase. This happens because these two pairs of functions describe states from different representations of the $SU_L(N) \times SU_R(N)$: while $\chi^{(i)}_{ab;1}$ and $\chi^{(i)}_{ab;4}$ correspond to the $(N, \bar{N}) \oplus (\bar{N}, N)$ representation, $\chi^{(i)}_{ab;2}$ and $\chi^{(i)}_{ab;3}$ correspond to the $(N \times \bar{N}, 1) \oplus (1, N \times \bar{N})$ representation.

We begin the analysis of these equations by expressing the functions $\chi^{(p)}_{ab;\mu}, \chi^{(p)}_{ab;\lambda\mu}$ through the function $\chi^{(p)}_{ab;1}$ by using Eqs.(9.88), (9.89) and by expressing the functions $\chi^{(s)}_{ab;\mu}, \chi^{(s)}_{ab;\lambda\mu}$ through the function $\chi^{(s)}_{ab;1}$ by using Eqs.(9.91), (9.92):

$$\chi^{(p)}_{ab;\mu} = \frac{\bar{m}_{ab}P_\mu + \Delta m_{ab}q_\mu}{\left(\frac{P^2}{4} - q^2 + m_a m_b\right)}\chi^{(p)}_{ab;1} , \tag{9.95}$$

$$\chi^{(p)}_{ab;\lambda\mu} = \frac{i\left(P_\lambda q_\mu - P_\mu q_\lambda\right)}{2\left(\frac{P^2}{4} - q^2 + m_a m_b\right)}\chi^{(p)}_{ab;1} , \tag{9.96}$$

$$\chi^{(s)}_{ab;\mu} = \frac{4\bar{m}_{ab}q_\mu + \Delta m_{ab}P_\mu}{2\left(q^2 - \frac{P^2}{4} + m_a m_b\right)}\chi^{(s)}_{ab;1} , \tag{9.97}$$

$$\chi^{(s)}_{ab;\lambda\mu} = \frac{i\left(P_\lambda q_\mu - P_\mu q_\lambda\right)}{2\left(\frac{P^2}{4} - q^2 - m_a m_b\right)}\chi^{(s)}_{ab;1} . \tag{9.98}$$

Therefore Eq.(9.87) and Eq.(9.90) can be transformed into the following equations for the functions $\chi^{(p)}_{ab;1}$ and $\chi^{(s)}_{ab;1}$, respectively:

$$\begin{aligned}&\left(\frac{P^2}{4} - q^2 + m_a m_b\right)\chi^{(p)}_{ab:1}\left(p^2, q^2, Pq\right)\\&\qquad + \frac{\left[P^2q^2 - (Pq)^2 - \bar{m}^2_{ab}P^2 - 2\bar{m}_{ab}\Delta m_{ab}Pq - (\Delta m_{ab})^2 q^2\right]}{\left(\frac{P^2}{4} - q^2 + m_a m_b\right)}\\&\qquad \cdot \chi^{(p)}_{ab;1}(P^2, q^2, Pq) = -\frac{i\kappa}{\pi^2\Lambda^2}\int d^4k\chi^{(p)}_{ab;1}(P^2, k^2, Pk) ,\end{aligned} \tag{9.99}$$

$$\begin{aligned}&\left(\frac{P^2}{4} - q^2 - m_a m_b\right)\chi^{(s)}_{ab;1}\left(P^2, q^2, Pq\right)\\&\qquad + \frac{\left[P^2q^2 - (Pq)^2 - 4\bar{m}^2_{ab}q^2 + \frac{(\Delta m_{ab})^2}{4}P^2\right]}{\left(\frac{P^2}{4} - q^2 - m_a m_b\right)}\chi^{(s)}_{ab;1}\left(P^2, q^2 Pq\right)\\&\qquad = -\frac{i\kappa}{\pi^2\Lambda^2}\int d^4k\chi^{(s)}_{ab;1}(P^2, k^2, Pk) .\end{aligned} \tag{9.100}$$

The meaning of the relations (9.95) - (9.98) is rather simple. Let us introduce the amputated BS wave function

$$\tilde{\chi}^{(i)}_{ab} \equiv G^{-1}_{aa'}\left(q+\frac{P}{2}\right)\chi^{(i)}_{a'b'}G^{-1}_{b'b}\left(q-\frac{P}{2}\right) . \tag{9.101}$$

One can easily check that the relations (9.95) - (9.98) follow from the fact that $\tilde{\chi}^{(p)}_{ab}$ and $\tilde{\chi}^{(s)}_{ab}$ are

$$\tilde{\chi}^{(p)}_{ab} = \gamma_5 \tilde{\chi}^{(p)}_{ab;1} , \quad \tilde{\chi}^{(s)}_{ab} = \tilde{\chi}^{(s)}_{ab;1} , \tag{9.102}$$

that is, the amputated BS functions $\tilde{\chi}^{(p)}_{ab}$ and $\tilde{\chi}^{(s)}_{ab}$ each contain only one Lorentz invariant function, $\tilde{\chi}^{(p)}_{ab;1}$ and $\tilde{\chi}^{(s)}_{ab;1}$, respectively. Thus the $N_c \longrightarrow \infty$ limit simplifies enormously the BS equations.

Although the BS equations are essentially simplified in this limit, they are still rather complicated, thus demonstrating how complicated the dynamics of relativistic bound states is. In order to get an insight into this dynamics, we begin the analysis by considering the chiral limit with $\hat{m}^{(0)} = 0$ and $m_a = m_b = m_{\text{dyn}}$. Because of the Goldstone theorem, we expect that there should be N^2 pseudoscalar NG bosons. Let us consider Eq.(9.99) at $P_\mu = 0, m_a = m_b = m_{\text{dyn}}$:

$$\left(m^2_{\text{dyn}} - q^2\right)\chi^{(p)}_{ab;1}(q^2) = -\frac{i\kappa}{\pi^2\Lambda^2}\int d^4k\chi^{(p)}_{ab;1}(k^2) . \tag{9.103}$$

Comparing this equation with the SD equation (9.78), we see that they coincide if one identifies the amputated function $\tilde{\chi}^{(p)}_{ab;1}(q^2) = \left(m^2_{\text{dyn}} - q^2\right)\chi^{(p)}_{ab;1}$ with $cB_{\text{dyn}} = cm_{\text{dyn}}$, where c is a normalization constant. In fact, the identity (9.38) tells us that $c = 2F^{-1}$. Thus in agreement with the Goldstone theorem, at $\kappa > 1$ and $m_{\text{dyn}} \neq 0$, there are N^2 pseudoscalar NG bosons in the NJL model.

On the other hand, at $P_\mu = 0$, the BS equaiton for the function $\chi^{(s)}_{ab;1}(q^2)$ is

$$-\left(m^2_{\text{dyn}} + q^2\right)\chi^{(s)}_{ab;1}(q^2) + \frac{4m^2_{\text{dyn}}q^2}{q^2 + m^2_{\text{dyn}}}\chi^{(s)}_{ab;1}(q^2) = -\frac{i\kappa}{\pi^2\Lambda^2}\int \chi^{(s)}_{ab;1}(q^2) . \tag{9.104}$$

Let us rewrite this equation and Eq.(9.103) in the following form

$$\left(m^2_{\text{dyn}} - q^2\right)^2\chi^{(p)}_{ab;1}(q^2) = q^2\frac{i\kappa}{\pi^2\Lambda^2}\int d^4k\chi^{(p)}_{ab;1}(k^2) - m^2_{\text{dyn}}\frac{i\kappa}{\pi^2}\int d^4k\chi^{(p)}_{ab;1}(k^2) , \tag{9.105}$$

$$\left(m_{\text{dyn}}^2 - q^2\right)^2 \chi_{ab;1}^{(s)}(q^2) = q^2 \frac{i\kappa}{\pi^2\Lambda^2} \int d^4k \chi_{ab;1}^{(s)}(k^2) + m_{\text{dyn}}^2 \frac{i\kappa}{\pi^2} \int d^4k \chi_{ab;1}^{(s)}(k^2) \,. \quad (9.106)$$

In Euclidean space, they are:

$$\left(m_{\text{dyn}}^2 + q^2\right)^2 \chi_{ab;1}^{(p)}(q^2) = q^2 \frac{\kappa}{\pi^2\Lambda^2} \int d^4k \chi_{ab;1}^{(p)}(k^2) + m_{\text{dyn}}^2 \frac{\kappa}{\pi^2} \int d^4k \chi_{ab;1}^{(p)}(k^2) \quad (9.107)$$

$$\left(m_{\text{dyn}}^2 + q^2\right)^2 \chi_{ab;1}^{(s)}(q^2) = q^2 \frac{\kappa}{\pi^2\Lambda^2} \int d^4k \chi_{ab;1}^{(s)}(k^2) - m_{\text{dyn}}^2 \frac{\kappa}{\pi^2} \int d^4k \chi_{ab;1}^{(s)}(k^2) \quad (9.108)$$

The minus sign in front of the second term on the right hand side of Eq.(9.108) implies an additional repulsion (caused by the generation of the fermion dynamical mass) in the scalar channel as compared to the pseudoscalar one. Because of this repulsion, there are no nontrivial solutions to Eq.(9.108) and hence there are no massless scalars in this phase. We will estimate the mass of scalars below; it will be shown that $M_s^2 \simeq 4m_{\text{dyn}}^2$ at $m_{\text{dyn}}^2/\Lambda^2 \ll 1$.

We note that the appearance of the additional repulsion in the scalar channel is not accidental. It is connected with the point that in the nonrelativistic limit scalars are bound states with orbital angular momentum $l = 1$, while the orbital angular momentum of pseudoscalars is $l = 0$. The additional repulsion in the scalar channel is the relativistic image of this difference between scalar and pseudoscalar bound states.

Let us now analyse Eqs.(9.99) and (9.100) with $P^2 \neq 0$. We will consider the case with $\Delta m_{ab} \ll \bar{m}_{ab}$ when all terms with Δm_{ab} can be neglected. Now the equations take the following form in Euclidean space ($P_0 \longrightarrow iP_4$, $q_0 \longrightarrow iq_4$):

$$\begin{aligned} &\left(-\frac{P^2}{4} + q^2 + m_a m_b\right) \chi_{ab:1}^{(p)}\left(P^2, q^2, Pq\right) \\ &\qquad + \frac{\left[P^2q^2 - (Pq)^2 + \bar{m}_{ab}^2 P^2\right]}{\left(q^2 + m_a m_b - \frac{P^2}{4}\right)} \chi_{ab;1}^{(p)}\left(P^2, q^2, Pq\right) \\ &\qquad = \frac{\kappa}{\pi^2\Lambda^2} \int d^4k \chi_{ab;1}^{(p)}\left(P^2, k^2, Pk\right) \,, \end{aligned} \quad (9.109)$$

$$\begin{aligned} &\left(-\frac{P^2}{4} + q^2 - m_a m_b\right) \chi_{ab:1}^{(s)}\left(P^2, q^2, Pq\right) \\ &\qquad + \frac{\left[P^2q^2 - (Pq)^2 + 4\bar{m}_{ab}^2 q^2\right]}{\left(q^2 - m_a m_b - \frac{P^2}{4}\right)} \chi_{ab;1}^{(s)}\left(P^2, q^2, Pq\right) \\ &\qquad = \frac{\kappa}{\pi^2\Lambda^2} \int d^4k \chi_{ab;1}^{(s)}(P^2, k^2, Pk) \,. \end{aligned} \quad (9.110)$$

These equations are difficult to solve because of the dependence of $\chi^{(i)}_{ab;1}$ on the variable $Pq = |P||q|\cos\theta$. We expand the wave function $\chi^{(i)}_{ab;1}$ in Gegenbauer polynomials [9]:

$$\chi^{(i)}_{ab;1}\left(P^2, q^2, Pq\right) = \sum_{n=0}^{\infty} G^{(i)}_{n;ab}\left(P^2, q^2\right) C^1_n(\cos\theta) \,. \tag{9.111}$$

Substituting the expansion (9.111) into Eqs.(9.109), (9.110) and using the recurrence formula for Gegenbauer polynomials [9],

$$\begin{aligned}
\cos^2\theta C^1_0(\cos\theta) &= \frac{1}{4}\left(C^1_0(\cos\theta) + C^1_2(\cos\theta)\right) \,, \\
\cos^2\theta C^1_{2n}(\cos\theta) &= \frac{1}{4}\left(C^1_{2(n-1)}(\cos\theta) + 2C^1_{2n}(\cos\theta) + C^1_{2(n+1)}(\cos\theta)\right) \,, \quad n \geq 1 \,,
\end{aligned}$$

and some formulae for the angular integrations of C^1_{2n} [9], we obtain the following infinite chain of coupled equations:

$$\begin{aligned}
&\left[\left(-\frac{x^2}{4} + y^2 + \frac{m_a m_b}{\Lambda^2}\right) + \frac{x^2\left(y^2 + \frac{\bar{m}^2_{ab}}{\Lambda^2}\right)}{\left(y^2 - \frac{x^2}{4} + \frac{m_a m_b}{\Lambda^2}\right)}\right] G^{(p)}_{0;ab} \\
&\qquad - \frac{x^2 y^2}{4\left(y^2 - \frac{x^2}{4} + \frac{m_a m_b}{\Lambda^2}\right)}\left(G^{(p)}_{0;ab} + G^{(p)}_{2;ab}\right) \\
&\qquad = \kappa \int_0^1 dz^2 z^2 G^{(p)}_{0;ab}(x^2, z^2) \,,
\end{aligned} \tag{9.112}$$

$$\begin{aligned}
&\left[\left(-\frac{x^2}{4} + y^2 + \frac{m_a m_b}{\Lambda^2}\right) + \frac{x^2\left(y^2 + \frac{\bar{m}^2_{ab}}{\Lambda^2}\right)}{\left(y^2 - \frac{x^2}{4} + \frac{m_a m_b}{\Lambda^2}\right)}\right] G^{(p)}_{2n;ab} \\
&\qquad - \frac{x^2 y^2}{4\left(y^2 - \frac{x^2}{4} + \frac{m_a m_b}{\Lambda^2}\right)}\left(G^{(p)}_{2n-2;ab} + 2G^{(p)}_{2n;ab} + G^{(p)}_{2n+2;ab}\right) = 0, \; n \geq 1 \,,
\end{aligned} \tag{9.113}$$

and

$$\left[\left(-\frac{x^2}{4}+y^2-\frac{m_a m_b}{\Lambda^2}\right)+\frac{y^2\left(x^2+4\frac{\bar{m}_{ab}^2}{\Lambda^2}\right)}{\left(y^2-\frac{x^2}{4}-\frac{m_a m_b}{\Lambda^2}\right)}\right]G^{(s)}_{0;ab}$$

$$-\frac{x^2y^2}{4\left(y^2-\frac{x^2}{4}-\frac{m_a m_b}{\Lambda^2}\right)}\left(G^{(s)}_{0;ab}+G^{(s)}_{2;ab}\right) \tag{9.114}$$

$$=\kappa\int_0^1 dz^2 z^2 G^{(s)}_{0;ab}(x^2,z^2)\,,$$

$$\left[\left(-\frac{x^2}{4}+y^2-\frac{m_a m_b}{\Lambda^2}\right)+\frac{y^2\left(x^2+4\frac{\bar{m}_{ab}^2}{\Lambda^2}\right)}{\left(y^2-\frac{x^2}{4}-\frac{m_a m_b}{\Lambda^2}\right)}\right]G^{(s)}_{2n;ab}$$

$$-\frac{x^2y^2}{4\left(y^2-\frac{x^2}{4}+\frac{m_a m_b}{\Lambda^2}\right)}\left(G^{(s)}_{2n-2;ab}+2G^{(s)}_{2n;ab}+G^{(s)}_{2n+2;ab}\right)=0,\ n\geq 1\,, \tag{9.115}$$

where we introduced the dimensionless quantitis

$$x^2=P^2/\Lambda^2,\ y^2=q^2/\Lambda^2,\ z^2=k^2/\Lambda^2\,. \tag{9.116}$$

It is clear that in order to find solutions of these two infinite chains of coupled equations, one should find a way to truncate them.

We note that at $x^2=0$ ($P^2=0$), the wave function $\chi^{(i)}_{ab;1}$ is $\chi^{(i)}_{ab;1}=G^{(i)}_{0;ab}$, and all $G^{(i)}_{2n;ab}$ with $n\geq 1$ equal zero; also, Eqs.(9.113) and (9.115) imply that at $x^2\ll y^2$, the ratio $G^{(i)}_{2(n+1);ab}/G^{(i)}_{2n;ab}$ is

$$G^{(i)}_{2(n+1);ab}/G^{(i)}_{2n;ab}\sim x^2\,. \tag{9.117}$$

Therefore in first order in x^2, the terms with $G^{(i)}_{2;ab}$ in Eqs.(9.112), (9.114) can be neglected, and as a result, in this order, we get rather simple equations for $G^{(p)}_{0;ab}$ and $G^{(s)}_{0;ab}$.

However the main question is: why may we assume that $x^2\ll y^2$? Fortunately, in the NJL model, there does exist a dynamical regime in which this assumption is justified. The point is that when the coupling constant κ is close to the critical value $\kappa_c=1$, i.e., $0<\kappa-\kappa_c\ll 1$, the dynamcial fermion mass $m_{\rm dyn}$ is $m_{\rm dyn}\ll\Lambda$ (see Eq.(9.79)).

One may expect that in this near-critical regime, the masses of scalar and pseudoscalar bound states should be also much smaller than Λ, that is, $x^2 \ll 1$. We will show that this expectation is indeed fulfilled. As to the assumption $x^2 \ll y^2$ $(P^2/\Lambda^2 \ll q^2/\Lambda^2)$, it is equivalent to that assumption that in the NJL model with a near-critical coupling constant κ, the ultraviolet region with $q^2 \sim \Lambda^2$ is primarly responsible for the formation of the bound states. As we will see in Chapter 12, this indeed happens there. In fact, such a trunctation of the BS equations corresponds to a truncation of the expansion of the effective action in powers of the field derivatives; a "small" parameter in that expansion is $1/\ln(\Lambda^2/P^2)$ (see Sec. 12.11).

As result of the truncation, we get the following BS equations valid to the first order in x^2 and $\frac{m_a m_b}{\Lambda^2}$:

$$\left(q^2 + \frac{P^2}{2} + m_a m_b\right) \chi^{(p)}_{ab;1}(P^2, q^2) = \frac{\kappa}{\Lambda^2} \int_0^{\Lambda^2} dk^2 k^2 \chi^{(p)}_{ab;1}(P^2, k^2) \,, \quad (9.118)$$

$$\left(q^2 + \frac{P^2}{2} + 4\bar{m}^2_{ab} - m_a m_b\right) \chi^{(s)}_{ab;1}(P^2, q^2) = \frac{\kappa}{\Lambda^2} \int_0^{\Lambda^2} dk^2 k^2 \chi^{(s)}_{ab;1}(P^2, k^2) \,. \quad (9.119)$$

We would like to recall that these equations are valid in leading order of the $1/N_c$ expansion and in the nearcritical regime with $|\kappa - \kappa_c| \ll 1$, $\kappa_c = 1$.

With the substitution $\frac{P^2}{2} + m_a m_b \longrightarrow m^2_{\text{dyn}}$, Eq.(9.118) transforms to Eq.(9.103). Eq.(9.119) transforms to that equation by substituting $\frac{P^2}{2} + 4\bar{m}^2_{ab} - m_a m_b \longrightarrow m^2_{\text{dyn}}$. Therefore we can easily solve them.

Let us begin their analysis by considering the model with $\hat{m}^{(0)} = 0$, $m_a = m_b = m_{\text{dyn}}$. The spectrum of ${M^{(p)}}^2 = -P^2$ of pseudoscalar bound states and the spectrum of ${M^{(s)}}^2 = -P^2$ of scalar bound states are determined from the following equations

$$\kappa - 1 = \kappa \frac{\left(m^2_{\text{dyn}} - \frac{{M^{(p)}}^2}{2}\right)}{\Lambda^2} \ln \left(\frac{\Lambda^2 + m^2_{\text{dyn}} - \frac{{M^{(p)}}^2}{2}}{m^2_{\text{dyn}} - \frac{{M^{(p)}}^2}{2}} \right) \,, \quad (9.120)$$

$$\kappa - 1 = \kappa \frac{\left(3m^2_{\text{dyn}} - \frac{{M^{(s)}}^2}{2}\right)}{\Lambda^2} \ln \left(\frac{\Lambda^2 + 3m^2_{\text{dyn}} - \frac{{M^{(s)}}^2}{2}}{3m^2_{\text{dyn}} - \frac{{M^{(s)}}^2}{2}} \right) \,. \quad (9.121)$$

The dynamical mass $m_{\rm dyn}$ is a solution of the SD equation (9.84) with $m_a^{(0)} = 0$, $m_a = m_{\rm dyn}$. There is always the trivial solution $m_{\rm dyn} = 0$, and when $\kappa > 1$, there is also a nontrivial solution satisfying the equation

$$\kappa - 1 = \kappa \frac{m_{\rm dyn}^2}{\Lambda^2} \ln \frac{\Lambda^2 + m_{\rm dyn}^2}{m_{\rm dyn}^2} . \tag{9.122}$$

Comparing this equation with Eqs.(9.120), (9.121), we are led to the following conclusions:

a) *the symmetric phase*; $m_{\rm dyn} = 0$.

At $\kappa < 1$, the SD equation (9.122) admits only the trivial solution. In this case, Eqs.(9.120), (9.121) have solutions with complex $M^{(p)^2} = M^{(s)^2} \equiv (M - (i/2)\Gamma)^2$. For κ close to 1, keeping only leading terms in Γ/M, we find:

$$M^2 \simeq \frac{2(1-\kappa)}{\kappa \ln [\kappa/(1-\kappa)]} \Lambda^2 , \tag{9.123}$$

$$\frac{\Gamma}{M} \simeq \frac{\pi}{\ln [\kappa/(1-\kappa)]} . \tag{9.124}$$

These states are unstable since they decay into massless fermion-antifermion pairs. When $1 - \kappa \ll 1$, their width Γ is small. Thus in this case there are $2N^2$ spinless resonances in the symmetric phase.

Let us now consider the case with $\kappa > 1$. Eqs.(9.120), (9.121) with $m_{\rm dyn} = 0$ and $\kappa > 1$ lead to $2N^2$ scalar and pseudoscalar tachyons with

$$M^{(s)^2} = M^{(p)^2} < 0 . \tag{9.125}$$

Therefore at $\kappa > 1$, the symmetric phase is unstable;

b) *the asymmetric phase*; $m_{\rm dyn} \neq 0$.

The solution $m_{\rm dyn} \neq 0$ to Eq.(9.122) exists only if $\kappa > 1$. As we showed above, the symmetric phase is unstable at $\kappa > 1$: there are $2N^2$ tachyons with the same mass, $M^2 < 0$. Eqs.(9.120), (9.121) and (9.122) show that in the asymmetric phase, while the N^2 pseudoscalars become NG bosons, the N^2 scalars acquire the mass

$$M^{(s)} = 2m_{\rm dyn} . \tag{9.126}$$

This is the famous NJL relation. We emphasize that it is valid only if $N_c \longrightarrow \infty$ and $\kappa - 1 \ll 1$.

We would like to note that the BS equations for bound states (collective excitations) in the BCS theory of superconductivity can be analysed in the same way. In particular one can show that the gap μ in the spectrum of excitations connected with the fluctuations of the density of the superconducting current $n_s(x)$ (see Sec.2.10) is $\mu = 2|M_d|$, where $|M_d|$ is the gap ("nonrelativistic Majorana mass") in the electron spectrum.

The results we have just derived in the NJL model yield the dynamical realization of spontaneous chiral symmetry breaking through the formation of spinless bound states. One can say that the results yield a dynamical realization of the linear σ-model considered in Chapter 4.

In principle, in the same way, one can also study other bound states (as vector and axial-vector bosons, baryons, etc.) in the NJL model. The result is that in leading order in $1/N_c$ and at $|\kappa - 1| \ll 1$ there are no light bound states (with $M \ll \Lambda$) among them. We will clarify the reason for this in Sec.15.4.

Let us now show that the NJL model reproduces also the PCAC mass relation (9.51) for pseudoscalar bosons.

We consider the mass matrix $\hat{m}_{ab}^{(0)} = m_a^{(0)}\delta_{ab}$; the physical mass m_a is a solution of the SD equation (9.84).

Since the mass of pseudoscalars appears only in the combination $\left(-\frac{M_{ab}^{(p)^2}}{2} + m_a m_b\right)$ in Eq.(9.118), it is clear that the expression for $M_{ab}^{(p)^2}$ takes the form:

$$M_{ab}^{(p)^2} = 2m_a m_b - \Lambda^2 f(\kappa) , \tag{9.127}$$

where the function $f(\kappa)$ is determined from Eq. (9.118). Actually, we even do not need to know its explicit form: the point is that at $m_a^{(0)} = 0$, the mass m_a is equal to the dynamical mass m_{dyn}, and pseudoscalars are NG bosons with $M_{ab}^{(p)} = 0$. Therefore we find that

$$f(\kappa) = 2\frac{m_{\text{dyn}}^2}{\Lambda^2} . \tag{9.128}$$

This equaiton and Eq.(9.127) imply that

$$M_{ab}^{(p)^2} = 2\left(m_a m_b - m_{\text{dyn}}^2\right) . \tag{9.129}$$

Let us write the mass m_a as

$$m_a = m_{\rm dyn} + m_{{\rm cur};a} \ , \tag{9.130}$$

where, by definition, $m_{{\rm cur};a}$ is the current fermion mass; the relation between $m_{{\rm cur};a}$ and $m_a^{(0)}$ is determined from the SD equation (9.84).

We get the following mass relation from Eqs.(9.129) and (9.130):

$$M_{ab}^{(p)^2} = 2m_{\rm dyn}\left(m_{{\rm cur};a} + m_{{\rm cur};b}\right) + 2m_{{\rm cur};a}m_{{\rm cur};b} \ . \tag{9.131}$$

Let us show that in the limit $m_{{\rm cur};a}, m_{{\rm cur};b} \ll m_{\rm dyn}$, this relation is equivalent to the PCAC relation (9.51).

Let us first calculate the decay constant F. In the PCAC approximation, it is sufficient to calculate F in the chiral limit with $m_a = m_{\rm dyn}$. Eqs.(9.71) and (9.95) imply that

$$iFP_\mu = \frac{2N_c}{(2\pi)^4}P_\mu \int d^4q \frac{m_{\rm dyn}}{\left(m_{\rm dyn}^2 - q^2\right)}\chi_{ab;1}^{(p)}\left(M^{(p)^2}, q^2\right)\Big|_{M^{(p)^2}=0} \ . \tag{9.132}$$

Using Eq.(9.38), we find that this relation takes the following form in Euclidean region:

$$F^2 = \frac{N_c}{4\pi^2}\int_0^{\Lambda^2} dq^2 q^2 \frac{m_{\rm dyn}^2}{\left(m_{\rm dyn}^2 + q^2\right)^2} \simeq m_{\rm dyn}^2 \frac{N_c}{4\pi^2}\ln\frac{\Lambda^2}{m_{\rm dyn}^2} \ . \tag{9.133}$$

We will discuss the logarithmic divergence in F^2 in a moment.

The condensate $<\bar\psi\psi>$ is

$$<\bar\psi\psi> = -\frac{1}{N}\lim_{x\longrightarrow 0} trG(x) = -\frac{N_c}{4\pi^2}\int_0^{\Lambda^2} dq^2 \frac{q^2 m_{\rm dyn}}{q^2 + m_{\rm dyn}^2} \simeq -\frac{N_c}{4\pi^2}\Lambda^2 m_{\rm dyn} \ . \tag{9.134}$$

The last thing which we need for comparing Eqs.(9.51) and (9.131) is the relation between $m_a^{(0)}$ and $m_{{\rm cur};a}$. In first order in $m_{{\rm cur};a}$, we find the following relation from the SD equation (9.84):

$$m_a^{(0)} = \frac{2m_{{\rm cur};a}m_{\rm dyn}^2}{\Lambda^2}\ln\frac{\Lambda^2}{m_{\rm dyn}^2} \ . \tag{9.135}$$

Eqs.(9.133), (9.134) and (9.135) imply that the mass relations (9.51) and (9.131) are equivalent in first order in $m_{{\rm cur};a}$.

Let us now return to the logarithmic divergence in F^2 (9.133). At the first glance, this may appear surprising: the decay constant is an observable, and it should be finite in the continuum limit. To understant this point, let us consider the Goldberger-Treiman relation (see Sec.4.5):

$$g_Y = \frac{g_A m_{\text{dyn}}}{F} \ . \tag{9.136}$$

It is easy to check that $g_A = 1$ in leading order in $1/N_c$. Then Eqs.(9.133) and Eq.(9.136) imply that the Yukawa coupling g_Y, determining the interactions between fermions and NG bosons in the low energy region, goes to zero as $\Lambda \longrightarrow \infty$. This point shows that the NJL model should be considered as an effective theory describing the dynamics up to a scale of the order of Λ. Therefore we should keep cutoff Λ as a free paramter in the model. In more realistic theories, like QCD, the decay constant is finite in continuum limit. We will discuss this point in more detail in Chapter 12.

9.6. Why Does the Fermion Dynamical Mass Stabilize the Vacuum ?

At $\kappa > 1$, there are tachyons in the symmetric vacuum of the NJL model. Their presence implies that the vacuum is unstable. In fact, as was shown in Sec.7.7, a tachyon leads to an imaginary part of the vacuum energy density:

$$\text{Im}V = -\frac{M^4}{64\pi}, \ M^2 < 0 \ . \tag{9.137}$$

Why does the appearance of the dynamical mass remove tachyons and so stabilize the vacuum ?

The answer to this question is in fact given in Eq.(9.127):

$$M_{ab}^{(p)^2} = 2m_a m_b - \Lambda^2 f(\kappa) \ . \tag{9.138}$$

At $\kappa > 1$, the function $f(\kappa)$ is positive, and $m_a = 0$ implies that there are tachyons in the symmetric phase. Eq.(9.138) shows that the square of the mass of a bound state monotonically increases with increasing masses of its constituents. At $f(\kappa) = 2m_{\text{dyn}}^2/\Lambda^2$, the tachyons become massless NG bosons. Therefore there are no tachyons in the vacuum with spontaneous chiral symmetry breaking, and it is stable.

This process is illustrated in Fig.9.3. At $m_a = m_b = 0$, there are scarar and pseudoscalar tachyons with the same mass: $M^{(s)^2} = M^{(p)^2} < 0$. Both $M^{(s)^2}$ and $M^{(p)^2}$ increase with increasing $m = m_a = m_b$. At $m = m^{(s)}, M^{(s)^2}$ equals zero, however, $M^{(p)^2}$ is still negative at this value of m. Only at $m = m_{\text{dyn}}$, all the tachyons disappear from the spectrum: there are now the N^2 NG pseudoscalar bosons and the N^2 massive scalars with $M^{(s)^2} \simeq 4m_{\text{dyn}}$.

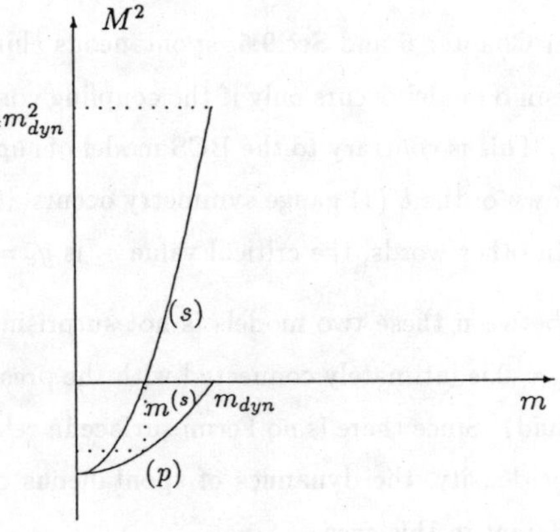

Fig. 9.3. An illustration of the behavior of M^2 as a function of the fermion mass m for scalar $((s))$ and pseudoscalar $((p))$ bound states in the Nambu-Jona-Lasinio model.

This consideration shows that the vacuum stabilization by generating dynamical fermion mass is a rather universal phenomenon.

Chapter 10

Critical Coupling Constant in the Problem of Dynamical Chiral Symmetry Breaking in Gauge Theories

10.1. Introduction

As was shown in Chapter 6 and Sec.9.5, spontaneous chiral symmetry breaking in the Nambu-Jona-Lasinio model occurs only if the coupling constant κ is larger than the critical value $\kappa_c = 1$. This is contrary to the BCS model of superconductivity where the spontaneous breakdown of the $U(1)$ gauge symmetry occurs at any positive value of the coupling constant. In other words, the critical value g_c is $g_c = 0$ in the BCS model.

This difference between these two models is not surprising. As was emphasized in Sec.2.8, the result $g_c = 0$ is intimately connected with the presence of a Fermi surface in a fermion gas (or liquid). Since there is no Fermi surface in relativistic field models with zero fermion number density, the dynamics of spontaneous chiral symmetry breaking should be rather different in this case.

This point can also be understood in the following way. Spontaneous chiral symmetry breaking leads to a nonzero fermion dynamical mass m_{dyn} and to the appearance of pseudoscalar massless NG bosons composed of a massive fermion and a massive antifermion. Therefore NG bosons are characterized by a large binding energy, $\Delta E = 2m_{dyn} - M^{(p)} = 2m_{dyn}$. It is natural that the dynamics producing such bound states should be rather strong.

Of course, the "strong dynamics" may imply not a large coupling constant but a specific form of the kernel in the BS equation for NG bosons. What we would like indicate here however is that the presence of the critical coupling in the problem of spontaneous chiral symmetry breaking is hardly surprising. The important point is to understand its dynamical origin and to estimate its value in concrete models.

In this chapter, we will consider qualitative features of the dynamics of spontaneous chiral symmetry breaking in vector-like gauge theories. We will show that a nonzero critical coupling constant appears naturally in this problem, and it admits a clear physical interpretation.

The history of the problem of the dynamical generation of a fermion mass in quantum gauge theories traces back to the paper of Landau, Abrikosov and Khalatnikov (LAK) [1], in which the running (in the modern language) fermion mass was considered in QED. In the approximation corresponding to the one-loop approximation for the renormalization group equations, they derived the following expression for the function $m(p^2) = B(p^2)/A(p^2)$ in Euclidean region (the fermion propagator $G(p) = \left[\hat{p}A(p^2) - B(p^2)\right]^{-1}$ with $\hat{p} = p_\mu \gamma^\mu$):

$$m(p^2) \underset{p^2 \gg m^2}{\simeq} m \left[1 - \frac{Ne^2}{3\pi} \ln \frac{p^2}{m^2}\right]^{9/4N} , \tag{10.1}$$

where m is the physical fermion mass, and N is the number of fermion flavors. The bare mass $m^{(0)}$ is given by $m(p^2)|_{p^2=\Lambda^2}$, Λ is ultraviolet cutoff, that is,

$$m^{(0)} = m \left[1 - \frac{Ne^2}{3\pi} \ln \frac{\Lambda^2}{m^2}\right]^{9/4N} = m \left(\frac{\alpha}{\alpha^{(0)}}\right)^{9/4N} , \tag{10.2}$$

where $\alpha \equiv \frac{e^2}{4\pi} \simeq \frac{1}{137}$, and the bare coupling $\alpha^{(0)}$ is

$$\alpha^{(0)} \simeq \frac{\alpha}{1 - \frac{4N\alpha}{3} \ln \frac{\Lambda^2}{m^2}} . \tag{10.3}$$

The approximation is valid at $\alpha^{(0)} \lesssim 1$.

The expression (10.2) shows that at $\alpha^{(0)} \sim 1$, the ratio $\frac{m^{(0)}}{m}$ is rather small if N is not too large. LAK wrote that "it can evidently be supposed that $\cdots$ the mass of an electron is of wholly electromagnetic origin". They also emphasized that this point might be established only beyond perturbative theory.

The further development of the problem was connected with the program of Baker, Johnson and Willey (BJW) [2, 3] and Adler [4] (see also Refs.[5, 6, 7]) known as finite QED. This was based on the result of Gell-Mann and Low [8] about the necessity of the existence of an ultraviolet stable fixed point $\alpha^{(0)} = \alpha_f$ for constructing local interacting QED. The BJW analogues of the relations (10.1), (10.2) are (at $\alpha_f \ll 1$)

$$m(p^2) \underset{p^2 \gg m^2}{\simeq} m \left(\frac{p^2}{m^2}\right)^{-\frac{3\alpha_f}{4\pi}} , \tag{10.4}$$

$$m^{(0)} \simeq m \left(\frac{\Lambda^2}{m^2}\right)^{-\frac{3\alpha_f}{4\pi}} \underset{\Lambda \longrightarrow \infty}{\longrightarrow} 0 \,. \tag{10.5}$$

Thus, the bare mass $m^{(0)}$ is necessary equal to zero in finite QED!

A problem with this program was in fact to show how one can overcome the Goldstone theorem implying that spontaneous chiral symmetry breaking in QED leads to the existence of a pseudoscalar electron-positron massless bound state (massless para-positronium!) which is certainly absent in nature. Baker, Johnson and Lee [3] (see also Ref.[6]) gave arguments showing that in the general case the equality $m^{(0)} = 0$ in local QED does not imply the conservation of the axial current. Therefore the Goldstone theorem may be avoided even at $m^{(0)} = 0$.

Thus at that time, the conventional wisdom was that there is no spontaneous chiral symmetry breaking in QED, even although the fermion mass m may be dynamical [3, 6, 7, 9].

The situation changed since the works of Maskawa and Nakajima [10] and Fomin and Miransky [11] (see also Ref.[12]). Maskawa and Nakajima studied the Schwinger-Dyson (SD) equation for the fermion propagator in the ladder approximation in QED with a finite cutoff Λ. They showed that there is a spontaneous chiral symmetry breaking solution at the values of the coupling constant α larger than the critical value $\alpha_c = \frac{\pi}{3}$ (in Landau gauge). Fomin and Miransky, using renormalization group arguments and analyzing Dirac's equation with the supercritical Coulomb field $-\frac{Ze^2}{4\pi r}$, $Z > Z_c \simeq 137$, put forward the hypothesis about the principal role of the supercritical Coulomb-like forces in triggering dynamical chiral symmetry breaking in gauge theories, in particular, they suggested that the vacuum rearrangement underlying this phenomenon is intimately connected with the vacuum rearrangement in the case of a charge fermion in the external supercritical Coulomb field.

The next important step was done by Fukuda and Kugo [13] who explicitly analyzed the solution of the SD equation for the fermion propagator in ladder QED. Fomin, Gusynin and Miransky [14, 15] analyzed the Bethe-Salpeter (BS) equations for spinless bound states in ladder QED and showed that the equations describe consistently spontaneous chiral symmetry breaking at supercritical $\alpha > \alpha_c = \pi/3$.

Thus, these studies revealed the existence of a new supercritical phase, at least in the ladder approximation, in QED. The present author put forward the hypothesis [16, 17] that the supercritical phase in QED exists beyond the ladder approximation. He in particular suggested that the renormalization structure in the supercritical phase should be different from that in perturbative phase; the critical value $\alpha = \alpha_c$ was treated as an ultraviolet stable fixed point defining the continuum limit in supercritical QED (see also the reviews [18]).

Bardeen, Leung and Love [19] indicated that because of a large anomalous dimension of the composite operators $\bar{\psi}\psi$ and $i\bar{\psi}\gamma_5\psi$ at $\alpha = \alpha_c$, the chiral invariant four-fermion operator

$$(\bar{\psi}\psi)^2 + (\bar{\psi}i\gamma_5\psi)^2 \tag{10.6}$$

becomes a marginally relevant one, i.e. its dynamical dimension becomes $d = 4$ at $\alpha = \alpha_c$. Therefore this operator may play important role in the dynamics of supercritical QED.*)

One should indicate that, in parallel with these analytical studies, QED was studied in lattice computer simulations. The Illinois group discovered the second order chiral phase transition at a nonzero $\alpha = \alpha_c$ in quenched lattice QED [20]. Then the systematic studies of Kogut, Dagotto Kocic and Hands et al. [21], Horowitz [22], the Edinburgh group [23] and the DESY group [24] showed the existence of such a phase transition in full QED, at least with a small ($N \leq 8$) number of fermion flavors. We will discuss these results in more detail in Sec. 14.6.

The important point is that the dynamics with critical and supercritical Coulomb-like forces found its applications in such important problems as chiral symmetry breaking in QCD and electroweak symmetry breaking. We will discuss these applications in Chapters 12 - 15. In this chapter, we will discuss the basic properties of this dynamics.

10.2. The Schwinger-Dyson equation in the Rainbow (Ladder) Approximation in QED

In Sec.8.4, we considered the two-loop effective action for composite operators in QED and derived the SD equation for the fermion propagator $G(p) = \left(\hat{p}A(p^2) - B(p^2)\right)^{-1}$. In Landau gauge, the equations for $A(p^2)$ and $B(p^2)$ take the following form in Euclidean

*) For discussions concerning the notion of a marginally relevant operator see Sec.11.2.

region:

$$A(p^2) = 1 , \tag{10.7}$$

$$B(p^2) = m^{(0)} + \frac{3\alpha}{4\pi} \int_0^{\Lambda^2} dq^2 \frac{q^2}{q^2 + B^2(q^2)} \left[\frac{\theta(p^2 - q^2)}{p^2} + \frac{\theta(q^2 - p^2)}{q^2} \right] B(q^2) , \tag{10.8}$$

where, for simplicity, we consider the same bare mass $m^{(0)}$ for all fermion flavors.

The graphical picture of the SD equation is shown in Fig.8.6. In the literature, the present approximation is called the rainbow (or ladder) approximation. As will be shown in Sec.10.5, the Landau gauge is most preferable in this approximation from the viewpoint of the Ward-Takahashi identities.

Differentiating equation (10.8) in p^2, we find that its solutions satisfy the differential equation

$$\frac{d}{dp^2} \left[p^2 \frac{dB(p^2)}{dp^2} \right] + \frac{3\alpha}{4\pi} \frac{B(p^2)}{(p^2 + B(p^2))} = 0 \tag{10.9}$$

and two boundary conditions:

$$\lim_{p^2 \to 0} \left((p^2)^2 \frac{dB(p^2)}{dp^2} \right) = 0 , \tag{10.10}$$

$$\lim_{p^2 \to \Lambda^2} \left[p^2 \frac{dB(p^2)}{dp^2} + B(p^2) \right] = m^{(0)} . \tag{10.11}$$

The nonlinear equation (10.9) can be solved only numerically [13]. However one can use the following linearized version of Eq.(10.9) which approximates very well the original equation in the whole Euclidean region [14, 18]:

$$\frac{d}{dp^2} \left[p^2 \frac{dB}{dp^2} \right] + \frac{3\alpha}{4\pi} \frac{B}{(p^2 + m^2)} = 0 , \tag{10.12}$$

where $m \equiv B(0)$. It is clear that this equation is a good approximation to Eq.(10.9) in both infrared $(p^2 \ll m^2)$ and ultraviolet $(p^2 \gg m^2)$ regions. The numerical analysis shows that it is a good approximation even at $p^2 \sim m^2$.

We emphasize however that it is a good approximation to Eq.(10.9) in Euclidean region only. The properties of these two equations are essentially different in Minkowski region. We shall return to this point in Sec.10.10.

The great advantage of Eq.(10.12) is that its general solution is expressed through the hypergeometric function [25]:

$$\begin{aligned} B(p^2) =& C_1 F\left(\frac{1+\omega}{2}, \frac{1-\omega}{2}, 2; -p^2/m^2\right) \\ &+ C_2 \Bigg[\left(\frac{p^2}{m^2}\right)^{-\frac{(1+\omega)}{2}} F\left(\frac{1+\omega}{2}, \frac{\omega-1}{2}, 1+\omega; -p^2/m^2\right) \\ &+ \left(\frac{p^2}{m^2}\right)^{-\frac{(1-\omega)}{2}} F\left(\frac{1-\omega}{2}, \frac{-\omega-1}{2}, 1-\omega; -p^2/m^2\right)\Bigg], \end{aligned} \tag{10.13}$$

where $\omega = (1 - 3\alpha/\pi)^{1/2}$. Using the expansion for the hypergeometric functions near $p^2 = 0$, one finds from Eq.(10.10) that $C_2 = 0$, and therefore $C_1 = m$, that is,

$$B(p^2) = mF\left(\frac{1+\omega}{2}, \frac{1-\omega}{2}, 2; -p^2/m^2\right). \tag{10.14}$$

The second boundary condition (10.11) yields a relation between $m, m^{(0)}$ and Λ. In order to find it, we need the asymptotic form of $B(p^2)$ (10.14):

$$\begin{aligned} B(p^2) \underset{p^2 \gg m^2}{\simeq} & m \Bigg[\frac{\Gamma(\omega)}{\Gamma\left(\frac{\omega+1}{2}\right)\Gamma\left(\frac{\omega+3}{2}\right)} \left(\frac{p^2}{m^2}\right)^{\frac{\omega-1}{2}} \\ & + \frac{\Gamma(-\omega)}{\Gamma\left(\frac{1-\omega}{2}\right)\Gamma\left(\frac{3-\omega}{2}\right)} \left(\frac{p^2}{m^2}\right)^{\frac{-(\omega+1)}{2}} \Bigg]; at\ \alpha < \frac{\pi}{3}, \end{aligned} \tag{10.15}$$

$$B(p^2) \underset{p^2 \gg m^2}{\simeq} \frac{2}{\pi} m \left(\frac{p^2}{m^2}\right)^{-\frac{1}{2}} \left(\ln\frac{p^2}{m^2} + 2(2\ln 2 - 1)\right); at\ \alpha = \frac{\pi}{3}, \tag{10.16}$$

$$B(p^2) \underset{p^2 \gg m^2}{\simeq} m \left(\frac{8\operatorname{cth}\frac{\pi\tilde{\omega}}{2}}{\pi\tilde{\omega}(\tilde{\omega}^2+1)}\right)^{1/2} \left(\frac{p^2}{m^2}\right)^{-\frac{1}{2}} \sin\left(\frac{\tilde{\omega}}{2}\ln\frac{p^2}{m^2} + \tilde{\omega}(2\ln 2 - 1)\right); at\ \alpha > \frac{\pi}{3}, \tag{10.17}$$

where $\tilde{\omega} = \left(\frac{3\alpha}{\pi} - 1\right)^{1/2}$.

Substituting these expressions in Eq.(10.11), we lead to the following conclusions:

a) $\alpha < \alpha_c = \frac{\pi}{3}$.

The relation between $m, m^{(0)}$ and Λ is

$$m^{(0)} \simeq m \frac{\Gamma(\omega)}{\Gamma\left(\frac{\omega+1}{2}\right)\Gamma\left(\frac{\omega+3}{2}\right)} \left(\frac{\Lambda}{m}\right)^{\omega-1} . \tag{10.18}$$

At $\alpha \ll 1$, it transforms to the BJW relation (10.5). This is not surprising since there is no charge renormalization in the ladder approximation (the renormalization constant $Z_3 = 1; \alpha^{(0)} = Z_3^{-1}\alpha$), and therefore each value of α can be considered as an ultraviolet stable fixed point (in fact it is true only at $\alpha < \alpha_c = \frac{\pi}{3}$; the case of supercritical $\alpha, \alpha > \alpha_c$, will be discussed below).

The relation (10.18) implies that at finite Λ, if $m^{(0)} = 0$, the physical mass is also equal to zero. Therefore there is no spontaneous chiral symmetry breaking at $\alpha < \alpha_c$ and finite cutoff Λ. One the other hand, the bare mass $m^{(0)}$ goes to zero as $\Lambda \longrightarrow \infty$, and therefore the physical mass m can be considered as purely dynamical in this continuum limit. Does spontaneous chiral symmetry breaking take place in the continuum limit $\Lambda \longrightarrow \infty$? We will show in Sec.10.7 that the answer to this question is negative;

b) $\alpha = \alpha_c = \frac{\pi}{3}$.

The relation between $m, m^{(0)}$ and Λ is now

$$m^{(0)} \simeq m\frac{4}{\pi}\left(\ln\frac{\Lambda}{m} + 2\ln 2 - 1\right)\left(\frac{\Lambda}{m}\right)^{-1} . \tag{10.19}$$

As in the case of $\alpha < \alpha_c$, there is no spontaneous chiral symmetry breaking at $\alpha = \alpha_c$ and finite Λ;

c) $\alpha > \alpha_c = \frac{\pi}{3}$.

Substituting the expression (10.17) for $B(p^2)$ in Eq.(10.11), we find the relation:

$$m^{(0)} \simeq \frac{m^2}{\Lambda}\left(\frac{2\mathrm{cth}\frac{\pi\tilde{\omega}}{2}}{\pi\tilde{\omega}}\right)^{1/2} \sin\left(\tilde{\omega}\ln\frac{\Lambda}{m} + \Sigma(\tilde{\omega})\right) , \tag{10.20}$$

where $\Sigma(\tilde{\omega}) = \arg\frac{\Gamma(1+i\tilde{\omega})}{\Gamma^2\left(\frac{1+i\tilde{\omega}}{2}\right)}$. Let us consider the chiral limit with $m^{(0)} = 0$. In this case we find the following relation from Eq.(10.20):

$$\sin\left(\tilde{\omega}\ln\frac{\Lambda}{m_{dyn}} + \Sigma(\tilde{\omega})\right) = 0 . \tag{10.21}$$

This equation has an infinite number of nontrivial solutions for the dynamical mass

m_{dyn}. At the near-critical values of α, satisfying the condition $\alpha - \alpha_c \ll 1$, they are

$$m_{dyn}^{(n)} \simeq 4\Lambda \exp\left(-\pi n/\tilde{\omega}\right), \; n = 1, 2, \cdots . \tag{10.22}$$

Thus, at $\alpha > \alpha_c = \frac{\pi}{3}$, there is spontaneous chiral symmetry breaking in ladder QED. As we will see in the next section, only the largest value $m_{dyn} = m_{dyn}^{(1)}$ leads to the stable vacuum; we will also clarify the physical meaning of the solutions $m_{dyn}^{(n)}$ with $n \geq 2$ there.

Let us indicate the following characteristic points of the solution with $\alpha > \alpha_c$:

1) in the ladder approximation, the renormalization constant of the electromagnetic field, Z_3, is equal to one. In perturbation theory, this would imply that in the chiral limit with $m^{(0)} \equiv 0$, the theory is scale invariant. However it does not happen for the nonperturbative solution (10.22): indeed, there is a new nonperturbative mass divergence at $\alpha > \alpha_c$, and the ultraviolet cutoff Λ (or m_{dyn} which is proportional to Λ) spoils the scale symmetry.

What will happen in the theory as $\Lambda \longrightarrow \infty$? Eq.(10.22) with $n = 1$ implies that the dynamical fermion mass m_{dyn} remains finite in this limit only if the coupling constant α goes to $\alpha_c = \pi/3$:

$$\alpha = \alpha_c + \frac{\pi^2 \alpha_c}{\ln^2\left(4\Lambda/m_{dyn}\right)} \underset{\Lambda \to \infty}{\longrightarrow} \alpha_c = \frac{\pi}{3} \, . \tag{10.23}$$

From the point of view of the theory of phase transitions [26], the critical coupling α_c corresponds to the continuous (second order) phase transition. It separates two phases: the symmetric phase with $\alpha < \alpha_c$ and the phase with spontaneous chiral symmetry breaking at $\alpha > \alpha_c$. The phase transition is continuous since the order parameter $m_{dyn} \simeq 4\Lambda \exp\left(-\pi/\left(\frac{3\alpha}{\pi} - 1\right)^{1/2}\right)$ continuously goes to zero as $\alpha \longrightarrow \alpha_c = \pi/3$ from the side of the asymmetric phase.*)

We will show that the scale symmetry remains broken here even in the continuum limit with $\Lambda \longrightarrow \infty$ and m_{dyn} being finite. The critical coupling constant $\alpha_c = \pi/3$ may be interpreted as an ultraviolet stable fixed point.

---

*) We will discuss this point in more detail in Chapter 11.

2) The characteristic feature of the supercritical solution (10.17) is the presence of oscillations in the function B_{dyn}. Since the Bethe-Salpeter function of NG bosons $\chi^\alpha(q^2)$ at $P^\mu = 0$ is expressed through B_{dyn} (see Eq.(9.38)),

$$\chi^\alpha(q^2) = \lambda^\alpha \gamma_5 \frac{B_{dyn}(q^2)}{F(q^2 + B^2_{dyn}(q^2))}, \tag{10.24}$$

this in turn implies that there are also the oscillations in the wave function of NG bosons. As we will see in Sec.10.4, the oscillations is a manifestation of the field theoretical analogue of the "fall to the centre" (collapse) phenomenon well-known in quantum mechanics [27].

10.3. The Bethe-Salpeter Equations in Ladder QED

In Sec.9.5, we analyzed the BS equations for spineless fermion-antifermion bound states in the Nambu-Jona-Lasinio (NJL) model. In this section, we will consider these equations in ladder QED. Our main aim here is to show that ladder QED with the supercritical coupling constant α yields a dynamical realization of spontaneous chiral symmetry breaking scenario considered before in the linear σ-model (see Chapter 4).

The BS wave functions for pseudoscalar and scalar bosons composed of the a-th fermion and the b-th antifermion $(a, b = 1, 2, \cdots, N)$ are:

$$\chi^{(p)}_{ab}(q,P) = \frac{1}{\sqrt{2}}\left[\chi^{(p)}_{ab;1} + \chi^{(p)}_{ab;2}\hat{P} + \chi^{(p)}_{ab;3}\hat{q} + \chi^{(p)}_{ab;4}\sigma^{\mu\nu}(P_\mu q_\nu - P_\nu q_\mu)\right]\gamma_5\,, \tag{10.25}$$

$$\chi^{(s)}_{ab}(q,P) = \frac{1}{\sqrt{2}}\left[\chi^{(s)}_{ab;1} + \chi^{(s)}_{ab;2}\hat{P} + \chi^{(s)}_{ab;3}\hat{q} + \chi^{(s)}_{ab;4}\sigma^{\mu\nu}(P_\mu q_\nu - P_\nu q_\mu)\right], \tag{10.26}$$

where $\chi^{(p)}_{ab;i}$, $\chi^{(s)}_{ab;i}$ depends on P^2, q^2, Pq (as in Eq. (9.41), the factor $\frac{1}{\sqrt{2}}$ is introduced here for convenience).

The expression (8.113) for the BS kernel in the two-loop approximation in QED includes two terms corresponding to the two terms in the effective action (8.59). The first term corresponds to the photon exchange forces (ladder approximation); the second one corresponds to the annihilation forces (see Fig.8.5). As was already indicated in Sec.8.4, the annihilation forces do not contribute to the SD equation. Since at $P^\mu = 0$, the BS wave function of NG bosons is expressed through B_{dyn} (see Eq.(10.24)), the annihilation term does not contribute to the equation for this wave function either. However, it certainly contributes to the BS equations for massive bound states. Nevertheless, we will ignore this contribution in the following and consider only the exchange forces corresponding to the ladder approximation. Graphical representations of the BS equations with the kernel (8.113) and the ladder kernel are shown in Fig.10.1 and Fig.10.2, respectively.

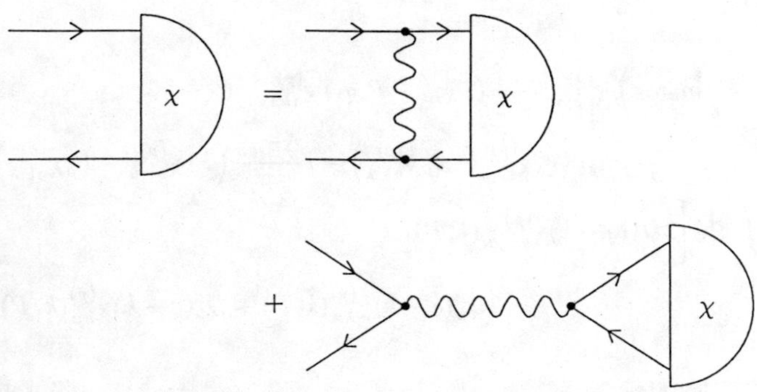

Fig. 10.1. Graphical representation of the homogeneous Bethe-Salpeter equation with the kernel (8.113) in QED.

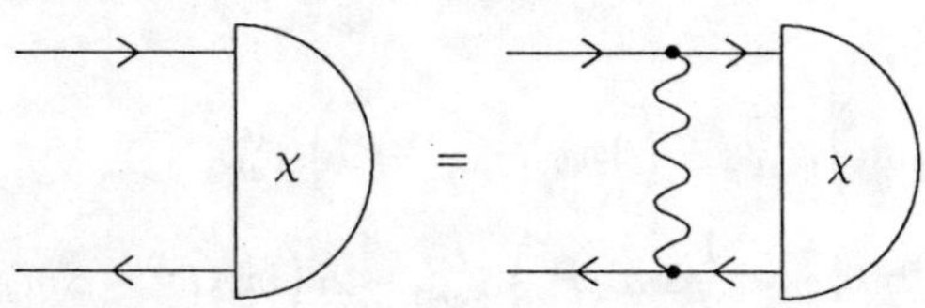

Fig. 10.2. Graphical representation of the homogeneous Bethe-Salpeter equation in the ladder approximation in QED.

The ladder BS equations for the wave functions (10.25), (10.26) take the following form (compare with Eqs.(9.87) - (9.92)):

a) pseudoscalar ($J^{PC} = 0^{-+}$) states:

$$\begin{aligned}
&\left(\frac{P^2}{4} - q^2 + m_a m_b\right)\chi^{(p)}_{ab;1} - (\bar{m}_{ab}P^\mu + \Delta m_{ab} q^\mu)\,\chi^{(p)}_{ab;\mu} \\
&\qquad - 2i\left(P^2q^2 - (Pq)^2\right)\chi^{(p)}_{ab;4} = \frac{i\alpha}{4\pi^3}\int d^4k D^\mu_\mu(q-k)\chi^{(p)}_{ab;1}(k,P)\,,
\end{aligned} \tag{10.27}$$

$$\begin{aligned}
&\left(\frac{P^2}{4} - q^2 - m_a m_b\right)\chi^{(p)}_{ab;\mu} + \left(2q_\mu q^\lambda - \frac{1}{2}P_\mu P^\lambda\right)\chi^{(p)}_{ab;\lambda} \\
&\qquad + (\bar{m}_{ab}P_\mu + \Delta m_{ab}q_\mu)\,\chi^{(p)}_{ab;1} - i\left(4\bar{m}_{ab}P^\lambda + \Delta m_{ab}q^\lambda\right)\chi^{(p)}_{ab;\mu\lambda} \\
&= \frac{i\alpha}{4\pi^3}\int d^4k\left[g^\lambda_\mu D^\nu_\nu(q-k) + 2D^\lambda_\mu(q-k)\right]\chi^{(p)}_{ab;\lambda}(k,P)\,,
\end{aligned} \tag{10.28}$$

$$\begin{aligned}
&\left(\frac{P^2}{4} - q^2 - m_a m_b\right)\chi^{(p)}_{ab;\lambda\mu} - \frac{i}{2}(P_\lambda q_\mu - P_\mu q_\lambda)\,\chi^{(p)}_{ab;1} \\
&\qquad - i\bar{m}_{ab}\left(q_\lambda\chi^{(p)}_{ab;\mu} - q_\mu\chi^{(p)}_{ab;\lambda}\right) - i\frac{\Delta m_{ab}}{4}\left(P_\lambda\chi^{(p)}_{ab;\mu} - P_\mu\chi^{(p)}_{ab;\lambda}\right) \\
&= \frac{-i\alpha}{4\pi^3}\int d^4k\Big[2D^\nu_\lambda(q-k)\chi^{(p)}_{ab;\mu\nu}(k,P) \\
&\qquad - 2D^\nu_\mu(q-k)\chi^{(p)}_{\lambda\nu}(k,P) + D^\nu_\nu(q-k)\chi^{(p)}_{\lambda\mu}(k,P)\Big]\,;
\end{aligned} \tag{10.29}$$

b) scalar (0^{++}) states:

$$\begin{aligned}
&\left(\frac{P^2}{4} - q^2 - m_a m_b\right)\chi^{(s)}_{ab;1} + \left(2\bar{m}_{ab}q^\mu - \frac{\Delta m_{ab}}{2}P^\mu\right)\chi^{(s)}_{ab;\mu} \\
&\qquad + 2i\left((Pq)^2 - P^2q^2\right)\chi^{(s)}_{ab;4} = \frac{i\alpha}{4\pi^3}\int d^4k D^\mu_\mu(q-k)\chi^{(s)}_{ab;1}(k,P)\,,
\end{aligned} \tag{10.30}$$

$$\begin{aligned}
&\left(\frac{P^2}{4} - q^2 + m_a m_b\right)\chi^{(s)}_{ab;\mu} + \left(2q_\mu q^\lambda - \frac{1}{2}P_\mu P^\lambda\right)\chi^{(s)}_{ab;\lambda} \\
&\qquad - \left(2\bar{m}_{ab}q_\mu - \frac{1}{2}\Delta m_{ab}P_\mu\right)\chi^{(s)}_{ab;1} - 2i\left(\bar{m}_{ab}P^\lambda - \Delta m_{ab}q^\lambda\right)\chi^{(s)}_{\lambda\mu} \\
&= \frac{i\alpha}{4\pi^3}\int d^4k\left[g^\lambda_\mu D^\nu_\nu(q-k) + 2D^\lambda_\mu(q-k)\right]\chi^{(p)}_{ab;\lambda}(k,P)\,,
\end{aligned} \tag{10.31}$$

$$\begin{aligned}
&\left(\frac{P^2}{4} - q^2 + m_a m_b\right)\chi^{(s)}_{ab;\lambda\mu} - \frac{i}{2}(P_\lambda q_\mu - P_\mu q_\lambda)\,\chi^{(s)}_{ab;1} \\
&\qquad + \frac{i}{2}\bar{m}_{ab}\left(P_\lambda\chi^{(s)}_{ab;\mu} - P_\mu\chi^{(s)}_{ab;\lambda}\right) - \frac{i}{2}\Delta m_{ab}\left(q_\lambda\chi^{(s)}_{ab;\mu} - q_\mu\chi^{(s)}_{ab;\lambda}\right) \\
&= \frac{-i\alpha}{4\pi^3}\int d^4k\Big[2D^\nu_\lambda(q-k)\chi^{(s)}_{\mu\nu}(k,P) \\
&\qquad - 2D^\nu_\mu(q-k)\chi^{(s)}_{\lambda\nu}(k,P) + D^\nu_\nu(q-k)\chi^{(s)}_{\lambda\mu}(k,P)\Big]\,.
\end{aligned} \tag{10.32}$$

Here we use the same notations as in Eqs.(9.93), (9.94). The photon propagator is

$$D_{\mu\nu}(q) = \left(g_{\mu\nu} - \frac{q_\mu q_\nu}{q^2}\right)\frac{1}{q^2} + \xi^{-1}\frac{q_\mu q_\nu}{(q^2)^2}\,. \tag{10.33}$$

As was already indicated above, the Landau gauge ($\xi \longrightarrow \infty$) is the most preferable in the ladder approximation (for more details see Sec.10.5). However, we consider here the general covariant gauge; the reason for this will become clear below.

The system of equations (10.27) - (10.32) is much more complicated than the system of equations (9.87) - (9.92) in the NJL model. In that case, the functions $\chi^{(i)}_{ab;\mu}, \chi^{(i)}_{ab;\mu\nu}, i =$ p or s, are expressed through the function $\chi^{(i)}_{ab;1}$ and this enormously simplifies the problem. Unfortunately this can not be done in ladder QED and therefore the analysis of the QED BS equations is essentially more complicated [14, 15].

Let us first consider the equations for pseudoscalars. They essentially simplifies at $P^\mu = 0$ (massless NG bosons). In particular, the equation for the invariant function $\chi^{(p)}_{ab;1}$ is isolated in this case:

$$\left(m^2_{dyn} - q^2\right)\chi^{(p)}_{ab;1}(q^2) = \frac{i\alpha\left(3+\xi^{-1}\right)}{4\pi^3}\int\frac{d^4k}{(q-k)^2}\chi^{(p)}_{ab;1}(k^2)\,, \tag{10.34}$$

where $m_a = m_b = m_{dyn}$ in the chiral limit. With the substitution $\left(m^2_{dyn} - q^2\right)\chi^{(p)}_{ab;1}(q^2)$ $\longrightarrow B(q^2)$, $\alpha\left(3+\xi^{-1}\right) \longrightarrow 3\alpha$, this equation transforms in the linearized version of the SD equation (10.8) with $m^{(0)} = 0$, and therefore we can use the results of the analysis of that equation.

The general solution to Eq.(10.34) takes the following form in Euclidean region:

$$\chi^{(p)}_{ab;1}(q^2) = C\left(q^2 + m^2_{dyn}\right)^{-1} F\left(\frac{1+\omega(\xi)}{2}, \frac{1-\omega(\xi)}{2}, 2; -q^2/m^2_{dyn}\right)\,. \tag{10.35}$$

At $\Lambda < \infty$, the nontrivial solutions exist only if $\alpha > \alpha_c(\xi) = \pi/\left(3+\xi^{-1}\right)$. At $\alpha > \alpha_c(\xi)$, there are an infinite number of the solutions with m_{dyn} defined from the equation

$$m^{(n)}_{dyn} \simeq 4\Lambda\exp\left(-\pi n/\tilde{\omega}(\xi)\right),\; n = 1, 2, \cdots\,, \tag{10.36}$$

where

$$\tilde{\omega}(\xi) = \left(\frac{\alpha}{\alpha_c(\xi)} - 1\right)^{1/2}\,. \tag{10.37}$$

Thus, massless NG bosons appear at the values of $m_{dyn} = m^{(n)}_{dyn}$ from the relation

(10.36). As we will see, only the largest value $m_{dyn}^{(1)}$ leads to the stable vacuum.

Let us turn now to the analysis of the system of equations (10.27)-(10.29) at $P^\mu \neq 0$. As we have already indicated, the Landau gauge with $\xi^{-1} = 0$ is the most preferable in the ladder approximation (see Sec.10.5). We note, however, that in this approximation the form of the BS equation for NG bosons is the same in all covariant gauges, and the only distinction is in the coefficient of the coupling constant α: as $\xi^{-1} = 0 \longrightarrow \xi^{-1}$, the coupling constant $\alpha \longrightarrow \alpha(1+1/3\xi)$ (see Eq.(10.34)). Therefore, in this approximation, the physical picture of dynamical chiral symmetry breaking should be similar in the gauges with $|\xi| \leq 1$, in which the critical value $\alpha_c(\xi)$ is of order one. We will use this fact in the following.

The system of equations (10.27)-(10.29) simplifies in the Feynman gauge with $\xi = 1$: the right-hand side of Eq.(10.29) vanishes and the equation becomes purely algebraic in this gauge. In the chiral limit $\left(P^\mu = 0, m_a = m_{dyn}\right)$, the system has a solution of the form

$$\chi_{ab;4}^{(p)} = -\frac{i\chi_{ab;1}^{(p)}}{2\left(q^2 + m_{dyn}^2\right)}, \quad \chi_2^{(p)} = \chi_3^{(p)} = 0 \; . \tag{10.38}$$

At $P^\mu \neq 0$, we will use the approach used before in analyzing the BS equations in the NJL model, that is, we will expand the wave functions in a double perturbative series in $|P|$ and m_a.

Considering Eq.(10.28) as an inhomogeneous equation for the function $\chi_{ab;\mu}^{(p)} = P_\mu \chi_{ab;2}^{(p)} + q_\mu \chi_3^{(p)}$, we find that

$$\chi_{ab;\mu}^{(p)} \sim \left(P_\mu \bar{m}_{ab} + q_\mu \Delta m_{ab}\right) \chi_{ab;1}^{(p)} - i\left(4\bar{m}_{ab} q^\lambda + i\Delta m_{ab} P^\lambda\right) \chi_{\mu\lambda}^{(p)} \; . \tag{10.39}$$

We assume now that $\Delta m_{ab} \ll \bar{m}_{ab}$ (the smallness of the breakdown of the $SU_V(N)$). Then, we find from Eqs.(10.38), (10.39) that in the first order in $|P|^2$ and $m_a m_b$, equation (10.27) reduces to the following equation (in Euclidean region):

$$\left[\frac{3}{4}P^2 + q^2 + m_a m_b - \frac{(Pq)^2}{q^2}\right] \chi_{ab;1}^{(p)}\left(P^2, q^2, Pq\right) = \frac{\alpha}{\pi^3} \int^{\Lambda} \frac{d^4k}{(q-k)^2} \chi_{ab;1}^{(p)}\left(P^2, k^2, Pk\right) \; . \tag{10.40}$$

Let us show that in this order, we may take the function $\chi_{ab;1}^{(p)}$ to be independent of Pq. In order to show this, we expand the kernel $(q-k)^{-2}$ and the function $\chi_{ab;1}^{(p)}$ in

Gegenbauer polynomials C_n^1 [28]:

$$(q-k)^{-2} = \frac{\theta(1-t)}{q^2}\sum_{n=0}^{\infty} t^n C_n^1(\cos\varphi) + \frac{\theta(t-1)}{k^2}\sum_{n=0}^{\infty} t^{-n} C_n^1(\cos\varphi) , \tag{10.41}$$

$$\chi_{ab;1}^{(p)}(P^2, q^2, Pq) = \sum_{n=0}^{\infty} G_{ab;n}^{(p)}(P^2, q^2) C_n^1(\cos\theta) , \tag{10.42}$$

where

$$t = k^2/q^2, \ \cos\varphi = \frac{q\cdot k}{|q||k|}, \ \cos\theta = \frac{P\cdot q}{|P||q|} . \tag{10.43}$$

Substituting the expressions (10.41), (10.42) in Eq.(10.40) and using the addition theorem for Gegenbauer polynomials [28], we obtain the following infinite chain of coupled equations (compare with Eqs.(9.112), (9.113));

$$\begin{aligned} &\left(\frac{3}{4}P^2 + q^2 + m_a m_b\right) G_{ab;0}^{(p)} - \frac{1}{4}P^2\left(G_{ab;0}^{(p)} + G_{ab;2}^{(p)}\right) \\ &= \frac{\alpha}{\pi}\int_0^{\Lambda^2} dk^2 K_0(q^2, k^2) G_{ab;0}^{(p)}(p^2, k^2) , \end{aligned} \tag{10.44}$$

$$\begin{aligned} &\left(\frac{3}{4}P^2 + q^2 + m_a m_b\right) G_{ab;n}^{(p)} - \frac{1}{4}P^2\left(G_{ab;n-2}^{(p)} + 2G_{ab;n}^{(p)} + G_{ab;n+2}^{(p)}\right) \\ &= \frac{\alpha}{\pi}\int_0^{\Lambda^2} dk^2 K_n(q^2, k^2) G_{ab;n}^{(p)}(P^2, k^2); n \geq 2 , \end{aligned} \tag{10.45}$$

where

$$K_n(q^2, k^2) = \frac{k^2}{n+1}\left[\frac{\theta(q^2-k^2)}{q^2}\left(\frac{k^2}{q^2}\right)^{n/2} + \frac{\theta(k^2-q^2)}{k^2}\left(\frac{q^2}{k^2}\right)^{n/2}\right] . \tag{10.46}$$

In the chiral limit, $P^2 = 0$, the function $\chi_{ab;1}^{(p)} = G_{ab;0}^{(p)}$, and all $G_{ab;n}^{(p)}$ with $n \geq 0$ vanish. Considering Eq.(10.45) as an inhomogeneous integral equation for harmonics

$G^{(p)}_{ab;n}$ with $n \geq 2$, we find that

$$G^{(p)}_{ab;n}/G^{(p)}_{ab;n-2} \sim P^2 \,. \tag{10.47}$$

Therefore, to first order in P^2, the term $P^2 G^{(p)}_{ab;2}$ in Eq.(10.44) can be neglected, and we are lead to the equation

$$\begin{gathered}\left(\frac{1}{2}P^2 + q^2 + m_a m_b\right)\chi^{(p)}_{ab;1}(P^2, q^2) \\ = \frac{\alpha}{\pi}\int^{\Lambda^2} dk^2 \left[\frac{\theta(q^2 - k^2)}{q^2} + \frac{\theta(k^2 - q^2)}{k^2}\right] k^2 \chi^{(p)}_{ab;1}(P^2, k^2)\,.\end{gathered} \tag{10.48}$$

The left-hand side of this equation coincides with the left-hand side of Eq.(9.118) in the NJL model, but the integral kernel here is essentially different.

How well is this approximation ? At present, the only way to answer this question is to use a numerical analysis of the BS equations. Results of such an analysis are encouraging [29].

Let us discuss the physical consequences of equation (10.48). With the substitution $\frac{P^2}{2} + m_a m_b \longrightarrow m^2_{dyn}$, this equation transforms into Eq.(10.34) with $\xi = 1$, which is in turn equivalent to the linearized version of the SD equation (10.8) with $m^{(0)} = 0$. Therefore we can use the results of the analysis of the latter.

As follows from that analysis, the mass spectrum of pseudoscalars is determined by Eq.(10.22), in which $m^2_{dyn} \longrightarrow m_a m_b - \frac{M^{(p)^2}_{ab}}{2}$, where $M^{(p)^2}_{ab} = -P^2$, and $\tilde{\omega} \longrightarrow \tilde{\omega}(\xi)|_{\xi=1} = \left(\frac{4\alpha}{\pi} - 1\right)^{1/2}$. Therefore we get:

$$M^{(p)^2}_{ab;n} = 2m_a m_b - \Lambda^2 f^{(n)}(\alpha)\,, \tag{10.49}$$

where

$$f^{(n)}(\alpha) = 32\exp\left(\frac{-2\pi n}{\left(\frac{4\alpha}{\pi} - 1\right)^{1/2}}\right), n = 1, 2, \cdots\,. \tag{10.50}$$

Let us discuss this relation in the following three cases:

1) $m_a = m_b = 0$.

This case corresponds to the symmetric (massless) phase. Eq.(10.49) implies that at $\alpha > \alpha_c$, there are tachyons in this phase (in fact, an infinite number of tachyons). Therefore at supercritical values of α, the symmetric phase is unstable.

2) $\hat{m}^{(0)} = 0; m_a = m_{dyn}$.

This case corresponds to the phase with spontaneous chiral symmetry breaking in chiral invariant QED. Because of the Goldstone theorem (see Sec.3.4), there must be massless NG bosons in this phase. The condition $M_{ab}^{(p)} = 0$ results in the equation for the dynamical mass m_{dyn}:

$$m_{dyn}^{(n)^2} = \frac{1}{2}\Lambda^2 f^{(n)}(\alpha),\ n = 1, 2, \cdots . \tag{10.51}$$

Each value $m_{dyn}^{(n)}$ defines its own fermion propagator and, therefore, its own vacuum. However, as follows from Eq.(10.49), only the largest value $m_{dyn}^{(1)}$ leads to the stable vacuum: at $m_{dyn} = m_{dyn}^{(n)}$ with $n \geq 2$, the bound states with $M_{ab;i}^{(p)^2} = \Lambda^2\left(f^{(n)}(\alpha) - f^{(i)}(\alpha)\right)$, $i < n$, remain tachyonic.

In the stable vacuum with $m_{dyn} = m_{dyn}^{(1)}$, there are an infinite number of radial excitations for each NG boson. Their masses are

$$M_{ab;n}^{(p)^2} = \Lambda^2\left(f^{(n)}(\alpha) - f^{(1)}(\alpha)\right), n = 1, 2, \cdots . \tag{10.52}$$

Of course, their presence is a manifestation of the long range character of Coulomb forces.

3) $\hat{m}^{(0)} \neq 0, m_a = m_{dyn} + m_{\mathrm{cur};a}$.

This case corresponds to the PCAC dynamics. The mass $m_{\mathrm{cur};a}$ is the current mass of the a-th fermion. By proceeding in the same way as in the case of the NJL model, we derive the same PCAC mass formula (see Eq.(9.131)):

$$M_{ab}^{(p)^2} = 2m_{dyn}\left(m_{\mathrm{cur};a} + m_{\mathrm{cur};b}\right) + 2m_{\mathrm{cur};a}m_{\mathrm{cur};b}\ , \tag{10.53}$$

where $M_{ab}^{(p)} \equiv M_{ab;1}^{(p)}$ (see Eq.(10.52)).

Thus, the essential difference of the dynamics of spontaneous chiral symmetry breaking in QED as compared to the NJL model is the spectrum of the radial excitations of NG bosons reflecting the long range character of Coulomb forces.

Let us consider now scalar (0^{++}) bound states.

First of all, because of the same reason as in the case of the NJL model (see Sec.9.5), the BS equations for scalars coincide with the BS equations for pseudoscalars in the symmetric phase. Therefore at $\alpha > \alpha_c$, there are an infinite number of scalar tachyons in the symmetric phase.

Unfortunately, in QED, unlike the NJL model, we cannot solve the BS equations for massive scalars, even to first order in P^2 and $m_a m_b$. In the NJL model, the functions $\chi^{(s)}_{ab;\lambda}$ and $\chi^{(s)}_{ab;\lambda\mu}$ are expressed through the function $\chi^{(s)}_{ab;1}$ from Eqs.(9.91), (9.92). However we cannot do this in QED. We cannot use also the trick used in the equations for pseudoscalars above: although Eq.(10.32) becomes purely algebraic in Feynman gauge, and therefore the function $\chi^{(s)}_{ab;\lambda\mu}$ can be expressed through $\chi^{(s)}_{ab;1}$ and $\chi^{(s)}_{ab;\mu}$, we cannot neglect the contribution of $\chi^{(s)}_{ab;\mu}$ in Eq.(10.30), due to the presence of the term $2\bar{m}_{ab}q^\mu\chi^{(s)}_{ab;\mu}$ there. As a result, massive scalars are described by a system of two integral equations even in Feynman gauge. Nobody has yet succeeded in solving them.

However, we still can show that in the asymmetric phase, with spontaneous chiral symmetry breaking, scalars are heavier than pseudoscalars. In order to show this, we will follow the way already used in the NJL model in Sec.9.5 (see Eqs.(9.105), (9.106)), that is, we will compare the BS equations for massless scalars and pseudoscalars.

First of all, we note that at $P^\mu = 0$, the BS wave function for scalars has the following general structure:

$$\chi^{(s)}_{ab} = \frac{1}{\sqrt{2}}\left[\chi^{(s)}_{ab;1} + \hat{q}\chi^{(s)}_{ab;3}\right] . \tag{10.54}$$

At $P^\mu = 0$, the system of Eqs.(10.30)-(10.32) is reduced to the following pair of equations in Landau gauge:

$$\begin{aligned}&\left(q^2 - m^2_{dyn}\right)\chi^{(s)}_{ab;1} - 2m_{dyn}q^2\chi^{(s)}_{ab;3}\\ &\qquad = \frac{3\alpha}{4\pi}\int\limits_0^{\Lambda^2} dk^2\left(\frac{\theta(q^2-k^2)}{q^2} + \frac{\theta(k^2-q^2)}{k^2}\right)k^2\chi^{(s)}_{ab:;1}(k^2) ,\end{aligned} \tag{10.55}$$

$$\left(q^2 - m^2_{dyn}\right)\chi^{(s)}_{ab;3} + 2m_{dyn}\chi^{(s)}_{ab;1} = 0 . \tag{10.56}$$

The purely algebraic character of Eq.(10.56) is the consequence of the identity

$$\int d^4k\left[\delta_{\mu\nu}(q-k)^2-(q-k)_\mu(q-k)_\nu\right](q-k)^{-4}\gamma_\mu\hat{k}f(k^2)\gamma_\nu=0\,, \tag{10.57}$$

which follows from formulae for the angle integrations in Euclidean space (see Sec.B.4 in Appendix B). Eqs.(10.55), (10.56) lead to the following equation:

$$\left(q^2+m_{dyn}^2\right)^2\chi_{ab;1}^{(s)}=q^2K\left[\chi_{ab;1}^{(s)}\right]-m_{dyn}^2K\left[\chi_{ab;1}^{(s)}\right]\,, \tag{10.58}$$

where

$$K\left[f(q^2)\right]\equiv\frac{3\alpha}{4\pi}\int\limits_0^{\Lambda^2}dk^2\left[\frac{\theta(q^2-k^2)}{q^2}+\frac{\theta(k^2-q^2)}{k^2}\right]k^2f(k^2)\,. \tag{10.59}$$

Let us compare equation (10.58) with equation (10.34) for NG bosons which can be rewritten in the form ($\xi^{-1}=0$):

$$\left(q^2+m_{dyn}^2\right)^2\chi_{ab;1}^{(p)}=q^2K\left[\chi_{ab;1}^{(p)}\right]+m_{dyn}^2K\left[\chi_{ab;1}^{(p)}\right]\,. \tag{10.60}$$

The minus sign in front of the second term on the right hand side of Eq.(10.58) implies an additional repulsion in the scalar channel as compared to the pseudoscalar one. This in turn implies that there are no nontrivial solutions to Eq.(10.58) describing massless scalars. Therefore scalars are heavier than pseudoscalars. Of course, as in the case of the NJL model, the physical reason for this is that scalars are bound states with the orbital angular momentum $l=1$, while the orbital angular momentum of pseudoscalar bound states is zero.

10.4. The Supercritical Dynamics and the Collapse Phenomenon

At the end of Sec.10.2, we already indicated that the characteristic property of the BS wave function for NG bosons in ladder QED is the occurrence of oscillations. Let us discuss this point in more detail.

In Euclidean region, the wave function for NG bosons takes the form (see Eqs.(10.17), (10.24)):

$$\chi^\alpha = \lambda^\alpha \gamma_5 \frac{B_{dyn}(q^2)}{F\left(q^2 + B^2_{dyn}(q^2)\right)} , \tag{10.61}$$

$$B_{dyn}(q^2) \underset{p^2 \gg m^2_{dyn}}{\simeq} m_{dyn} \left(\frac{8\mathrm{cth}\frac{\pi\tilde{\omega}}{2}}{\pi\tilde{\omega}(\tilde{\omega}^2+1)} \right)^{1/2} \left(\frac{q^2}{m^2_{dyn}} \right)^{-\frac{1}{2}} \cdot \sin\left(\frac{\tilde{\omega}}{2} \ln \frac{p^2}{m^2_{dyn}} + \tilde{\omega}(2\ln 2 - 1) \right) \tag{10.62}$$

with $\tilde{\omega} = \left(\frac{\alpha}{\alpha_c} - 1\right)^{1/2}$. If we consider m_{dyn} as a free parameter here, then at $\Lambda = \infty$ and any finite value of m_{dyn}, this wave function has an infinite number of zeroes. This is a typical manifestation of the collapse ("fall to the centre") phenomenon which is well known in quantum mechanics: in this case, there is no ground state (vacuum) in a system [27]. Indeed, the wave function of the ground state has no zeroes: wave functions with zeroes describe excitations. The wave function (10.61) describes a massless bound state, $M^2 = 0$. In relativistic theory M^2 plays the role of energy. The presence of an infinite number of zeroes in this wave function implies that at any finite value of m_{dyn}, there should be an infinite number of states with $M^2 < 0$, that is, tachyons. But that is exactly what we have already encountered in ladder QED. Indeed, let us rewrite the mass relation (10.49) as

$$M_n^{(p)^2} = 2m^2_{dyn} - 32\Lambda^2 \exp\left(\frac{-2\pi n}{\sqrt{\frac{\alpha}{\alpha_c} - 1}} \right) , \tag{10.63}$$

where m_{dyn} is treated now as a free parameter in the BS equation. It is clear that as $\Lambda \longrightarrow \infty$, there are an infinite number of tachyons at any finite value of m_{dyn}. Of course, this reflects the point that at $\alpha > \alpha_c$, on the stable vacuum, the dynamical mass m_{dyn} goes to infinity as $\Lambda \longrightarrow \infty$. This in turn implies that at $\alpha > \alpha_c$, the continuum theory does not exist as a well defined theory in the Fock space.

In nonrelativistic quantum mechanics, the collapse phenomenon occurs for singular attractive potentials $U(r)$ which obey the inequality $\lim_{r\to 0} \left[-8mr^2 U(r)\right] > 1$ [27]. In particular it occurs for the potential $U(r) = -\beta/r^2$ with $\beta > 1/8m$. We will discuss here the results concerning the form of the wave function and the energy spectrum in that problem.

First of all, let us replace the initial potential $-\beta/r^2$ by the potential

$$U_{r_0}(r) = -\beta\left[\theta(r-r_0)r^{-2} + \theta(r_0-r)r_0^{-2}\right] . \tag{10.64}$$

The inverse radius r_0^{-1} plays the role of ultraviolet cutoff Λ. In the conventional situation, the results of the problem with the potential $U_{r_0}(r)$ would coincide with the results of the problem with the initial potential as $r_0 \longrightarrow 0$. However, as we will show, at the supercritical values $\beta > 1/8m$, this is not the case.

The Schrödinger equation with the potential (10.64) leads to the following equation for the radial function $R_l(r)$:

$$\frac{1}{r^2}\frac{d}{dr}\left(r^2\frac{dR}{dr}\right) + \gamma R/r^2 = E,\ r > r_0 , \tag{10.65}$$

and

$$\frac{1}{r^2}\frac{d}{dr}\left(r^2\frac{dR}{dr}\right) + \gamma R/r_0^2 = E,\ r < r_0 , \tag{10.66}$$

where

$$\gamma = 2m\beta - l(l+1) , \tag{10.67}$$

and l is the orbital angular momentum.

We are interested in the behavior of the solutions at small distances, that is, at $mr \ll 1$ (the radius r_0 is chosen to satisfy the condition $mr_0 \ll 1$). At $r > r_0$, the general solution of Eq.(10.65) takes the following form at short distances:

$$R = Ar^{s_+} + Br^{s_-} , \tag{10.68}$$

$$s_\pm = -\frac{1}{2} \pm \left(\frac{1}{4} - \gamma\right)^{1/2} , \tag{10.69}$$

where A and B are some constants. At $r < r_0$, the solution which is finite at the origin has the form

$$R = C\frac{\sin kr}{r},\ k = \sqrt{\gamma}/r_0 . \tag{10.70}$$

Equating the logarithmic derivative of these two solutions at $r = r_0$, we find that

$$B/A = \text{constant} \cdot r_0^{2\left(\frac{1}{4}-\gamma\right)^{1/2}} . \tag{10.71}$$

The character of the limit $r_0 \longrightarrow 0$ is essentially different at subcritical ($\beta < 1/8m$) and supercritical ($\beta > 1/8m$) values of β. Indeed, at $\beta < 1/8m$, the parameter γ is real and it is smaller than $\frac{1}{4}$. Therefore the ratio B/A goes to zero as $r_0 \longrightarrow 0$. Thus, at $\beta < 1/8m$, we have the regular situation when the wave function is expressed through a less singular (at $r \longrightarrow 0$) solution:

$$R(r) = Ar^{s_+} \; . \tag{10.72}$$

However, at $\beta > 1/8m$, the parameter γ becomes larger than $\frac{1}{4}$ for at least one solution with $l = 0$. Then, the parameters $s_\pm$ becomes complex for this solution, and the ratio B/A does not tend to any definite limit as $r_0 \longrightarrow 0$. The solution can be written in the form:

$$R = \text{constant} \cdot r^{-1/2} \cos\left(\sqrt{\left(\gamma - \frac{1}{4}\right)} \log(r/r_0) + \text{constant}\right) \; . \tag{10.73}$$

Moreover, one can show [30] that there is no discrete spectrum at $\beta < 1/8m$, and that at $\beta > 1/8m$, the discrete spectrum corresponding to the solution (10.73) takes the following form:

$$E^{(n)} \sim -r_0^{-1} \exp\left(\frac{2\pi n}{\nu}\right) , \quad n = 1, 2, \cdots \; , \tag{10.74}$$

where $\nu = \left(\gamma - \frac{1}{4}\right)^{1/2}$.

The similarity between Eqs.(10.15), (10.17), (10.63) and Eqs.(10.68), (10.73), (10.74), respectively, is striking. In particular, as $r_0 \longrightarrow 0$, the number of zeroes of the function (10.73) goes to infinity. This of course reflects the point that the energy $E^{(1)}$ of the ground state goes to $-\infty$ as $r_0 \longrightarrow 0$, that is, there is no vacuum state at $\beta > 1/8m$ and $r_0 = 0$.

As is known, in quantum mechanics, a particle connected with a discrete level is moving in those regions of space where $E < U(r)$ [27]. Therefore $E^{(1)} = -\infty$ implies that the "ground state" corresponds to a particle at the origin $r = 0$, that is, the particle "falls" to the centre.

The existence of such a phenomenon for $U(r) \sim r^{-2}$ and more singular potentials follows in fact from the uncertainty principle. Indeed, the hamiltonian is

$$H = \frac{\hat{p}^2}{2m} + U(r) \; . \tag{10.75}$$

Let us consider a particle with a wave function localized in a small region of radius r_0. Then the uncertainty in the value of the momentum is of the order of r_0^{-1}. The average energy of the particle is

$$\bar{E} \sim \frac{1}{mr_0^2} + U(r) \,. \tag{10.76}$$

Therefore if $U(r) = -\beta r^{-s}$ with $s > 2$ or $s = 2$ and β is order of $\frac{1}{m}$ or larger, the sum (10.76) goes to $-\infty$ as $r_0 \longrightarrow 0$.

In relativistic quantum physics, since the kinetic energy $E_k = \left(p^2 + m^2\right)^{1/2} \underset{p \gg m}{\simeq} \frac{1}{r_0}$, the collapse phenomenon occurs already for the supercritical Coulomb potential [31]. The second important difference of relativistic theory from nonrelativistic one is that the processes of particle pair creation are allowed there; therefore stationary levels with $E < -m$ cannot exist in relativistic quantum systems. Let us consider the Dirac equation with the supercritical Coulomb potential $-\frac{\alpha}{r}, \alpha > 1$, in more detail. As was pointed out in Ref.[11], this problem is intimately connected with spontaneous chiral symmetry breaking in QED.

Since we are interested in spontaneous chiral symmetry breaking, the most interesting case is the case of a light fermion: $|E| \gg m$, and, in particular, $m = 0$. This problem is considered in Appendix D. Here we quote the main results. The energy $E^{(n)}$ of $nS_{1/2}$-levels is:

$$E^{(n)} \simeq E_0^{(n)} - \frac{m}{2} - i\frac{m^2}{50|E_0^{(n)}|} \,, \tag{10.77}$$

where $E_0^{(n)}$ is the energy of massless fermion:

$$\begin{gathered} E_0^{(n)} = ar_0^{-1}(\sin\varphi - i\cos\varphi)\exp\left[\frac{-\pi n}{\sqrt{\alpha^2-1}}\right] , \\ a \simeq 0.4, \ \varphi = -\frac{\pi}{2}\text{cth}\pi \simeq -\frac{\pi}{2}\cdot 1.004 \,, \end{gathered} \tag{10.78}$$

and r_0 appears in the modified potential

$$V_{r_0} = -\alpha\left[\theta(r-r_0)r^{-1} + \theta(r_0 - r)r_0^{-1}\right] . \tag{10.79}$$

The radial wave function $F_2(r)$ (see Eq.(D.3)) takes the following form at small r:

$$F_2(r) \simeq r^{1/2}\left(\frac{2\pi}{\nu\sinh 2\pi\nu}\right)^{1/2}\frac{1}{\Gamma(i)}\cos\left[\nu\ln r - \arg\Gamma(2i\nu)\right] \,, \tag{10.80}$$

where $\nu = (m^2 - E^2)^{1/2}$. We note that $ReE_0^{(n)} < 0$. According to the conventional interpretation of the levels with $ReE^{(n)} < 0$, these levels determines the antifermion state energy

$$E_{an}^{(n)} = -E^{(n)}, \ Re E_{an}^{(n)} > 0, \ \mathrm{Im} E_{an}^{(n)} < 0 \tag{10.81}$$

corresponding to an outgoing antifermion wave. The appearance of such quasi-stationary levels is interpreted as instability with respect to the spontaneous creation of fermion-antifermion pairs from the vacuum [31]. The created fermion couples to the centre thus shielding its charge, while the antifermion goes to infinity; the process is repeated until the charge of the centre is reduced to a subcritical value.

The expression (10.77) for the energy indicates however another possibility for the vacuum rearrangement: to generate spontaneously a fermion mass. Indeed, this expression shows that the imaginary part $\mathrm{Im}E^{(n)} > 0$ decreases with increasing the mass. Since the Dirac equation with the Coulomb potential is an one-particle problem, it cannot be used for the description of the dynamics of spontaneous chiral symmetry breaking. As we showed in Sec.10.3, one has to use the BS equations for this purpose. However the deep connection between these two problems is evident. In particular, it is clear that the tachyons in an unstable symmetric phase are the field theoretical analogues of the quasi-stationary levels (10.77).

Thus, at $\alpha > \alpha_c$, a new nonperturbative divergence occurs in ladder QED: the fermion mass m diverges as $m \sim \Lambda$. As is known, ultraviolet divergences are intimately connected with the properties of field theories with respect to scale transformations. We will consider this issue in Sec.10.8.

10.5. The Gauge Invariance in the Problem of Dynamical Mass Generation in QED

The ladder (rainbow) approximation is not gauge invariant. In particular, as we have already seen, the value of the critical coupling $\alpha_c(\xi)$ depends sensitively on the gauge parameter $\xi : \alpha_c(\xi) = \pi/(3 + \xi^{-1})$.

Why can we trust the results in this approximation ? And if we can, what is the most preferable gauge in this approximation ?

Let us recall that the gauge invariance is reflected in the corresponding Ward-Takahashi (WT) identities. If they are satisfied in some approximation in a certain gauge, then we can use the gauge transformation law for Green's function to rewrite them in an arbitrary gauge (such a transformation law was established by Landau and Khalatnikov in QED [32], and was extended to non-Abelian gauge theories by 't Hooft and Veltman [33]). The transformation law guarantees that the WT identities will take place in all other gauges, although the approximation may change.

Let us look at the WT identity (9.33) for Green's function $\Gamma_\mu(q,k)$,

$$G(q)\Gamma_\mu(q,k)G(k) = \int dx dy \exp[iqx - iky]\,\tilde{\Gamma}_\mu(x,y) \, , \tag{10.82}$$

$$\tilde{\Gamma}_\mu(x,y) = \langle 0|T\jmath_\mu(0)\psi(x)\bar{\psi}(y)|0\rangle \, . \tag{10.83}$$

It is:

$$(q-k)^\mu \Gamma_\mu(q,k) = -G^{-1}(q) + G^{-1}(k) = -\hat{q}A(q^2) + \hat{k}A(k^2) + B(q^2) - B(k^2) \, . \tag{10.84}$$

In the ladder approximation, the vertex $\Gamma_\mu(q,k)$ is just the bare one, $\Gamma_\mu(q,k) = -\gamma_\mu$. This fact and the WT identity imply that the Landau gauge is preferable in this approximation. Indeed, as we know, the function $A(q^2)$ is equal to one in this gauge in the ladder approximation (see Eq.10.7). Therefore, up to a correction connected with the mass function $B(q^2)$, the WT identity is satisfied in Landau gauge.

One may expect that this correction is small. Indeed, $B(q^2) - B(k^2) = 0$ at $q^2 = k^2$, and the function $B(q^2)$ decreases rather rapidly with momentum, $B(q^2) \sim (q^2)^{-1/2}$ at $\alpha \simeq \alpha_c$. Therefore one can expect that by transforming properly the rainbow SD equation in Landau gauge to other gauges, we get the same (or nearly the same) values for gauge invariant quantities, although the approximation will change.

This program was realized by Appelquist, Lane and Mahanta [34]. In particular, they showed that the value α_c is nearly the same in all covariant gauges.

Another approach is based on improving the ansatz for the vertex in such a way that the WT identity (10.84) is exactly satisfied [35 - 40]. For example, Atkinson, Johnson and Stam [35] and Kondo [38] used the following ansatz for the longitudinal part of Γ_l^μ:

$$\Gamma_l^\mu(q,k) = \frac{(q-k)^\mu}{(q-k)^2}\left(G^{-1}(k) - G^{-1}(q)\right) \, . \tag{10.85}$$

The transverse part of Γ^μ was chosen to satisfy the multiplicative renormalizability condition for Green's functions. They showed that with this ansatz, the critical coupling constant and dynamical mass are gauge independent, at least in leading order in $\lambda = \alpha/4\pi$.

The problem with the ansatz (10.85) is however that this expression for Γ^μ contains a kinematic singularity at $(q-k)^2 = 0$.

It is only recently that Curtis and Pennington have suggested an ansatz for Γ^μ which is free from kinematic singularities and which satisfies the WT identity and the multiplicative renormalizability [39, 40]. It also reproduces the results of low orders in perturbation theory.

Their ansatz is:

$$\begin{aligned}\Gamma^\mu(q,k) = & \frac{-1}{2}\left(A(q^2)+A(k^2)\right)\gamma^\mu + \frac{1}{2}\left(A(k^2)-A(q^2)\right)\cdot\frac{(q+k)^\mu(\hat{q}+\hat{k})}{(q^2-k^2)} \\ & + \frac{1}{2}\left(A(k^2)-A(q^2)\right)\frac{\left[\gamma^\mu(q^2-k^2)-(q+k)^\mu(\hat{q}-\hat{k})\right]}{d(q,k)} \\ & + \left(B(q^2)-B(k^2)\right)\frac{(q+k)^\mu}{q^2-k^2}\,,\end{aligned} \tag{10.86}$$

where

$$d(q,k) = \frac{\left(q^2-k^2\right)^2 + \left[\left(\frac{B(q^2)}{A(q^2)}\right)^2 + \left(\frac{B(k^2)}{A(k^2)}\right)^2\right]^2}{q^2+k^2}\,. \tag{10.87}$$

With this vertex, the SD equations for the functions $A(q^2)$ and $B(q^2)$ take a rather complicated form [39, 40]. The equations have been analyzed by two groups. Atkinson, Gusynin and Maris [41], using some approximations to simplify the equations, found the critical value $\alpha \simeq 0.91 \div 0.98$ in Landau gauge, which is rather close to the critical value $\alpha_c = \frac{\pi}{3}$ in the ladder (rainbow) approximation. Curtis and Pennington analyzed the equations numerically in three different gauges with $\xi^{-1} = 0$ (Landau gauge), $\xi^{-1} = 1$ (Feynman gauge), and $\xi^{-1} = 3$ (Yennie gauge) [40]. They found that the values of the critical coupling α_c and dynamical mass m_{dyn} are (to the numerical accuracy) independent of $\xi; \alpha_c \simeq 0.92$.

Thus, at present, there are strong evidences that the physical picture of dynamical chiral symmetry breaking in the ladder approximation in Landau gauge reproduces qualitatively correctly this phenomenon in the full quenched QED, i.e. in the approximation in which all diagrams without fermion loops are taken into account (for general arguments in support of this conclusion see also Ref.[42]). As we will see in Chapter 14, the quenched approximation in QED is relevant for some scenarios of dynamic electroweak symmetry breaking.

10.6. The Critical Coupling Constant in Dynamical Symmetry Breaking Problem in Asymptotically Free Theories

Up to now we have considered dynamical chiral symmetry breaking in QED. It is important that, with some modifications, the dynamics with a supercritical coupling constant may also be relevant for asymptotically free gauge theories as QCD and technicolor models of dynamic electroweak symmetry breaking (see Chapters 12-14).

In asymptotically free gauge theories, the running coupling constant decreases with increasing the momentum. In the one-loop approximation, it is [43]:

$$\bar{g}^2(q^2) = \frac{2}{b\ln(q^2/M^2)},$$

where the characteristic infrared scale parameter M is a free parameter determined from experiment (in QCD, $M \equiv \Lambda_{QCD} \sim 300MeV$), and b is the first coefficient of the β function. Therefore it is naturally that at some scale $q^2 = \Lambda_\chi^2$, the value $g^2(\Lambda_\chi^2)$ becomes equal to a critical value g_c^2, and the dynamics in the infrared region with $q^2 \lesssim \Lambda_\chi^2$ is responsible for dynamic symmetry breaking of some symmetries, as the chiral symmetry in QCD.

There are essentially two possibilities. The first one is that the scale Λ_χ in a dynamics is $\Lambda_\chi \sim M$. The second possibility is a dynamics with $\Lambda_\chi \gg M$. For example, in QCD the scale Λ_χ is apparently only few times larger than M, while in some scenarios of dynamical electroweak symmetry breaking Λ may be essentially larger than M (see Chapters 12 and 14).

Thus, in asymptotically free theories, Λ_χ is a physical finite scale determined by the dynamics. However, in those cases when $\Lambda_\chi \gg M$, the dynamics of quenched QED may serve as a toy model for such dynamics, as it happens in the so called walking technicolor for example (see Chapter14).

10.7. The Conservation Current Condition and Proper Definition of Equations for Green's Functions. The Solution of the Goldstein Problem

In quenched QED, by definition, the renormalization constant of photon field Z_3 is $Z_3 = 1$. In the rainbow (ladder) approximation, in Landau gauge, the renormalization constant of a fermion field, Z_2, and the renormalization constant of the vertex, Z_1, are also equal to one. Therefore in this approximation, the only perturbative divergences in QED are those connected with the renormalization of the fermion mass, $m^{(0)} = Z_m m$. Since the bare mass $m^{(0)}$ equals zero in the chiral limit, one may expect that in this limit, there is no need for a regularization in ladder QED, and one can consider the equations for Green's functions without cutoff Λ from the very begining.

However, as we will show below, following Refs.[17, 44], the situation in QED, and in gauge theories generally, is in fact more subtle.

Let us consider again the SD equation (10.8) for the mass function $B(p^2)$. This equation is equivalent to the differential equation (10.9) with the boundary conditions (10.10) and (10.11). Let us consider the boundary condition (10.11) at $m^{(0)} = 0$ and $\Lambda \longrightarrow \infty$:

$$\lim_{p^2 \to \infty} \left[p^2 \frac{dB(p^2)}{dp^2} + B(p^2) \right] = 0 \, . \tag{10.88}$$

Eqs. (10.15) - (10.17) imply that this condition is satisfied by the function (10.14) at any positive value of α. Thus at $\Lambda = \infty$ and $m^{(0)} = 0$, the SD equation has nontrivial solutions for $B(p^2)$ at any $\alpha > 0$.

Does it mean that dynamical chiral symmetry breaking in ladder continuum QED occurs at arbitrary small coupling α ? We will show that the answer to this question is "no".

We begin the analysis by considering the equation of motion for the axial currents $j^{\alpha}_{5\mu} = \bar{\psi}\gamma_\mu\gamma_5 \frac{\lambda^\alpha}{2}\psi$ (see Eq.(9.7)):

$$\partial^\mu j^{\alpha}_{5\mu} = im^{(0)}\bar{\psi}\gamma_5\lambda^\alpha\psi \, . \tag{10.89}$$

The composite operator $j^{\alpha}_5 \equiv \bar{\psi}\gamma_5\lambda^\alpha\psi$ is a singular object, i.e. its matrix elements depend on cutoff Λ. Let us introduce the renormalized operator $j_{5\mu}^{(\sigma)^\alpha}$,

$$j_5^{(\sigma)^\alpha} = Z_m^{(\sigma)}(\Lambda) j_5^\alpha \tag{10.90}$$

connected with a subtraction point σ. The renormalization constant $Z_m^{(\sigma)}$ is renormalization-scheme dependent but it is important that it always can be chosen to coincide with the renormalization constant of the fermion mass [5]:

$$m^{(0)} = Z_m^{(\sigma)}(\Lambda) m_{\text{cur}}^{(\sigma)} , \tag{10.91}$$

where $m_{\text{cur}}^{(\sigma)}$ is the renormalization (current) mass connected with the subtraction point σ.

The central point in calculating $Z_m^{(\sigma)}$ is the definition of the current mass $m_{\text{cur}}^{(\sigma)}$, which itself is of course renormalization-scheme dependent.

Let us begin by considering the case of subcritical $\alpha < \alpha_c$. We will use the momentum-space subtraction scheme (MOM) [43]. In this scheme, $Z_m^{(\sigma)}$ is defined in the following way. Let us first rewrite the mass function $m(p^2) = B(p^2)/A(p^2)$ as

$$m(p^2) = \tilde{m}_{\text{cur}}^{(\sigma)} \tilde{f}^{(\sigma)}(p^2) , \tag{10.92}$$

where, by definition, the function $\tilde{f}^{(\sigma)}(p^2)$ satisfies the normalization condition

$$\tilde{f}^{(\sigma)}(p^2)|_{p^2=\sigma^2} = 1 \tag{10.93}$$

(we recall that we work in Euclidean space). Then the current mass $\tilde{m}_{\text{cur}}^{(\sigma)}$ is

$$\tilde{m}_{\text{cur}}^{(\sigma)} = m(\sigma^2) , \tag{10.94}$$

and the renormalization constant $\tilde{Z}_m^{(\sigma)}$ is

$$\tilde{Z}_m^{(\sigma)} = m^{(0)}/\tilde{m}_{\text{cur}}^{(\sigma)} . \tag{10.95}$$

The renormalization constant $Z_m^{(\sigma)}$ is defined now as

$$Z_m^{(\sigma)} \equiv \lim_{\tilde{m}_{\text{cur}}^{(\sigma)} \to 0} \tilde{Z}_m^{(\sigma)} , \tag{10.96}$$

and the current mass $m_{\rm cur}^{(\sigma)}$ is

$$m_{\rm cur}^{(\sigma)} \equiv m^{(0)} Z_m^{(\sigma)^{-1}} . \tag{10.97}$$

The mass function $m(p^2)$ (10.92) can be written as

$$m(p^2) = m_{\rm cur}^{(\sigma)} f^{(\sigma)}(p^2) \tag{10.98}$$

with $f^{(\sigma)}(p^2) = Z_m^{(\sigma)} \tilde{Z}_m^{(\sigma)^{-1}} \tilde{f}^{(\sigma)}(p^2)$. The ratio $\tilde{Z}_m^{(\sigma)}/Z_m^{(\sigma)}$ is finite as $\Lambda \longrightarrow \infty$.

Let us calculate $Z_m^{(\sigma)}$ in the rainbow approximation in the Landau gauge. Eq.(10.7) implies that in this case, the mass function $m(p^2)$ equals $B(p^2)$. Then, using Eqs.(10.11), (10.14), (10.15) and (10.96), one can check that

$$Z_m^{(\sigma)} = (\Lambda^2/\sigma^2)^{\frac{\omega-1}{2}} . \tag{10.99}$$

at $\alpha < \alpha_c \left(\omega = (1-\alpha/\alpha_c)^{1/2} < 1\right)$.

Let us now return to the question of whether spontaneous chiral symmetry breaking may or may not occur at $\alpha < \alpha_c$ in the continuum limit.

By definition, spontaneous chiral symmetry breaking implies the conservation of axial currents. Therefore in the continuum limit we must get:

$$\lim_{\Lambda\to\infty} \partial^\mu j_{5\mu}^\alpha = \lim_{\Lambda\to\infty} m^{(0)}(\Lambda) j_5^\alpha = \lim_{\Lambda\to\infty} m^{(0)}(\Lambda) Z_m^{(\sigma)^{-1}}(\Lambda) j_5^{(\sigma)^\alpha} = 0 . \tag{10.100}$$

This equation implies that the *necessary and sufficient* condition for the conservation of axial currents in the continuum limit is [17, 44]

$$\lim_{\Lambda\to\infty} m^{(0)}(\Lambda) Z_m^{(\sigma)^{-1}}(\Lambda) = 0 . \tag{10.101}$$

Although this condition looks as almost evident, we will see that it is helpful in clarifying some subtle issues in the problem of spontaneous chiral symmetry breaking in gauge theories. The point is that in gauge theories, the inverse renormalization constant $Z_m^{(\sigma)^{-1}}$ diverges as $\Lambda \longrightarrow \infty$. Therefore this condition is essentially stronger that $\lim_{\Lambda\to\infty} m^{(0)}(\Lambda) = 0$. In particular, the condition shows that the equality of the bare mass $m^{(0)}$ to zero in continuum theories is not sufficient to guarantee the conservation of axial currents.

Let us apply the condition (10.101) to ladder QED with $\alpha < \alpha_c$. Substituting the expression (10.15) for $B(p^2)$ in the boundary condition (10.11), we find that for $m^{(0)}$ satisfying the condition (10.101) with $Z_m^{(\sigma)}$ from Eq.(10.99), only the trivial solution $B(p^2) = 0$ satisfies this boundary condition in the continuum limit. Thus, there is no spontaneous chiral symmetry breaking in continuum ladder QED at $\alpha < \alpha_c$.

Let us consider now supercritical values of α. Since there is a dynamical piece in the fermion mass connected with spontaneous chiral symmetry breaking, we should modify the definition of the renormalization constant $Z_m^{(\sigma)}$. The point is that there is no renormalization of the dynamical mass m_{dyn}; renormalizations occur only for the piece of m connected with the explicit breakdown of chiral symmetry.*)

Let us define the current mass m_{cur} as

$$m_{\text{cur}} \equiv m - m_{dyn} \; , \tag{10.102}$$

where m_{dyn} corresponds to the solution of the SD equation at $m^{(0)}(\Lambda) = 0$. Let us define further the current mass function $m_{\text{cur}}(p^2)$ as

$$m_{\text{cur}}(p^2) \equiv m(p^2) - m_{dyn}(p^2) \; , \tag{10.103}$$

where $m_{dyn}(p^2)$ is the solution of the SD equation with $m^{(0)}(\Lambda) = 0$. Then we define the renormalization constant $Z_m^{(\sigma)}$ in the same way as before but with the mass function $m(p^2)$ replaced by the mass function $m_{\text{cur}}(p^2)$. Of course, we can realize this procedure consistently only at those values of the coupling constant at which the continuum limit is well defined (i.e. at ultraviolet fixed points, in the language of the renormalization group [26]).

As we already showed in Sec.10.2, in rainbow QED the dynamical mass m_{dyn} remains finite in the continuum limit only if the coupling constant α goes to the critical value $\alpha_c = \pi/3$ as

$$\alpha = \alpha_c + \frac{\pi^2 \alpha_c}{\ln^2 \left(4\Lambda/m_{dyn}\right)} \underset{\Lambda \to \infty}{\longrightarrow} \alpha_c = \pi/3 \tag{10.104}$$

(see Eq.(10.23)). Therefore we must calculate $Z_m^{(\sigma)}$ near the critical point $\alpha_c = \pi/3$.

---

*) Indeed, the mass renormalizations imply that the bare mass $m^{(0)}$ is chosen as an appropriate function of cutoff Λ (or, in the dimensional regularization, of the parameter $\frac{1}{\epsilon} = \frac{2}{4-n}$) to remove mass divergences. Since in the chiral limit $m^{(0)} \equiv 0$, the dynamical mass m_{dyn} is not affected by the renormalizations.

Substituting the expression (10.17) for $B(p^2)$ in the boundary condition (10.11) and using the procedure outlined above, we find that

$$Z_m^{(\sigma)} = (\sigma/\Lambda) \ . \tag{10.105}$$

Now it is easy to check that in the continuum limit, the condition (10.101) leads to the solution

$$B_{dyn}(p^2) = m_{dyn} F\left(\frac{1}{2}, \frac{1}{2}, 2; -p^2/m_{dyn}^2\right) \tag{10.106}$$

with the fixed value of $\alpha, \alpha = \alpha_c = \pi/3$. Since the asymptotic of this function is (see Eq.(10.16))

$$B_{dyn}(p^2) \underset{p^2 \gg m_{dyn}^2}{\simeq} \frac{2}{\pi} m_{dyn} \left(\frac{p^2}{m_{dyn}^2}\right)^{-1/2} \left(\ln \frac{p^2}{m_{dyn}^2} + 2(2\ln 2 - 1)\right) , \tag{10.107}$$

we see that oscillations occurring at $\alpha > \alpha_c$ disappear in the continuum limit $\Lambda \longrightarrow \infty, \alpha \longrightarrow \alpha_c$. This in turn implies that there are no zeroes in the BS function of NG bosons, i.e. massless bound states are states with the lowest value of M^2 in the pseudoscalar channel, and there are no tachyons with $M^2 < 0$ (compare with the discussion in the preceding section).

What we have shown is that the condition of the conservation of currents plays the important role in the proper definition of equations for Green's functions. In quantum mechanics, the Schrödinger equation becomes properly defined only if it is supplemented with two boundary conditions at $r = 0$ and $r = \infty$. The conservation of currents leads to a boundary condition at large momentum, i.e. it is an analogue of the quantum mechanical condition at $r = 0$.

So far we have considered only axial currents. But the same consideration is valid also for vector currents. It is easy to check that in QED with, for example, two fermion flavors, $a = 1, 2$, the rainbow SD equations without cutoff ($\Lambda = \infty$) and $\hat{m}^{(0)} = 0$ admit solutions with an arbitrary ratio m_1/m_2, where m_a is the mass of the a-th fermion, i.e. they admit solutions with the breakdown of the vector (isospin) symmetry. However, taking into account Eq.(9.6), we find that the condition of the conservation of vector currents takes the form:

$$\lim_{\Lambda\to\infty}\left[\lambda^\alpha,\hat{m}^{(0)}\right]Z_m^{(\sigma)^{-1}}(\Lambda)=0\,. \qquad (10.108)$$

Then, it is easy to show that the solutions with $m_1/m_2 \neq 1$ violate this condition, i.e. they correspond to explicit (and not to spontaneous) isospin symmetry breaking.

Of course the condition (10.108) and the corresponding condition for axial currents,

$$\lim_{\Lambda\to\infty}\left\{\lambda^\alpha,\hat{m}^{(0)}\right\}Z_m^{(0)^{-1}}(\Lambda)=0 \qquad (10.109)$$

are always satisfied with $\hat{m}^{(0)}(\Lambda)\equiv 0$ (the condition (10.108) is also satisfied with $\hat{m}^{(0)}_{ab}(\Lambda)=m^{(0)}(\Lambda)\delta_{ab}$). Therefore the most convenient way for getting a right solution with spontaneous symmetry breaking is to start with a theory with a regularizing cutoff Λ preserving the symmetry; to solve the equations there, and at the end, to remove cutoff in the solution (if it exists at all).

Thus we conclude that introducing cutoff in field theory is not just a reflection of our inability to work with complicated equations but a necessary procedure to define the equations properly.

The present approach leads to the solution of the problem known as the Goldstein problem in the literature [45]. Goldstein was first to study the ladder BS equation for massless electron-positron pseudoscalar bound states (parapositronium) in QED. He considered the equations without cutoff and of course found that nontrivial solutions to them exist at all positive values of the coupling constant α. A paradoxical situation occurs: *massless* parapositronium exists at arbitrary small coupling in QED!

However, as follows from the present consideration, if the subtleties of the transition $\Lambda \longrightarrow \infty$ are taken into account, one will come to the picture with the fixed value of the coupling constant, $\alpha=\alpha_c$, considered above.

10.8. The Breakdown of Scale Symmetry in QED. Composite Operators with Large Anomalous Dimensions

In the classical limit, QED with massless fermions is a scale invariant theory. However, in quantum QED, ultraviolet divergences and their subsequent renormalizations spoil this simmetry (for a review see Refs.[26, 46]).

The current connected with scale transformations is called the dilatation current, $\mathfrak{D}_\mu$. Its divergence is equal to the trace of the energy-momentum tensor θ^μ_ν [46]:

$$\partial^\mu \mathfrak{D}_\mu = \theta^\mu_\mu \ . \tag{10.110}$$

The quantum number connected with the dilatation generator $\mathfrak{D} = \int d^3x \mathfrak{D}_0(x)$ is called the dynamical dimension:

$$-i\left[\mathfrak{D},\ \varphi(x)\right] = \left(d_\varphi + x^\mu \partial_\mu\right)\varphi(x),$$

where d_φ is the dynamical dimension of the field φ. In the case of free massless fields, when the scale symmetry is exact, the dynamical dimension d_φ coincides with the canonical (engineering) dimension $d_{c;\varphi}$ of φ. However, for interacting fields, d_φ is usually different from $d_{c;\varphi}$. In this case, one says that the anomalous dimension of the operator φ, $\gamma_\varphi \equiv \mid d_{c;\varphi} - d_\varphi \mid$, is non-zero. Since in quantum theory, $\mathfrak{D}_\mu$ and θ^μ_μ are composite operators, care should be taken to define them properly. Usually such properly defined operators are written with the symbol N, for example, $N(\mathfrak{D}_\mu), N(\theta^\mu_\nu)$, etc. The definition of the N-operation is different in different models. We will return to this point in a moment.

In vector-like gauge theories, the following relation for $N(\partial^\mu \mathfrak{D}_\mu)$ takes place in all orders of perturbation theory [47]:

$$N\left(\partial^\mu \mathfrak{D}_\mu\right) = N\left(\theta^\mu_\mu\right) = \lim_{\Lambda\to\infty}\left[\frac{\beta(\alpha)}{4\alpha} N\left(F_{\mu\nu}F^{\mu\nu}\right)^{(\sigma)} + \sum_{a=1}^{N}\left(1+\gamma_m(\alpha)\right) m^{(\sigma)}_{\mathrm{cur};a} N\left(\bar{\psi}_a\psi_a\right)^{(\sigma)}\right] . \tag{10.111}$$

Here the renormalization group (RG) β-function is

$$\beta(\alpha) = \frac{\partial \alpha}{\partial \ln \Lambda} \, , \tag{10.112}$$

the anomalous dimension γ_m of the composite operator $N\left(\bar{\psi}_a\psi_a\right)^{(\sigma)}$ is

$$\gamma_m(\alpha) = -\frac{\partial \ln Z_m^{(\sigma)}}{\partial \ln \Lambda}, \tag{10.113}$$

and renormalized quantities $m_{\mathrm{cur};a}^{(\sigma)}, N\left(\bar{\psi}_a\psi_a\right)^{(\sigma)}$ are defined in Eq.(10.90), (10.91) (the summation in color indices is understood in Eq.(10.111).

In perturbative theory, operators $N(\hat{A})$, $\hat{A} = \theta_\mu^\mu$, $F_{\mu\nu}F^{\mu\nu}$ or $\bar{\psi}_a\psi_a$, are defined as

$$N(\hat{A}) = \hat{A} - \langle 0|\hat{A}|0\rangle\,. \tag{10.114}$$

However, the definition of $N(\hat{A})$ in the full theory is rather subtle. For example, in QCD, it is not appropriate to subtract the nonperturbative part of $\langle 0|\hat{A}|0\rangle$ connected with hadronic interactions. Therefore in this case one subtracts only its perturbative part [48]:

$$N(\hat{A}) = \hat{A} - \langle 0|\hat{A}|0\rangle_{pert}\,. \tag{10.115}$$

Unfortunately, such a separation of perturbative and nonperturbative contributions is somewhat ambiguous, and it leads to some difficulties [49]. We will return to this point in Sec. 12.15.

We also note that although the quantities $N\left(F_{\mu\nu}F^{\mu\nu}\right)^{(\sigma)}, N\left(\bar{\psi}_a\psi_a\right)^{(\sigma)}, m_{\mathrm{cur};a}^{(\sigma)}$ depend on the scale parameter σ, the operator $N(\theta_\mu^\mu)$, and therefore the combinations of the operators appearing on the right-hand side of Eq.(10.111), are observables independent of σ. At $m_{\mathrm{cur};a}^{(\sigma)} = 0$, in the rainbow approximation, the scale symmetry is exact in perturbative QED. However, in the supercritical phase, because of spontaneous chiral symmetry breaking, fermions acquire a dynamical mass m_{dyn}, and therefore the scale symmetry should be broken. Is the scale symmetry broken explicitly or spontaneously there ? In the latter case, the conservation of the dilatation current would imply the existence of a massless NG scalar, dilaton. We will see that the breakdown of the scale symmetry is explicit in the supercritical phase. But it is important that near $\alpha = \alpha_c$, one can treat the scale symmetry as an approximate one. In particular, it is appropriate to consider the dynamical dimension for operators, which is the quantum number connected with the scale transformations. The situation resembles that in QCD, however, there is an essential difference between these two dynamics: while in QCD,because of asymptotic freedom, anomalous dimensions of all operators are small [43] (i.e. their dynamical dimensions are close to their canonical (engineering) ones), in ladder QED, the critical value α_c is rather big, and therefore some operators may acquire large anomalous dimensions. As we will see, this leads to some important consequences.

What is an appropriate definition of the composite operators $N(\theta^\mu_\nu), N(F_{\mu\nu}F^{\mu\nu})$ and $N(\bar{\psi}_a\psi_a)$ in the present problem ? We will define these operators as the canonical operators minus the contribution of their vacuum expectation values calculated in free massless theory:

$$N_0(\theta^\mu_\nu) \equiv \theta^\mu_\nu - \langle 0|(\theta^\mu_\nu)_{free}|0\rangle\big|_{m_a^{(cur)}=0} \,, \tag{10.116}$$

$$N_0\left(F_{\mu\nu}F^{\mu\nu}\right) \equiv F_{\mu\nu}F^{\mu\nu} - \langle 0|(F_{\mu\nu}F^{\mu\nu})_{free}|0\rangle|_{m_a^{cur}=0} \,, \tag{10.117}$$

$$N_0(\bar{\psi}_a\psi_a) \equiv \bar{\psi}_a\psi_a \tag{10.118}$$

(of course, $\langle 0|\bar{\psi}_a\psi_a|0\rangle$ is equal to zero in free massless theory). The necessity of the subtraction of the vacuum expectation value $\langle 0|(\theta^\mu_\nu)_{free}|0\rangle|_{m_a^{cur}=0}$ from θ^μ_ν is dictated by the point that the scale symmetry is exact in free massless theory, and therefore the dilatation current $\mathfrak{D}_\mu$ is conserved, that is, $\partial_\mu \mathfrak{D}^\mu_{free}|_{m_a^{cur}=0} = N_0(\theta^\mu_\mu) = 0$ there. The nontrivial point is that, as we will show below, this single subtraction in Eqs.(10.116), (10.117) provides finiteness of the operators in Eq.(10.111).

In the supercritical phase, the continuum limit $\Lambda \to \infty$ with m_{dyn} being finite can be arranged only if the coupling constant α depends on cutoff Λ as is shown in Eq.(10.104). Therefore, following Bardeen, Leung and Love [19], it is convenient to introduce the β-function

$$\beta = \frac{\partial \alpha}{\partial \ln \Lambda} = -\frac{2}{3}\left(\frac{\alpha}{\alpha_c} - 1\right)^{3/2} \,, \quad \alpha \geq \alpha_c \,. \tag{10.119}$$

Since $\beta(\alpha)$ is negative at $\alpha > \alpha_c$, its zero an $\alpha = \alpha_c$ corresponds to an ultraviolet stable fixed point. Is this β-function genuine ? This question is natural since at $\alpha < \alpha_c$ the perturbative β-function equals zero in this approximation. The answer to the question is "it is": we will show, following Ref.[50], that the β-function (10.119) governs scale symmetry breaking in the supercritical phase.

What the authors of Ref.[50] showed is that at $\alpha > \alpha_c$ the vacuum expectation values of the operators (10.116)-(10.118) satisfy the relation (10.111) with the β-function defined in Eq.(10.119) and the anomalous dimension

$$\gamma_m = -\frac{\partial \ln Z_m^{(\sigma)}}{\partial \ln \Lambda} = 1 \tag{10.120}$$

(see Eq.(10.105). For simplicity, we will consider here only the most interesting case with $m_{\text{cur};a} = 0$.

Our aim is to show that

$$\langle 0|N_0(\theta^\mu_\mu)|0\rangle = \lim_{\substack{\Lambda\to\infty \\ (\alpha\to\alpha_c+0)}} \frac{\beta(\alpha)}{4\alpha}\langle 0|N_0(F_{\mu\nu}F^{\mu\nu})|0\rangle \neq 0\ . \tag{10.121}$$

First of all, we note that because of the Lorentz invariance, we get the following relation:

$$\langle 0|N_0(\theta^\mu_\mu)|0\rangle = 4\langle 0|N_0(\theta^0_0)|0\rangle = 4V(\bar{G})\ , \tag{10.122}$$

where V is the effective potential for composite operators, considered in Chapter 8, and the fermion propagator $\bar{G}$ is its stationary point (i.e., $\bar{G}$ is a solution of the Schwinger-Dyson equation).

In the chiral limit, the expression for $V(G)$ takes the following form in Landau gauge (see Eq.(8.60)):

$$V(G) = V(B) = -\frac{N}{4\pi^2}\left\{\int_0^{\Lambda^2} duu\left[\frac{1}{2}\ln\left(1+\frac{B^2(u)}{u}\right) - \frac{B^2(u)}{u+B^2(u)}\right]\right.$$
$$\left. + \frac{3\alpha}{8\pi}\int_0^{\Lambda^2}\frac{duuB(u)}{u+B^2(u)}\int_0^{\Lambda}\frac{dvvB(v)}{v+B^2(v)}K(u,v)\right\}\ , \tag{10.123}$$

where $u \equiv p^2, v = k^2$ and $K(u,v) = \theta(u-v)/u + \theta(v-u)/v$. Let us show that

$$V(\bar{B}) = \frac{1}{2}\frac{\partial V(B)}{\partial \ln \Lambda^2}|_{B=\bar{B}}\ , \tag{10.124}$$

where $\bar{B}$ satisfies the Schwinger-Dyson equation

$$B(u) = \frac{3\alpha}{4\pi}\int_0^{\Lambda^2} dv K(u,v)\frac{vB(v)}{v+B^2(v)} . \tag{10.125}$$

The relation (10.124) follows from the fact that $\bar{B}(u)$ is a sationary point of the effective potential. Indeed, considering the variations of V with respect to the transformations $\bar{B} \longrightarrow B^{(s)} = s\bar{B}(u/s)$ $(V(B^{(s)},\Lambda^2) = s^4 V(\bar{B},\Lambda^2/s^2))$, we find at the stationary point:

$$\frac{dV(B^{(s)},\Lambda^2)}{ds}|_{s=1} = \int du \frac{\delta V}{\delta B^{(s)}}|_{B^{(s)}=\bar{B}}\frac{dB^{(s)}}{ds}|_{s=1} = 0 . \tag{10.126}$$

Using the expression (10.123) for $V(B)$, it is easy to check that Eq.(10.126) leads to the relation (10.124).

The relation (10.124) together with the relation

$$\bar{B}(\Lambda^2) = \frac{3\alpha}{4\pi\Lambda^2}\int_0^{\Lambda^2} \frac{du u\bar{B}(u)}{u+\bar{B}^2(u)} , \tag{10.127}$$

following from Eq.(10.125), allow to calculate $\langle 0|N_0(\theta^\mu_\mu)|0\rangle$. Indeed:

$$\begin{aligned}\langle 0|N_0(\theta^\mu_\mu)|0\rangle &= \lim_{\substack{\Lambda\to\infty \\ (\alpha\to\alpha_c+0)}} 4V(\bar{B}) = \lim_{\substack{\Lambda\to\infty \\ (\alpha\to\alpha_c+0)}} 2\frac{\partial V(\bar{B})}{\partial \ln \Lambda^2} \\ &= -\lim_{\substack{\Lambda\to\infty \\ (\alpha\to\alpha_c+0)}} \frac{N}{4\pi^2}\Lambda^4 \ln\left(1+\frac{\bar{B}^2(\Lambda^2)}{\Lambda}\right) = -\frac{4N}{\pi^4}m^4_{dyn} .\end{aligned} \tag{10.128}$$

Here we used the expression for $\bar{B}(p^2)$ in Eq.(10.17).

The vacuum expectation value $\langle 0|N_0(F_{\mu\nu}F^{\mu\nu})|0\rangle$ can be calculated by using the well-known relation

$$\langle 0|N_0(F_{\mu\nu}F^{\mu\nu})|0\rangle = -4\alpha\frac{\partial V(\alpha,\Lambda)}{\partial\alpha} \tag{10.129}$$

following from the path integral representation for the vacuum energy density $V(\bar{B})$. It can be also calculated directly; in the present approximation, it is given by the contribution of the diagram in Fig.10.3.

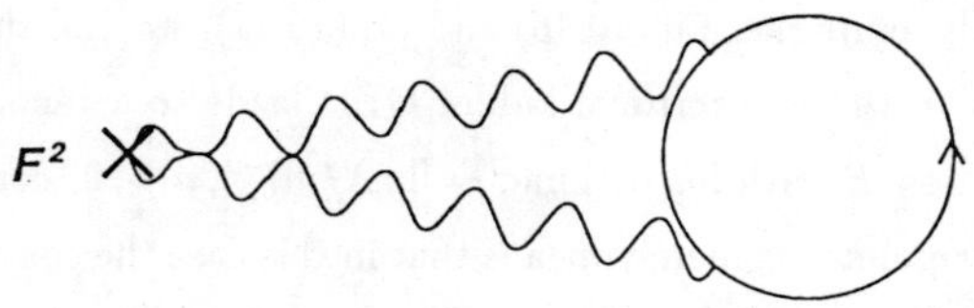

Fig. 10.3.

The result of the calculations leads to the desired equality (10.121).

The following comments are in order.

We note that the calculated contribution into $\langle 0|N_0(\theta^\mu_\mu)|0\rangle$ is purely nonperturbative. Therefore it is appropriate to call it a nonpertubative scale anomaly. The nonperturbative scale anomaly occurs because of the presence of the nonperturbative mass divergence in the supercritical phase. We emphasize that the origin of this divergence is different from that of loop divergences in perturbative theory: the mass divergence (10.22) occurs due to a singular character (at small distances) of the exchange interactions with a strong α, which is harmless in perturbative theory. Since such divergences occur already in quantum mechanics (see Sec.10.4), it is natural to call them the quantum-mechanical divergences. We will discuss the origin of the nonperturbative scale anomaly in the framework of the theory of critical phenomena in Sec. 11.6.

As was indicated in Refs.[16, 17], the collapse phenomenon in the supercritical phase in QED has common features with phenomena in the supercritical phases of such two-dimensional models as the Thirring model and the sine-Gordon model. This issue has been recently clarified in the work [51].

The explicit violation of scale symmetry implies that there should not be very light scalars in quenched QED; as in the NJL model, one may expect that the mass of the lightest scalar is of the order of m_{dyn}. A numerical analysis of the BS equations for scalars confirms this expectation[29].

In perturbative renormalizations, the following relations are valid:

$$\Gamma(\{p\},\mu,\alpha) = Z(\Lambda/\mu)\Gamma^{(0)}(\{p\},\Lambda,\alpha^{(0)}) + (\textit{power corrections } p/\Lambda, \mu/\Lambda, etc.). \tag{10.130}$$

On the other hand, comparing Eq.(10.16) and Eq.(10.17), we find that the renormalization (10.104) of α in the supercritical ladder QED leads to a renormalization relation for the mass function B with logarithmic $\sim [\ln \Lambda/m]^{-n}, n > 0$, corrections. From the mathematical view point this merely means that in this case the convergence to a continuum theory is slower than in perturbative QED or in QCD. However, from the physical viewpoint it reflects a deep distinction between these two dynamics. In particular, small power corrections in Eq.(10.130) imply decoupling of high energy dynamics from the low energy region [52].*) The situation can be different in field theories with a rather strong bare coupling constant. For example, the anomalous dimension γ_m of the composite operators $\bar{\psi}\lambda^\alpha\psi$ and $i\bar{\psi}\lambda^\alpha\gamma_5\psi$ is equals one at $\alpha = \alpha_c$ (see Eq.(10.120)), i.e. their dynamical dimension $d = 3 - \gamma_m$ equals two. As was pointed out by Bardeen, Leung and Love [19], this implies that the four-fermion chiral invariant operator

$$\frac{1}{\Lambda^2}\sum_\alpha [(\bar{\psi}\lambda^\alpha\psi)^2 + (\bar{\psi}i\gamma_5\lambda^\alpha\psi)^2] \tag{10.131}$$

acquires the dynamical dimension $d = 4$ and it acts as a renormalizable operator. In order words, strong Coulomb-like interactions amplify short-distance ($r \sim \Lambda^{-1}$) interactions in a way that they manifest themselves without any suppression at large scales. In the language of the renormalization group, it means that such a strongly interacting dynamics transform formally irrelevant operators, posessing canonical (engineering)) dimensions $d_c > 4$ into relevant ones.**) We will consider these issues in detail in Chapter 11.

From the formal point of view, it implies that QED with a large bare coupling constant is not a closed theory, and it should be supplemented by those perturbatively irrelevant interactions which become relevant due to the strong QED interaction. We will also see in Chapters 11, 14, 15 that the phenomenon of the amplification of formally irrelevant interactions find interesting applications in model building, in particular, in models of electroweak symmetry breaking.

---

*) For a more detailed discussion of the decoupling effect (the decoupling theorem[52]), see Sec. 13.8.

**) The notion of relevant and irrelevant operators is discussed in Sec.11.2.

10.9. Spontaneous Chiral Symmetry Breaking in QED Beyond the Quenched Approximation

Up to now we have considered spontaneous chiral symmetry breaking in quenched QED, in which the contribution of diagrams with fermion loops is neglected. In this section, we will discuss the dynamics beyond the quenched approximation in QED.

The first heuristic consideration of the effects of fermion loops for spontaneous chiral symmetry breaking dynamics in QED was done in Ref.[17]. It was argued there that spontaneous chiral symmetry breaking at supercritical values of $\alpha > \alpha_c \sim 1$ should survive the screening effect connected with fermion loops; also, speculations about the possibility of the existence of a nontrivial continuum QED with the bare coupling constant $\alpha^{(0)} = \alpha_c$ were considered.

The first systematic studies of this problem were done in Refs.[53 - 56]. They were essentially improved in the further works [57]. The main conclusions of these studies are the following:

a) at $\alpha > \alpha_c \sim 1$, there is spontaneous chiral symmetry breaking even in QED beyond the quenched approximation; the critical value α_c corresponds to the second order (continuous) phase transition;

b) as a result of including fermion loops, the essential singularity at $\alpha = \alpha_c$ in the expression (10.22) for m_{dyn} is replaced by a square root singularity:

$$m_{dyn} \sim \Lambda(\alpha - \alpha_c)^{1/2} \ . \tag{10.132}$$

We will discuss the physical meaning of these results below.

Let us consider the effect of including the fermion loops by following the work of Gusynin [55] who used a method known as the bifurcation approximation in the literature.

First, let us consider this very useful approximation introduced by Atkinson and Johnson [58].

In Sec.10.2, we considered the linearized approximation to Eq.(10.8), in which the function $B^2(p^2)$ in the denominator is replaced by m^2. This allowed us to express its approximation solution through the hypergeometric functions.

The idea of the bifurcation approximation is the following. If we are interested in the solution near the critical point $\alpha_c, \alpha - \alpha_c \ll 1$, where the dynamical mass is very small, then $B^2(q^2)$ in the denominator $(q^2 + B^2(q^2))$ in Eq.(10.8) may be neglected everywhere except the infrared region around $q^2 = 0$. Therefore, in this case, one may expect that a good approximation is that one in which the denominator $(q^2 + B^2(q^2))$ is replaced by just q^2 but m^2 is introduced as an infrared cutoff:

$$B(p^2) = m^{(0)} + \frac{3\alpha}{4\pi} \int_{m^2}^{\Lambda^2} dq^2 \left[\frac{\theta(p^2 - q^2)}{p^2} + \frac{\theta(q^2 - p^2)}{p^2} \right] B(q^2) \, . \tag{10.133}$$

This trick simplifies the analysis enormously. In particular, one can easily check that the general solution of this equation is expressed through elementary functions:

$$B(p^2) = C_1 (p^2)^{-\frac{1}{2}[1-(1-\frac{\alpha}{\alpha_c})^{1/2}]} + C_2 (p^2)^{-\frac{1}{2}[1+(1-\frac{\alpha}{\alpha_c})^{1/2}]} \, ; \alpha_c = \pi/3 \, . \tag{10.134}$$

The approximation is called the bifurcation approximation since the value $\alpha = \alpha_c$, at which a second (nontrivial) solution to the equation occurs, is called a bifurcation point in the mathematical literature.

Let us now consider the SD equation with the bare vertex and $A(p^2) = 1$ but including the vacuum polarization effects (in Landau gauge):

$$B(p^2) = \frac{3\alpha^{(0)}}{4\pi^3} \int d^4q \frac{B(q^2)}{q^2 + B^2(q^2)} \frac{1}{1 - \Pi((p-q)^2)} \frac{1}{(p-q)^2} \tag{10.135}$$

(we work in Euclidean region), where $\alpha^{(0)}$ is the bare coupling and $\Pi(q^2)$ is the polarization operator in the photon propagator:

$$D_{\mu\nu}(q) = \left(g_{\mu\nu} - \frac{q_\mu q_\nu}{q^2} \right) \frac{1}{q^2(1 - \Pi(q^2))} \, . \tag{10.136}$$

Having using the substitution $\Pi\left((p-q)^2\right) \longrightarrow \theta(p^2 - q^2)\Pi(p^2) + \theta(q^2 - p^2)\Pi(q^2)$ one can integrate over the angular variables in Eq.(10.135) and obtain

$$B(u) = \int_0^{\Lambda^2} \frac{dv v B(v)}{v + B^2(v)} K(u, v), \; u = p^2, \; v = q^2 \, , \tag{10.137}$$

where in the one-loop approximation, the kernel K is

$$K(u,v) = \frac{g(u)}{u}\theta(u-v) + \frac{g(v)}{v}\theta(v-u) \, , \tag{10.138}$$

$$g(u) = \frac{3\alpha^{(0)}(\Lambda)}{4\pi\left(1 + \frac{\alpha^{(0)}(\Lambda)N}{3\pi}\ln\frac{\Lambda^2}{u}\right)} \, . \tag{10.139}$$

Let us now use the bifurcation approximation in equation (10.137). As a result, we obtain the following equation:

$$B(u) = \int\limits_{m^2_{dyn}}^{\Lambda^2} dv B(v)K(u,v), \; B(m^2_{dyn}) \equiv m_{dyn} \, . \tag{10.140}$$

It is convenient to introduce the variable $z = 9/4Ng(u)$. Then one can check that equation (10.140) is equivalent to the differential equation

$$\frac{d^2B}{dz^2} + \left(\frac{2}{z} + \frac{z}{z-1}\right)\frac{dB}{dz} + \frac{9}{4N}\frac{z-1}{z^2}B = 0 \tag{10.141}$$

with the boundary conditions

$$\frac{dB}{dz}\Big|_{z=z_0=\frac{3\pi}{\alpha^{(0)}(\Lambda)N}+\ln\frac{\Lambda^2}{m^2_{dyn}}} \, , \tag{10.142}$$

$$\left[\frac{dB}{dz} + \left(\frac{1}{z} - 1\right)B\right]_{z=\bar{z}=\frac{3\pi}{\alpha^{(0)}(\Lambda)N}} \tag{10.143}$$

(compare with Eqs.(10.9) - (10.11)).

Eq.(10.141) is a linear second order differential equation with three singular points $z = 0, 1$ and ∞. The points $z = 0, 1$ are regular singular points, and $z = \infty$ is an irregular one. Because of this, the solutions of this equations cannot be expressed through known special functions[25]. However, if N is not very large ($N \leq 2$), the equation can be further simplified. Indeed, since the variable z changes within $z_0 \geq z \geq$

$3\pi/\alpha^{(0)}(\Lambda)N$, then at $\frac{3\pi}{\alpha^{(0)}(\Lambda)N} \gg 1$, the variable $z \gg 1$ and simplifying Eq.(10.141) by retaining only $\frac{1}{z}$ and constant terms at the derivative $\frac{dB}{dz}$, we get the following equation:

$$\frac{d^2B}{dz^2} + \left(\frac{1}{z} - 1\right)\frac{dB}{dz} + \frac{9}{4N}\left(\frac{z-1}{z^2}\right)B = 0 \,. \tag{10.144}$$

Of course, the consistency of the assumption $\frac{3\pi}{\alpha^{(0)}(\Lambda)N} \gg 1$ implies that the critical coupling α_c should not be very large. As we will see, $\alpha_c \simeq 2$ at $N = 1$, and therefore the ratio $\frac{3\pi}{\alpha_c N}$ is reasonably large in this case.

The general solution of equation (10.144) takes the form

$$B(z) = z^\gamma[C_1\Phi(a, c; z) + C_2\Psi(a, c; z)] \,, \tag{10.145}$$

where Φ and Ψ are the confluent hypergeometric functions [25]; $\gamma = 3/2N^{1/2}, a = \gamma - \gamma^2, c = 1 + 2\gamma$.

We omit the analysis of the boundary conditions (10.142), (10.143) for the solution (10.145) [55] and indicate here only the main results.

The critical coupling α_c is $\alpha_c \simeq 1.95$ at $N = 1$. At $\alpha^{(0)}(\Lambda) > \alpha_c$, there is spontaneous chiral symmetry breaking. The dynamical mass is

$$m_{dyn} \simeq \Lambda(\alpha^{(0)} - \alpha_c)^{1/2} \,. \tag{10.146}$$

Thus, there is a second order phase transition at $\alpha^{(0)} = \alpha_c$: the order parameter m_{dyn} continuosly goes to zero as $\alpha^{(0)} \longrightarrow \alpha_c$ from the side of the supercritical values of $\alpha^{(0)}$.

The relation between m_{dyn} and Λ (10.146), $m_{dyn} = \Lambda(\alpha^{(0)} - \alpha_c)^\nu$ with $\nu = 1/2$, corresponds to the scaling law in the mean-field theory [26] (for a more detail see Secs. 11.3, 11.6). This in turn implies that as $\Lambda \longrightarrow \infty$ with m_{dyn} being finite, the value of the bare coupling constant in the continuum theory is fixed:

$$\alpha^{(0)} = \alpha_c + \frac{m_{dyn}^2}{\Lambda^2} \underset{\Lambda \to \infty}{\longrightarrow} \alpha_c \,, \tag{10.147}$$

and it describes non-interacting fermions, photon and $2N^2$ spineless states: N^2 massless pseudoscalars (NG bosons) and N^2 massive scalars. More precisely, in such a theory coupling constants, describing interactions at low ($p \ll \Lambda$) momenta, go slowly (logarithmically) to zero as cutoff Λ goes to infinity ($g \sim (\ln p/\Lambda)^{-a}$, $a > 0$). Of course, at finite cutoff Λ (considered as a scale of a "new" physics) such a theory may serve as an effective theory describing low energy interactions of all these particles.

A more sophisticated analysis in the one-loop approximation in QED, which takes into account the SD equations for the vertex and for the photon propagator, leads to similar conclusions [56, 57]. It is however clear that since there is no reliable perturbation scheme in strong coupling QED, all these studies are not conclusive. At present, our understanding of the dynamics of supercritical QED is still far from being complete. The main problem is connected not with just absence of an adequate approximation for studying such a nonperturbative dynamics but also with the point that, as was already indicated in Sec.10.8, supercritical QED is not a closed theory: there might be a lot of formally irrelevant operators which, because of strong short distance QED interactions, become relevant ones, and therefore adding them to QED lagrangian may change the dynamics.

We will return to the discussion of this point in Chapter 11 and Sec. 14.6.

10.10. Nonperturbative Dynamics in QED and Beyond: Miscellaneous

In this section, we will review some problems in the nonperturbative dynamics in QED (and beyond) intimately connected with the problem of spontaneous chiral symmetry breaking.

1) *Complex branch-point singularities in the fermion propagator.*

So far we have analyzed the rainbow SD equation for the fermion propagator by using its linearized version. As we already indicated in Sec.10.2, this approximation is rather good in Euclidean region. However, the situation in Minkowski region is essentially different. Let us discuss this point in more detail.

In the linearized approximation, the solution of the SD equation for the mass function $B(p^2)$ is expressed through the hypergeometric function (see Eq.(10.14)),

$$B(p^2) = mF\left(\frac{1+\omega}{2}, \frac{1-\omega}{2}, 2; -p^2/m^2\right) . \tag{10.148}$$

This function has a branch-point in the time-like region with a threshold at $p^2 = m^2$. This is what one should expect: the branch-point corresponds to one-fermion plus many-photon intermediate states. However, it turns out that the analytic properties of the solution of the nonlinear SD equation are absolutely different.

This problem was first considered by Fukuda and Kugo [13]. By using numerical methods, they showed that the solution of the nonlinear equation does not have singularities at real p^2. Their hypothesis was that the fermion propagator is an entire function, and therefore the solution might be appropriate for describing the confinement phenomenon.

However, Dragovic, Mavlo and Filippov [59] showed later, again by numerical methods, that an expansion of the propagator in powers of of p^2 has a finite radius of convergence. This implies that the propagator has at least one singularity. At last, Atkinson and Blatt [60], and Alekseev, Arbuzov and Rodionov [61] showed the existence of complex singularities in the p^2-plane in the fermion propagator. More precisely, they found a pair of complex-conjugate branch-points at any positive value of the coupling constant α.

Needless to say, the existence of such singularities are forbidden by the general principles of quantum field theory, such as locality, Lorentz invariance, etc.. It seems that the only appropriate explanation of their existence is that the rainbow approximation to the SD equation is not fully consistent and that an appropriate improving this approximation will remove the complex singularities.

Up to now this problem is still unsolved.

Does this imply that we should not trust any result in the rainbow approximation if it leads to such undesirable singularities ? Our viewpoint is that this does not. The point is that the analytic properties of functions are a global characteristics: a small variation of a function in some region may drastically change its analytic properties in another. Therefore the analytic properties of Green's functions are very sensitive to the truncation of SD equations. On the other side, such characteristics of a theory as a critical coupling constant in spontaneous chiral symmetry breaking phenomenon, anomalous dimensions of composite operators, etc. might be much less sensitive to the truncation operation. The important point is that, as we argued above in this chapter, spontaneous chiral symmetry breaking in supercritical QED admits a rather clear physical interpretation.

Nevertheless, it is certainly very desirable to clarify the issue with the complex singularities. In particular, it would be very interesting to check the analytical properties of the solution of the SD equation for the fermion propagator with the Curtis-Pennington ansatz for the vertex considered in Sec.10.5. This ansatz restores the gauge invariance. But does it cure the desease with the complex singularities ?

2) *Abnormal Solutions of the BS equation.*

As is known, the ladder BS equations have pathological features such as the appearance of ghost solutions (introducing the indefinite metrics) or tachyonic solutions (for a recent review see Refs.[62, 63]). This problem relates not just to QED but to many field theoretical models.

In Sec.10.7, we discussed the Goldstein problem and suggested its solution. That discussion may be helpful for the understanding of the general situation concerning the BS equation problems indicated above.

The usual treatment of the ladder BS equation is as the relativistic analogue of the Schrödinger equation. In particular, a particle mass is considered as a free parameter there. But such a treatment is somewhat misleading. The BS equation is just one of an infinite number of equations for Green's functions. Also, a mass parameter in field theory cannot always be treated as a free one: it is determined from the particle propagator, which itself satisfies the Schwinger-Dyson equation. For example, as we saw in Sec.10.2, in rainbow QED with a supercritical value of the coupling constant, the fermion mass is always large, $m \geq m_{dyn} \sim \Lambda$, and one cannot consider freely small values for m.

This point was essence of the Goldstein problem: one cannot consistently treat the BS equation by isolating it from other equations for Green's functions, in particular, from the SD equation for propagators. One may expect that pathological features of some solutions of the ladder BS equation just reffect those inconsistencies.

The second important point in solving the Goldstein problem was the choice of the right boundary condition. Of course, this point is primary important for the Schrödinger equation either: it is not enough to solve the equation; it is important to select the "right" solutions satisfying the boundary conditions at $r = 0$ and $r \longrightarrow \infty$. In the case of the Schrödinger equation these conditions are uniquely defined by the probabilistic interpretation of quantum mechanics. The situation in field theory is far less definite. Boundary conditions may be different in different problems (in the case of the Goldstein problem, the boundary condition was determined from the condition of the axial current conservation for example). But it seems that as a general rule, introducing a proper regularization must play the important role for the consistent treatment of the nonperturbative dynamics in field theory.

Of course, in reality, in order to get concrete results, we must always truncate equations for Green's functions. There is no algorithm how one should do the truncation. Clearly, this is a kind of art, where the physical intuition is very helpful. The lessons we have learned from the experience of working with nonperturbative QED (even although in a rather crude approximation) may in turn help the intuition.

3) *Chiral Symmetry Breaking in Coulomb gauge.*

In this book, only covariant gauges are used for studying chiral symmetry breaking in gauge theories. Some authors studied this phenomenon in the Coulomb gauge [64 - 67]. This gauge is convenient for constructing the vacuum state in the broken phase (which is similar to constructing the vacuum in the NJL model (see Chapter 6)). However, it is rather difficult to satisfy the WT identities, beyond perturbation theory, in this gauge.

Chapter 11

Spontaneous Symmetry Breaking and Critical Phenomena

11.1. Introduction

In the previous two chapters, where we described dynamical chiral symmetry breaking in the Nambu-Jona-Lasinio (NJL) model and QED, we often used such words as "second order phase transition", "symmetric and asymmetric phases", "critical point" and so on, which are normaly used in the description of critical phenomena in statistical physics.

A deep relation between the existence of a continuum limit in quantum field theory (QFT) and critical phenomena was established by Wilson [1, 2]. He made use of a fact that a partition function of a (regularized) euclidean QFT can be considered as a statistical sum in the corresponding classical field theory. The crucial observation is that the continuum limit of the QFT corresponds to the neighbourhood of a second order phase transition in the statistical system. Indeed, suppose Λ is an ultraviolet cutoff in the QFT, say $\Lambda = 1/a$ where a is the lattice spacing of the discretized space-time. Then, in the neighbouhood of a second order phase transition, the correlation length ξ in the system is very large compared to a [3]. In the QFT the correlation length is defined as the inverse mass m^{-1}. Therefore, in this neighbourhood, the ratio $\xi/a \sim \Lambda/m$ is very large, and $\Lambda/m \to \infty$ at the critical point of the second order phase transition. This in turn implies that there is a possibility to arrange a continuum limit $\Lambda/m \to \infty$ in the theory by tuning the bare parameters as functions of Λ in such a way that the physical quantities, like m, remain finite as $\Lambda/m \to \infty$. This tuning procedure is nothing else but a renormalization procedure in QFT.

Therefore continuum limit is defined near a critical hypersurface in the parameter space of a theory. A special and very important subset of the critical hypersurfaces is formed by conformally invariant points where the theory is conformally invariant. The conformal transformations include the Lorentz transformations ($SO(D)$ transformations in D dimensional Euclidean space), the scale (dilatation) transformations and the special conformal transformations (for a review see the books [3]). In the vicinity of conformal points, local operators are characterized by a dynamical dimension, which is the quantum number connected with the generator of scale transformations (see Sec.11.3). The mathematical tool for the description of the dynamics near conformal points is the renormalization group [2-4]. Conformal points coincide with the fixed points of the renormalization group equations considered first by Gell-Mann and Low long ago [5].

Thus a critical hypersurface with the infinite correlation length ξ is a hypersurface of a second order phase transition. According to the Landau theory of phase transitions [6], continuous (second order) phase transitions are intimately connected with spontaneous breakdown of some symmetries. Therefore it is not surprising that dynamical symmetry breaking is described as a critical phenomenon.

In the present chapter, we will discuss these points by considering, as an illustration, some low dimensional models with spontaneous symmetry breaking. The advantage of this is that critical phenomena in dimensions $D < 4$ are rather well understood. Therefore, low dimensional models can serve as a very useful laboratory for studying these issues. They, in particular, will help us in better understanding the dynamics in the NJL model and QED already discussed in Chapter 6, 9 and 10.

From the viewpoint of the problem of generating dynamical fermion mass, the most interesting low dimensional models are models with four-fermion interactions. They were first studied by Wilson [7] and Gross and Neveu [8] in the framework of $1/N_c$ expansion (N_c is the number of fermion colors). It was suggested by these authors and Parisi [9] that although the models are perturbatively nonrenormalizable at $D > 2$, they become renormalizable at some value of the four-fermion coupling constant. This value corresponds to a second order phase transition. The proof of the renormalizability of these models to all orders in the $1/N_c$ expansion was given by Rosenstein, Warr and Park [10, 11].

The phenomenon underlying the renormalizability of these models is the transmutation of the dynamical dimension of some operators. This phenomenon, which is interesting in itself, occurs also in some four dimensional models used in the description of electroweak symmetry breaking. This is the main reason we are going to discuss this phenomenon in some detail.

11.2. Renormalizations and Relevant, Marginal and Irrelevant Interactions

Let us consider an action in D dimensional space-time:

$$S = \int d^D x \sum_i g_i^{(0)} O_i(x) \,, \tag{11.1}$$

where $O_i(x)$ are some local operators composed of elementary fields $\varphi_a(x)$.

As we know from text-books (see for example Ref.[4]), the theory with the action (11.1) is perturbatively renormalizable if the canonical (engineering) dimensions d_{ci} of the operators O_i are equal to or less than D (d_{ci} are expressed in the energy units).

Operators O_i with $d_{ci} < D$ describe the so called superrenormalizable interactions, and are called relevant operators. Operators with $d_{ci} = D$ lead to strictly renormalizable interactions, and are called marginal operators. Now what about the operators O_i with $d_{ci} > D$ which lead to nonrenormalizable interactions ? They seem to be the most difficult operators to work with, so it may seem surprising that they are called irrelevant operators !

This terminology, introduced by Wilson [1, 2], will become intelligible if the old renormalization procedure is reformulated à la Wilson by using the notion of the low energy effective action [1, 2, 12, 13].*)

The initial point of Wilson's approach is to introduce a cutoff Λ in the action (11.1):

$$S \longrightarrow S_\Lambda \,. \tag{11.2}$$

Now the action S_Λ is considered as an effective action describing interactions at scale $E \ll \Lambda$. One may imagine the derivation of S_Λ from some more fundamental action S_f in the following way. Let us divide the fields in the path integral with the action S_f into high-and low-energy parts :

$$\varphi_a = \varphi_{a,h} + \varphi_{a,l} \,, \tag{11.3}$$

where the energy E is $E > \Lambda$ for $\varphi_{a,h}$ and $E < \Lambda$ for $\varphi_{a,l}$. The effective action S_Λ is

*) As will be clear from the following discussion, this effective action should be distinguished from the effective action (generating functional for one-particle irreducible Green's functions) considered in Chapter 7.

defined as

$$\int \mathcal{D}(\varphi_l) \exp(iS_\Lambda(\varphi_l)) = \int \mathcal{D}(\varphi_l)\mathcal{D}(\varphi_h) \exp(iS_f(\varphi_l, \varphi_n)) , \tag{11.4}$$

i.e.

$$\exp iS_\Lambda(\varphi_l) = \int \mathcal{D}(\varphi_h) \exp(iS_f(\varphi_l, \varphi_h)) . \tag{11.5}$$

The action S_Λ is nonlocal on the scale $x \lesssim 1/\Lambda$. However, because the remaining fields are of energy $E < \Lambda$, we can expand S_Λ in terms of local operators O_i :

$$S_\Lambda = \int d^D x \sum_i g_i^{(0)} O_i , \tag{11.6}$$

i.e. the action S_Λ is approximately local in the low energy region with $E \ll \Lambda$. The dimension of the coupling $g_i^{(0)}$ is $d_{g_i^{(0)}} = D - d_{ci}$, where d_{ci} is the canonical dimension of the operator O_i. The crucial Wilson assumption is that the dimensionless coupling constants

$$\lambda_i^{(0)} \equiv \Lambda^{d_{ci}-D} g_i^{(0)} \tag{11.7}$$

have generic values $\lambda_i^{(0)} \sim 1$ for nonrenormalizable interactions.

Now nonrenormalizable operators O_i, with $d_{ci} > D$, appear in the action S_Λ with a negative power of the cutoff, i.e. they are suppressed. Of course, they lead to new divergent diagrams in the theory, however, the important point is that all these divergences can be removed by counter terms which have the form of relevant or marginal operators. *In other words, nonrenormalizable operators lead to new divergences but not to new types of divergences.* This is the reason why they are called irrelevant operators.

Let us illustrate this important point by considering some examples.

Let us first recall the main facts concerning divergent diagrams in D dimensional field theory [4].

Consider a generic diagram G. The index of the diagram $\omega(G)$ is defined as

$$\omega(G) = \sum_{l_{\rm int}} (r_l + 2) - D(n-1) , \tag{11.8}$$

where n is the number of vertices in the diagram, the summation is taken over all internal lines, and r_l is the highest power of momentum in the numerator of the free propagator of a particle corresponding to the l-th line.*) If $\omega(G) < 0$, the diagram is superficially convergent, if $\omega(G) \geq 0$, it is divergent. Let us now introduce the maximum index of the i-th vertex, $\omega_i^{\max}$:

$$\omega_i^{\max} = \frac{1}{2}\sum_l (r_l + 2) - D\,, \tag{11.9}$$

where the summation is taken over all lines entering the vertex. Then, the index of the diagram $\omega(G)$ is

$$\omega(G) = \sum_i \omega_i^{\max} - \frac{1}{2}\sum_{lext}(r_l + 2) + D\,, \tag{11.10}$$

where $\sum\limits_{lext}$ corresponds to the summation over all external lines in the diagram.

As the first example, let us consider a four dimensional scalar theory with the interaction of the form

$$\mathcal{L}_{\text{int}} = \frac{\lambda_4^{(0)}}{4!}\varphi^{(0)^4} + \frac{\lambda_6^{(0)}}{6!\Lambda^2}\varphi^{(0)^6}\,, \tag{11.11}$$

where $\lambda_4^{(0)}$, $\lambda_6^{(0)}$ and $\varphi^{(0)}$ are bare quantities. The index $\omega_4^{\max}$ of the four-scalar vertex is $\omega_4^{\max} = 0$; the index $\omega_6^{\max}$ of the six-scalar vertex is $\omega_6^{\max} = 2$. However, because of the factor Λ^{-2}, this vertex acts as a vertex with $\tilde{\omega}_6^{\max} = 0$. Let us look at the self energy diagram in Fig.11.1. Its index $\omega(G)$ is

$$\omega(G) = 2\tilde{\omega}_6^{\max} - \frac{1}{2}\sum_{lext}(r_l + 2) + 4 = 2\,, \tag{11.12}$$

i.e. it coincides with $\omega(G)$ of the self energy diagram in Fig.11.2 in the renormalizable theory with $\lambda_6^{(0)} = 0$. The divergences in both diagrams are removed by the mass counter term and by the counter term connected with the renormalization of the field $\varphi^{(0)}$: $\frac{1}{2}(m - m^{(0)})Z_2\varphi^2$ and $\frac{1}{2}(Z_2 - 1)\partial^\mu\varphi\partial_\mu\varphi$, where $\varphi = Z_2^{-1/2}\varphi^{(0)}$ is the renormalized field.

---

*) That is, r_l is the highest power of the momentum in the polynomial $Z(p)$ in the propagator $\frac{Z(p)}{m^2 - p^2 - i\epsilon}$.

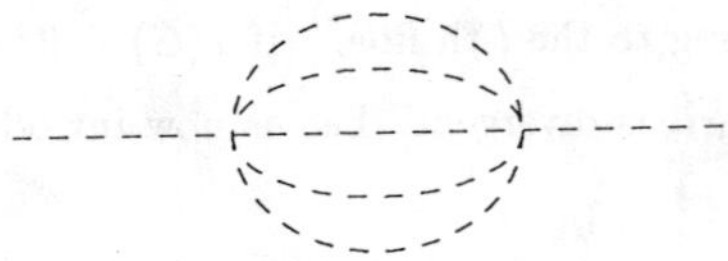

Fig. 11.1. A self energy diagram in ϕ^6 theory. A dashed line corresponds to the propagator of ϕ.

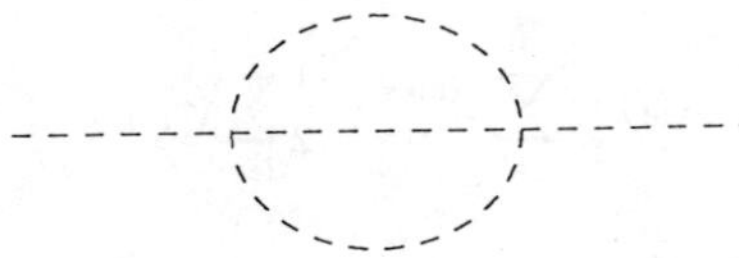

Fig. 11.2. A self energy diagram in ϕ^4 theory.

Thus, the irrelevant operator $\frac{\lambda_6^{(0)}}{6!\Lambda^2}\varphi^{(0)^6}$ does not lead to new types of divergences in the theory, although it leads to new divergent diagrams as the one shown in Fig.11.1. In other words, irrelevant operators change the coefficients of the counter terms but they do not introduce new types of the counter terms.

Another example is QED with additional four-fermion interactions:

$$\mathcal{L}_4 = \frac{g}{\Lambda^2}[(\bar{\psi}\psi)^2 - (\bar{\psi}\gamma_5\psi)^2] \,. \tag{11.13}$$

The $\omega_4^{\max}$ is $\omega_4^{\max} = 2$ but the effective $\tilde{\omega}_4^{\max}$ is $\tilde{\omega}_4^{\max} = 0$. One can again check that this vertex does not lead to new types of counter terms in QED.

Since the dynamics in the infrared region ($E \ll \Lambda$) is defined by renormalized parameters, it is insensitive to irrelevant operators: their effect is absorbed by suitable modifications of bare couplings (i.e. by counter terms) of relevant and marginal interactions. As a result, the effect of irrelevant interactions in the infrared region is suppressed by factors $\left(\frac{E}{\Lambda}\right)^n$ with $n > 0$.

It is instructive to discuss the effects of irrelevant and relevant interactions in the framework of the renormalization group. As is known, in renormalizable theories, higher vertices, such as the $\Gamma^{(6)}$ proper vertex in φ^4 theory for example, are determined uniquely in terms of renormalized couplings of relevant and marginal operators. Their insensitivity to irrelevant interactions can be interpreted as the presence of infrared stable fixed points in a theory. As an illustration, let us consider the following simple renormalization group equation for the running coupling $\lambda_6(\sigma)$, of the φ^6-operator (σ is a renormalization group scale):

$$\frac{d\lambda_6(\sigma)}{d\ln\sigma} = a\lambda_6 + b;\ a > 0 \tag{11.14}$$

(in fact, one can check that $a = 2$ in the theory with the lagrangian (11.11)). The β-function $\beta_6 = a\lambda_6 + b$ has a zero at $\lambda_6 = -b/a$. Since $\frac{d\beta_6}{d\lambda_6} = a > 0$, it is an infrared stable fixed point. The general solution of equation (11.14) takes the following form:

$$\lambda_6(\sigma) = -\frac{b}{a} + \left(\frac{\sigma}{\Lambda}\right)^a \left(\lambda_6^{(0)} + \frac{b}{a}\right);\ \lambda_6^{(0)} \equiv \lambda_6(\Lambda)\ . \tag{11.15}$$

Therefore at $\frac{\sigma}{\Lambda} \ll 1$, the value $\lambda_6(\sigma)$ is approximately universal, $\lambda_6(\sigma) \simeq -\frac{b}{a}$, and it is insensitive to the vale of the bare coupling $\lambda_6^{(0)}$. In fact, at $a = 2$, the effect of $\lambda_6^{(0)}$ is suppressed by the factor $\left(\frac{\sigma}{\Lambda}\right)^2$ in the infrared region.

Unlike irrelevant interactions, relevant interactions are very sensitive to the choice of bare parameters. Let us illustrate this point by considering the renormalization group equation for the square of the mass of scalar field, $m^2(\sigma)$:

$$\frac{ds(\sigma)}{d\ln\sigma} = -2s + b\ , \tag{11.16}$$

where $s(\sigma) \equiv \frac{m^2(\sigma)}{\sigma^2}$. The β-function $\beta_s = -2s + b$ has a zero at $s = s_0 = \frac{b}{2}$. Since $\frac{d\beta}{ds}|_{s=s_0} = -2 < 0$, it is an ultraviolet stable fixed point. The solution of equation (11.16) is

$$s(\sigma) = \frac{b}{2} + \left(\frac{\Lambda}{\sigma}\right)^2 \left(s(\Lambda) - \frac{b}{2}\right)\ , \tag{11.17}$$

i.e.

$$m^2(\sigma) = \frac{b}{2}\sigma^2 + \left(m^{(0)^2} - \frac{b}{2}\Lambda^2\right)\ , \tag{11.18}$$

where $m^{(0)} = m(\Lambda)$ is the bare mass. Eqs.(11.17), (11.18) imply that in order to get a physical mass $m \simeq m(\sigma)|_{\sigma=m}$ to be much lighter than cutoff Λ, $s(\Lambda)$ $(m^{(0)^2})$ must be tuned very precisely to the ultraviolet stable fixed point $s_0 = \frac{b}{2}$ $(m^{(0)^2} = \frac{b}{2}\Lambda^2)$:

$$\begin{aligned} s(\Lambda) - s_0 &\lesssim \frac{m^2}{\Lambda^2} \ll 1 \\ m^{(0)^2} - \frac{b}{2}\Lambda^2 &\lesssim m^2 \ll \Lambda^2 \, . \end{aligned} \tag{11.19}$$

What we have encountered now is the famous fine tuning problem for relevant interactions. As we shall see in the next sections, such ultraviolet fixed points correspond to critical points separating different phases in the space of coupling constants.

Marginal interactions are intermediate between relevant and irrelevant interactions. Usually running coupling constants connected with them depend logarithmically on the scale σ. In this case the amount of fine tuning of the bare parameters is much less than for relevant interactions. For example, in gauge theories, we have the following type of the relation for the running coupling constant $\alpha(\sigma)$:

$$\frac{1}{\alpha(\sigma)} = \frac{1}{\alpha^{(0)}} + C \ln \frac{\sigma}{\Lambda} \, , \tag{11.20}$$

where $\alpha^{(0)} \equiv \alpha(\Lambda)$, and $C > 0$ in asymptotically free gauge theories. The value $\alpha = 0$ is an ultraviolet stable fixed point. One can see that even at a large ratio $\frac{\Lambda}{\sigma}$, say $\frac{\Lambda}{\sigma} \sim e^{10}$, and reasonable values of $\alpha(\sigma)$, say $\alpha(\sigma) \sim 0.1 \div 1$, the amount of fine tuning of $\alpha^{(0)}$ is not very large:

$$\alpha^{(0)^{-1}} - 10C \sim 1 - 10 \, , \tag{11.21}$$

and therefore these values of $\alpha(\sigma)$ are not unnatural.

The irrelevant interactions are intimately connected with the problem of the interplay between dynamics of heavy and light particles, which is described by the Appelquist-Carazzone (decoupling) theorem. We will discuss this theorem in Sec.13.8.

In conclusion, we would like to emphasize that the consideration in the present section is essentially based on the perturbative dynamics; beyond the perturbative theory, the situation can be essentially different: at sufficiently strong couplings, some operators may acquire large anomalous dimensions and, as a result, some of perturbatively irrelevant operators may become marginal or even relevant (and vice versa). This phenomenon is used in some interesting applications, in particular, in describing the mechanisms of electroweak symmetry breaking (see Chapters 14, 15). In the next section, we will consider examples of such a phenomenon.

11.3. Transmutation of Dynamical Dimensions of Composite Operators at Strong Coupling

The quantum number connected with the dilatation operator $\mathfrak{D}$ generating scale transformations is called the dynamical dimension:

$$-i[\mathfrak{D},\ \varphi(x)] = (d_\varphi + x^\mu \partial_\mu)\varphi(x)\ , \tag{11.22}$$

where d_φ is the dynamical dimension of the field φ [3]. In the case of free massless fields, when the scale symmetry is exact, the dynamical dimension d_φ coincides with the canonical (engineering) dimension $d_{c;\varphi}$. Since dynamical dimensions of fields govern the ultraviolet asymptotics of their Green's functions, changing the values of d_φ may result in switching the operators between relevant, marginal and irrelevant ones.

The best known example of this phenomenon is given by the $O(N)$ invariant σ-model in dimensions with $2 < D \leqslant 4$ [14]. However, from the viewpoint of the problem of generating fermion mass, a more interesting example is given by the theory with four-fermion interactions in dimensions with $2 < D \leqslant 4$ [7 - 11, 15, 16]. Here we will discuss this model in detail. At the end of this section, we will also indicate the main results concerning the $O(N)$ σ-model.

Let us consider the following action for a set of N_c massless fermion fields:

$$S(\bar{\psi}, \psi) = \int d^D x \left[i\bar{\psi}\gamma^\mu \partial_\mu \psi + \frac{G^{(0)}}{2}(\bar{\psi}\psi)^2 \right] . \tag{11.23}$$

Henceforth we will discuss $D = 3$; the results concerning the general $D \leq 4$ will be considered in Sec.11.5.

First, we recall the properties of spinors in $D = 3$ dimensions [11, 17]. A spinorial representation of the Lorentz group $SO(2,1)$ is provided by two-component spinors. The Dirac algebra $\{\gamma_\mu, \gamma_\nu\} = 2g_{\mu\nu}$ has the following two-dimensional representation:

$$\gamma^0 = \sigma_3,\ \gamma^1 = i\sigma_1,\ \gamma^2 = i\sigma_2\ , \tag{11.24}$$

where σ_i are the Pauli matrices. We note that there is no 2×2 matrix which anticommutes with all these γ^μ. Therefore there is no analogue of the chiral symmetry in $(2+1)$ dimensional space which forbids generating fermion mass.

One can however introduce [17] a flavor symmetry playing the role of the chiral symmetry if one considers four-fermion spinors assigned to a reducible representation of the Dirac algebra with the following γ matrices:

$$\gamma^0 = \begin{pmatrix} \sigma_3 & 0 \\ 0 & -\sigma_3 \end{pmatrix}, \; \gamma^1 = \begin{pmatrix} i\sigma_1 & 0 \\ 0 & -i\sigma_1 \end{pmatrix}, \; \gamma_2 = \begin{pmatrix} i\sigma_2 & 0 \\ 0 & -i\sigma_2 \end{pmatrix} . \tag{11.25}$$

Now there are two 4×4 matrices,

$$\gamma^3 = \begin{pmatrix} 0 & I \\ I & 0 \end{pmatrix}, \; \gamma^5 = i \begin{pmatrix} 0 & I \\ -I & 0 \end{pmatrix}, \tag{11.26}$$

that anticommute with γ^μ matrices (11.25). The massless free theory is invariant under the "chiral" transformations

$$\psi \longrightarrow e^{i\varphi\gamma^3}\psi, \; \psi \longrightarrow e^{i\theta\gamma^5}\psi . \tag{11.27}$$

For each four-component spinor, there is a global $U(2)$ symmetry with generators

$$I, \; \gamma^3, \; \gamma^5, \; \frac{i}{2}[\gamma^3, \; \gamma^5] . \tag{11.28}$$

Therefore in a theory with N_c four-component spinors, the full symmetry is $U(2N_c)$. A mass term $m\bar{\psi}\psi$ breaks this symmetry to the subgroup $SU(N_c) \times SU(N_c) \times U(1) \times U(1)$, where the generators of the $U(1) \times U(1)$ are I and $\frac{i}{2}[\gamma^3, \; \gamma^5]$.

The action (11.23) has the same continuous $SU(N_c) \times SU(N_c) \times U(1) \times U(1)$ symmetry as the action of free fermions with mass term $m\bar{\psi}\psi$. However it also has a Z_2 discrete symmetry: $\psi \longrightarrow \gamma_5\psi$. Generating fermion mass will spontaneously break this symmetry. Since the symmetry is discrete, no NG bosons will appear in the system.

Of course, we can modify the action (11.23) by replacing $\frac{G^{(0)}}{2}(\bar{\psi}\psi)^2$ by

$$G^{(0)}\left[\left(\bar{\psi}\frac{\lambda^\alpha}{2}\psi\right)^2 - \left(\bar{\psi}\gamma_5\frac{\lambda^\alpha}{2}\psi\right)^2\right], \tag{11.29}$$

where $\alpha = 0, 1, \cdots, N^2 - 1$. In this case, generating fermion mass will spontaneously break the continuous $U(N) \times U(N)$ symmetry to the diagonal $U(N)$, and N^2 NG bosons will appear. However, for our purposes, it is sufficient to consider the simpler model (11.23).

It is convenient to introduce the dimensionless coupling constant

$$\kappa \equiv \frac{2N_c\Lambda}{\pi^2}G^{(0)} . \tag{11.30}$$

Then, the action (11.23) becomes

$$S(\psi,\bar{\psi}) = \int d^3x \left[i\bar{\psi}\gamma^\mu\partial_\mu\psi + \frac{\pi^2\kappa}{4N_c\Lambda}(\bar{\psi}\psi)^2 \right] . \tag{11.31}$$

In $D = 3$, this interaction is nonrenormalizable and the operator $(\bar{\psi}\psi)^2$ is irrelevant. As it follows from the discussion in the previous section, this implies that in perturbative theory, as $\Lambda \longrightarrow \infty$ and $\kappa \sim 0(1)$, this theory becomes equivalent to the theory of free massless fermions. We will however show that, beyond perturbation theory, for values of κ close to the critical value $\kappa_c = 1$, it becomes equivalent to the renormalizable Yukawa model with the action

$$S_Y = \int d^3x \left[i\bar{\psi}\gamma^\mu\partial_\mu\psi + \frac{1}{2}\partial^\mu\varphi\partial_\mu\varphi - \frac{1}{2}{\mu^{(0)}}^2\varphi^2 - g^{(0)}\varphi\bar{\psi}\psi \right] , \tag{11.32}$$

in which the bare parameters ${\mu^{(0)}}^2$ and $g^{(0)}$ take certain values [15, 16]. The action is invariant under the following Z_2 discrete transformations

$$\psi \longrightarrow \gamma_5\psi, \ \varphi \longrightarrow -\varphi . \tag{11.33}$$

The model (11.31) can be analyzed in the same way as the NJL model in Chapter 6. Introducing an auxiliary field φ and putting $G^{(0)} = {g^{(0)}}^2/{\mu^{(0)}}^2$, we rewrite the path integral for the partition function $Z \equiv Z(J)|_{J=0}$ of the model (11.31) as

$$Z = \int \mathcal{D}(\psi)\mathcal{D}(\bar{\psi})\mathcal{D}(\varphi)\exp\left[i\int d^3x\mathcal{L}(x) \right] , \tag{11.34}$$

where

$$\mathcal{L}(x) = i\bar{\psi}\gamma^\mu\partial_\mu\psi - g^{(0)}\varphi\bar{\psi}\psi - \frac{{\mu^{(0)}}^2}{2}\varphi^2 . \tag{11.35}$$

Integrating over N_c fermion fields in the path integral, we get the effective action for the field $\varphi(x)$:

$$S(\varphi) = \int d^3x \left[-\frac{1}{2}\mu^{(0)^2}\varphi^2 - iN_c \text{TrLn}\left(i\gamma^\mu\partial_\mu - g^{(0)}\varphi \right) \right] . \qquad (11.36)$$

In the limit $N_c \longrightarrow \infty$, the path integral with the action (11.36) is dominated by stationary points of the action,

$$\frac{\delta S(\varphi)}{\delta\varphi(x)} = 0 . \qquad (11.37)$$

In particular, the vacuum expectation value $v = \langle 0|\varphi|0\rangle$ satisfies the following equation (compare with Eq.(6.29)):

$$\mu^{(0)^2} v = \frac{4iN_c g^{(0)^2} v}{(2\pi)^3} \int d^3p \frac{1}{p^2 - g^{(0)^2} v^2 + i\epsilon} , \qquad (11.38)$$

which is equivalent to

$$\frac{1}{G^{(0)}} = \frac{2N_c\Lambda}{\pi^2} - |m_{dyn}| , \qquad (11.39)$$

where $m_{dyn} = g^{(0)}v$. This equation implies that the model has a critical point $G_c^{(0)} = \frac{\pi^2}{2N_c\Lambda}$ (i.e. $\kappa_c = 1$). At $\kappa < \kappa_c = 1$, the model is in the symmetric phase where $v = 0$ and the discrete Z_2 symmetry (11.33) is exact; at $\kappa > 1$, the model is in the asymmetric phase, with $v \neq 0, m_{dyn} \neq 0$, where the Z_2 symmetry is spontaneously broken. Since v goes to zero continuously as $\kappa \longrightarrow \kappa_c = 1$ from the side of the broken phase, the critical point corresponds to a continuous (second order) phase transition.

From the viewpoint of the renormalization group, the critical value $\kappa_c = 1$ can be interpreted as an ultraviolet stable fixed point. Indeed, Eq.(11.39) can be rewritten as

$$\kappa^{-1} = 1 - \frac{\pi^2}{2N_c}\frac{|m_{dyn}|}{\Lambda} . \qquad (11.40)$$

Therefore the β-function β_κ is

$$\beta_\kappa = \frac{d\kappa}{d\ln\Lambda} = \kappa(1-\kappa) . \qquad (11.41)$$

This β-function has two zeroes: $\kappa = 0$ and $\kappa = 1$. Since the first zero corresponds to the solution with $\frac{|m_{dyn}|}{\Lambda} \longrightarrow \infty$, it is irrelevant for the continuum theories. The second zero corresponds to an ultraviolet stable fixed point: indeed, $\frac{d\beta_\kappa}{d\kappa}|_{\kappa=1} = -1 < 0$. We will show that this fixed point defines an interesting continuum theory.

Let us return to the Yukawa model (11.32). The effective action $S_Y(\varphi)$ for the field φ can be obtained in the same way as for the model (11.35). We get:

$$S_Y(\varphi) = \int d^3x \left[\frac{1}{2}\partial_\mu\varphi\partial^\mu\varphi - \frac{1}{2}\mu^{(0)^2}\varphi^2 - iN_c \mathrm{TrLn}\left(i\gamma^\mu\partial_\mu - g^{(0)}\varphi\right)\right] . \quad (11.42)$$

Thus, the only difference here as compared to the action (11.36) is the appearance of the bare kinetic term $\frac{1}{2}\partial_\mu\varphi\partial^\mu\varphi$ in the action $S_Y(\varphi)$. In particular, Eq.(11.38) for v, which can be rewritten as

$$\mu^{(0)^2} = \frac{N_c g^{(0)^2}}{\pi}\left(\frac{2\Lambda}{\pi} - |m_{dyn}|\right) , \; m_{dyn} = g^{(0)}v , \quad (11.43)$$

is also valid in this model. Since the Yukawa model has two free parameters, $g^{(0)}$ and $\mu^{(0)^2}$, Eq.(11.43) with $m_{dyn} = 0$ defines now the critical line of second order phase transition points:

$$\mu^{(0)^2} = \frac{2N_c g^{(0)^2}}{\pi^2}\Lambda . \quad (11.44)$$

At $\mu^{(0)^2} > \frac{2N_c g^{(0)^2}}{\pi^2}\Lambda$, the model is in the symmetric phase where the Z_2 symmetry (11.33) is exact; at $\mu^{(0)^2} < \frac{2N_c g^{(0)^2}}{\pi^2}\Lambda$, the model is in the phase where the Z_2 symmetry is spontaneously broken.

At the critical line (or critical point $\kappa_c = 1$ in the model (11.31)), the ratio $\frac{m}{\Lambda} \longrightarrow 0$, where m is the characteristic mass scale in the theory ($m \sim m_{dyn} \sim \mu_\varphi$ in the broken phase, and $m \sim \mu_\varphi$ in the symmetric phase; μ_φ is the mass of the φ-particle). Since the correlation length of the dynamics ξ is $\xi \sim m^{-1}$, in the scaling region near the critical line, ξ is very large: $\xi \gg a \equiv \Lambda^{-1}$.

Thus, the critical line (critical point) is the place in the $\mu^{(0)^2} g^{(0)^2}$-plane (κ-line) where the continuum limit $\Lambda/m \longrightarrow \infty$ can be defined. Let us show that the theory with the four-fermion interaction and $\kappa = \kappa_c = 1$ is equivalent to the Yukawa model in the strong critical regime, in which $g^{(0)^2} \longrightarrow \infty$, and $\mu^{(0)^2}$ and $g^{(0)^2}$ lie at the critical line (11.44).

Let us begin by considering the propagator of the field φ in the Yukawa model:

$$G_\varphi(x) = i\langle 0|T\varphi(x)\varphi(0)|0\rangle = \frac{1}{(2\pi)^3}\int d^3p e^{-ipx} G_\varphi(p) . \quad (11.45)$$

For simplicity, we will consider the dynamics only in the broken phase. The case of the symmetric phase can be considered in a similar way; some special features of that case will be indicated below.

To leading order in $1/N_c$, the contribution to $G_\varphi(p)$ is given by the diagrams in Fig.11.3.

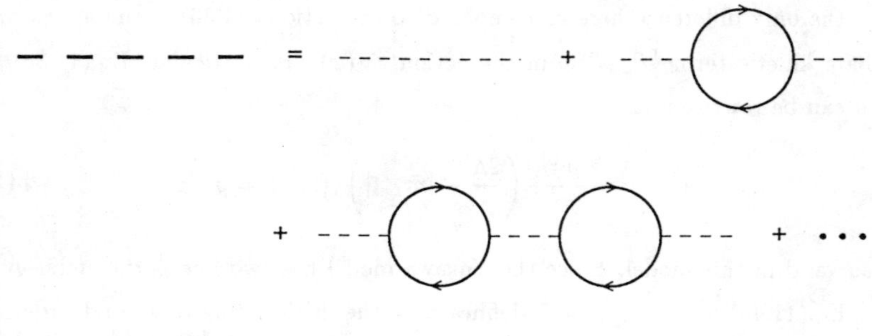

Fig. 11.3. Diagrams contributing to $G_\varphi(p)$ at leading order in $1/N_c$. A solid line correspond to the fermion propagator; a dashed thick line corresponds to the full propagator $G_\varphi(p)$; a dashed thin line corresponds to the free propagator of φ

We find that it takes the following form in Euclidean region (near the critical line (11.44)):

$$G_\varphi(p) = \left[p^2 + 2N_c g^{(0)^2}\left(p^2 + 4m_{dyn}^2\right)\left(I(p) - I_\Lambda(p)\right)\right]^{-1}, \tag{11.46}$$

where

$$I(p) = \frac{1}{(2\pi)^3}\int d^3k \frac{1}{\left(k^2 + m_{dyn}^2\right)\left((k+p)^2 + m_{dyn}^2\right)} = \frac{arctan\sqrt{p^2/4m_{dyn}^2}}{4\pi p}, \tag{11.47}$$

$$I_\Lambda(p) = \frac{1}{(2\pi)^3}\int d^3k \frac{1}{(k^2+\Lambda^2)((k+p)^2+\Lambda^2)} \simeq \frac{1}{8\pi\Lambda}. \tag{11.48}$$

One can see that the ultraviolet asymptotics of $G_\varphi(p)$ is essentially different in the following two cases:

a) weak coupling: $g^{(0)^2} = const; g^{(0)^2}/\Lambda \ll 1$.

In this case, $G_\varphi(p)$ has a large momentum behavior which is characteristic for superrenormalizable theories:

$$G_\varphi(p) \longrightarrow \frac{1}{p^2}, \tag{11.49}$$

i.e. the theory is asymptotically free and the Born term survives in the large momentum limit;

b) strong coupling:

$$g^{(0)^2} \equiv N_c^{-1} \Lambda \tilde{g}^{(0)^2} \longrightarrow \infty \,, \tag{11.50}$$

where $\tilde{g}^{(0)^2}$ is a dimensionless coupling constant.

In this case, the ultraviolet asymptotics of $G_\varphi(p)$ takes the form

$$G_\varphi(p) \longrightarrow \frac{4}{\tilde{g}^{(0)2} \Lambda} \frac{1}{p} \,. \tag{11.51}$$

From the viewpoint of the renormalization group, such a change of the behavior of $G_\varphi(p)$ means that the dynamical dimension $d_\varphi = \frac{1}{2}$ of the field at the weak coupling (equal to the canonical (engineering) dimension of the field φ) is replaced by the dynamical dimension $d_\varphi = 1$ at the strong coupling. Indeed, Eq.(11.22) implies that in a scale invariant theory the propagator $G_\varphi(x)$ satisfies the relation:

$$\begin{aligned} G_\varphi(x) =& i\langle 0|T\varphi(x)\varphi(0)|0\rangle = i\langle 0|TU_D^{-1}(s)\varphi(x)U_D(s)U_D^{-1}(s)\varphi(0)U_D(s)|0\rangle \\ =& s^{2d\varphi} i\langle 0|T\varphi(sx)\varphi(0)|0\rangle = s^{2d\varphi} G_\varphi(sx) \,, \end{aligned} \tag{11.52}$$

where $U_D(s) = \exp(iD \ln s)$ is the operator of the scale transformations:

$$U_D^{-1}(s)\varphi(x)U_D(s) = s^{2d\varphi}\varphi(sx) \,. \tag{11.53}$$

Therefore,

$$G_\varphi(p) \sim p^{2d\varphi - D} = p^{2d\varphi - 3} \tag{11.54}$$

in 3 dimensional space-time. In particular, $G_\varphi(p) \sim p^{-1}$ corresponds to $d_\varphi = 1$.

In fact, because of ultraviolet divergences, the real situation is somewhat more complicated. Let us discuss this in more detail.

First of all, we note that $G_\varphi(p)$ is a bare propagator. Let us define a renormalizable propagator $G_\varphi^{(\sigma)}(p)$:

$$G_\varphi(p) = Z_\varphi^{(\sigma)} G_\varphi^{(\sigma)}(p) \,, \tag{11.55}$$

where σ is the renormalization group scale. The bare mass $\mu^{(0)}$ (and therefore the mass counter term) is determined from the relation (11.43).*) It will be convenient to define all other counter terms in the theory with massless fermions ($m_{dyn} = 0$). In particular, the renormalization constant $Z_\varphi^{(\sigma)}$ is determined from the following normalization condition

$$\frac{d\bar{\Gamma}_\varphi^{(\sigma)}}{dp^2}|_{p^2=\sigma^2} = -1\ , \tag{11.56}$$

where $\bar{\Gamma}_\varphi^{(\sigma)}(p) = -\bar{G}_\varphi^{(\sigma)^{-1}}$ is the two-particle proper vertex (see Eq.(7.16)), and "bar" means that Green's functions correspond to the massless theory with $m_{dyn} = 0$.

Then, Eqs.(11.46), (11.55) and (11.56) imply that

$$Z_\varphi^{(\sigma)} = \left(1 + \frac{\tilde{g}^{(0)^2}}{8\pi}\frac{\Lambda}{\sigma} - \frac{\tilde{g}^{(0)^2}}{4\pi}\right)^{-1}\ , \tag{11.57}$$

where (we recall)

$$\tilde{g}^{(0)^2} \equiv N_c \Lambda^{-1} g^{(0)^2}\ . \tag{11.58}$$

Let us define the renormalized coupling constant $\tilde{g}^2(\sigma)$ as

$$\sigma \tilde{g}^2(\sigma) = \Lambda Z_\varphi^{(\sigma)} Z_\psi^{(\sigma)^2} Z_1^{(\sigma)^{-2}} \tilde{g}^{(0)^2}\ , \tag{11.59}$$

where $Z_\psi^{(\sigma)}$ is the renormalization constant of the fermion field and $Z_1^{(\sigma)}$ is the renormalization constant of the $\varphi\bar{\psi}\psi$-vertex. Since n-particle Green's functions $G_\varphi^{(n)}(p_1, \cdots, p_{n-1})$ satisfy the multiplicative renormalization relations

$$G_\varphi^{(n)} = (Z_\varphi^{(\sigma)})^{\frac{n}{2}} G_\varphi^{(\sigma);(n)}\ , \tag{11.60}$$

and since $\left(\frac{d}{d\ln\sigma} G_\varphi^{(n)}\right) = 0$, the renormalized Green's functions $G_\varphi^{(\sigma);(n)}$ satisfy the

---

*) As was already indicated above, we consider here only the nearcritial dynamics from the side of the broken phase. In the unbroken phase, with $m_{dyn} = 0$, the relation (11.43) is replaced by the following one:

$$\mu^{(0)^2} = \frac{N_c g^{(0)^2}}{\pi}\left(\frac{2\Lambda}{\pi} - t\right),$$

where t determines the width of the unstable φ-resonance decaying into massless fermion-antifermion pairs [11, 15] (compare with the discussion in Sec.9.5).

Callan-Symanzik equation:

$$\left[\frac{\partial}{\partial \ln \sigma} + \beta_{\tilde{g}^2}\frac{\partial}{\partial \tilde{g}^2} + n\gamma_\varphi\right] G_\varphi^{(\sigma);(n)}(p) = 0 , \tag{11.61}$$

where the β-function $\beta_{\tilde{g}^2}$ is defined as

$$\beta_{\tilde{g}^2} = \frac{\partial \tilde{g}^{(2)}(\tilde{g}^{(0)^2},\ \Lambda/\sigma)}{\partial \ln \sigma} , \tag{11.62}$$

and the anomalous dimension γ_φ is

$$\gamma_\varphi = \frac{1}{2}\frac{\partial \ln Z_\varphi^{(\sigma)}(\tilde{g}^{(0)^2},\ \Lambda/\sigma)}{\partial \ln \sigma} . \tag{11.63}$$

The dynamical dimension d_φ is

$$d_\varphi = d_{c;\varphi} + \gamma_\varphi , \tag{11.64}$$

where $d_{c;\varphi} = 1/2$ is the canonical dimension of scalar field in 3 dimensions.

One can easily check that there are no corrections to the fermion propagator and the vertex function in leading order in $1/N_c$. Therefore $Z_\psi^{(\sigma)} = Z_1^{(\sigma)} = 1$, and we find from Eqs.(11.57), (11.59), (11.62), (11.63) that

$$\beta_{\tilde{g}^2} = -\tilde{g}^2 + \frac{1}{8\pi}\tilde{g}^4 , \tag{11.65}$$

$$\gamma_\varphi = \frac{1}{16\pi}\tilde{g}^2 , \tag{11.66}$$

$$\gamma_\psi = \frac{1}{2}\frac{\partial \ln Z_\psi^{(\sigma)}}{\partial \ln \sigma} = 0 . \tag{11.67}$$

The β-function (11.65) has two zeroes: $\tilde{g}^2 = 0$ and $\tilde{g}^2 = 8\pi$. Since $\frac{d\beta_{\tilde{g}^2}}{d\tilde{g}^2}|_{\tilde{g}^2=0} = -1 < 0$, the first zero is an ultraviolet stable fixed point; it corresponds to the weak coupling regime where all anomalous dimensions are equal to zero. The second zero is an infrared fixed point: indeed, $\frac{d\beta_{\tilde{g}^2}}{d\tilde{g}^2}|_{\tilde{g}^2=8\pi} = 1 > 0$. The physical meaning of this fixed point becomes clear from Eq.(11.59) which is now:

$$\left(1 - \frac{\tilde{g}^2}{8\pi}\right) + \frac{\sigma}{\Lambda}\left(\frac{1}{4\pi} - \tilde{g}^{(0)^{-2}}\right)\tilde{g}^2 = 0\,. \tag{11.68}$$

This equation shows that as (σ/Λ) goes to zero, which corresponds to the critical point, $\tilde{g}^{(2)}$ goes to the fixed point $\tilde{g}^{(2)} = 8\pi$ independently of the value of the bare coupling constant $\tilde{g}^{(0)^2}$.

Since the dimensionful coupling constant $g^{(0)^2} = \Lambda\tilde{g}^{(0)^2}$ goes to infinity with $\Lambda \longrightarrow \infty$ (at $\tilde{g}^{(0)^2} \neq 0$), the infrared fixed point corresponds to the strong coupling regime. The anomalous dimension γ_φ is $\gamma_\varphi = \frac{1}{16\pi}\tilde{g}^2 = \frac{1}{2}$ at this fixed point. Therefore the dynamical dimension d_φ is

$$d_\varphi = \tfrac{1}{2} + \gamma_\varphi = 1,$$

and because of Eq.(11.61), we find that

$$G_\varphi^{(\sigma)}(p) \sim p^{-1}\,, \tag{11.69}$$

which agrees with the relation (11.51).

Now it becomes rather evident that in the critical strong regime, the Yukawa model is equivalent to the model with four-fermion interactions at $\kappa = \kappa_c = 1$. In order to show this, we note that at the critical line (11.44), the ratio of $g^{(0)^2}\Lambda$ and $\mu^{(0)^2}$ is fixed. Therefore in the critical strong limit, when $g^{(0)^2} = \tilde{g}^{(0)^2}\Lambda \longrightarrow \infty$, $\mu^{(0)^2}$ also goes to infinity. As a result, the kinetic term of the φ field becomes negligible compared to the mass term and can be dropped in this limit in the action. This leads us to the theory with four-fermion interactions after the auxiliary field φ is integrated out. The coupling constant κ in this theory is automatically fixed: $\kappa = \kappa_c = 1$.

In more formal language, one can say that since the dynamical dimension of the kinetic operator $\partial_\mu\varphi\partial^\mu\varphi$ equals 4 in the critical strong regime, it becomes an irrelevant operator and can be dropped.

To see the equivalence between these two models in a more explicit form, let us calculate the propagator

$$G_{\bar{\psi}\psi}(x) = i\langle 0|T\bar{\psi}\psi(x)\bar{\psi}\psi(0)|0\rangle \tag{11.70}$$

of the composite field $\bar{\psi}\psi$ in the Yukawa model.

In leading order in $1/N_c$, the contribution to the connected Green's function $G_{\bar{\psi}\psi}$ is given by the diagrams in Fig.11.4. It takes the following form in Euclidean space:

$$G_{\bar{\psi}\psi}(p) = N_c \frac{\left(p^2 + 4N_c g^{(0)^2} \jmath\right)\left[4\jmath - 2\left(p^2 + 4m_{dyn}^2\right) I(p)\right]}{p^2 + 2N_c g^{(0)^2}\left(p^2 + 4m_{dyn}^2\right) I(p)} , \tag{11.71}$$

where $I(p)$ is given in Eq.(11.47), and $\jmath$ is

$$\jmath = \frac{1}{(2\pi)^3} \int d^3p \frac{1}{p^2 + m_{dyn}^2} . \tag{11.72}$$

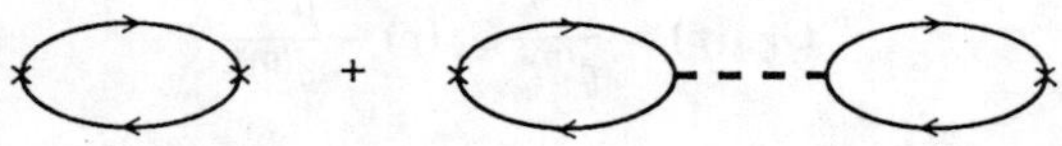

Fig. 11.4. Diagrams contributing to the $\bar{\psi}\psi$ propagator at leading order in $1/N_c$. A solid line denotes the fermion propagator and a thick dashed line denotes the full scalar propagator.

At strong coupling, when $g^{(0)^2} \longrightarrow \infty$, $G_{\bar{\psi}\psi}(p)$ is

$$G_{\bar{\psi}\psi}(p) = N_c \frac{8\jmath^2}{\left(p^2 + 4m_{dyn}^2\right) I(p)} - 4N_c \jmath . \tag{11.73}$$

This expression of course coincides with that of the propagator in the theory with the four-fermion interaction at $\kappa = \kappa_c = 1$.

We note that Eqs.(11.46) and (11.73) lead to the following relation:

$$G_{\bar{\psi}\psi}(p) = 16 N_c^2 \jmath^2 g^{(0)^2} G_\varphi(p) - 4N_c \jmath . \tag{11.74}$$

In other words, the composite operator $\bar{\psi}\psi$ essentially acts as the scalar field φ. Let us make this statement more precise.

We consider the generating functional $W(J) = \frac{1}{i} \ln Z(J)$ for Green's functions of the auxiliary field φ in the theory with the lagrangian (11.35):

$$Z(J) = \int \mathcal{D}(\psi)\mathcal{D}(\bar{\psi})\mathcal{D}(\varphi) \exp\left[i \int d^3x \left(\mathcal{L}(x) + J(x)\varphi(x)\right)\right] . \tag{11.75}$$

Integrating over φ in $Z(J)$, we find that, up to some inessential normalization factor,

$Z(J)$ is

$$Z(J) \sim \int \mathcal{D}(\psi)\mathcal{D}(\bar{\psi})\mathcal{D}(\psi) \exp\left[i \int d^3x \left(\mathcal{L}_4(x) - \frac{g^{(0)}}{\mu^{(0)^2}} J(x)\bar{\psi}(x)\psi(x) + \frac{1}{2\mu^{(0)^2}} J^2(x)\right)\right], \tag{11.76}$$

where $\mathcal{L}_4$ is the lagrangian density of the four-fermion interaction model (11.23). This equation leads to the following relation for $G_{\bar{\psi}\psi}(p)$:

$$G_{\bar{\psi}\psi}(p) = \frac{\mu^{(0)^4}}{g^{(0)^2}} G_\varphi(p) - \frac{\mu^{(0)^2}}{g^{(0)^2}} . \tag{11.77}$$

Since the relation (11.38) can be rewritten as

$$\frac{\mu^{(0)^2}}{g^{(0)^2}} = 4N_c \mathcal{J} , \tag{11.78}$$

we find that Eq.(11.77) is equivalent to Eq.(11.74).

Eq.(11.76) also implies that for Green's functions including only one composite field $\bar{\psi}\psi$ and any number of elementary fields ψ and $\bar{\psi}$, we have the following relation:

$$\begin{aligned} &\langle 0|T\bar{\psi}(x)\psi(x)\psi(y_1)\cdots\psi(y_n)\bar{\psi}(z_1)\cdots\bar{\psi}(z_n)|0\rangle \\ =&\frac{\mu^{(0)^2}}{g^{(0)}}\langle 0|T\varphi(x)\psi(y_1)\cdots\psi(y_n)\bar{\psi}(z_1)\cdots\bar{\psi}(z_n)|0\rangle . \end{aligned} \tag{11.79}$$

Therefore it is appropriate to define the renormalization operator $(\bar{\psi}\psi)^{(\sigma)}$ in such a way that it will coincide with the renormalized field $\sigma^{3/2}\varphi^{(\sigma)} = \sigma^{3/2} Z_\varphi^{(\sigma)^{-1/2}} \varphi$:

$$(\bar{\psi}\psi)^{(\sigma)} = Z_{\bar{\psi}\psi}^{(\sigma)^{-1/2}} (\bar{\psi}\psi) = \sigma^{3/2}\varphi^{(\sigma)} , \tag{11.80}$$

where

$$Z_{\bar{\psi}\psi}^{(\sigma)} = \sigma^{-3} \left(\frac{\mu^{(0)^2}}{g^{(0)}}\right)^2 Z_\varphi^{(\sigma)} \simeq \frac{32N_c}{\pi^3} \left(\frac{\Lambda}{\sigma}\right)^2 \tag{11.81}$$

(we used Eqs.(11.44), (11.57) and (11.58) in the derivation of this relation).

Then, one can easily check that in the continuum limit, *all* renormalized Green's functions of $(\bar{\psi}\psi)^{(\sigma)}$ coincide with those Green's functions in which $(\bar{\psi}\psi)^{(\sigma)} \longrightarrow \sigma^{3/2}\varphi^{(\sigma)}$. In other words, the renormalized composite operator $(\bar{\psi}\psi)^{(\sigma)}$ equals the scalar operator $\sigma^{3/2}\varphi^{(\sigma)}$. In particular, its dynamical dimension $d_{\bar{\psi}\psi}$ equals $d_\varphi = 1$.

Thus, at strong coupling in the Yukawa model, two different operators $\bar{\psi}\psi$ and φ (having in particular different canonical dimensions) have the same dynamical dimension $d_{\bar{\psi}\psi} = d_\varphi = 1$. This in turn implies that they acquire large anomalous dimensions:

$$\gamma_{\bar{\psi}\psi} \equiv d_{c;\bar{\psi}\psi} - d_{\bar{\psi}\psi} = 1 \, , \tag{11.82}$$

$$\gamma_\varphi \equiv d_\varphi - d_{c;\varphi} = 1 \tag{11.83}$$

(this is our convention for the definition of the anomalous dimensions of these operators).

At strong coupling, the renormalized operators $(\bar{\psi}\psi)^{(\sigma)}$ and $\sigma^{3/2}\varphi^{(\sigma)}$ are equivalent, and this is the main reason of the equivalence of the Yukawa model and the model with four-fermion interactions in the strong critical regime.

The equivalence of the Yukawa model and the model with four-fermion interactions was shown in all orders in the $1/N_c$ expansion [10, 11]. This equivalence is in turn the basis of the proof of the renormalizability of the four-fermion interaction model in all orders of the $1/N_c$ expansion in 3 dimensional space-time. The central point of the proof [10, 11] is that Green's functions in the Yukawa model are multiplicatively renormalizable in the critical strong coupling limit. The latter is based on that fact that physical couplings in the model are determined by infrared fixed points, and therefore they remain finite in this limit.

In the conclusion of this section, we will briefly discuss a similar phenomenon which takes place in the $O(N)$ invariant σ-model [14].

The lagrangian densities of the $O(N)$ invariant linear and nonlinear σ-models are:

$$\mathcal{L}_l = \frac{1}{2}\left(\partial^\mu \varphi^a \partial_\mu \varphi^a\right) + \frac{1}{2}\mu^{(0)^2}\varphi^a\varphi^a + \frac{\lambda^{(0)}}{4!}(\varphi^a\varphi^a)^2 \, , \tag{11.84}$$

$$\mathcal{L}_n = \frac{1}{2F}\left[\left(\partial^\mu \pi^a \partial_\mu \pi^a\right) + \sigma\left(\pi^{a^2} - 1\right)\right] \, , \tag{11.85}$$

where the Lagrange multiplier $\sigma(x)$ enforces the constraint $(\pi^a)^2 = 1; a = 1, 2 \cdots, N$. The nonlinear σ-model is perturbatively nonrenormalizable.

In leading order in $1/N$, the analysis of these models is similar to that of the Yukawa model and the four-fermion interaction model considered above. We find that in $(2+1)$ dimensional space-time, the critical line in the linear σ-model is:

$$\frac{\mu^{(0)^2}}{\lambda^{(0)}} = \frac{N\Lambda}{12\pi^2} . \tag{11.86}$$

It separates two phases: at $\frac{\mu^{(0)^2}}{\lambda^{(0)}} < \frac{N\Lambda}{12\pi^2}$, the model is in the symmetric phase with the $O(N)$ symmetry unbroken; at $\frac{\mu^{(0)^2}}{\lambda^{(0)}} > \frac{N\Lambda}{12\pi^2}$, the model is in the phase where the $O(N)$ is spontaneously broken to $O(N-1)$.

The critical point in the nonlinear σ-model is

$$F_c^{-1} = \frac{N\Lambda}{2\pi^2} . \tag{11.87}$$

In the strong coupling limit, when $\lambda^{(0)} \longrightarrow \infty$, the critical linear σ-model $\left(\text{with} \frac{\mu^{(0)^2}}{\lambda^{(0)}} = \frac{N\Lambda}{12\pi^2}\right)$ becomes equivalent to the critical nonlinear σ-model with $F^{-1} = F_c^{-1} = \frac{N\Lambda}{2\pi^2}$.

The basic reason of this equivalence is connected with the fact that near the critical point $F = F_c$, the composite operator $\sigma^2(x)$ acquires a large dynamical dimension, $d_{\sigma^2} = 4$, i.e. it becomes irrelevant operator. Since it is irrelevant, we can add it to $\mathcal{L}_n$ without changing the behavior of the model in the infrared region with $p^2/\Lambda^2 \ll 1$:

$$\mathcal{L}_n \longrightarrow \tilde{\mathcal{L}}_n = \frac{1}{2F}\left[(\partial^\mu \pi^a \partial_\mu \pi^a) + \sigma(\pi^{a2} - 1) - \frac{1}{2\mu^{(0)^2}}\sigma^2\right] . \tag{11.88}$$

Integrating over the field $\sigma(x)$ in the path integral we come to the linear σ-model with $\lambda^{(0)} = 6F\mu^{(0)^2}$.

The proof of the equivalence of these two models in all orders of the $1/N$ expansion was given in Ref.[18].

In the conclusion of this section, we would like to interpret these results from the viewpoint of low energy effective lagrangians.

The equivalence between the Yukawa model and the four-fermion interaction model, and between the linear and nonlinear σ-models is exact in the strict continuum limit, $p/\Lambda \longrightarrow 0$. At small but finite p/Λ, the equivalence can be interpreted as the derivation of the low energy effective lagrangians for the two "fundamental" theories, the nonlinear σ-model and the four-fermion interaction model: at $p/\Lambda \ll 1$, the linear σ-model is the low energy effective theory of the former, and the Yukawa model is the effective theory of the latter. The φ-partial can be interpreted as a fermion-antifermion bound state, and the σ-particle can be considered as a $\pi - \pi$ bound state.

The concept of effective theories is quite useful and general. Cutoff Λ is considered as a scale of a "new" physics in particle physics and as some microscopical scale (as the Debye frequency (see Sec.2.8)) in condensed matter physics. In particular, in this way, one can derive the Ginzburg-Landau (GL) model from the BCS theory of superconductivity (see Sec.2.9). The GL model is the low energy effective theory of the latter.

In Chapters 12, 14 - 15, we will apply this concept to the description of the nonperturbative dynamics in QCD and to models of electroweak symmetry breaking. For some recent applications of the effective theories of the four-fermion interaction models to condensed matter systems, see Ref.[19].

11.4. Universality, Critical Exponents and Fine Tuning

The Yukawa model and the model with four-fermion interactions are in general case essentially different models. However in the critical dynamical regime described in the previous section they become equivalent.

This is an example of the universality of critical phenomena: different models may have the same critical dynamics. In this case one says that they relate to the same universality class.

The number of universality classes is essentially smaller than the number of different models. Important characteristics of universality are critical exponents [3, 20]. Let us discuss them in more detail.

Critical exponents describe the approach to criticality of such quantities as the order parameter, the effective potential, the particle masses, etc.. To be concrete let us define the critical exponents $\alpha, \beta, \gamma, \nu$ and δ by using notations in the four-fermion interaction model:

$$V \sim (\Delta\kappa)^{2-\alpha} \text{ at } m^{(0)} = 0 \,, \tag{11.89}$$

$$\langle 0|\bar{\psi}\psi|0\rangle \sim (\Delta\kappa)^{\beta} \text{ at } m^{(0)} = 0 \,, \tag{11.90}$$

$$\frac{\partial(\langle 0|\bar{\psi}\psi|0\rangle)}{\partial m^{(0)}}|_{m^{(0)}=0} \sim (\Delta\kappa)^{-\gamma} \,, \tag{11.91}$$

$$\mu_{\varphi} \sim (\Delta\kappa)^{\nu} \text{ at } m^{(0)} = 0 \,, \tag{11.92}$$

$$\langle 0|\bar{\psi}\psi|0\rangle \sim [m^{(0)}]^{1/\delta} \text{ at } \kappa = \kappa_c \,, \tag{11.93}$$

where $\Delta\kappa = \kappa - \kappa_c$, $m^{(0)}$ is the bare mass of fermions, and the potential $V(\varphi_c)$ is $V(\varphi_c) = (\int d^3x)^{-1} S(\varphi_c)$, where $S(\varphi_c)$ is the action (11.36) calculated at $\varphi = \varphi_c$ independent of x. The potential is:

$$\begin{aligned} V =& \frac{1}{2}\mu^{(0)^2}\varphi_c^2 - \frac{N_c}{\pi^2}\int dp \cdot p^2 \ln\left(1 + \frac{g^{(0)^2}\varphi_c^2}{p^2}\right) \\ =& \frac{1}{2G^{(0)}}\tilde{\varphi}_c^2 - \frac{N_c}{\pi^2}\int dp \cdot p^2 \ln\left(1 + \frac{\tilde{\varphi}_c^2}{p^2}\right) \,, \end{aligned} \tag{11.94}$$

where $\tilde{\varphi}_c = g^{(0)}\varphi_c$, $G^{(0)} = \frac{g^{(0)^2}}{\mu^{(0)^2}} = \frac{\pi^2}{2N_c\Lambda}\kappa$.

The calculation of the critical exponents in leading order in $1/N_c$ is straightforward. In order to calculate β, γ and δ, we must first solve the Schwinger-Dyson equation for the fermion mass m,

$$m = m^{(0)} + \frac{2N_c G^{(0)} m}{\pi^2}\int dp \frac{p^2}{p^2 + m^2} \tag{11.95}$$

(compare with Eqs.(8.55), (11.38)), and then calculate the condensate

$$\langle 0|\bar{\psi}\psi|0\rangle = -i \lim_{x\to 0} tr(G(x)) = -\frac{2N_c}{\pi^2}\int dp \frac{p^2 m}{p^2 + m^2} \,. \tag{11.96}$$

The results are

$$\beta = 1, \;\; \gamma = 1, \;\; \delta = 2 \,. \tag{11.97}$$

The exponent ν is secured by noting that as $m^{(0)} \to 0$, the mass μ_{φ} is $\mu_{\varphi} \sim m$. We find

that

$$\nu = 1\,. \tag{11.98}$$

The exponent α is calculated from the expression (11.94) for V. It is

$$\alpha = -1\,. \tag{11.99}$$

For completeness, we also indicate the critical exponents to order $1/N_c$ calculated in Ref.[21]:

$$\begin{aligned}\alpha &= -1 - \frac{8}{\pi^2 N_c},\ \beta = 1,\ \gamma = 1 + \frac{8}{\pi^2 N_c},\\ \delta &= 2 + \frac{8}{\pi^2 N_c},\ \nu = 1 + \frac{8}{3\pi^2 N_c}\,.\end{aligned} \tag{11.100}$$

One can check that they satisfy the so called hyperscaling relations which follow from the requirement of having a correletation length $\xi \sim m^{-1}$ as the only relevant scale in the critical dynamics [22, 3, 20]:

$$2 - \alpha = D\nu,\ 2\beta + \gamma = 2 - \alpha,\ 2\beta\delta - \gamma = 2 - \alpha\,. \tag{11.101}$$

Different universality classes have different critical exponents. Therefore the dependence of the exponents on N_c in Eq.(11.100) implies that the theories with different N_c belong to different universality classes. In principle there could be even a crossover in the behavior at some $N_c = \bar{N}_c$ so that theories with low $N_c < \bar{N}_c$ might have a dynamics essentially different from that in theories with $N_c > \bar{N}_c$. At present this possibility is still open. Recent lattice computer simulations of these theories show that $1/N_c$ expansion is accurate at least for $N_c \geq 12$ [21].

Another interesting point is that the values of the critical exponents in Eq. (11.100) are essentially different from the values of mean-field exponents,

$$\alpha = -2,\ \beta = \frac{1}{2},\ \gamma = 1,\ \delta = 3,\ \nu = \frac{1}{2}\,. \tag{11.102}$$

The latter corresponds to the critical dynamics with the trivial S-matrix [3, 20]. This unambiguously establishes that at $D = 3$, the continuum theories with four-fermion interactions are nontrivial theories with S-matrix different from $S = 1$.

Let us return to the fine tuning problem. As we already indicated in Sec.11.2, infrared fixed points lead to low energy dynamics which are stable under the variations of bare parameters. On the other hand, ultraviolet fixed points define low energy dynamics which are sensitive to the variations of bare parameters.

Let us discuss the fine tuning problem in the critical region of the four-fermion interaction model and the Yukawa model, where these two models are equivalent.

At $\kappa \simeq \kappa_c = 1$, the β-function β_κ in the four-fermion interaction model is (see Eq.(11.41)):

$$\beta_\kappa \simeq 1 - \kappa \ . \tag{11.103}$$

It corresponds to the mass relation (11.40):

$$\kappa^{-1} = 1 - \frac{\pi^2}{2N_c}\frac{|m_{dyn}|}{\Lambda} \ . \tag{11.104}$$

This relation shows that in order to get a light m_{dyn}, with $|m_{dyn}|/\Lambda \ll 1$, the coupling constant κ must be tuned very precisely to the ultraviolet stable fixed point $\kappa_c = 1$.

On the other hand, the β-function $\beta_{\tilde{g}^2}$ (11.65) in the Yukawa model has a zero at $\tilde{g}^2 = 8\pi$ corresponding to an infrared fixed point. As was already shown in Sec.11.3 (see Eq.(11.68)), this zero defines the running coupling constant $\tilde{g}^2(\sigma)$ in the low energy region (with $\sigma/\Lambda \ll 1$) which is close to the fixed point 8π and it is insensitive to the value of the bare coupling constant $\tilde{g}^{(0)^2}$.

So it may seem that there is no fine tuning problem in the Yukawa model. However, this is not so. The point is that there is the mass parameter $\mu^{(0)^2}$ in that model, and in the critical region, it satisfies the relation (11.43):

$$\frac{\mu^{(0)^2}}{\Lambda^2} = \frac{2\tilde{g}^{(0)^2}}{\pi^2}\left(1 - \frac{\pi^2}{2N_c}\frac{|m_{dyn}|}{\Lambda}\right) \ . \tag{11.105}$$

Comparing this relation with Eq.(11.104), we see that the quantity $\frac{\pi^2 \mu^{(0)^2}}{2\tilde{g}^{(0)^2}\Lambda^2}$ plays here the role of κ^{-1}. Thus, we conclude that the fine tuning of the coupling constant κ is just equivalent to the fine tuning of the mass parameter $\mu^{(0)^2}$ in the Yukawa model.

The appearance of new relevant operators in the action of the models makes the fine tuning problem even more severe. Let us for example add mass term $-m^{(0)}\bar{\psi}\psi$ in the lagrangian density (11.31). Then at $\kappa = \kappa_c = 1$, we find the following relation from the Schwinger-Dyson equation (11.95):

$$m^{(0)} = \frac{\pi}{2} \frac{m_{\text{cur}}|m_{\text{cur}}|}{\Lambda}, \tag{11.106}$$

where $m = m_{\text{cur}}$ at $\kappa = \kappa_c$. This relation shows that the bare mass $m^{(0)}$ must be tuned very precisely to zero in order to get $m_{\text{cur}} \ll \Lambda$. In fact, we can introduce the β-function

$$\beta_{\tilde{m}^{(0)}} = \frac{\partial \tilde{m}^{(0)}}{\partial \ln \Lambda} = -2\tilde{m}^{(0)} \tag{11.107}$$

for the dimensionless mass $\tilde{m}^{(0)} \equiv \frac{m^{(0)}}{\Lambda}$. One can see that $\beta_{\tilde{m}^{(0)}}$ has an ultraviolet stable fixed point at $\tilde{m}^{(0)} = 0$.

Thus we see that the fine tuning problem is a very general phenomenon in the critical dynamics. In some models, it can be essentially reduced by imposing some symmetries as, for example, supersymmetry [23]. However, as we will discuss in Chapters 14, 15, there can be other possibilities to deal with the fine tunning, and even to use it.

11.5. The Four-Fermion Interaction Model in D dimensional space-time ($2 \leq D \leq 4$)

Up to now we have discussed the four-fermion interaction dynamics in 3 dimensional space-time. This analysis can be extended to theories in D dimensions, $2 \leq D \leq 4$ [24, 21]. In this section we will just indicate the results of the analysis in Ref.[21].

In leading order in $1/N_c$, the Schwinger-Dyson equation takes the following form:

$$m_{dyn} = \frac{4N_c' G^{(0)}}{(2\pi)^D} \int d^D p \frac{m_{dyn}}{p^2 + m_{dyn}^2}, \tag{11.108}$$

where $N_c' \equiv \frac{N_c}{4} trI$, and I is the identity matrix in the space of the Dirac matrices in D dimensions.

The critical coupling $G_c^{(0)}$ is determined from the equation

$$1 = \frac{4N_c' G_c^{(0)}}{(2\pi)^D} \int d^D p \frac{1}{p^2} \, . \tag{11.109}$$

Eq.(11.108) is equivalent to the following one:

$$\frac{1}{G^{(0)}} = \frac{1}{G_c^{(0)}} - CN_c' m_{dyn}^{D-2} \, , \tag{11.110}$$

where

$$C = \frac{8\Gamma\left(2 - \frac{D}{2}\right)}{(4\pi)^{D/2}(D-2)} \tag{11.111}$$

(in the derivation of Eq.(11.111), we used Eq.(C.6) in Appendix C). It is convenient to introduce the following dimensionless coupling constant *)

$$\kappa \equiv (D-2)\frac{G^{(0)}}{G_c^{(0)}} \, . \tag{11.112}$$

Then, Eq.(11.110) takes the form:

$$\frac{1}{\kappa} = \frac{1}{D-2}\left(1 - CN_c' G_c^{(0)} m_{dyn}^{D-2}\right) \, . \tag{11.113}$$

The β-function β_κ is

$$\beta_\kappa = \frac{\partial\kappa}{\partial \ln\Lambda} = (D-2)\kappa - \kappa^2 \tag{11.114}$$

(compare with Eq.(11.41)).**) It has a zero at

$$\kappa = \kappa_c = D - 2 \tag{11.115}$$

corresponding to an ultraviolet stable fixed point. The two cases, $D = 2$ and $D = 4$, deserve a separate discussion.

---

*) This κ coincides with κ in Eq.(11.30) but it is different (by the factor 2) from κ defined in Eq.(6.15) at $D = 4$.

**) In accord with the Wilson theory (see Sec.11.2), our definition of the integral in Eq.(11.109) is such that $G_c^{(0)^{-1}} \sim \Lambda^{(D-2)}$.

a) $D = 4$.

In this case, since $\Gamma\left(2-\frac{D}{2}\right) \simeq \frac{1}{2-\frac{D}{2}}$ at $D \simeq 4$, the coefficient C in Eq.(11.111) has a pole singularity. The Schwinger-Dyson equation (11.110) is now:

$$m_{dyn}^2 \frac{1}{2-\frac{D}{2}} = \frac{4\pi^2}{N_c G_c^{(0)}} - \frac{4\pi^2}{N_c G^{(0)}} \ . \tag{11.116}$$

Comparing this equation with the Schwinger-Dyson equation (6.14) in the NJL model obtained by using a sharp cutoff Λ in the integrals, we find that they are equivalent if we use the following identification:

$$\begin{aligned} \frac{1}{2-\frac{D}{2}} &\longrightarrow \ln \frac{\Lambda^2}{m_{dyn}^2} \\ \frac{4\pi^2}{N_c G_c^{(0)}} &\longrightarrow \Lambda^2 \ . \end{aligned} \tag{11.117}$$

b) $D = 2$.

In this case, the critical coupling $\kappa_c = 0$, and spontaneous Z_2 symmetry breaking takes place at all $\kappa > 0$.

In fact, at $D = 2$, the coupling constant $G^{(0)}$ is dimensionless and the model is perturbatively renormalizable.

The Schwinger-Dyson equation (11.108) leads to the following relations at $D = 2$:

$$G^{(0)} = \frac{2\pi}{N_c \ln \frac{\Lambda^2}{m_{dyn}^2}} \ , \tag{11.118}$$

$$m_{dyn}^2 = \Lambda^2 \exp\left(-\frac{2\pi}{N_c G^{(0)}}\right) \ . \tag{11.119}$$

Thus, $G^{(0)} \longrightarrow 0$ as $\Lambda \longrightarrow \infty$, i.e. the model is asymptotically free [8]. Another characteristic point in the model is an essential singularity at $G^{(0)} = 0$ in the expression (11.119) for m_{dyn}^2.

Let us return to the general D. To leading order in $1/N_c$, the critical exponents were calculated in Ref.[21] (compare with Eq.(11.100)):

$$\begin{aligned}\alpha =&\frac{D-4}{D-2},\ \beta=\frac{1}{D-2},\ \gamma=1\ ,\\ \delta =&D-1,\ \nu=\frac{1}{D-2}\ .\end{aligned} \tag{11.120}$$

We note that at $D = 2$, the exponents α, β and ν become infinite. This point reflects the essential singularity at $G^{(0)} = 0$ in the expression (11.119) for m^2_{dyn}. On the other hand, at $D = 4$, the critical exponents (11.120) coincide with the mean-field exponents (11.102). This point reflects the fact that at $D = 4$, in the continuum limit, the Yukawa model and the four-fermion interaction model become free theories. We will return to this point in Sec.14.6.

11.6. Running Parameters and Renormalization of Bare Parameters. More About the Phase Transition in QED

After getting an experience in critical phenomena, it is appropriate to discuss some problems connected with renormalizations in quantum field theory from this viewpoint. In this section we will discuss two of them: the connection between bare and running parameters, and the renormalizations in the supercritical QED.

In four dimensional gauge theories with fermions, there is a simple connection between the form of running parameters, such as running coupling constants, running masses, etc., and the form of their bare counterparts as functions of cutoff Λ.

For example, in the one-loop approximation, the running coupling has the following form:

$$\frac{1}{\alpha(p)} = \frac{1}{\alpha(\sigma)} + C\ln\frac{p}{\sigma}\ , \tag{11.121}$$

where $C > 0$ ($C < 0$) in asymptotically free (non-asymptotically free) gauge theories. The bare coupling constant $\alpha^{(0)}$ is

$$\frac{1}{\alpha^{(0)}} = \frac{1}{\alpha(\Lambda)} = \frac{1}{\alpha(\sigma)} + C\ln\frac{\Lambda}{\sigma}\ . \tag{11.122}$$

In the same approximation, the running fermion mass takes the following form (see

Sec.12.2):

$$m(p) = m(\sigma)\left(\frac{\alpha(p)}{\alpha(\sigma)}\right)^{c_m}. \tag{11.123}$$

And the bare mass $m^{(0)}$ is

$$m^{(0)} = m(p)|_{p=\Lambda} = m(\sigma)\left(\frac{\alpha(\Lambda)}{\alpha(\sigma)}\right)^{c_m}. \tag{11.124}$$

Such a simple correspondence follows of course from the fact that the bare parameters are given by the values of the running functions at $p = \Lambda$.

However, this is not an universal rule in field theory. Let us look at the relation between the bare mass $m^{(0)}$ and the physical mass m in the four-fermion interaction models.

In leading order in $1/N_c$, the mass function $m(p^2)$ is just a constant, $m(p^2) = m$. But at the critical value of $\kappa, \kappa = \kappa_c$, we find that the bare mass $m^{(0)}$ is not equal to $m(p^2)|_{p^2=\Lambda^2} = m$ but rather

$$m^{(0)} \sim \frac{m|m|}{\Lambda} \tag{11.125}$$

at $D = 3$ (see Eq.(11.106)), and

$$m^{(0)} \sim \frac{m^3}{\Lambda^2} \tag{11.126}$$

at $D = 4$ (the relation (11.126) follows directly from Eq.(9.84)).

These relations show a nonperturbative enhancement of the fermion mass: the bare mass is enhanced by the factor $\left(\frac{\Lambda}{m}\right)^{D-2}$*).

This example demonstrates that the relations between the running parameters and their bare counterparts are in general case rather delicate. From the viewpoint of the theory of critical phenomena, the renormalization of bare parameters means their fine tuning to the critical values. In particular, it is not difficult to convince oneself that the

*) As we will see in Chapter 15, this enhancement mechanism is used in some models for the description of the mass spectrum of quarks and leptons.

parameter γ_m in the relation,

$$m \sim m^{(0)} \left(\frac{\Lambda}{m}\right)^{\gamma_m} , \tag{11.127}$$

representing the relations (11.125) and (11.126) in the four-fermion interaction models, is expressed through the critical exponent δ (see Eq.(11.93)):

$$\gamma_m = \frac{\delta(D-1)-1}{1+\delta} \tag{11.128}$$

Since δ is a universal characteristics of a critical phenomenon, γ_m is also a universal parameter. Therefore the relation (11.127) expresses a genuine physical effect.

We conclude that both the renormalizations of bare parameters and the form of running parameters are determined from the equations for Green's functions. However the relations between bare and running parameters are in general rather nontrivial.

Let us discuss the chiral phase transition in QED (considered in Chapter 10) from this viewpoint.

We begin by considering the ladder approximation. In this approximation, the running coupling $\alpha(p^2)$ is just a constant, $\alpha(p^2) = \alpha = \alpha^{(0)}$. However, as we showed in Sec.10.2, there is a nonperturbative mass divergence at $\alpha > \alpha_c = \pi/3$. Although the renormalization constant Z_3 is $Z_3 = 1$ in this approximation, the β-function connected with the nonperturbative renormalization of α is nonzero (see Eq.(10.119)):

$$\beta(\alpha) = -\frac{2}{3}\left(\frac{\alpha}{\alpha_c} - 1\right)^{3/2} . \tag{11.129}$$

From the viewpoint of the theory of critical phenomena, the renormalization of α is the fine tuning of α to the critical value $\alpha = \alpha_c$. The critical value defines a universality class different from that of perturbative ladder QED. In particular, in this universality class, new degrees of freedom (chiral composite fields $\pi^\alpha \sim \bar{\psi}\gamma_5\lambda^\alpha\psi$ and $\sigma^\alpha \sim \bar{\psi}\lambda^\alpha\psi$) appear. In fact, the renormalization (fine tuning) of α provides finiteness of the mass of σ^α in the continuum limit.

As we showed in Sec.10.8, the β-function (11.129) leads to nonperturbative breakdown of scale symmetry in the ladder approximation. From the viewpoint of the effective theory, which corresponds to this universality class, this fact is not surprising:

the nonperturbative mass term of σ particles violates the scale symmetry. In Sec.14.6, we will discuss this effective theory in more detail. In particular, the critical exponents corresponding to this universality class will be calculated.

Let us discuss now the chiral phase transition in the one-loop QED considered in Sec.10.9. The expression for the running coupling $\alpha(p^2)$ is just the standard one (see Eq.(10.139)). The critical coupling $\alpha^{(0)} = \alpha_c \simeq 2$ defines an universality class different from that of perturbative QED at $\alpha^{(0)} < \alpha_c$. At $\alpha^{(0)} = \alpha_c$, the universality class of QED (in this approximation) corresponds to the mean-field σ-model (plus the photon field).

Thus, the chiral phase transition in QED can be well understood in the framework of the theory of critical phenomena.

We will return to these issues in Sec.14.6.

11.7. The Second Coleman Theorem: There are no Goldstone Bosons in Two Dimensions

Our discussion of critical phenomena in D dimensional systems would be incomplete if we did not discuss the problem of the spontaneous breakdown of continuous symmetries at $D = 2$. In fact, the general statement concerning this problem is very simple: there is no spontaneous breakdown of continuous symmetries at $D = 2$.

This theorem for condensed matter systems was proved by Mermin and Wagner [25] and extended to relativistic field theory by Coleman [26].

Rather than to consider the rigorous proof of the theorem, we will discuss its physical meaning.

As is well-known, the ultraviolet divergences become less severe with decreasing D.*) However, the infrared divergences become stronger as D decreases. They are in fact the reason of impossibility of the spontaneous breakdown of continuous symmetries at $D = 2$.

The essence of the Coleman theorem is the following. Because of the Goldstone theorem (see Sec.3.4), there would be massless scalars if a continuous symmetry was

*) Actually, this was motivation for studying spontaneous chiral symmetry breaking at $D = 2$ for the authors of Ref.[27]. This paper considering a two dimensional model with four-fermion interaction appeared just after the Nambu and Jona-Lasinio paper.

spontaneously broken. But in a two dimensional space-time it is not possible to construct a massless scalar field operator satisfying the axioms of relativistic field theory (as positivity of the Fock space metric, locality, etc.). In order to see this, let us consider the two-point function

$$\langle 0|\varphi(x)\varphi(0)|0\rangle = \int \frac{d^2p}{2\pi}\theta(p^0)\delta(p^2)e^{-ipx} = \int_0^\infty \frac{dp^0}{2\pi p^0}\cos(p^0x^1)e^{-ip^0x^0} . \quad (11.130)$$

It is divergent at the origin at all x^0. This reflects the point that a δ-function singularity on the light-cone is forbidden in two dimensions.

The presence of NG bosons provides the validity of the chiral Ward-Takahashi (WT) identities (considered in Sec.9.2) even although some symmetry breaking Green's functions are nonzero. But there are no NG bosons at $D = 2$.

This result also admits the following simple interpretation. Spontaneous symmetry breaking implies that there is an order parameter $v = \langle 0|\varphi|0\rangle$ describing the condensate in the vacuum. There are of course quantum fluctuations around v. It is clear that the notion of the order parameter is meaningful only if the fluctuations are not very strong. But Eq.(11.130) implies that NG bosons would lead to infinitely strong fluctuations. So we conclude that the notion of the order parameter connected with a continuous symmetry is physically meaningless at $D = 2$.

Of course, discrete symmetries, as the Z_2 in the four-fermion interaction model, can be spontaneously broken at $D = 2$.

One should also understand that the Coleman theorem does not forbid the dynamical generation of particle masses at $D = 2$. As an example of what may happen at $D = 2$, let us consider the $U(N)$ invariant Gross-Neveu model [8]:

$$S = \int d^2x \left[\bar{\psi}_a i\gamma^\mu\partial_\mu\psi_a + G^{(0)}\left(\left(\bar{\psi}_a\psi_a\right)^2 - \left(\bar{\psi}_a\gamma_5\psi_a\right)^2\right)\right] , \quad (11.131)$$

where $a = 1, 2, \cdots, N_c$. The action possesses the $U(1)$ chiral symmetry, $\psi_a \longrightarrow e^{i\theta\gamma_5}\psi_a$. By using the auxiliary field method, used already in Chapter 6 and in this chapter, one can show that in leading order in $1/N_c$, the vacuum expectation value $\langle 0|\bar{\psi}\psi|0\rangle$ is nonzero, and fermions acquire a dynamical mass. This, in the contradiction with Coleman's theorem, implies the spontaneous breakdown of the $U(1)$ chiral symmetry.

A resolution of this puzzle was suggested by Witten [28]. He argued that in the next to leading order in $1/N_c$, the order parameter equals zero, and Green's function $\langle 0|\bar{\psi}(x)\psi(x)\bar{\psi}(0)\psi(0)|0\rangle$ is

$$\langle 0|\bar{\psi}(x)\psi(x)\bar{\psi}(0)\psi(0)|0\rangle \sim |x|^{-1/N_c} \tag{11.132}$$

for large $|x|$. This behavior indeed implies that the order parameter is zero:

$$\langle 0|\bar{\psi}(x)\psi(x)\bar{\psi}(0)\psi(0)|0\rangle \underset{|x|\longrightarrow\infty}{=} \langle 0|\bar{\psi}(x)\psi(x)|0\rangle\langle 0|\bar{\psi}(0)\psi(0)|0\rangle = 0\ . \tag{11.133}$$

However, the Green's function (11.132) goes to zero rather slowly as $|x| \longrightarrow \infty$. Moreover, in this order, fermions remain massive. A "right" fermion field $\tilde{\psi}(x)$ describing them is chiral invariant.

The kind of behavior in Eq.(11.132) corresponds to the Berezinski- Kasterlitz-Thouless (BKT) phase transition discovered first in the two-dimensional XY model in statistical physics [29]. While in the conventional phase transition, the behavior of this Green's function would be

$$\langle 0|\bar{\psi}(x)\psi(x)\bar{\psi}(0)\psi(0)|0\rangle \underset{|x|\to\infty}{\sim} |x|^{-2} \tag{11.134}$$

in the symmetric phase, and

$$\langle 0|\bar{\psi}(x)\psi(x)\bar{\psi}(0)\psi(0)|0\rangle \underset{|x|\longrightarrow\infty}{\sim} const \neq 0 \tag{11.135}$$

in the broken phase, in the case of the BKT phase transition, the relation (11.135) is replaced by the relation (11.132), although Eq.(11.134) is still valid.

The relation (11.132) implies that although the order parameter equals zero, there are long rangle correlations in the BKT phase. Thus, at $D = 2$, this phase replaces the phase with spontaneous symmetry breaking existing at $D > 2$.

Chapter 12

Dynamical Chiral Symmetry Breaking in QCD

12.1. Introduction

Hadron dynamics gives the most important example of the phenomenon of dynamical symmetry breaking in particle physics. In fact, the concept of spontaneous symmetry breaking in particle physics itself first appeared there. In their classical papers, Nambu [1], Nambu and Jona-Lasinio [2], and Goldstone [3] introduced this concept being motivated by the experimental fact that the pions are light compared to other hadrons.

The microscopic theory of the hadron interactions is quantum chromodynamics (QCD). Since a lot of experimental facts have left no doubts on the reality of spontaneous chiral symmetry breaking in the dynamics of hadrons (see Chapter 4), this phenomenon has necessarily to be realized in QCD.

QCD is a vector-like gauge theory (see Sec.9.1). Its lagrangian density takes the following form :

$$\mathcal{L} = -\frac{1}{4} G^K_{\mu\nu} G^{\mu\nu K} + i\bar{\psi}\gamma^\mu D_\mu \psi - \bar{\psi}\hat{m}^{(0)}\psi \ , \tag{12.1}$$

where

$$G^K_{\mu\nu} = \partial_\mu G^K_\nu - \partial_\nu G^K_\mu - g^{(0)} f_{KST} G^S_\mu G^T_\nu \ , \tag{12.2}$$

$$(D_\mu \psi)_{al} = \partial_\mu \psi_{al} + \frac{i}{2} g^{(0)} t^K_{lm} \psi_{am} G^K_\mu \ , \tag{12.3}$$

$$\left(\hat{m}^{(0)}\psi\right)_{al} = m^{(0)}_{ab}\psi_{bl} + i\gamma_5 \tilde{m}^{(0)}_{ab}\psi_{bl} \ . \tag{12.4}$$

Here f_{KST} are the structure constants of the gauge (color) group $SU_c(3)$, $K, S, T = 1, 2, \cdots, 8$, the matrices t^K_{lm} define the fundamental representation of the $SU_c(3)$ ($trt^K t^K = 2$), and a, b are the flavor indices of quark fields ; $a, b = 1, \cdots, N$. QCD describes the interaction between quarks (ψ_{al}) and gluons $\left(G^K_\mu\right)$; $g^{(0)}$ is the bare gauge coupling, and $\hat{m}^{(0)}$ is the bare mass matrix for quarks.

Although in the real world the color group is $SU_c(3)$, it is often convenient to consider QCD with the color group $SU(N_c)$, with N_c arbitrary. The point is that the dynamics simplifies in the limit $N_c \longrightarrow \infty$, and the $1/N_c$ expansion may still be reliable at $N_c = 3$.

The chiral limit with $\hat{m}^{(0)} = 0$ is very special. As we already showed in Sec.9.2, in this limit, the currents

$$j^{\alpha}_{\mu} = \bar{\psi}\frac{\lambda^{\alpha}}{2}\gamma_{\mu}\psi; \alpha = 0, 1, \cdots, N^2 - 1 , \tag{12.5}$$

$$j^{\alpha}_{5\mu} = \bar{\psi}\frac{\lambda^{\alpha}}{2}\gamma_{\mu}\gamma_5\psi; \alpha = 1, \cdots, N^2 - 1 \tag{12.6}$$

connected with the chiral group $SU_L(N) \times SU_R(N) \times U_V(1)$ are conserved. The singlet axial current

$$j^{0}_{5\mu} = \bar{\psi}\frac{\lambda^{0}}{2}\gamma_{\mu}\gamma_5\psi \tag{12.7}$$

is not conserved even in the chiral limit because of the Adler-Bell-Jackiw (ABJ) anomaly [4, 5] :

$$\partial^{\mu} j^{0}_{5\mu} = -\left(\frac{N}{2}\right)^{1/2} \frac{g^2}{32\pi^2}\epsilon^{\mu\nu\delta\sigma}G^{K}_{\mu\nu}G^{K}_{\delta\sigma} . \tag{12.8}$$

As was shown in Sec.9.2, if the chiral group is spontaneously broken to the $SU_V(N) \times U_V(1)$, the quark propagator takes the following form in the chiral limit:

$$G^{-1}_{al,bm}(p) = \delta_{ab}\delta_{lm}(\hat{p}A(p^2) - B_{dyn}(p^2)) , \tag{12.9}$$

where $B_{dyn}(p^2)$ is the dynamical quark mass function.

If the bare mass matrix $\hat{m}^{(0)}$ is not zero, the chiral symmetry is explicitly broken, but if the bare masses of some of the quark flavors are small, we are led to the dynamics of the partially conserved axial currents (PCAC) considered in Sec.4.5. This is what happens with three lightest quarks u, d and s in nature. The corresponding approximate chiral symmetry is $SU_L(3) \times SU_R(3) \times U_V(1)$. A signature of the PCAC dynamics is the lightness of eight pseudoscalar mesons $\pi^0, \pi^{\pm}, K^0, \bar{K}^0, K^{\pm}$ and η composed of these three quarks. They are pseudo-NG (Nambu-Goldstone) bosons. The ninth pseudoscalar meson η' is essentially heavier because of the ABJ anomaly (12.8)(see Secs.12.8 and 12.12).

In many respects, the description of dynamical chiral symmetry breaking in QCD reduces to the description of the dynamics of pseudoscalar mesons. Since, as we already showed in Sec.9.2, the Bethe-Salpeter wave function of NG bosons is expressed through $B_{dyn}(p^2)$ (see Eq.(9.38)), it is important to find the form of the dynamical mass function of quarks.

The whole program of the realization of spontaneous chiral symmetry breaking in QCD can be divided into two parts. The first part is the description of the microscopic dynamics which is responsible for this phenomenon. The second ("macroscopic") part is the derivation of the low energy effective action describing the interactions of light hadrons.

Although at present this program is far from being completed, some key points of the dynamics of chiral symmetry breaking in QCD are already rather well understood.

In the present chapter, we will consider the issues connected both with the microscopic dynamics and with the low energy effective action in QCD.

12.2. Perturbative Dynamics in QCD

In this section we recall the basic points of perturbative dynamics in QCD [6].

QCD is an asymptotically free theory [7]. In the one-loop approximation, its running coupling constant takes the form (in Euclidean region) :

$$\bar{g}^2(p^2) = \frac{2}{b\ln\left(p^2/\Lambda_{QCD}^2\right)} , \tag{12.10}$$

where b is the first coefficient of the QCD β function,

$$\begin{aligned} \beta(g) &= -\frac{b}{2}g^3 , \\ b &= \frac{11N_c - 2N}{24\pi^2} , \end{aligned} \tag{12.11}$$

and Λ_{QCD} is the infrared scale parameter in QCD ($\Lambda_{QCD} \sim 300MeV$ in the real QCD with $N_c = 3$).

We will assume that the bare mass matrix $\hat{m}^{(0)}$ (12.4) takes the form :

$$\hat{m}_{ab}^{(0)} = \delta_{ab}m_a^{(0)} \tag{12.12}$$

(as we already noted in Sec.9.3, the general mass matrix $\hat{m}^{(0)}$ can be reduced to the form (12.12) by using an appropriate chiral transformation of fermion fields).*) It is

---

*) There are however subtle points connected with the ABJ anomaly (12.8) (see.Sec.12.12).

convenient to use the Landau gauge for the description of the quark propagator

$$G^{(\sigma)^{-1}}_{al,bm}(p) = \delta_{ab}\delta_{lm}\left(\hat{p}A^{(\sigma)}_a(p^2) - B^{(\sigma)}_a(p^2)\right) , \tag{12.13}$$

where the renormalized propagator $G^{(\sigma)}$ is

$$G^{(\sigma)} = Z_2^{(\sigma)^{-1}} G , \tag{12.14}$$

and $Z_2^{(\sigma)}$ is the renormalization constant of quark fields :

$$\psi^{(\sigma)} = Z_2^{(\sigma)^{-1/2}} \psi \tag{12.15}$$

(see Eq.(9.57)). The point is that $Z_2^{(\sigma)} = 1$ in the one-loop approximation in this gauge, and the form of the propagator simplifies in this case. In particular, the ultraviolet asymptotics of $A^{(\sigma)}_a(p^2)$ is

$$A^{(\sigma)}_a(p^2) \underset{p^2\to\infty}{\longrightarrow} 1 \tag{12.16}$$

in this approximation. The ultraviolet asymptotics of $B^{(\sigma)}_{\text{cur};a}(p^2)$, where $B^{(\sigma)}_{\text{cur};a}(p^2)$ is the part of $B^{(\sigma)}_a$ connected with explicit chiral symmetry breaking, is

$$B^{(\sigma)}_{\text{cur};a}(p^2) \underset{p^2\to\infty}{\longrightarrow} m^{(\sigma)}_{\text{cur};a}\left(\frac{\ln p^2/\Lambda^2_{QCD}}{\ln \sigma^2/\Lambda^2_{QCD}}\right)^{-c_m} , \tag{12.17}$$

where

$$c_m = \frac{3(N_c^2-1)}{16\pi^2 b N_c} . \tag{12.18}$$

The current mass $m^{(\sigma)}_{\text{cur};a}$ is expressed through the bare mass $m^{(0)}_a$ as

$$m^{(\sigma)}_{\text{cur};a} = Z_m^{(\sigma)^{-1}} m^{(0)}_a \tag{12.19}$$

with

$$Z_m^{(\sigma)} = \left(\frac{\ln \sigma^2/\Lambda^2_{QCD}}{\ln \Lambda^2/\Lambda^2_{QCD}}\right)^{c_m} . \tag{12.20}$$

Eqs.(12.17), (12.19) and (12.20) imply that

$$B^{(\sigma)}_{\text{cur};a} = B^{(\sigma')}_{\text{cur};a} \tag{12.21}$$

in this approximation, i.e. the function $B^{(\sigma)}_{\text{cur};a}$ is independent of σ, and is a renormalization group invariant. This is of course a consequence of the equality $Z_2^{(\sigma)} = 1$. In the general case, the mass functions

$$m_a(p^2) \equiv \frac{B_a^{(\sigma)}(p^2)}{A_a^{(\sigma)}(p^2)} \text{ and } m_{\text{cur};a}(p^2) \equiv \frac{B_{\text{cur};a}^{(\sigma)}(p^2)}{A_a^{(\sigma)}(p^2)} \tag{12.22}$$

(and not $B_a^{(\sigma)}$ and $B_{\text{cur};a}^{(\sigma)}$) are independent of σ. The expressions for the ultraviolet asymptotics of $A_a^{(\sigma)}(p^2)$ and $B_{\text{cur};a}^{(\sigma)}(p^2)$ in other covariant gauges can be found in the review [8].

12.3. The Ultraviolet Asymptotics of the Dynamical Mass Function of Quarks

In the previous section we considered the ultraviolet asymptotics of the piece of the mass function of quarks connected with the explicit breakdown of the chiral symmetry. Here we will analyze the asymptotics of the dynamical mass function of quarks $B_{dyn}^{(\sigma)}(p^2)$.

The simplest way [9, 10] to obtain it is to use the relation (9.52) for the condensate $\langle\bar{\psi}\psi\rangle \equiv \lim_{x\to 0} \frac{1}{N} tr\langle 0|T\bar{\psi}(0)\psi(x)|0\rangle$ (in Minkowski space):

$$\langle\bar{\psi}\psi\rangle = -\lim_{x\to 0} \frac{i}{N} trG(x) = -\frac{4iN_c}{(2\pi)^4}\int d^4p \frac{B_{dyn}(p^2)}{p^2A^2(p^2) - B_{dyn}^2(p^2)} \,. \tag{12.23}$$

All quantities in this relation are bare (unrenormalized). Eq.(12.14) implies that

$$A(p^2) = Z_2^{(\sigma)^{-1}} A^{(\sigma)}(p^2) \,, \tag{12.24}$$

$$B_{dyn}(p^2) = Z_2^{(\sigma)^{-1}} B_{dyn}^{(\sigma)}(p^2) \,. \tag{12.25}$$

As we already showed in Sec.9.3, the renormalization constant of the composite operator $\bar{\psi}\psi$ coincides with $Z_m^{(\sigma)}$ in Eq.(12.19) :

$$\bar{\psi}\psi = Z_m^{(\sigma)^{-1}} (\bar{\psi}\psi)^{(\sigma)} \,. \tag{12.26}$$

Since $Z_2^{(\sigma)} = 1$ in Landau gauge (in the one-loop approximation), we find from Eqs. (12.23), (12.26) that

$$Z_m^{(\sigma)^{-1}} \langle(\bar{\psi}\psi)^{(\sigma)}\rangle = \left(\frac{\ln \Lambda^2/\Lambda_{QCD}^2}{\ln \sigma^2/\Lambda_{QCD}^2}\right)^{c_m} \langle(\bar{\psi}\psi)^{(\sigma)}\rangle \simeq -\frac{N_c}{4\pi^2}\int^{\Lambda^2} dp^2 B_{dyn}^{(\sigma)}(p^2) \,. \tag{12.27}$$

Here we used Eqs.(9.66), (12.16) and (12.20), and the integral in this relation is written in Euclidean region.

Eq.(12.27) implies that

$$B_{dyn}^{(\sigma)}(p^2) \underset{p^2 \gg \Lambda_{QCD}^2}{\longrightarrow} -\frac{4\pi^2 c_m}{N_c} \langle(\bar{\psi}\psi)_{inv}\rangle \frac{1}{p^2} \left(\ln p^2/\Lambda_{QCD}^2\right)^{c_m - 1} , \tag{12.28}$$

where the renormalization group invariant condensate $\langle(\bar{\psi}\psi)_{inv}\rangle$ is

$$\langle(\bar{\psi}\psi)_{inv}\rangle \equiv \frac{\langle(\bar{\psi}\psi)^{(\sigma)}\rangle}{\left(\ln \sigma^2/\Lambda_{QCD}^2\right)^{c_m}} . \tag{12.29}$$

In accordance with $Z_2^{(\sigma)} = 1$, Eq.(12.28) shows that $B_{dyn}^{(\sigma)}$ is independent of σ in this approximation (in the general case, $m_{dyn}(p^2) = B_{dyn}^{(\sigma)}(p^2)/A^{(\sigma)}(p^2)$ is of course independent of σ).

Comparing Eq.(12.17) and Eq.(12.28), we see that the dynamical mass function $B_{dyn}^{(\sigma)}(p^2)$ decreases as $p^2 \longrightarrow \infty$ much more rapidly than the function $B_{\mathrm{cur};a}^{(\sigma)}(p^2)$ connected with explicit chiral symmetry breaking. This point can be understood if we, following Lane and Politzer [11, 12], apply the operator product expansion (OPE) [13] to the bilocal operator $T\psi^{(\sigma)}(x)\psi^{(\sigma)}(0)$:

$$T\psi^{(\sigma)}(x)\bar{\psi}^{(\sigma)}(0) \sim \left[i\gamma^\mu \partial_\mu C_0^{(\sigma)}(x^2) + m_{\mathrm{cur}}^{(\sigma)} C_1^{(\sigma)}(x^2)\right] I + C_2^{(\sigma)}(x^2)(\bar{\psi}\psi)^{(\sigma)} + \cdots , \tag{12.30}$$

where I denotes the identity operator. For simplicity, we consider here the case with the same current mass for all quark flavors.

At short distances (high energy), just these low dimensional operators dominate in the OPE of $T\psi^{(\sigma)}(x)\psi^{(\sigma)}(0)$. Let us compare the Fourier transform of the vacuum expectation value of the expressions in Eq.(12.30) with the ultraviolet asymptotics of the quark propagator (12.13),

$$G^{(\sigma)}(p) \longrightarrow \frac{1}{A^{(\sigma)}(p^2)} \left[\frac{\hat{p}}{p^2} + \frac{1}{p^2} m(p^2) + \cdots\right] . \tag{12.31}$$

Then we get :

$$\frac{-i}{p^2 A^{(\sigma)}(p^2)} = C_0^{(\sigma)}(p^2) , \tag{12.32}$$

$$\frac{i}{p^2 A^{(\sigma)}(p^2)} m(p^2) = m_{\text{cur}}^{(\sigma)} C_1^{(\sigma)}(p^2) + \langle (\bar{\psi}\psi)^{(\sigma)} \rangle C_2^{(\sigma)}(p^2) \,, \tag{12.33}$$

where

$$C_i^{(\sigma)}(p^2) = \int d^4x e^{ipx} C_i^{(\sigma)}(x^2) \,. \tag{12.34}$$

If we try to use the naive dimensional analysis, then, just comparing the dimensions of quantities on both sides of Eq.(12.33), we would find that

$$\begin{aligned} C_1^{(\sigma)}(p^2) &\sim \frac{1}{p^2} \,, \\ C_2^{(\sigma)}(p^2) &\sim \frac{1}{(p^2)^2} \,, \end{aligned} \tag{12.35}$$

i.e. the function $C_2^{(\sigma)}(p^2)$ would decrease more rapidly than $C_1^{(\sigma)}(p^2)$ because of the fact that while the dimension of $m_{\text{cur}}^{(\sigma)}$ is one, the dimension of $(\bar{\psi}\psi)^{(\sigma)}$ is three. Although in reality the situation is somewhat more complicated, this simplified picture yields a correct qualitative explanation of the reason of a rapid decreasing of the dynamical mass function.

The full analysis of the behavior of the coefficients $C_i^{(\sigma)}(p^2)$ is based on the renormalization-group equations. Since the bilocal operator $T\psi(x)\psi(0) = Z_2^{(\sigma)} T\psi^{(\sigma)}(x)\psi^{(\sigma)}(0)$ is independent of σ, we get :

$$\sigma \frac{d}{d\sigma} \left[Z_2^{(\sigma)} T\psi^{(\sigma)}(x)\psi^{(\sigma)}(0) \right] = 0 \,. \tag{12.36}$$

This in turn leads to the following equations for $C_i^{(\sigma)}$ in OPE (12.30) :

$$\left[\sigma \frac{\partial}{\partial \sigma} + \beta(g) \frac{\partial}{\partial g} - \gamma_m(g) \frac{\partial}{\partial \ln m_{\text{cur}}^{(\sigma)}} + 2\gamma_\psi \right] C_0^{(\sigma)}(p^2, g) = 0 \,, \tag{12.37}$$

$$\left[\sigma \frac{\partial}{\partial \sigma} + \beta(g) \frac{\partial}{\partial g} - \gamma_m(g) \frac{\partial}{\partial \ln m_{\text{cur}}^{(\sigma)}} + 2\gamma_\psi - \gamma_m \right] C_1^{(\sigma)}(p^2, g) = 0 \,, \tag{12.38}$$

$$\left[\sigma \frac{\partial}{\partial \sigma} + \beta(g) \frac{\partial}{\partial g} - \gamma_m(g) \frac{\partial}{\partial \ln m_{\text{cur}}^{(\sigma)}} + 2\gamma_\psi + \gamma_m \right] C_2^{(\sigma)}(p^2, g) = 0 \,, \tag{12.39}$$

where

$$g \equiv g^{(\sigma)} , \tag{12.40}$$

$$\gamma_\psi = \frac{1}{2}\frac{\partial \ln Z_2^{(\sigma)}}{\partial \ln \sigma} = 0 , \tag{12.41}$$

$$\gamma_m = \frac{\partial \ln Z_m^{(\sigma)}}{\partial \ln \sigma} = bc_m g^2 = \frac{3(N_c^2-1)}{16\pi^2 N_c} g^2 . \tag{12.42}$$

In the calculations of γ_ψ and γ_m, we used Eqs.(12.18), (12.20) and the equality $Z_2^{(\sigma)} = 1$, in this approximation.

Equations (12.37) - (12.39) can be solved in the same way as the renormalization-group equations in Sec.7.6. In particular, for this purpose, we need to know the boundary conditions at $p^2 = \sigma^2$ in Euclidean region. To find them, we consider the short distance limit ($x \to 0$) in the Green's function $\langle 0|T\psi(x)\bar{\psi}(0)\psi(y)\bar{\psi}(z)|0\rangle$. We get the following coefficient functions in the lowest order approximation (in Euclidean region):

$$C_0^{(\sigma)}(p^2, g) = \frac{i}{p^2} , \tag{12.43}$$

$$C_1^{(\sigma)}(p^2, g) = \frac{-i}{p^2} , \tag{12.44}$$

$$C_2^{(\sigma)}(p^2, g) = \frac{3i}{4}\frac{N_c^2-1}{2N_c^2}\frac{g^2}{(p^2)^2} . \tag{12.45}$$

The solutions of Eqs.(12.37) - (12.39) must coincide with the functions in Eqs.(12.43) - (12.45) at $p^2 = \sigma^2$. As a result, we are lead to the following solutions for $A^{(\sigma)}(p^2)$ and $m(p^2) = B^{(\sigma)}(p^2)/A^{(\sigma)}(p^2)$:

$$A^{(\sigma)}(p^2) \longrightarrow 1 , \tag{12.46}$$

$$m(p^2) \longrightarrow m_{\rm cur}^{(\sigma)} \left(\frac{\ln \sigma^2/\Lambda_{QCD}^2}{\ln p^2/\Lambda_{QCD}^2}\right)^{c_m} - \frac{4\pi^2 c_m}{N_c}\langle(\bar{\psi}\psi)_{inv}\rangle \frac{1}{p^2}\left(\ln p^2/\Lambda_{QCD}^2\right)^{c_m - 1} . \tag{12.47}$$

These expressions are of course in agreement with those in Eqs.(12.16), (12.17) and (12.28).

The following comments are in order.

1) The comparison of Eq.(12.47) and Eq.(12.35) shows that the behavior of the composite operator $\bar{\psi}\psi$ in OPE in QCD is similar to that in free theory: up to small logarithmic corrections, one can assign the dynamical dimension $d_{\bar{\psi}\psi} = 3$ to this operator. This point reflects the property of asymptotic freedom in QCD.

2) The derivation of the expression (12.47) does not prove of course that spontaneous chiral symmetry breaking occurs in QCD. What this expression shows is that *if* spontaneous chiral symmetry breaking takes place in QCD ($\langle\bar{\psi}\psi\rangle \neq 0$), then the asymptotics of the dynamical mass function $m_{dyn}(p^2)$ takes the form (12.47) with $m_{\text{cur}}^{(\sigma)} = 0$. We will discuss the dynamical issues of spontaneous chiral symmetry breaking in QCD in Secs.12.6 - 12.7.

3) So far we have worked in Landau gauge. Of course, there is nothing wrong with using a concrete gauge to describe dynamics in QCD. One can choose the vacuum expectation value $\langle 0|(\bar{\psi}\psi)^{(\sigma)}|0\rangle$ as a gauge invariant order parameter characterizing spontaneous chiral symmetry breaking. This paramter is expressed through the functions $A^{(\sigma)}(p^2)$ and $m_{dyn}(p^2)$ which are not gauge invariant themselves. Eq.(12.23) implies that the condition $m_{dyn}(p^2) \neq 0$ in Landau gauge is sufficient for a proof of spontaneous chiral symmetry breaking in QCD.

12.4. The Exact Relation for the Mass of Pseudoscalar Mesons. Hadrons are Loose Bound States

In Sec.9.3, the exact relation for the mass of pseudoscalar bosons was derived. It is

$$M_{ab}^2 = -\frac{2iN_c}{(2\pi)^4 F_{ab}} \left(m_a^{(0)} + m_b^{(0)}\right) \int d^4q \chi_{ab;1}^{(p)} , \tag{12.48}$$

where M_{ab} is the mass of a pseudoscalar boson (meson) composed of the a-th quark and the b-th antiquark, F_{ab} is the decay constant,

$$\langle 0|\bar{\psi}_a(0)\gamma_\mu\gamma_5\psi_b(0)|P\rangle = i\sqrt{2}F_{ab}P_\mu , \tag{12.49}$$

and $\chi_{ab;1}^{(p)}$ is a Lorentz invariant function appearing in the expansion of the Bethe-Salpeter (BS) wave function (see Eq.(9.41)) :

$$\chi_{ab}^{(p)}(q,P) = \frac{1}{\sqrt{2}}\left[\chi_{ab;1}^{(p)} + \chi_{ab;2}^{(p)}\hat{P} + \chi_{ab;3}^{(p)}\hat{q} + \chi_{ab;4}^{(p)}\sigma^{\mu\nu}\left(P_\mu q_\nu - P_\nu q_\mu\right)\right]\gamma_5 . \tag{12.50}$$

As was also shown in that section, in leading order in $m_a^{(0)}, m_b^{(0)}$, this relation reduces to the PCAC relation

$$M_{ab}^{(p)^2} = -\frac{\langle\bar{\psi}\psi\rangle}{F^2}\left(m_a^{(0)} + m_b^{(0)}\right), \tag{12.51}$$

where F is $F \equiv F_{ab}$ in the chiral limit ($F = F_\pi \simeq 93MeV$ in QCD).

The relation (12.48) leads to the following constraint for the ultraviolet asymptotics of the BS wave function $\chi_{ab;1}^{(p)}$ in Landau gauge (see Eq.(9.64)) :

$$\chi_{ab;1}^{(p)} \sim \chi_{(\sigma);ab;1}^{(p)} \underset{q^2 \gg \Lambda_{QCD}^2}{\sim} (q^2)^{-2}\left(\ln q^2/\Lambda_{QCD}^2\right)^{c_m - 1}. \tag{12.52}$$

This relation takes place at any values of the current masses $m_{\text{cur};a}^{(\sigma)}$. Let us discuss this point in more detail.

In the chiral limit with $m_{\text{cur};a}^{(\sigma)} = 0$, the BS wave function $\chi_{ab;1}^{(p)}$ is expressed through $B_{dyn}(q^2)$ (see Eq.(9.38)). In particular, in this limit

$$\chi_{ab;1}^{(p)}(q^2) \underset{q^2 \gg \Lambda_{QCD}^2}{\sim} (q^2)^{-1} B_{dyn}(q^2). \tag{12.53}$$

However, at $m_{\text{cur};a}^{(\sigma)} \neq 0$, the situation changes drastically : while a new, slowly changing with q^2, piece appears in $B_a(q^2)$,

$$B_a(q^2) \sim m_{\text{cur};a}^{(\sigma)} \left(\frac{\ln \sigma^2/\Lambda_{QCD}^2}{\ln q^2/\Lambda_{QCD}^2}\right)^{c_m} \tag{12.54}$$

(see Eq.(12.47)), the wave function $\chi_{ab;1}^{(p)}$ still satisfies the relation (12.52) as before.*)

As is known, the BS wave function defines the form factors of the corresponding bound state. Eq.(12.52) implies that the form factors of mesons rapidly decrease with increasing the momentum at any value of $m_{\text{cur};a}^{(\sigma)}$, i.e. mesons are loose bound states insensitive to $m_{\text{cur};a}^{(\sigma)}$ [14]. On the other hand, the mass functions of quarks sensitively depend on $m_{\text{cur};a}^{(\sigma)}$.

---

*) Of course, a constant C in the asymptotic relation $\chi_{ab;1}^{(p)} \to C(q^2)^{-2}\left(\ln q^2/\Lambda_{QCD}^2\right)^{c_m - 1}$ should depend on the current masses. However, it is important that the functional form of this asymptotics is independent of them.

In order to understand the situation better, let us return to OPE (12.30). The slowly decreasing piece in the mass function is connected with the term

$$m_{\mathrm{cur};a}^{(\sigma)} C_1^{(\sigma)}(x^2) I \tag{12.55}$$

in OPE (12.30). But while this term does give contribution to the propagator

$$G^{(\sigma)}(x) = -i\langle 0|T\psi^{(\sigma)}(x)\bar{\psi}^{(\sigma)}(0)|0\rangle \;, \tag{12.56}$$

it does not give any contribution to the BS wave function $\chi_{(\sigma)}^{(i)} = \langle 0|T\psi(x)\bar{\psi}(0)|P\rangle$ [14, 15]. Indeed,

$$\langle 0|m_{\mathrm{cur};a}^{(\sigma)} C_1^{(\sigma)}(x^2) I|P\rangle = 0 \;. \tag{12.57}$$

In other words, the slowly decreasing piece in the quark mass function (the so called irregular asymptotics) is connected with vacuum fluctuations which completely decouple from the nonperturbative dynamics forming hadrons. *Mesons are loose bound states at any value of* $m_{\mathrm{cur};a}^{(\sigma)}$. One should expect that this is a general property of all hadrons. This implies a certain universality of the dynamics for both light and heavy hadrons.

12.5. The Equation for the Mass Function of Quarks

In this section, following Refs.[9 - 11, 16], we will analyse the equation for the function $B_a(q^2)$ in the propagator of quarks.

By using the WT identity (9.33) at $q = k$ and the BS equation (9.40), we get the following equation for $B_a(q^2)$:

$$(\gamma_5)_{nm} B_a(q^2) = (\gamma_5)_{nm} m_a^{(0)} + \frac{1}{(2\pi)^4} \int d^4p K_{mn,n'm'}(q,p) \left[G(p)\gamma_5 B_a(p^2) G(p)\right]_{n'm'} \;, \tag{12.58}$$

where the kernel $K_{mn,n'm'}(q,p)$ is

$$K_{mn,n'm'}(q,p) \equiv K_{mn,n'm'}\left(\eta_y q + \eta_x k, p; q-k\right)|_{q=k} \tag{12.59}$$

(see Eq.(9.40)). Here n, m represent both Dirac and color indices.

In the lowest order approximation, the kernel is (compare with Eq.(8.113)) :

$$\begin{aligned}K_{mn,n'm'}(q,p) &= -\frac{i}{4}g^{(0)^2}(t^K\gamma^\mu)_{nn'}(t^K\gamma^\nu)_{m'm}D_{\mu\nu}(q-p) \\ &= -ig^{(0)^2}\frac{N_c^2-1}{2N_c}(\gamma^\mu)_{nn'}(\gamma^\nu)_{m'm}D_{\mu\nu}(q-p) \ .\end{aligned} \tag{12.60}$$

Here we used Eqs.(B.4) and (B.7) in Appendix B in calculating the Casimir invariant $C_2(f)$:

$$\delta_{ll'}C_2(f) = \frac{1}{4}\sum_{K=1}^{N_c^2-1}(t^Kt^K)_{ll'} = \delta_{ll'}\frac{N_c^2-1}{2N_c} \ . \tag{12.61}$$

Introducing the renormalized functions

$$B_a^{(\sigma)} = Z_2^{(\sigma)}B_a, \ G^{(\sigma)} = Z_2^{(\sigma)^{-1}}G, \ K_{mn,n'm'}^{(\sigma)} = Z_2^{(\sigma)^2}K_{mn,n'm'} \ , \tag{12.62}$$

we get the equation

$$\begin{aligned}(\gamma_5)_{nm}B_a^{(\sigma)}(q^2) &= (\gamma_5)_{nm}Z_2^{(\sigma)}m_a^{(0)} \\ &+ \frac{1}{(2\pi)^4}\int d^4pK_{mn,n'm'}^{(\sigma)}(q,p)\left[G^{(\sigma)}(p)\gamma_5B_a^{(\sigma)}(p^2)G^{(\sigma)}(p)\right]_{n'm'} \ .\end{aligned} \tag{12.63}$$

Let us consider the ultraviolet asymptotics of the solutions of this equation in Euclidean region. We will use the Landau gauge.

Since the asymptotics of the functions $A_a^{(\sigma)}$ and $K_{mn,n'm'}^{(\sigma)}$ are insensitive to the mass term, they should not be changed significantly if the nonperturbative dynamics of spontaneous chiral symmetry breaking is taken into account. Therefore, in the one-loop approximation, one can take for them the expressions following from the renormalization group analysis (see Sec.12.2 and Eq.(12.60)) :

$$A_a^{(\sigma)}(q^2) = 1 \ , \tag{12.64}$$

$$K_{mn,n'm'}^{(\sigma)}(q,p) = -i\bar{g}^2((q-p)^2)\frac{N_c^2-1}{2N_c}\gamma_{nn'}^\mu\gamma_{m'm}^\nu D_{\mu\nu}(q-p) \ , \tag{12.65}$$

where

$$\bar{g}^2(q^2) = \frac{2}{b\ln(q^2/\Lambda_{QCD}^2)} \ , \tag{12.66}$$

$$D_{\mu\nu}(q) = \left(g_{\mu\nu} - \frac{q_\mu q_\nu}{q^2}\right)\frac{1}{q^2} \ . \tag{12.67}$$

We will also assume the standard renormalization group arguments showing that the region with $p^2, (p-q)^2 \gg \Lambda^2_{QCD}$ gives the main contribution to the integral on the right-hand side of Eq.(12.63) as $q^2 \to \infty$. In this region,

$$G^{(\sigma)}(p) \simeq \frac{1}{\hat{p}A^{(\sigma)}(p^2)} \simeq \frac{1}{\hat{p}} \,. \tag{12.68}$$

From this relation and from Eqs.(12.64), (12.65), we find the asymptotic form of equation (12.63):

$$B_a^{(\sigma)}(q^2) = m_a^{(0)} + \frac{3(N_c^2-1)}{2(2\pi)^4 N_c} \int^{\Lambda} d^4p \frac{\bar{g}^2((q-p)^2)}{(q-p)^2} \frac{B^{(\sigma)}(p^2)}{p^2} \tag{12.69}$$

($Z_2^{(\sigma)} = 1$ in this approximation).

Although this is the equation for the renormalized function $B_a^{(\sigma)}(q^2)$, we retained cutoff Λ here. In accordance with the analysis in Sec.10.7, this point will be crucial for selecting the right solution in the chiral limit.

In the region with $p^2, (p-q)^2 \gg \Lambda^2_{QCD}$, in the leading logarithmic approximation, the function $\bar{g}^2((q-p)^2)$ can be substituted by the following function :

$$\bar{g}^2((q-p)^2) = \theta(q^2-p^2)\bar{g}^2(q^2) + \theta(p^2-q^2)\bar{g}^2(p^2) \,. \tag{12.70}$$

As a result, we come to the following asymptotic equation [9, 10, 16]

$$B_a^{(\sigma)}(q^2) = m_a^{(0)} + \frac{3(N_c^2-1)}{32\pi^2 N_c} \int_{q^2}^{\Lambda^2} \frac{\bar{g}^2(p^2) B_a^{(\sigma)}(p^2)}{p^2} dp^2 + \frac{3(N_c^2-1)\bar{g}^2(q^2)}{32\pi^2 N_c q^2} \int_{q_0^2}^{q^2} B_a^{(\sigma)}(p^2) dp^2 \tag{12.71}$$

(the infrared cutoff q_0^2 remains undetermined in this asymptotic equation).

It is easy to verify that the solutions to Eq.(12.71) satisfy the second order differential equation,

$$\frac{32\pi^2 N_c}{3(N_c^2-1)} \frac{d}{dq^2} \left\{ \left[\frac{d}{dq^2} \left(\frac{\bar{g}^2(q^2)}{q^2} \right) \right]^{-1} \frac{dB_a^{(\sigma)}(q^2)}{dq^2} \right\} = B_a^{(\sigma)}(q^2) \,, \tag{12.72}$$

and the boundary condition :

$$\left[B_a^{(\sigma)}(q^2) - \frac{\bar{g}^2(q^2)}{q^2 \frac{d}{dq^2}\left(\frac{\bar{g}^2(q^2)}{q^2}\right)} \frac{dB_a^{(\sigma)}(q^2)}{dq^2} \right] |_{q^2=\Lambda^2} = m_a^{(0)} \,. \tag{12.73}$$

For our purposes, it is sufficient to know that the general solution to Eq.(12.72) takes the form

$$B_a^{(\sigma)}(q^2) = C_1 B_{a1}^{(\sigma)}(q^2) + C_2 B_{a2}^{(\sigma)}(q^2) , \tag{12.74}$$

where (as can be verified by direct substitution) the functions $B_{ai}^{(\sigma)}$ have the following ultraviolet asymptotics :

$$B_{a1}^{(\sigma)}(q^2) \sim (\ln q^2/\Lambda_{QCD}^2)^{-c_m} , \tag{12.75}$$

$$B_{a2}^{(\sigma)}(q^2) \sim \frac{1}{q^2}(\ln q^2/\Lambda_{QCD}^2)^{c_m-1} \tag{12.76}$$

with c_m defined in Eq.(12.18). From the boundary condition (12.73) we get :

$$C_1 = -\frac{c_m C_2}{\Lambda^2}\left(\ln\frac{\Lambda^2}{\Lambda_{QCD}^2}\right)^{2(c_m-1)} + m_a^{(0)}\left(\ln\frac{\Lambda^2}{\Lambda_{QCD}^2}\right)^{c_m} . \tag{12.77}$$

Let us consider the following two cases.

a) Spontaneous chiral symmetry breaking: $\lim\limits_{\Lambda\to 0}\partial^\mu j_{5\mu}^\alpha = 0$.

As was shown in Sec.10.7, the necessary and sufficient condition of the conservation of the axial currents $j_{5\mu}^\alpha$ in the continuum limit is

$$\lim_{\Lambda\to\infty} m_a^{(0)}(\Lambda) Z_m^{(\sigma)^{-1}}(\Lambda) = 0 . \tag{12.78}$$

With $Z_m^{(\sigma)}$ given in Eq.(12.20), this condition takes the form :

$$\lim_{\Lambda\to\infty} m_a^{(0)}(\Lambda)\left(\ln\Lambda^2\right)^{c_m} = 0 . \tag{12.79}$$

This equation implies that C_1 in Eq.(12.77) goes to zero as $\Lambda\to\infty$. Therefore the asymptotics of the dynamical mass function is

$$m_{dyn}(q^2) = \frac{B_{dyn}^{(\sigma)}(q^2)}{A^{(\sigma)}(q^2)} \longrightarrow C\frac{1}{q^2}(\ln q^2/\Lambda_{QCD}^2)^{c_m-1} . \tag{12.80}$$

As in the previous section, the constant C can be expressed through the renormalization group invariant condensate $\langle(\bar\psi\psi)_{inv}\rangle$ (see Eq.(12.29)), and it is :

$$C = -\frac{4\pi^2 c_m}{N_c}\langle(\bar\psi\psi)_{inv}\rangle . \tag{12.81}$$

Thus, the condition of the conservation of the axial currents selects the rapidly decreasing (at large q^2) solution (called the regular solution in the literature) for the dynamical mass function of quarks.

We emphasize that the equality $C_1 = 0$ was proved as a result of the choice of the procedure of the transition to the continuum theory with $\Lambda = \infty$. The point is that the bare mass $m^{(0)}(\Lambda) = Z_m^{(\sigma)}(\Lambda) m_{\rm cur}^{(\sigma)}$ goes to zero as $\Lambda \to \infty$ even if $m_{\rm cur}^{(\sigma)} \neq 0$, i.e. if the axial currents are not conserved. Therefore if the transition $\Lambda \to \infty$ was directly performed in Eqs.(12.71), (12.73), then the ratio C_1/C_2 would be arbitrary : indeed, both functions $B_{a1}^{(\sigma)}$ and $B_{a2}^{(\sigma)}$ satisfy the boundary condition (12.73) with $\Lambda = \infty$ and $m^{(0)} = 0$. This reflects the fact that (as was already indicated in Sec.10.7) the vanishing of the bare mass, $m^{(0)} = \lim\limits_{\Lambda\to\infty} m^{(0)}(\Lambda) = 0$, does not ensure the conservation of axial currents. The essential point is that the choice of the transition procedure to the continuum theory is determined uniquely by the physical content of the problem, i.e. by the axial current conservation [10].

Although our final result for the asymptotics of $m_{dyn}(q^2)$ just reproduces the result which was already obtained in the previous section by using OPE, the present derivation also gives some insight in the dynamics of spontaneous chiral symmetry breaking in QCD. In particular, a simple examination of equation (12.71) with $m_a^{(0)} = 0$,

$$B^{(\sigma)}(q^2) = \frac{3(N_c^2-1)}{32\pi^2 N_c} \int_{q^2}^{\Lambda^2} \frac{\bar{g}^2(p^2) B^{(\sigma)}(p^2)}{p^2} dp^2 + \frac{3(N_c^2-1)}{32\pi^2 N_c q^2} \bar{g}^2(q^2) \int_{q_0^2}^{q^2} B^{(\sigma)}(p^2) dp^2 \ , \quad (12.82)$$

shows the following characteristic point : if only the first term (describing the interactions in the high energy region with $p^2 > q^2$) was retained on the right-hand side of this equation, its solution would be the function $C_1 B_1^{(\sigma)}(q^2)$ with the irregular asymptotics; on the other hand, if only the second term was retained there, the solution would be the "right" function $C_2 B_2(q^2)$. This fact indicates that the dynamics in the infrared region is responsible for spontaneous chiral symmetry breaking in QCD. In particular, this implies that in order to describe this phenomenon, it is important to guess the behavior of the running coupling $\bar{g}^2(q^2)$ at low momenta. We will return to this point in the next section.

b) PCAC : $\lim\limits_{\Lambda\to\infty} \partial^\mu j_{5\mu}^\alpha = \frac{i}{2}\bar{\psi}\left\{\hat{m}_{\rm cur}^{(\sigma)}, \lambda^\alpha\right\}\gamma_5\psi \neq 0;\ m_{{\rm cur};a}^{(\sigma)} = Z_m^{(\sigma)^{-1}} m_a^{(0)} \neq 0.$

In this case the boundary condition (12.77) leads to the relation (compare with Eq.(12.47)) :

$$m(q^2) \to \frac{C}{q^2}\left(\ln q^2/\Lambda_{QCD}^2\right)^{c_m - 1} + m_{{\rm cur};a}^{(\sigma)} \left(\frac{\ln q^2/\Lambda_{QCD}^2}{\ln \sigma^2/\Lambda_{QCD}^2}\right)^{-c_m} . \quad (12.83)$$

12.6. Two Scales in QCD Nonperturbative Dynamics. Dynamics Underlying Spontaneous Chiral Symmetry Breaking in QCD

As was indicated at the end of the previous section, the dynamics responsible for spontaneous chiral symmetry breaking in QCD is formed at low momenta. It is important to guess the characteristic features of this nonperturbative dynamics. In particular, it is important to estimate the characteristic momentum scale of the dynamics. This in turn will help us in constructing a simple model of this phenomenon.

Since QCD is a real theory of real interactions, its dynamics is necessary rich and complicated. Therefore, model building in QCD is of primary importance. The aim of this model building is to construct simple and, at the same time, reasonably well working models describing nonperturbative dynamics. Different nonperturbative phenomena can be described by different models. This is in fact the reason why the models can be relatively simple : QCD is complicated because it describes a lot of nonperturbative phenomena (including nuclear physics); on the other hand, a concrete model describes only a small piece of the QCD dynamics, and it may therefore be relatively simple.

Our aim here is to construct models describing the dynamics of spontaneous chiral symmetry breaking of hadrons. This implies that the models have to describe the dynamics of *light* hadrons, such as pseudoscalar mesons, but they may be irrelevant for the description of the dynamics of heavier hadrons.

At present, there exists a whole "industry" studying spontaneous chiral symmetry breaking in hadronic physics. Our aim is not to review the whole subject but rather to discuss the key points of this dynamics. The discussion will be based on a particular class of models which are rather simple and elaborated. In particular, the models lead to a low energy effective action describing rather well the dynamics of light hadrons.

We begin by considering the approach to the chiral dynamics in QCD elaborated by the Kiev group [17, 18, 19]. Similar models were also considered by Goldman and Haymaker [20, 21], and by Finger and Mandula [22].

Let us recall that dynamics of chiral symmetry breaking is intimately connected with the dynamics forming NG bosons (pseudoscalar mesons in QCD). These hadrons are very specific. Our initial assumption is that the dynamics forming these light hadrons is insensitive to confinement forces responsible for the formation of heavy (unstable) hadrons. The physical picture underlying this hypothesis is that the size of pseudoscalar mesons is rather small, and they are insensitive to confinement forces dominating at larger distances.

This assumption implies a certain factorization of the dynamics of pseudoscalar mesons from the dynamics of the rest of hadrons. It is clear that this is a simplified picture of real QCD. However, as we will see later, it leads to a rather successful description of the chiral dynamics in QCD.

The concrete model based on this assumption is the following. We assume that the chiral dynamics is formed at momenta q satisfying $\delta^2 \lesssim q^2 \lesssim \Lambda_\chi^2$, where $\delta^2 \gtrsim \Lambda_{QCD} \sim 300MeV$ and $\Lambda_\chi \sim 1GeV$. Thus the scale Λ_χ is a factor of three or so larger than the confinement scale Λ_{QCD}.

Let us now return to equation (12.63) for the quark mass function. As we know (see Sec.9.2), at $m_a^{(0)} = 0$, this equation is equivalent to the BS equation for NG bosons. The crucial point is how should one model the running coupling in the region with q^2 satisfying $\delta^2 \lesssim q^2 \lesssim \Lambda_\chi^2$. Our model for $\bar{g}^2(q^2)$ in that region will be the simplest one : we will take $\bar{g}^2(q^2)$ as just a constant g_s^2 which equals to some averaged value of $\bar{g}^2(q^2)$ there. It is clear that this hypothesis implies that $\bar{g}^2(q^2)$ should be a rather smooth function of q^2 in that region.

As we have already known, the dynamical mass function, and therefore the BS wave function of NG bosons, decreases rapidly at $q^2 \gtrsim \Lambda_\chi^2 > \Lambda_{QCD}^2$. Therefore it seems reasonable to assume the following simple ansatz for $m_{dyn}(q^2)$:

$$m_{dyn} = 0 \text{ at } q^2 \gtrsim \Lambda_\chi^2 \tag{12.84}$$

and

$$m_{dyn}(q^2) = m_{dyn} \text{ at } \delta^2 \lesssim q^2 \lesssim \Lambda_\chi^2 \tag{12.85}$$

where m_{dyn} is an averaged value of $m_{dyn}(q^2)$ in that region.

Using now in this model the kernel (12.65) with the running coupling $\bar{g}^2$ described above, we find that the model reduces to the simple models considered in Chapters 6, 9, and 10. Indeed, as the infrared cutoff δ goes to zero, we recover the ladder (rainbow) QED with $\alpha = \frac{(N_c^2-1)}{8\pi N_c} g_s^2$ and cutoff $\Lambda = \Lambda_\chi$. On the other hand, since the infrared cutoff δ plays the role of an effective gluon mass, we find that if δ is not much smaller than the scale Λ_χ, the model is just a Nambu-Jona-Lasinio (NJL)-like model.

In particular, it is clear that in both cases spontaneous chiral symmetry breaking occurs only if the coupling constant g_s^2 exceeds a critical value $g_c^2 > 0$.

Let us add few more comments about the infrared cutoff δ. As we already noted above, it plays the role of an effective gluon mass. The necessity of the appearance of such a mass in QCD was advocated by Cornwall [23] (see also earlier papers [24, 25]). It may also happen that the cutoff δ is not very important for describing NG bosons. As we showed in Sec.10.3, the only essential difference between the NJL dynamics and the Coulomb-like dynamics is in the type of the spectrum of the radial excitations of NG bosons : while there are no such radial excitations in the NJL model, the Coulomb-like long range forces lead to an infinite number of them. However, since the dynamics responsible for the formation of the excitations is connected with rather long distances where both models are hardly reliable, this difference is not in fact essential. The point is that the supercritical dynamics forming NG bosons in these two models are rather similar.

The program of an improvement of the model considered above by introducing a running coupling of the form (12.70) in the model was initiated in Refs.[9, 16, 19]. In particular, Higashijima [16, 26] used the form (12.70) with the following ansatz for the running coupling :

$$\bar{g}^2(q^2) = \frac{2}{b}\frac{1}{\ln\left((q^2+q_c^2)/\Lambda_{QCD}^2\right)}, \tag{12.86}$$

where q_c^2 is a free parameter (see also Refs.[27, 28]). The numerical analysis of the equation (12.58) with the BS kernel (12.65) and the ansatz (12.70), (12.86) for the running coupling leads to the picture agreeing with the simple model considered above. If q_c is rather low, i.e. $\bar{g}^2(q^2)$ is rather large at $q^2 \sim q_c^2 \gtrsim \Lambda_{QCD}^2$, spontaneous chiral symmetry breaking occurs. The mass function $m_{dyn}(q^2)$ is of course rapidly decreasing at large q^2, and is a rather smooth function in the infrared region.

Numerous phenomenological applications of this model were considered by the Florence-Geneve group [28].

We also note that the hypothesis about the two different scales in QCD, Λ_χ and Λ_{QCD} with $\Lambda_\chi > \Lambda_{QCD}$, was also considered by Callan, Dashen and Gross in the framework of instanton dynamics [29] (see also Ref.[30]). It was further developed by Shuryak [31, 32], and by Dyakonov and Petrov [33, 34]. It is interesting that this approach eventually also leads to an effective non-local NJL-like model describing the dynamics of pseudoscalar mesons.

We also note that there exists a class of models based on the hypothesis that confinement forces are primarily responsible for triggering spontaneous chiral symmetry breaking in QCD [35 - 37]. Although this possibility is still certainly viable, these models have not yet lead to sufficiently riach phenomenological consequences which would allow one to examine such a scenario in a serious way.

12.7. More about the Dynamics of Chiral Symmetry Breaking in QCD

In the previous section, we described the main characteristics of the models of spontaneous chiral symmetry breaking in QCD based on the hypothesis of the existence of the two scales in the QCD nonperturbative dynamics. Here we will discuss these models in more detail.

In this class of models, the confinement forces are ignored in the description of the dynamics of spontaneous chiral symmetry breaking. The models are based on the following ansatz for the BS kernel (compare with Eqs.(12.60), (12.65)):

$$K_{mn,n'm'}(q,p) = -\frac{i}{4}\bar{g}^2(q,p)(t^K\gamma^\mu)_{nn'}(t^K\gamma^\nu)_{m'm}D_{\mu\nu}(q-p) . \tag{12.87}$$

Different models use different ansätze for the running coupling $\bar{g}(q,p)$ and gluon propagator $D_{\mu\nu}(q-p)$.*)

We note that the corresponding Schwinger-Dyson (SD) equation for the fermion (quark) propagator takes the form (compare with Eq.(8.61)) :

$$G_a^{-1}(q) = \hat{p} - m_a^{(0)} + \frac{i}{(2\pi)^4}\frac{N_c^2-1}{2N_c}\int d^4p\bar{g}^2(q,p)D_{\mu\nu}(q-p)\gamma^\mu G_a(p)\gamma^\nu . \tag{12.88}$$

The important point in the model building is to construct a simple model which, at the same time, would be consistent with such basic properties of QCD as asymptotic freedom, the Ward-Takahashi (WT) identities, etc..

*) Here we will not consider the models based on the dynamics of instantons [29 - 34]. As we already noted in the previous section, these models lead to non-local NJL-like effective action.

The simplest, and very popular, model is the extended Nambu-Jona-Lasinio (ENJL) model [39 - 42]. Its "derivation" proceeds in the following way. First, we assume that gluons acquire an effective mass $M_g \lesssim \Lambda_\chi$, where Λ_χ is the scale of spontaneous chiral symmetry breaking. Then, as in the case of the low energy effective theory of electroweak interactions, we use for the propagator $D_{\mu\nu}(q-p)$

$$D_{\mu\nu}(q-p) = \frac{-1}{M_g^2} g_{\mu\nu} \simeq -\frac{1}{\Lambda_\chi^2} g_{\mu\nu} \,. \tag{12.89}$$

The running coupling constant $\bar{g}^2(q,p)$ is chosen as $\bar{g}^2(q,p) = g_s^2\theta(\Lambda_\chi^2 - (q-p)^2)$, where here q, p are Euclidean momenta. Thus, $\bar{g}^2$ is just a constant in the low energy region.

The model is described by the following interaction lagrangian :

$$\mathcal{L}_{int}(x) = -\frac{g_s^2}{\Lambda_\chi^2} \bar{\psi}(x) \frac{t^K}{2} \gamma^\mu \psi(x) \bar{\psi}(x) \frac{t^K}{2} \gamma_\mu \psi(x) \,. \tag{12.90}$$

Using Fierz identities (see Sec.B.3 in Appendix B) and neglecting subleading terms in $1/N_c$, we come to the ENJL model :

$$\mathcal{L}_{ENJL} = i\bar{\psi}(\hat{\partial} - \hat{m}^{(0)}) + \mathcal{L}^{S,P} + \mathcal{L}^{V,A} \,, \tag{12.91}$$

where

$$\mathcal{L}^{S,P} = G_s^{(0)} \sum_{\alpha=0}^{N^2-1} \left\{ \left(\bar{\psi} \frac{\lambda^\alpha}{2} \psi \right)^2 + \left(\bar{\psi} \frac{\lambda^\alpha}{2} i\gamma_5 \psi \right)^2 \right\} \,, \tag{12.92}$$

$$\mathcal{L}^{V,A} = -G_V^{(0)} \sum_{\alpha=0}^{N^2-1} \left\{ \left(\bar{\psi} \gamma^\mu \frac{\lambda^\alpha}{2} \psi \right)^2 + \left(\bar{\psi} \gamma^\mu \gamma_5 \frac{\lambda^\alpha}{2} \psi \right)^2 \right\} \,, \tag{12.93}$$

with

$$G_s^{(0)} = 2G_V^{(0)} = \frac{g_s^2}{\Lambda_\chi^2} \,. \tag{12.94}$$

In practice, the coupling constants G_S and G_V are treated as independent free parameters. This is reasonable since some dynamical effects in QCD may certainly change the relation (12.94) based on the simple approximation (12.87) with $D_{\mu\nu}$ from Eq.(12.89).

A nice feature of this model is its simplicity. We will also see in Sec.12.13 that it describes the dynamics of light mesons rather well. However, it is clear that the model simplifies some important points of the QCD dynamics. In particular, the sharp cut-off Λ_χ removes completely the ultraviolet "tail" of the dynamics in the model. What we have in the ENJL model is the realization of the light meson dominance hypothesis in Green's functions of such composite operators as $\bar{\psi}\lambda^\alpha\psi, \bar{\psi}\gamma_5\lambda^\alpha\psi,\ \bar{\psi}\gamma_\mu\lambda^\alpha\psi$ and $\bar{\psi}\gamma_\mu\gamma_5\lambda^\alpha\psi$ (for a more detailed discussion of this point see Secs.12.10, 12.15).

Other popular models are the models based on the so called improved ladder approximation. In this approximation, one uses the BS kernel (12.87) with a free gluon propagator and with the following ansätze for $\bar{g}^2(q,p)$:

$$\bar{g}^2(q,p) = \bar{g}^2((q-p)^2) \tag{12.95}$$

or

$$\bar{g}^2(q,p) = \theta(q^2-p^2)\bar{g}^2(q^2) + \theta(p^2-q^2)\bar{g}^2(p^2) \tag{12.96}$$

(in Euclidean region). We emphasize that while the ansatz (12.96) was used in the ultraviolet region in asymptotic equations in the previous section, here it is used in the whole dynamical region. We will return to this point later.

The function $\bar{g}^2(q,p)$ in Eqs.(12.95), (12.96) is chosen in such a way that as $\max(q^2,p^2)\to\infty$, it coincides with the one-loop expression (12.10) :

$$\bar{g}^2(q,p) \longrightarrow \frac{2}{b\ln\left(\frac{\max(q^2,p^2)}{\Lambda_{QCD}^2}\right)}\,. \tag{12.97}$$

These models improve the ENJL model in that they correctly reproduce the ultraviolet behavior of the BS wave functions (form factors) of hadrons. The important point in these models is the choice of the "best" gauge. As in the case of the ladder approximation in QED, the "best" means here the gauge in which the appropriate WT identities are satisfied.

However, the situation with the choice of the gauge in the case of non-Abelian gauge theories, such as QCD, is more complicated. The point is that the Slavnov-Taylor (ST) identity takes place for the fermion-antifermion-gluon vertex in these theories [43]. In covariant gauges, the ST identity includes a contribution connected with the Faddeev-Popov ghost fields. Unfortunately, at present, nobody can use this identity beyond perturbative theory.

A heuristic approach to this problem is just to neglect the ghost contribution both in ST identity and in equations for Green's functions in the improved ladder approximation. This corresponds to abelianizing the theory: the dynamics of the ghost fields and that of self-interaction of gluon fields manifest themselves only in the running coupling $\bar{g}^2$ in this approximation. At present, the only justification of this approach is that it leads to reasonable results.

Let us discuss now the concrete results obtained in the improved ladder approximation with the ansätze (12.95) and (12.96) for the running coupling.

We begin by considering the SD equation (12.88). It leads to the following two equations for the functions $A_a(q^2)$ and $B_a(q^2)$ in Landau gauge :

$$A_a(q^2) = 1 - i\frac{N_c^2-1}{2N_c}\int \frac{d^4p}{(2\pi)^4}\frac{3A_a(p^2)}{A_a^2(p^2)p^2 - B_a^2(p^2)}\frac{\bar{g}^2(q,p)}{(q-p)^2}\left[\frac{qp}{q^2} - \frac{2}{3} + \frac{2}{3}\frac{(q\cdot(p-q))^2}{q^2(p-q)^2}\right], \tag{12.98}$$

$$B_a(q^2) = m_a^{(0)} - i\frac{N_c^2-1}{2N_c}\int \frac{d^4p}{(2\pi)^4}\frac{3B_a(p^2)}{A_a^2(p^2)p^2 - B_a^2(p^2)}\frac{\bar{g}^2(q,p)}{(q-p)^2}. \tag{12.99}$$

In the case of the ansatz (12.96), because of the relation (10.57), we find, after the angle integration, that $A_a(q^2) = 1$. However, $A_a(q^2)$ is not equal to one for the ansatz (12.95). More precisely, the situation is the following. There are two forms of $\bar{g}^2((q-p)^2)$ used in the concrete calculations :

$$\bar{g}^2(k^2) = \frac{2}{b}\frac{1}{\ln\left((k^2+k_c^2)/\Lambda_{QCD}^2\right)} \tag{12.100}$$

(see.Eq.(12.86)) or

$$\bar{g}^2(k^2) = \frac{2}{b}\left[\theta(k_c^2-k^2)\frac{1}{\ln k_c^2/\Lambda_{QCD}^2} + \theta(k^2-k_c^2)\frac{1}{\ln k^2/\Lambda_{QCD}^2}\right]. \tag{12.101}$$

(the equations are written in Euclidean space; k_c^2 is a free parameter). The system of equations (12.98), (12.99) with the ansatz $\bar{g}^2(q,p) = \bar{g}^2((q-p)^2)$, where $\bar{g}^2(k^2)$, $k^2 = (q-p)^2$, is given in Eq.(12.100) or in Eq.(12.101), was numerically analysed by

Williams, Krein and Roberts [44] and by Jain and Munczek [45]. They found that although $A_a(q^2) \longrightarrow 1$ as $q^2 \gg \Lambda^2_{QCD}$, it is closer to 1.5 - 2 in the infrared region. Since in this approximation one uses the bare quark-antiquark-gluon vertex, this result implies that the WT identity $Z_1^{(\sigma)} = Z_2^{(\sigma)}$ is violated.

There are few possibilities to react to this result. One is just to ignore the violation of the WT identity : after all, it is not a "right" identity in QCD, and, as was noted above, the right one, the ST identity, cannot be practically used in this nonperturbative problem.

This viewpoint was adopted by the authors of Refs.[44 - 46]. In particular, Jain and Munczek [46] modified the form (12.100) for

$$\bar{g}^2(k^2) = \frac{2}{b}\frac{1}{\ln\left((k^2+k_c^2)/\Lambda^2_{QCD}\right)} + \eta k^2\delta^4(k) + ak^2\left(1-\frac{2k^2}{3k_0^2}\right)e^{-k^2/k_0^2} \qquad (12.102)$$

(η, a, k_0^2 are new free parameters) and applied the BS equations with such a kernel to describe properties of pseudoscalar $\bar{q}q$ bound states composed both from light (u, d, s) and heavy (c, b) quarks and antiquarks. They calculated bound-state masses, leptonic decay constants, and wave functions of light-light, light-heavy, and heavy-heavy $\bar{q}q$ bound states and found that the results agree rather well with experimental data.*)

The principal result of these studies is that, because of a rapid decreasing of the BS wave functions of bound states in QCD (independently of the values of the masses of their constituents q and $\bar{q}$), the main characteristics of the bound states are determined in the infrared region with $p^2 \lesssim 1(GeV)^2$, i.e. both relativistic and nonrelativistic hadronic bound states are loose ones (compare with the discussion in Sec.12.4).

Another viewpoint concerning the role of the WT identity in the improved ladder approximation was adopted by the Kyoto group [47 - 49]. These authors believe that even although the ST identity does not lead to the exact relation $Z_2^{(\sigma)} = Z_1^{(\sigma)}$ in QCD (as the WT identity does in QED), the relation $Z_2^{(\sigma)} = 1$ must be satisfied in a consistent improved ladder approximation. Their, somewhat hand-waving, argument goes as follows [49] : the running coupling $\bar{g}^2(\sigma)$ is, by definition, the coupling strength between fields normalized with weight one at the scale σ ; since on the other hand,

---

*) The second and third terms in the expression (12.102) for $\bar{g}^2(k^2)$ are essential for the dynamics of light-heavy and heavy-heavy $\bar{q}q$ bound states.

in the improved ladder approximation the running coupling is used at all scales, the consistency of the approximation implies using quark and gluon fields satisfying $Z_2^{(\sigma)} = 1$ and $Z_3^{(\sigma)} = 1$ independent of the scale σ. This requires $A(q^2) = 1$ (since the free form gluon propagator is used in this approximation, the condition $Z_3^{(\sigma)} = 1$ is automatic).

The simplest way to satisfy $A(q^2) = 1$ is to use the ansatz (12.96) in Landau gauge. This ansatz was used by Aoki, Kugo and Mitchard [48] for the description of the spectrum of the lowest lying pseudoscalar, scalar, vector and axial-vector mesons. The results agree with experimental data rather well.

However as was pointed out by Jain and Munczek [50], the ansatz (12.96) violates the WT identities for the axial currents (see Eq.(9.34)). In order to get the improved ladder approximation with $A(q^2) = 1$ and in which, at the same time, the axial WT identities were satisfied, Kugo and Mitchard [49] suggested to use a new (non-local) gauge in which $A(q^2) = 1$ takes place even for the ansatz (12.95).*)

The gluon propagator takes the following form in this gauge :

$$D_{\mu\nu}^{KL}(p) = \delta_{KL}\left(\delta_{\mu\nu} - \eta(p^2)\frac{p_\mu p_\nu}{p^2}\right)\frac{1}{p^2}, \tag{12.103}$$

where

$$\eta(p^2) = \frac{2}{(p^2)^2 \bar{g}^2(p^2)} \int_0^{p^2} dx \left[x\bar{g}^2(x) - x^2(\bar{g}^2(x))'\right] \tag{12.104}$$

(these equations are written in Euclidean region).**) It is easy to check that for $\bar{g}^2(p^2)$ from Eq.(12.100) or Eq.(12.101), $\eta(p^2)$ approaches the Landau gauge value $\eta(p^2) = 1$ both at $p^2 = 0$ and $p^2 \to \infty$.

The numerical analysis of the SD and BS equations in this gauge with the ansatz (12.95), (12.101) has shown [49] that the results are similar to those obtained earlier in Landau gauge with the ansatz (12.96), (12.101).

The present discussion reflects the general situation in the QCD model building : there are a lot of competing models based on similar assumptions but differing in detail,

---

*) They also showed that this ansatz automatically provides the validity of the axial WT identities in the improved ladder approximation.

**) This gauge belongs to a class of gauges with $A(q^2) = 1$ which was described earlier in Ref.[51].

and all of them work quite well ! This of course reflects our still rather poor understanding of the nonperturbative dynamics in QCD. On the other hand, the success of these models in the description of the dynamics of hadrons indicates that we apparently are on the right track. We will discuss possible reasons of this success in Sec.12.15.

In this situation, it is important to select the key features of the chiral dynamics in QCD from model dependent ones. It would be also useful to compare the results obtained in these models with those obtained in other approaches to the nonperturbative dynamics in QCD, such as QCD sum rules [52] for example (see Sec.12.15).

In the next sections, we will consider the simplest phenomenological consequences obtained in these models.

12.8. Masses of Pseudoscalar Mesons

The PCAC mass relation for pseudoscalar mesons takes the form (see Eq.(12.51)) :

$$M_{ab}^{(p)^2} = -\frac{\langle(\bar{\psi}\psi)^{(\sigma)}\rangle}{F^2}\left(m_{\text{cur;a}}^{(\sigma)} + m_{\text{cur;b}}^{(\sigma)}\right) , \tag{12.105}$$

where the decay constant F is determined from the decay process $\pi^- \longrightarrow \mu + \bar{\nu}_\mu$, and is $F \simeq 93.3MeV$.

We will consider the nine pseudoscalar mesons composed of the three lightest quarks and antiquarks u, d, s and $\bar{u}, \bar{d}, \bar{s}$.

If we use the $SU_V(3)$-covariant wave functions for pseudoscalar mesons,

$$\begin{aligned}
\pi^+ &= |u\bar{d}\rangle, \ \pi^- = |d\bar{u}\rangle , \\
\pi^0 &= \frac{1}{\sqrt{2}}\left(|u\bar{u}\rangle - |d\bar{d}\rangle\right), \ K^0 = |d\bar{s}\rangle, \ \bar{K}^0 = |s\bar{d}\rangle , \\
K^- &= |s\bar{u}\rangle, \ K^+ = |u\bar{s}\rangle , \\
\eta_8 &= \frac{1}{\sqrt{6}}\left(|u\bar{u}\rangle + |d\bar{d}\rangle - 2|s\bar{s}\rangle\right) , \\
\eta_1 &= \frac{1}{\sqrt{3}}\left(|u\bar{u}\rangle + |d\bar{d}\rangle + |s\bar{s}\rangle\right) ,
\end{aligned} \tag{12.106}$$

then Eq.(12.105) will yield the following mass relations

$$\begin{aligned}
M^2(\pi^\pm) &= -\frac{\langle(\bar{\psi}\psi)^{(\sigma)}\rangle}{F^2}\left(m_{\text{cur};u}^{(\sigma)} + m_{\text{cur};d}^{(\sigma)}\right) \\
M^2(\pi^0) &= -\frac{\langle(\bar{\psi}\psi)^{(\sigma)}\rangle}{F^2}\left(m_{\text{cur};u}^{(\sigma)} + m_{\text{cur};d}^{(\sigma)}\right) ,
\end{aligned} \tag{12.107}$$

$$M^2(K^\pm) = -\frac{\langle(\bar\psi\psi)^{(\sigma)}\rangle}{F^2}\left(m^{(\sigma)}_{\text{cur};u} + m^{(\sigma)}_{\text{cur};s}\right)$$
$$M^2(K^0) = M^2(\bar K^0) = -\frac{\langle(\bar\psi\psi)^{(\sigma)}\rangle}{F^2}\left(m^{(\sigma)}_{\text{cur};d} + m^{(\sigma)}_{\text{cur};s}\right)\,, \tag{12.108}$$

$$M^2(\eta_8) = -\frac{\langle(\bar\psi\psi)^{(\sigma)}\rangle}{3F^2}\left(m^{(\sigma)}_{\text{cur};u} + m^{(\sigma)}_{\text{cur};d} + 4m^{(\sigma)}_{\text{cur};s}\right)$$
$$M^2(\eta_1) = -\frac{2\langle(\bar\psi\psi)^{(\sigma)}\rangle}{3F^2}\left(m^{(\sigma)}_{\text{cur};u} + m^{(\sigma)}_{\text{cur};d} + m^{(\sigma)}_{\text{cur};s}\right)\,. \tag{12.109}$$

Because of the ABJ anomaly (12.8), the relations (12.109) for the masses of η_1 and η_8 should be essentially modified. We will return to this point later. Now we will analyse the four relations (12.107), (12.108).

Using the known relations $M^2_{em}(\pi^0) = M^2_{em}(K^0) = M^2_{em}(\bar K^0) = 0$, $M^2_{em}(\pi^\pm) = M^2_{em}(K^\pm)$ [53] to extract the contribution of the electromagnetic interactions into the meson masses, we get the following ratios for the quark masses [54] :

$$\frac{m^{(\sigma)}_{\text{cur};d}}{m^{(\sigma)}_{\text{cur};u}} = \frac{M^2(K^0) - M^2(K^\dagger) + M^2(\pi^\dagger)}{2M^2(\pi^0) + M^2(K^\dagger) - M^2(K^0) - M^2(\pi^\dagger)} \simeq 1.8\,, \tag{12.110}$$

$$\frac{m^{(\sigma)}_{\text{cur};s}}{m^{(\sigma)}_{\text{cur};d}} = \frac{M^2(K^0) + M^2(K^\dagger) - M^2(\pi^\dagger)}{M^2(K^0) - M^2(K^\dagger) + M^2(\pi^\dagger))} \simeq 20.1\,. \tag{12.111}$$

Since the renormalization constant $Z^{(\sigma)}_m$ is canceled in these ratios, they are independent of the renormalization-group scale σ.

The current masses $m^{(\sigma)}_{\text{cur};a}$ are usually evaluated at $\sigma = 1GeV$. For the s quark, $m_{\text{cur};s} \equiv m^{(\sigma)}_{\text{cur};s}|_{\sigma=1GeV}$ extracted from experimental data is $m_{\text{cur};s} \simeq 199 \pm 33MeV$ [55]. Then the relations (12.110) and (12.111) yield $m_{\text{cur};u} \simeq 5.6{\pm}1.1MeV$ and $m_{\text{cur};d} \simeq 9.9{\pm}1.1MeV$. From Eq.(12.89) we also find the condensate $\langle(\bar\psi\psi)\rangle_1 \equiv \langle(\bar\psi\psi)^{(\sigma)}\rangle|_{\sigma=1GeV}$:

$$\langle(\bar\psi\psi)\rangle_1 \simeq -(207-228)^3 MeV^3\,. \tag{12.112}$$

Let us look at these mass relations in the models considered in the previous section. They have the same form both in the NJL-like model (the infrared cutoff $\delta \gtrsim \Lambda_{QCD}$)

and in the QED-like model (with $\delta = 0$) :

$$M_{ab}^{(p)^2} \simeq 2m_{dyn}\left(m_{\text{cur};a} + m_{\text{cur};b}\right) \tag{12.113}$$

(see Eqs.(9.131), (10.53)). Comparing this relation with Eq.(12.105), we find

$$\langle(\bar{\psi}\psi)\rangle_1 \simeq -2m_{dyn}F^2 = -(173 - 240)MeV^3 \tag{12.114}$$

if we take the standard value for $m_{dyn}, m_{dyn} = (300 - 400)MeV$. Thus, the models lead to a reasonable value for the chiral condensate.

Let us now discuss the mass relations for η_8 and η_1 mesons. Because of the ABJ anomaly, the dynamics of these mesons is rather specific. Since the pioneering paper of ‘t Hooft [56], this problem has been extensively discussed in the literature. For our purposes, the most appropriate approach is a phenomenological approach considered in Refs.[57, 58]. The main point of this approach is the old observation [59] that the $SU(3)$ singlet η_1 can mix with two-gluon bound states (glueballs). This in turn leads to the modification of the mass relation (12.109) for η_1 :

$$M^2(\eta_1) = -\frac{2\langle(\bar{\psi}\psi)^{(\sigma)}}{3F^2}\left(m_{\text{cur};u}^{(\sigma)} + m_{\text{cur};d}^{(\sigma)} + m_{\text{cur};s}^{(\sigma)}\right) + \lambda_\eta \ , \tag{12.115}$$

where λ_η is a parameter which can be in principle calculated in QCD ; we will treat it, following Ref.[58], as a free phenomenological parameter.

At $\lambda_\eta \neq 0$, the perturbative mass term

$$\Delta L = - \sum_{a=u,d,s} m_a^{(0)}\bar{\psi}_a\psi_a \tag{12.116}$$

yields the following mass matrix for η_8 and η_1 mesons in the basis (12.106) (we neglect the isospin violation) :

$$\begin{pmatrix} M^2(\eta_8) & M^2(\eta_{81}) \\ M^2(\eta_{18}) & M^2(\eta_1) \end{pmatrix} = \begin{pmatrix} \frac{4}{3}M^2(K) - \frac{1}{3}M^2(\pi) & -\frac{2\sqrt{2}}{3}\left(M^2(K) - M^2(\pi)\right) \\ -\frac{2\sqrt{2}}{3}\left(M^2(K) - M^2(\pi)\right) & \frac{2}{3}M^2(K) + \frac{1}{3}M^2(\pi) + \lambda_\eta \end{pmatrix} . \tag{12.117}$$

We determine the parameter λ_η from the condition that the sum $M^2(\eta) + M^2(\eta')$, where η and η' are the eigenstates of the matrix (12.117), is to be equal to the experimental value $1.22(GeV)^2$. Then, the diagonalization of this matrix yields : $M^2(\eta) \simeq$

$0.27(GeV)^2$, $M^2(\eta') \simeq 0.95(GeV)^2$, $\theta \simeq -14^0$, where θ is the mixing angle : $|\eta\rangle = \cos|\eta_8\rangle - \sin\theta|\eta_1\rangle$, $|\eta'\rangle = \cos\theta|\eta_1\rangle + \sin\theta|\eta_8\rangle$. These results agree rather well with the experimental values : $M^2(\eta) \simeq 0.3(GeV)^2$, $M^2(\eta') \simeq 0.92(GeV)^2$ and $\theta \simeq -20^0$ [55].

We will return to the problem of η' meson in Sec.12.12.

12.9. The Pagels-Stokar formula for the Decay Constant

In Sec.9.4, we derived the exact relation for the decay constant. In the chiral limit, it is (see Eq.(9.75)) :

$$F = \frac{N_c}{32\pi^2}\int duu\left(4\chi_2^{(p)}(u) - u\chi_3^{(p)}(u)\right) , \tag{12.118}$$

where $u = q^2$, and the functions $\chi_i^{(p)}(u)$ are defined in Eq.(9.74).

In this section, we will derive the approximate form of this relation known as the Pagels-Stokar (PS) formula in the literature [60].

The PS formula corresponds to the approximation with

$$G^{-1}(p) = \hat{p} - B_{dyn}(p^2) \tag{12.119}$$

and

$$\tilde{\chi}^{(p);\alpha}(q,P) \equiv G^{-1}\left(q+\frac{P}{2}\right)\chi^{(p);\alpha}(q,P)G^{-1}\left(q-\frac{P}{2}\right) = \lambda^\alpha\gamma_5\frac{B_{dyn}(q^2)}{F} \tag{12.120}$$

(see Eqs.(9.38) and (9.101)). Here $\chi^{(p);\alpha}(q,P)$ is the BS wave function of massless pseudoscalar mesons ($P^2 = 0$) (see Eq.(9.37)). We will discuss the meaning of this approximation at the end of this section.

Eqs.(12.119) and (12.120) imply that the functions $\chi_i^{(p)}$ in the expansion

$$\chi^{(p);\alpha}(q,P) = \left[\chi_1^{(p)} + \chi_2^{(p)}\hat{P} + \chi_3^{(p)}\hat{q} + \chi_4^{(p)}\sigma^{\mu\nu}(P_\mu q_\nu - P_\nu q_\mu)\right]\frac{\lambda^\alpha}{2}\gamma_5 \tag{12.121}$$

can be expressed through B_{dyn}. At $P_\mu = 0$, in Euclidean space, we obtain :

$$\chi_1^{(p)}(q^2) = \frac{2}{F}\frac{1}{(q^2 + B_{dyn}^2(q^2))}B_{dyn}(q^2) , \tag{12.122}$$

$$\chi_2^{(p)}(q^2) = \frac{2}{F}\frac{1}{(q^2+B_{dyn}^2(q^2))^2}B_{dyn}^2(q^2)\ , \tag{12.123}$$

$$\chi_3^{(p)}(q^2) \equiv \lim_{P_\mu\to 0}\frac{\chi_3^{(p)}(q^2,Pq)}{Pq} = \frac{4}{F}\frac{1}{(q^2+B_{dyn}^2(q^2))^2}B'_{dyn}(q^2)B_{dyn}(q^2)\ , \tag{12.124}$$

$$\chi_4^{(p)}(q^2) = \frac{2i}{F}\frac{1}{\left(q^2+B_{dyn}^2(q^2)\right)^2}B_{dyn}(q^2)\ . \tag{12.125}$$

Substituting the expressions (12.123) and (12.124) for $\chi_2^{(p)}$ and $\chi_3^{(p)}$ into Eq.(12.118), we obtain the PS formula :

$$F^2 = \frac{N_c}{8\pi^2}\int_0^\infty duu\frac{B_{dyn}(u)\left[2B_{dyn}(u)-uB'_{dyn}(u)\right]}{\left(u+B_{dyn}^2(u)\right)^2}\ . \tag{12.126}$$

Let us discuss the assumptions (12.119) and (12.120), which the PS formula is based on, in the improved ladder approximation with the BS kernel given in Eq.(12.87). The running coupling constant takes the form (12.95) or (12.96) with $\bar{g}^2(k^2)$ defined in Eq.(12.100) or Eq.(12.101).

First of all, it is clear that in order to get $A(q^2)=1$, one should work in Landau gauge with the ansatz (12.96), or in non-local gauge (12.103) with the ansatz (12.95).

The assumption (12.120) is equivalent to that in the expansion

$$\tilde{\chi}^\alpha(q,P) = \left[\tilde{\chi}_1^{(p)} + \tilde{\chi}_2^{(p)}\hat{P} + \tilde{\chi}_3^{(p)}(Pq)\hat{q} + \tilde{\chi}_4^{(p)}\sigma^{\mu\nu}(P_\mu q_\nu - P_\nu q_\mu)\right]\frac{\lambda^\alpha}{2}\gamma_5\ , \tag{12.127}$$

the function $\tilde{\chi}_2^{(p)}$, $\tilde{\chi}_3^{(p)}$ a $\tilde{\chi}_4^{(p)}$ become equal zero as $P_\mu \longrightarrow 0$.

In the improved ladder approximation, the BS equations for $\tilde{\chi}^\alpha$ were studied in Ref.[47]. The equations for the functions $\tilde{\chi}_i^{(p)}$, $i=2,3,4$, can be written in the following symbolical form at $P_\mu = 0$ (in Landau gauge) :

$$(1+\hat{g}^2 I[B_{dyn}])\begin{pmatrix}\tilde{\chi}_2^{(p)}\\ \tilde{\chi}_3^{(p)}\\ \tilde{\chi}_4^{(p)}\end{pmatrix} = \hat{g}^2 C[B_{dyn}]\ , \tag{12.128}$$

where I and C are functionals of B_{dyn}, and the "operator" $\hat{g}^2$ here represents $\bar{g}^2(v)$ in the integration region $\int_0^v$ and $\bar{g}^2(u)$ in $\int_v^{\Lambda^2} du$, respectively.

Considering Eq.(12.128) as an inhomogeneous equation for $\tilde{\chi}_i^{(p)}$ with a given B_{dyn}, we conclude that as $\hat{g}^2 \longrightarrow 0$, the functions $\tilde{\chi}_i^{(p)}$, with $i = 2, 3, 4$, are suppressed with respect to B_{dyn} (at $\hat{g}^2 = 0$, the only solution of this equation is $\tilde{\chi}_i^{(p)} = 0$).

Is this a justification of the PS formula ? Since spontaneous chiral symmetry breaking in QCD implies that $\bar{g}^2(q^2)$ is not small at low momenta, one can hardly consider this argument as a strong one. Moreover, beyond the improved ladder approximation, the situation becomes even more uncertain [45, 50].

Nevertheless, numerical studies of the PS formula indicates that it works not so badly [49]. Although, at present, there is not complete agreement concerning this formula in the literature, the PS formula deserves what it deserves: it is an useful heuristic relation.

12.10. The Low Energy Effective Action in the Extended Nambu-Jona-Lasinio Model: Preliminaries

So far we have considered the "microscopic" dynamics of spontaneous chiral symmetry breaking in QCD. The main objects which we have studied were the quark propagator and the BS wave function of mesons. In reality, however, we are often interested in a more direct description of hadronic dynamics. In particular at low energies, when one can neglect the structure of hadrons, we are interested in a low energy effective action in which hadrons are treated as structureless, point-like, particles.

In Chapter 5, we considered the general theory of effective lagrangians. In Sec.12.13, we will apply this theory to QCD. In this section, we begin the derivation of the low energy effective action in the ENJL model (12.91) - (12.93) [40, 41]. This will help us to establish the connection of the effective lagrangian with the microscopic dynamics in QCD.

As in the case of the NJL model considered in Chapter 6, we introduce the collective hadron fields in the ENJL model. But now, besides the field M describing scalars and pseudoscalars, we also introduce the collective fields V_μ and A_μ describing vector and axial-vector mesons. Thus, we rewrite the generating functional of the ENJL model as (compare with Eq.(6.17)):

$$Z(J) = C \int \mathcal{D}M\mathcal{D}M^\dagger\mathcal{D}V\mathcal{D}A \exp\left\{i\int d^4x\left[-\frac{1}{2G_s^{(0)}}tr\left(\left(M-\hat{m}^{(0)}\right)^\dagger\left(M-\hat{m}^{(0)}\right)\right)\right.\right.$$
$$\left.\left. + \frac{1}{2G_V^{(0)}}tr(V_\mu^2 + A_\mu^2) + F(J)\right]\right\} Z_f(M,V,A)\,, \tag{12.129}$$

where $F(J) = \sum_i J_i(x)\varphi_i(x)$ is a source term for some local operators, and

$$Z_f(M,V,A) = \int \mathcal{D}\psi\mathcal{D}\bar{\psi}\exp\left\{i\int d^4x\bar{\psi}i\mathfrak{D}\psi\right\} = C(\mathrm{Det}\, i\mathfrak{D})^{N_C} \tag{12.130}$$

with

$$i\mathfrak{D} = i\gamma^\mu\left(\partial_\mu - iV_\mu - iA_\mu\gamma_5\right) - M\frac{1-\gamma_5}{2} - M^\dagger\frac{1+\gamma_5}{2}\,. \tag{12.131}$$

(see Eq.(C.32) in Appendix C). We note that the field M includes the mass matrix $\hat{m}^{(0)}$.

The matrix fields M, V_μ and A_μ are (compare with Eq.(4.51)):

$$M = S + iP\ ;\ S = \sqrt{2}\sum_{\alpha=0}^{N^2-1} S^\alpha\frac{\lambda^\alpha}{2}\,,\ P = \sqrt{2}\sum_{\alpha=0}^{N^2-1} P^\alpha\frac{\lambda^\alpha}{2}\,, \tag{12.132}$$

$$V_\mu = \sum_{\alpha=0}^{N^2-1} V_\mu^\alpha\frac{\lambda^\alpha}{2}\,,\ A_\mu = \sum_{\alpha=0}^{N^2-1} A_\mu^\alpha\frac{\lambda^\alpha}{2}\,. \tag{12.133}$$

Thus, in the ENJL model, the derivation of the effective action for the collective fields M, V_μ and A_μ is reduced to the calculation of the determinant of the operator $i\mathfrak{D}$. This is a great advantage of this model as compared to other, more complicated, models, like, for example, those based on the improved ladder approximation (see Secs.12.6, 12.7). The point is that there exist regular methods to study the determinant of such operators [61 - 66].

The action $\bar{\psi}i\mathfrak{D}\psi$ is invariant under the following $U_L(N)\times U_R(N)$ gauge transformations :

$$\psi_L(x)\to\psi_L'(x) = L(x)\psi_L(x),\ \ \psi_R(x)\to\psi_R'(x) = R(x)\psi_R(x)\,, \tag{12.134}$$

$$A_{\mu L}(x) \to A'_{\mu L}(x) = L(x)A_{\mu L}(x)L^{\dagger}(x) + iL(x)\partial_{\mu}L^{\dagger}(x) \,, \tag{12.135}$$

$$A_{\mu R}(x) \to A'_{\mu R}(x) = R(x)A_{\mu R}(x)R^{\dagger}(x) + iR(x)\partial_{\mu}R^{\dagger}(x) \,, \tag{12.136}$$

$$M(x) \to M'(x) = L(x)M(x)R^{\dagger}(x) \,. \tag{12.137}$$

Here $A_{\mu L}$ and $A_{\mu R}$ are :

$$\begin{aligned} A_{\mu L} &= V_{\mu} + A_{\mu} \\ A_{\mu R} &= V_{\mu} - A_{\mu} \,. \end{aligned} \tag{12.138}$$

However, although the action is invariant under these transformations, the determinant Det$i\mathfrak{D}$ is not. The point is that Det$i\mathfrak{D}$ is not properly defined object unless it is regularized. As was first shown by Bardeen [67], the regularization may preserve only a part of the gauge $U_L(N) \times U_R(N)$ symmetry. In particular, there are regularization schemes preserving the $U_{L+R}(N) \equiv U_V(N)$ vector symmetry. They are most interesting for physical applications.

Let us consider an infinitesimal $U_L(N) \times U_R(N)$ gauge transformation in such regularization schemes :

$$\begin{aligned} L(x) &\simeq 1 + i\varphi(x) + i\omega(x) \\ R(x) &\simeq 1 + i\varphi(x) - i\omega(x) \,, \end{aligned} \tag{12.139}$$

where

$$\varphi(x) = \sum_{\alpha=0}^{N^2-1} \varphi^{\alpha}(x)\frac{\lambda^{\alpha}}{2}, \; \omega(x) = \sum_{\alpha=0}^{N^2-1} \omega^{\alpha}(x)\frac{\lambda^{2}}{2} \,. \tag{12.140}$$

Then, Eqs.(12.135) - (12.138) imply that the corresponding transformations of the fields are :

$$\begin{aligned} \delta V_{\mu} &= \partial_{\mu}\varphi + i[\varphi, V_{\mu}] + i[\omega, A_{\mu}] \,, \\ \delta A_{\mu} &= \partial_{\mu}\omega + i[\varphi, A_{\mu}] + i[\omega, V_{\mu}] \,, \\ \delta S &= i[\varphi, S] - \{\omega, P\} \,, \\ \delta P &= i[\varphi, P] + \{\omega, S\} \,. \end{aligned} \tag{12.141}$$

As was shown by Bardeen [67], the resulting change in $W_f(M, V, A) \equiv \frac{1}{i} \ln Z_f(M, V, A)$

is

$$\delta W_f = \frac{N_c}{16\pi^2}\int d^4x\epsilon^{\alpha\beta\mu\nu}tr\Big[\omega\Big(V_{\alpha\beta}V_{\mu\nu}+\frac{4}{3}D_\alpha A_\beta D_\mu A_\nu+\frac{2i}{3}\{V_{\alpha\beta},\, A_\mu A_\nu\} \\ +\frac{8i}{3}A_\mu V_{\alpha\beta}A_\nu+\frac{4}{3}A_\alpha A_\beta A_\mu A_\nu\Big)\Big]\ , \tag{12.142}$$

where

$$V_{\alpha\beta}=\partial_\alpha V_\beta-\partial_\beta V_\alpha-i[V_\alpha, V_\beta] \tag{12.143}$$

is the $U_V(N)$ - covariant vector field strength, and

$$D_\alpha A_\beta=\partial_\alpha A_\beta-i[V_\alpha, A_\beta] \tag{12.144}$$

is the $U_V(N)$ - convariant derivative of A_β.

We note that at $A_\mu=0, \delta W_f$ corresponding to the $U_5(1)\equiv U_{L-R}(1)$ transformation is defined by the ABJ anomaly,

$$\partial^\mu j^0_{5\mu}=-\sqrt{\frac{1}{2N}}\frac{N_c}{16\pi^2}\epsilon^{\alpha\beta\mu\nu}tr(V_{\alpha\beta}V_{\mu\nu}) \tag{12.145}$$

(compare with Eq.(9.9)). Actually, in the ENJL model, based on the approximate form (12.87) for the BS kernel, the *strong* ABJ anomaly (12.8) is neglected. We will return to this important point in Sec.12.12, but now, for a while, we will ignore the strong anomaly.

The action of the effective theory for the collective fields M, V_μ and A_μ is (see Eq.(12.129)) :

$$S(M,V_\mu,A_\mu)=\int d^4x\Big[-\frac{1}{2G_s^{(0)}}tr\Big(\Big(M-\hat{m}^{(0)}\Big)^\dagger\Big(M-\hat{m}^{(0)}\Big)\Big) \\ +\frac{1}{2G_V^{(0)}}tr\left(V_\mu^2+A_\mu^2\right)\Big]-iN_c\ln \mathrm{Det}i\mathfrak{D}\ , \tag{12.146}$$

where the operator $i\mathfrak{D}$ is given in Eq.(12.131). As $N_c\to\infty$, the path integral (12.129) is dominated by stationary points of this action which are defined from the classical equations of motions :

$$\frac{\delta S}{\delta M} = 0, \ \frac{\delta S}{\delta V_\mu} = 0, \ \frac{\delta S}{\delta A_\mu} = 0 \ . \tag{12.147}$$

Thus, in this limit, the problem is essentially reduced to the calculation of

$$\ln \mathrm{Det} i\mathfrak{D} = \mathrm{TrLn} i\mathfrak{D} \tag{12.148}$$

Since we are interested in a low energy effective action, we need to get an expansion of $\ln \mathrm{Det} i\mathfrak{D}$ as a series in powers of the derivatives of the collective fields M, V_μ and A_μ [64]. In Appendix E, we discuss a concrete method which is suitable for the present problem.

It is convenient to introduce a polar decomposition of the field M:

$$M(x) = \xi(x) Q(x) \xi(x) \ , \tag{12.149}$$

where (see Sec.5.2)

$$\xi(x) = \exp(i\pi(x)/F); \pi(x) = \sum_{\alpha=0}^{N^2-1} \pi^\alpha(x) \frac{\lambda^\alpha}{2} \ , \tag{12.150}$$

and

$$Q(x) = \sqrt{2} \sum_{\alpha=0}^{N^2-1} \sigma^\alpha(x) \frac{\lambda^\alpha}{2} \ . \tag{12.151}$$

The new fields $\pi^\alpha(x)$ and $\sigma^\alpha(x)$ can in principle be expressed through the fields $P^\alpha(x)$ and $S^\alpha(x)$ in Eq.(12.132).

Since the transformation of ξ under the $U_L(N) \times U_R(N)$ is (see Eq.(5.33)) :

$$\xi \longrightarrow L \xi U^\dagger = U \xi R^\dagger \ , \tag{12.152}$$

we find the following transformation for Q :

$$Q \longrightarrow U Q U^\dagger \ . \tag{12.153}$$

It is convenient to calculate $\ln \mathrm{Det} i\mathfrak{D}$ in Euclidean space. The effective action $S_E(M, V_\mu, A_\mu)$ is defined as

$$\exp(-S_E) = \exp(iS) ,$$

i.e.

$$S_E = -\,iS = \int d^4x_E \left[\frac{1}{2G_s^{(0)}} tr \left(\left(M - \hat{m}^{(0)} \right)^\dagger \left(M - \hat{m}^{(0)} \right) \right) + \frac{1}{2G_V^{(0)}} tr \left(V_\mu^2 + A_\mu^2 \right) \right] - N_c \ln \mathrm{Det} i\mathfrak{D} , \tag{12.154}$$

where $d^4x_E = id^4x(x_4 = ix_0)$. Lorentz vectors ∂_μ, V_μ and A_μ are continued as $\partial_4 = -i\partial_0$, $V_4 = -iV_0$, $A_4 = -iA_0$, and $V_\mu^2 \equiv \sum_{\mu=1}^{4} V_\mu^2$, etc..

The operator $i\mathfrak{D}$ (12.131) takes the following form in Euclidean space :

$$i\mathfrak{D} = -i\gamma_\mu(\partial_\mu - iV_\mu - iA_\mu\gamma_5) - M\frac{1-\gamma_5}{2} - M^\dagger\frac{1+\gamma_5}{2} , \tag{12.155}$$

where the Euclidean γ-matrices are : $\gamma_4 = -i\gamma^0$, $\gamma_k = \gamma^k (k = 1,2,3)$. They are antihermitian. The γ_5-matrix does not change : $\gamma_5 = -i\gamma^0\gamma^1\gamma^2\gamma^3 = -\gamma_1\gamma_2\gamma_3\gamma_4$.

12.11. The Derivation of the Low Energy Effective Action in the Extended Nambu-Jona-Lasinio Model

The determinant of the operator $i\mathfrak{D}$ (12.155) is a complex quantity. In order to find the phase of Det$i\mathfrak{D}$, that is $\mathrm{Im}W_f^E = -\ln Z_f = -N_c \ln \mathrm{Det} i\mathfrak{D} + constant$, where

$$Z_f(M,V,A) = \int \mathcal{D}\psi\mathcal{D}\bar{\psi} \exp\left\{ \int d^4x_E \bar{\psi} i\mathfrak{D}\psi \right\} \tag{12.156}$$

in Euclidean space (see Eq.(12.130)), we will use the Bardeen relation (12.142). In Euclidean space, it takes the following form :

$$\delta W_f^E = \int d^4x_E tr[\omega(x)\Omega(x)] , \tag{12.157}$$

where

$$\Omega(x) = i\frac{N_c}{16\pi^2}\epsilon_{\alpha\beta\mu\nu}\left[V_{\alpha\beta}V_{\mu\nu} + \frac{4}{3}D_\alpha A_\beta D_\mu A_\nu + \frac{2i}{3}\{V_{\alpha\beta}, A_\mu A_\nu\} + \frac{8i}{3}A_\mu V_{\alpha\beta}A_\nu + \frac{4}{3}A_\alpha A_\beta A_\mu A_\nu \right] \tag{12.158}$$

(in the derivation of this relation from Eq.(12.142), we took into account that $\delta W_f^E = -i\delta W_f$, $d^4x_E = id^4x$, ∂_4, V_4, A_4 are $-i\partial_0$, $-iV_0$, $-iA_0$, and $\epsilon_{1234} = -\epsilon^{1230} = 1$).

A functional W^E_{WZW} which correctly reproduces the relation (12.157) was constructed by Wess and Zumino [68]. The geometrical meaning of the functional was established by Witten [69]. It is called the Wess-Zumino-Witten (WZW) action in the literature.

The important point for us will be that since both W^E_{WZW} and W^E_f satisfy the relation (12.157), their difference W^E_f - W^E_{WZW} is invariant with respect to local gauge transformations in Eq.(12.141).

The Wess-Zumino construction of W^E_{WZW} proceeds as follows [68]. Let us introduce the generator of the infinitesimal chiral transformations (12.141) $D(\omega)$ with $\varphi = 0,\ \omega \neq 0$. Then, Eq.(12.141) implies that

$$\begin{aligned} D(\omega)F(M, V_\mu, A_\mu) = \int d^4x_E \Big\{ tr \Big[i[\omega, A_\mu]\frac{\delta F}{\delta V_\mu(x)} + D_\mu\omega\frac{\delta F}{\delta A_\mu(x)} \\ - \{\omega, P\}\frac{\delta F}{\delta S(x)} + \{\omega, S\}\frac{\delta F}{\delta P(x)}\Big]\Big\} , \end{aligned} \tag{12.159}$$

where $D_\mu\omega = \partial_\mu\omega + i[\omega, V_\mu]$, and F is some function of M, V_μ and A_μ. As one can easily check, the exponent $\exp[D(\omega)]$ generates the global gauge transformations (12.135) - (12.137) with

$$L(x) = R^\dagger(x) = \exp(i\omega(x)) , \tag{12.160}$$

i.e.

$$e^{D(\omega)}A_{\mu L}(x) = e^{i\omega(x)}\left(i\partial_\mu + A_{\mu L}\right)e^{-i\omega(x)} , \tag{12.161}$$

$$e^{D(\omega)}A_{\mu R}(x) = e^{-i\omega(x)}\left(i\partial_\mu + A_{\mu R}\right)e^{i\omega(x)} , \tag{12.162}$$

$$e^{D(\omega)}M(x) = e^{i\omega(x)}Me^{i\omega(x)} , \tag{12.163}$$

where $A_{\mu L}$ and $A_{\mu R}$ are defined in Eq.(12.138).

Let us choose $e^{i\omega(x)} = \xi^\dagger(x) = (\Sigma^\dagger)^{1/2}$, where here $\xi(x)$ is from Eq.(12.149). The WZW action is constructed as follows. Let us consider a functional $W^E_{WZW}(\Sigma, V_\mu, A_\mu)$

which depends only on V_μ, A_μ and $\Sigma = \xi^2$ and satisfies the condition

$$W^E_{WZW}(I, V_\mu, A_\mu) = 0 \ , \tag{12.164}$$

where I is the identity matrix. Then, Eq.(12.157) implies that

$$D(\omega)W^E_{WZW} = \int d^4x_E tr\left[\omega(x)\Omega(x)\right] \ . \tag{12.165}$$

Choosing $e^{2i\omega(x)} = \Sigma^\dagger(x)$ and taking into account Eqs.(12.163), (12.164), we find that

$$e^{D(\omega)}W^E_{WZW}(\Sigma, V_\mu, A_\mu) = 0 \ , \tag{12.166}$$

i.e.

$$W^E_{WZW}(\Sigma, V_\mu, A_\mu) = \sum_{n=1}^{\infty}\frac{1}{n!}\int d^4x_E tr\left\{\omega(x)[D(\omega)]^{n-1}\Omega(x)\right\} \ , \tag{12.167}$$

where $\Omega(x)$ is defined in Eq.(12.158). The remarkable property of W^E_{WZW} is that it is purely imaginary.

Since $e^{2i\omega(x)} = \Sigma^\dagger(x) = \exp(-2i\pi(x)/F)$, see Eq.(5.48), the WZW action describes interactions of NG bosons with vector and axial - vector fields. The first term in the series (12.167) contains in particular the piece

$$-i\frac{N_c F^{-1}}{16\pi^2}\epsilon_{\alpha\beta\mu\nu}\int d^4x_E tr\left\{\pi(x)V_{\alpha\beta}V_{\mu\nu}\right\} \tag{12.168}$$

describing the anomalous coupling of pseudoscalar mesons to vector mesons.

Let us show that the WZW action yields the dominant contribution (in the low energy region) to the phase of the functional W^E_f.

Let us consider the difference

$$\Delta W^E_f(M, \Sigma, V_\mu, A_\mu) = W^E_f(M, V_\mu, A_\mu) - W^E_{WZW}(\Sigma, V_\mu, A_\mu) \ . \tag{12.169}$$

Since it is invariant under the gauge transformations (12.35) - (12.137), we get the following relation :

$$\begin{aligned}\Delta W_f^E(M,\Sigma,V_\mu,A_\mu) &= \Delta W_f^E\left(Q,V_\mu^{\xi^\dagger},A_\mu^{\xi^\dagger}\right)\\ &= W_f^E\left(Q,V_\mu^{\xi^\dagger},A_\mu^{\xi^\dagger}\right) - W_{WZW}^E\left(I,V_\mu^{\xi^\dagger},A_\mu^{\xi^\dagger}\right) = W_f^E\left(Q,V_\mu^{\xi^\dagger},A_\mu^{\xi^\dagger}\right) ,\end{aligned} \tag{12.170}$$

where

$$\begin{aligned}A_{\mu L}^{\xi^\dagger} &= \xi^\dagger\left(i\partial_\mu + A_{\mu L}\right)\xi ,\\ A_{\mu R}^{\xi^\dagger} &= \xi\left(i\partial_\mu + A_{\mu R}\right)\xi^\dagger ,\end{aligned} \tag{12.171}$$

and we used the condition (12.164).

In the low energy region, in the phase with spontaneous chiral symmetry breaking, we can approximate the hermitian operator $Q(x)$ by its vacuum expectation value (see Eqs.(12.132) and (6.29)):

$$Q(x) \sim \langle 0|Q(x)|0\rangle = v\frac{\lambda^0}{\sqrt{2}} = m_{dyn} I , \tag{12.172}$$

where m_{dyn} is the dynamical quark mass.

Let us look at the determinant of the operator

$$i\mathfrak{D}^{\xi^\dagger} = -i\gamma_\mu\left(\partial_\mu - iV_\mu^{\xi^\dagger} - iA_\mu^{\xi^\dagger}\gamma_5\right) - m_{dyn} . \tag{12.173}$$

Because γ_5 anticommutes with γ_μ, non-vanishing eigenvalues of the operator $i\mathfrak{D}_0^{\xi^\dagger} = -i\gamma_\mu\left(\partial_\mu - iV_\mu^{\xi^\dagger} - iA_\mu^{\xi^\dagger}\gamma_5\right)$ occur in pairs of opposite sign :

$$\begin{aligned}i\mathfrak{D}_0^{\xi^\dagger}\psi_r &= \lambda_r\psi_r ,\\ i\mathfrak{D}_0^{\xi^\dagger}\gamma_5\psi_r &= -\lambda_r\gamma_5\psi_r .\end{aligned} \tag{12.174}$$

The corresponding two eigenvalues of the operator $i\mathfrak{D}^{\xi^\dagger}$ (12.173) are

$$\tilde{\lambda}_r = \lambda_r - m_{dyn},\ \ \tilde{\lambda}_r' = -\lambda_r - m_{dyn} . \tag{12.175}$$

Therefore we get:

$$\ln \mathrm{Det} i\mathfrak{D}^{\xi^\dagger} = \sum_r \ln\left[\left(\lambda_r - m_{dyn}\right)\left(-\lambda_r - m_{dyn}\right)\right] . \tag{12.176}$$

The imaginary part of $\ln \mathrm{Det} i\mathfrak{D}^{\xi^\dagger}$ is

$$\mathrm{Im} \ln \mathrm{Det} i\mathfrak{D}^{\xi^\dagger} = -\sum_r \arctan\left(\frac{2v_r u_r}{m_{dyn}^2 + u_r^2 - u_r^2}\right), \tag{12.177}$$

where $\lambda_r = u_r + iv_r$. It vanishes as $m_{dyn} \to \infty$. This in turn implies that $\mathrm{Im} \ln \mathrm{Det} i\mathfrak{D}^{\xi^\dagger}$ is of $O\left(\frac{1}{m_{dyn}^2}\right)$, and therefore there is an additional power of p^2 in ΔW_f^E as compared to W_{WZW}^E which is independent of m_{dyn}.

Thus, the WZW action gives the leading contribution to the imaginary part of the action S_E in the low energy region.

Let us now consider the logarithm of the modulus of the determinant contributing into the real part of the action S_E in Eq.(12.154).

It is :

$$\ln |\mathrm{Det} i\mathfrak{D}| = -\frac{1}{2} \int\limits_{1/\Lambda^2}^{\infty} \frac{dt}{t} \mathrm{Tr} \exp\left(-t\mathfrak{D}^\dagger\mathfrak{D}\right). \tag{12.178}$$

In the derivation of this relation, we used the formula

$$\ln \lambda = -\int\limits_{\epsilon}^{\infty} \frac{dt}{t} e^{-\lambda t} + const., \tag{12.179}$$

for $\lambda > 0$.

The expression (12.178) corresponds to the proper-time regularization scheme preserving the invariance with respect to the vector $U_V(N)$ subgroup of the $U_L(N) \times U_R(N)$.

The concrete method for calculating $\ln \mathrm{Det} i\mathfrak{D}$ as a series in powers of the derivatives of the collective fields M, V_μ, A_μ is considered in Appendix E. Here we will discuss the results.

It is rather straightforward to calculate the effective potential $V = \left(\int d^4x_E\right)^{-1} S_E$, where here S_E is calculated with the fields M, V_μ and A_μ taken constant. The explicit expression for V can be easily obtained in the case of the $U_L(2) \times U_R(2)$ chiral symmetry (we will consider the general case in a moment). It is (compare with Eq.(12.154)):

$$V = \frac{1}{2G_s^{(0)}} tr\left(\left(M - \hat{m}^{(0)}\right)^\dagger \left(M - \hat{m}^{(0)}\right)\right) + \frac{1}{2G_V^{(0)}} tr\left(V_\mu^2 + A_\mu^2\right) \\ + \frac{N_c}{8\pi^2}\left\{C_1^2 \Gamma\left(-2, C_1/\Lambda^2\right) + C_2^2 \Gamma\left(-2, C_2/\Lambda^2\right)\right\}, \tag{12.180}$$

where

$$\Gamma(\alpha, x) = \int_x^\infty dt e^{-t} t^{\alpha-1} \tag{12.181}$$

is the incomplete gamma function [70], and

$$C_1 = \frac{1}{2}\sigma^\alpha\sigma^\alpha + \sigma^0(\boldsymbol{\sigma}^2)^{1/2}, \tag{12.182}$$

$$C_2 = \frac{1}{2}\sigma^\alpha\sigma^\alpha - \sigma^0(\boldsymbol{\sigma}^2)^{1/2} \tag{12.183}$$

with the fields σ^α defined in Eq.(12.151) ($\boldsymbol{\sigma} = (\sigma^1, \sigma^2, \sigma^3)$).

The incomplete gamma function $\Gamma(\alpha, x)$ is an analytical function of x at $\alpha > 0$, and it has poles or a logarithmic singularity at $x = 0$ for $\alpha = -n$ (n positive integer) and $\alpha = 0$, respectively. At $\alpha = -n$, it has the representation

$$\Gamma(-n, x) = \frac{(-1)^n}{n!}\left[\Gamma(0, x) - e^{-x}\sum_{m=0}^{n-1}(-1)^m \frac{m!}{x^{m+1}}\right], \tag{12.184}$$

and

$$\Gamma(0, x) \simeq -\ln x \tag{12.185}$$

at $x \ll 1$.

By using Eq.(12.184), we can rewrite the expression (12.180) for V as

$$V = \frac{1}{2G_s^{(0)}} tr\left(\left(M - \hat{m}^{(0)}\right)^\dagger \left(M - \hat{m}^{(0)}\right)\right) + \frac{1}{2G_V^{(0)}} tr\left(V_\mu^2 + A_\mu^2\right) \\ + \frac{N_c}{16\pi^2}\left\{C_1^2 \Gamma\left(0,\ C_1/\Lambda^2\right) + C_2^2 \Gamma\left(0,\ C_2/\Lambda^2\right) - \Lambda^2\left[C_1 \exp\left(-C_1^2/\Lambda^2\right) \right.\right. \\ \left.\left. + C_2 \exp\left(-C_2^2/\Lambda^2\right)\right] + \Lambda^4\left(e^{-C_1^2/\Lambda^2} + e^{-C_2^2/\Lambda^2}\right)\right\}. \tag{12.186}$$

The vacuum in the model is defined from the equation

$$\frac{\partial V}{\partial \pi^\alpha} = 0, \ \frac{\partial V}{\partial \sigma^\alpha} = 0\,. \tag{12.187}$$

The solutions of this equation coincide with those in the NJL model (see Chapters 6 and 9). In particular, at $\hat{m}^{(0)} = 0$, we find that spontaneous chiral symmetry breaking occurs at supercritical values of the coupling constant $G_s^{(0)}$: $\kappa \equiv \frac{G_s^{(0)} N_c \Lambda^2}{4\pi^2} > 1$. A vacuum solution takes the form

$$\pi^\alpha = 0\ ;\ \sigma^0 = \upsilon\ ,\ \sigma^1 = \sigma^2 = \sigma^3 = 0\ , \tag{12.188}$$

where υ is determined from the equation (compare with Eq.(6.14) in which $m_{dyn} = N^{-1/2}\upsilon$)

$$\kappa - 1 = \kappa \frac{\upsilon^2}{2\Lambda^2} \ln \frac{2\Lambda^2 + u^2}{\upsilon^2}\,. \tag{12.189}$$

We recall that $\upsilon = \sqrt{2} m_{dyn}$ at $N = 2$.

In the near-critical regime, with $\kappa - 1 \ll 1$ and $\frac{u^2}{\Lambda^2} \ll 1$, the expression for V essentially simplifies :

$$\begin{aligned} V \simeq & \frac{1}{2G_s^{(0)}} tr\left(\left(M - \hat{m}^{(0)}\right)^\dagger \left(M - \hat{m}^{(0)}\right)\right) + \frac{1}{2G_V^{(0)}} tr\left(V_\mu^2 + A_\mu^2\right) \\ & + \frac{N_c}{16\pi^2}\left\{\ln\frac{\Lambda^2}{m_{dyn}^2}\left[\frac{1}{2}(\sigma^\alpha\sigma^\alpha)^2 + 2(\sigma^0)^2\boldsymbol{\sigma}^2\right] - 2\Lambda^2\sigma^\alpha\sigma^\alpha + 2\Lambda^4\right\}\,, \end{aligned} \tag{12.190}$$

or *)

$$\begin{aligned} V \simeq tr\Big\{ & \frac{1}{2G_s^{(0)}}\left(M - \hat{m}^{(0)}\right)^\dagger \left(M - \hat{m}^{(0)}\right) + \frac{1}{2G_V^{(0)}}\left(V_\mu^2 + A_\mu^2\right) \\ & + \frac{N_c}{16\pi^2}\ln\frac{\Lambda^2}{m_{dyn}^2}\left(M^\dagger M M^\dagger M\right) - 2\Lambda^2\left(M^\dagger M\right) + 2\Lambda^4\Big\}\,. \end{aligned} \tag{12.191}$$

In other words, it takes a "renormalizable" form: the highest power of the fields equals 4 here.

---

*) In this approximation, the parameter μ^2 in $\ln \Lambda^2/\mu^2$ is not uniquely defined. What we can say is that it is of the order of a characteristic mass scale in the model: $\mu \sim \upsilon \sim m_{dyn}$.

In order to understand this fact more deeply, let us look at the terms with two (covariant) derivatives, i.e. at the kinetic term (see Appendix E):

$$\mathcal{L}_k^E = \frac{N_c}{16\pi^2} \ln \frac{\Lambda_\chi^2}{m_{dyn}^2} tr \left[\nabla_\mu S \nabla_\mu S + \nabla_\mu P \nabla_\mu P + \frac{1}{3} F_{\mu\nu}^V F_{\mu\nu}^V + \frac{1}{3} F_{\mu\nu}^A F_{\mu\nu}^A \right] , \tag{12.192}$$

where

$$F_{\mu\nu}^V = \partial_\mu V_\nu - \partial_\nu V_\mu - i[V_\mu, V_\nu] - i[A_\mu, A_\nu] , \tag{12.193}$$

$$F_{\mu\nu}^A = \partial_\mu A_\nu - \partial_\nu A_\mu - i[V_\mu, A_\nu] - i[A_\mu, V_\nu] , \tag{12.194}$$

and the covariant derivatives $\nabla_\mu S$, $\nabla_\mu P$ are :

$$\begin{aligned} \nabla_\mu S &= \partial_\mu S - i[V_\mu, S] + \{A_\mu, P\} , \\ \nabla_\mu P &= \partial_\mu P - i[V_\mu, P] - \{A_\mu, S\} , \end{aligned} \tag{12.195}$$

We note that these expressions for $\nabla_\mu S$ and $\nabla_\mu P$ follow from the expression for the covariant derivative

$$\nabla_\mu M = \partial_\mu M - iA_{L\mu} M + iMA_{R\mu} \tag{12.196}$$

We emphasize that the field strengths $F_{\mu\nu}^V$, $F_{\mu\nu}^A$ and covariant derivatives $\nabla_\mu S$, $\nabla_\mu P$ are covariant objects with respect to the whole chiral gauge group $U_L(N) \times U_R(N)$ and not only to its vector subgroup $U_V(N)$, as the quantities $V_{\alpha\beta}$ and $D_\alpha A_\beta$ in Eqs.(12.143), (12.144). It is not surprising : in the effective action, only the WZW term vilolets chiral invariance ; all other terms are invariant under the local $U_L(N) \times U_R(N)$ transformations.

By using the same approach as in Sec.4.4, we find the decay constant F from the action $S = iS_E = i \int d^4 x_E \left(\mathcal{L}_k^E + V \right)$. It is :

$$F = \frac{m_{dyn}}{2\pi} \left(N_c \ln \frac{\Lambda_\chi^2}{m_{dyn}^2} \right)^{1/2} . \tag{12.197}$$

We note that this expression coincides with that in Eq.(9.133) obtained from the Bethe-Salpeter (BS) equation in the NJL model.

We see that F diverges as $\frac{\Lambda_\chi^2}{m_{dyn}^2} \to \infty$. Then, we find, from the Goldberger-Treiman relation (4.50), that the Yukawa coupling constant $g_Y^2 = \frac{m_{dyn}^2}{F^2} \simeq \frac{4\pi^2}{N_c}\frac{1}{\ln\frac{\Lambda_\chi^2}{m_{dyn}^2}}$ goes to zero as $\frac{\Lambda_\chi^2}{m_{dyn}^2} \to \infty$. One can check that all other coupling constants, as those describing self-interactions of the collective fields, also logarithmically go to zero in this limit. The simplest way to check this is to note that besides the terms in the potential V (12.191) and in the kinetic term (12.192), all other terms in the effective action have finite (or zero) coefficients as $\Lambda_\chi^2/m_{dyn}^2 \to \infty$ (see Appendix E). Introducing the interpolating fields,

$$\begin{aligned}
S_{int}^\alpha &= \left(\frac{N_c}{8\pi^2}\ln\frac{\Lambda_\chi^2}{m_{dyn}^2}\right)^{1/2} S^\alpha\,, \\
\pi_{int}^\alpha &= \left(\frac{N_c}{8\pi^2}\ln\frac{\Lambda_\chi^2}{m_{dyn}^2}\right)^{1/2} \pi^\alpha\,, \\
V_{int}^\alpha &= \left(\frac{N_c}{24\pi^2}\ln\frac{\Lambda_\chi^2}{m_{dyn}^2}\right)^{1/2} V^\alpha\,, \\
A_{int}^\alpha &= \left(\frac{N_c}{24\pi^2}\ln\frac{\Lambda_\chi^2}{m_{dyn}^2}\right)^{1/2} A^\alpha\,,
\end{aligned} \tag{12.198}$$

which have the canonical kinetic terms, we find that all terms with higher powers of the fields and their derivatives are logarithmically suppressed.

In the strict continuum limit, the interactions are completely switched off in the model (in the present approximation). In the language of the theory of critical phenomena (see Sec.11.4), this means that the universality class of the four-dimensional ENJL model is that of the free theory.

This consideration also sheds light on the validity of the approximation used in the analysis of the BS equations in the NJL model in Sec.9.5. In that analysis, by truncating the BS equations, we found the relation

$$M = 2m_{dyn} \tag{12.199}$$

for the mass of the scalars in the chiral limit. One can easily check that the same result follows from the action $S = iS_E = i\int d^4x_E(\mathcal{L}_k^E + V)$. Thus, the truncated BS equations correspond to truncating the expansion of the effective action S_E in powers of the derivatives of the fields at second order. This approximation is justified in the near-critical regime with $\kappa - 1 \ll 1$ and $\ln \frac{\Lambda_\chi^2}{m_{dyn}^2} \gg 1$.

So far we have discussed the model with the $U_L(2) \times U_R(2)$ chiral symmetry. It is not difficult to convince oneself that Eqs.(12.191), (12.192), (12.198) and the main conclusions which we have come to above remain valid in the general case of the $U_L(N) \times U_R(N)$ chiral symmetry.

By using the expressions for V and $\mathcal{L}_k^E$ in Eqs.(12.191), (12.192), it is straightforward to analyze the spectrum of mesons in the ENJL model. For details, we refer the reader to Refs.[41, 42]. Here we will point out some characteristic points of this analysis.

In the most useful case, QCD with three light flavors, the chiral group is $U_L(3) \times U_R(3)$ and $N_c = 3$. In this case, the ENJL model contains six free parameters : $m_u^{(0)}, m_d^{(0)}, m_s^{(0)}, G_S^{(0)}, G_V^{(0)}$ and Λ_χ. To fix them one can use different sets of input data. For example, Ebert and Reinhardt [41], taking as input data $M_\pi, M_\rho, M_\varphi, F_\pi$ and the $\rho - \pi - \pi$ coupling constant $g_{\rho\pi\pi}$ and assuming that $m_u^{(0)} \simeq m_d^{(0)}$, find that $\Lambda_\chi \sim 1.4GeV$, $G_S^{(0)}/4\pi \sim 1.2\Lambda_\chi^{-2}$, $G_V^{(0)}/4\pi \sim 5\Lambda_\chi^{-2}$, $m_u^{(0)} \simeq m_d^{(0)} = 2MeV$, $m_s^{(0)} \simeq 82MeV$. Although the obtained values for the masses $m_u^{(0)}, m_d^{(0)}$, $m_s^{(0)}$ are visibly smaller than the values of $m_{\mathrm{cur};a}$ (relating to the scale $\sigma = 1GeV$) which were obtained in Sec.12.8, this discrepancy is not actually important. The point is that we have not yet understand the connection between the bare masses $m_a^{(0)}$ in the ENJL model and the current masses $m_{\mathrm{cur};a}$ in QCD [71, 72].

With these six parameters, the model describes the meson spectrum and the low energy dynamics of mesons pretty well [41].

An interesting point is the value $1.2\Lambda_\chi^{-2}$ for the coupling constant $G_S^{(0)}/4\pi$. The critical value for this coupling constant is

$$\Lambda_\chi^2 \left(G_S^{(0)}/4\pi\right)_c = \frac{\pi\kappa_c}{N_c}\Big|_{\substack{N_c=3\\ \kappa_c=1}} \simeq 1.05\ . \tag{12.200}$$

Thus the obtained value for $G_S^{(0)}/4\pi$ is rather close to its critical value, i.e. the dynamics of spontaneous chiral symmetry breaking in the model is near-critical. From the practical viewpoint, this implies that the coupling constants describing the interactions in the low energy region are not too strong. Let us for example look at the pion-quark Yukawa coupling constant g_Y. From the Goldberger-Treiman relation we find that $g_Y \sim \frac{m_{dyn}}{F_\pi} \sim$

3 with $m_{dyn} \sim 300MeV$ and $F_\pi \sim 93MeV$. Therefore the characteristic parameter in the loop expansion, $\frac{g_Y^2}{4\pi^2}$, is $\frac{g_Y^2}{4\pi^2} \sim 0.3$; it 10 times smaller than the value of the corresponding parameter for nucleons.

An interesting effect in the ENJL model is the $S - V_\mu$ and $P - A_\mu$ mixing. Let us look at $\mathcal{L}_k^E$ in Eq.(12.192). The covariant derivatives for S and P imply that such a mixing occurs in the model. The situation resembles that in gauge models with the Higgs mechanism considered in Secs.3.5, 3.6. However there is an essential difference between these two cases : while in gauge models, such a mixing can be removed by an appropriate choice of the gauge, in the ENJL model, the mass terms for V_μ and A_μ spoil the gauge $U_L(3) \times U_R(3)$ symmetry. Therefore in this case, the $S - V_\mu$ and $P - A_\mu$ mixing is a genuine physical effect. The $P - A_\mu$ mixing is especially important for applications. For a recent discussion of this effect see Ref.[42].

In conclusion, we note that the calculations of non-leading terms in the series of the derivative expansion of S_E is, although straightforward, rather involved. We will consider some of them in Sec.12.13.

12.12. The Strong $U_5(1)$ Anomaly. Effective Lagrangians Obeying the Anomalous Conservation Law

As was pointed out in Sec.12.10, the effects of the strong ABJ anomaly are neglected in the ENJL model. It is known [73, 74] that these effects can be coded by adding the θ-term in the QCD lagrangian density :

$$\mathcal{L}_{QCD}^{\theta} = \mathcal{L}_{QCD}^{\theta=0} + \theta \left(\frac{g^2}{64\pi^2} \right) \epsilon^{\mu\nu\delta\sigma} G_{\mu\nu}^K G_{\delta\sigma}^K \,, \tag{12.201}$$

where $\mathcal{L}_{QCD}^{\theta=0}$ is given Eq.(12.1), and the angle parameter $\theta (0 \leq \theta < 2\pi)$ defines different θ-vacua.

As we have already stated several times, by using an appropriate chiral transformation, the matrix $\hat{m}^{(0)}$ can be always reduced to the form $\hat{m}_{ab}^{(0)} = m_a^{(0)} \delta_{ab}$ with $m_a^{(0)} \geq 0$. This statement remains valid in the presence of the θ-term in Eq.(12.201) if under the chiral transformation (12.134), (12.139) θ transforms as [73, 74].

$$\theta \longrightarrow \theta + 2tr\omega = \theta + \sqrt{2N}\omega^0 \,. \tag{12.202}$$

This implies that the actual observable is not θ but the chiral invariant

$$\bar{\theta} \equiv \theta + \arg\det \hat{m}^{(0)} \tag{12.203}$$

(we recall that the mass term is $\bar{\psi}_L \hat{m}^{(0)} \psi_R + \bar{\psi}_R \left(\hat{m}^{(0)}\right)^\dagger \psi_L$).

We are interested in establishing the role of the strong anomaly in the low energy dynamics of mesons, in particular, the η' meson. The basic points in solving this problem were established by 't Hooft [56], Witten [57] and Veneziano [58]. A concrete effective chiral lagrangian for the field M obeying the anomalous conservation law (12.8) was constructed by Rosenzweig, Schechter and Trahern [75], Di Vecchia and Veneziano [76], and Nath and Arnowitt [77]. They showed that the anomaly modifies the effective potential and, to first order in $\hat{m}^{(0)}$, the potential takes the following form:

$$\begin{aligned} V(M,M^\dagger) =& V_0(M,M^\dagger) - A\, tr\left(M^\dagger \hat{m}^{(0)} + M\left(m^{(0)}\right)^\dagger\right) \\ & - \frac{F_\pi^2 M^2(P^0)}{16N}\left[\ln e^{i\theta}\det M - \ln e^{-i\theta}\det M^\dagger\right]^2 , \end{aligned} \tag{12.204}$$

where $V_0(M,M^\dagger)$ is a function which is invariant under the $U_L(N)\times U_R(N)$, $M(P^0)$ is the mass of P^0 meson (see Eq.(12.132)), and A is a real constant.*) We recall that $\det M$, being invariant with respect to the $SU_L(N)\times SU_R(N)\times U_V(1)$, is not invariant under the $U_5(1)$ transformations.

The main steps in establishing the form (12.204) are the following.

The effective lagrangian has to reproduce the anomalous conservation law (12.8) :

$$\partial^\mu j^0_{5\mu} = G\ ;\ G \equiv -\left(\frac{N}{2}\right)^{1/2} \frac{g^2}{32\pi^2}\epsilon^{\mu\nu\delta\beta} G^K_{\mu\nu} G^K_{\delta\beta}\ . \tag{12.205}$$

Let us consider the following lagrangian density (compare with Eq.(4.53)):

$$\begin{aligned} \mathcal{L} =& \frac{1}{2} tr\left[\partial_\mu M^\dagger \partial^\mu M\right] - V_0(M,M^\dagger) + CG^2 + \frac{i\sqrt{2}}{4\sqrt{N}} G\Big(\ln e^{i\theta}\det M \\ & - \ln e^{-i\theta}\det M^\dagger\Big) + A\, tr\left(M^\dagger \hat{m}^{(0)} + M\left(\hat{m}^{(0)}\right)^\dagger\right) , \end{aligned} \tag{12.206}$$

where C is another real constant.

*) Actually, as will be shown below, this is the simplest but not unique form for V obeying the anomalous conservation law.

The $U_5(1)$ current is (compare with Eqs.(4.56), (4.57)) :

$$\jmath^0_{5\mu} = \frac{i}{\sqrt{2N}} tr \left[M^\dagger \partial_\mu M - M \partial_\mu M^\dagger\right] , \tag{12.207}$$

so that

$$\partial^\mu \jmath^0_{5\mu} = \frac{i}{\sqrt{2N}} tr \left[M^\dagger \Box M - M \Box M^\dagger\right] . \tag{12.208}$$

The equations of motion for M and $M^\dagger$ are :

$$-\frac{1}{2}\Box M - \frac{\partial V_0}{\partial M^\dagger} - \frac{i\sqrt{2}}{4\sqrt{N}} G(M^\dagger)^{-1} + A\hat{m}^{(0)} = 0 , \tag{12.209}$$

$$-\frac{1}{2}\Box M^\dagger - \frac{\partial V_0}{\partial M} + \frac{i\sqrt{2}}{4\sqrt{N}} G(M)^{-1} + A\left(\hat{m}^{(0)}\right)^\dagger = 0 . \tag{12.210}$$

Taking into account the identity $tr\left(M^\dagger \frac{\partial V_0}{\partial M} - M\frac{\partial V_0}{\partial M^\dagger}\right) = 0$, we find the following relation from Eqs.(12.208) - (12.210) :

$$\partial^\mu \jmath^0_{5\mu} = G + iA\sqrt{\frac{2}{N}} tr\left(M^\dagger m^{(0)} - M\left(m^{(0)}\right)^\dagger\right) . \tag{12.211}$$

This relation coincides with the anomalous conservation equation in the chiral limit.

Since the gluonic field G occurs in the lagrangian as an auxiliary field, it can be integrated out in the path integral, or, what is the same, be expressed in terms of the fields $M, M^\dagger$ by using the equation of motion

$$2CG + \frac{i\sqrt{2}}{4\sqrt{N}}\left(\ln e^{i\theta}\det M - \ln e^{-i\theta}\det M^\dagger\right) = 0 . \tag{12.212}$$

Now the lagrangian density (12.206) takes the following form :

$$\begin{aligned}\mathcal{L} =& \frac{1}{2} tr\left[\partial_\mu M^\dagger \partial^\mu M\right] - V_0(M, M^\dagger) + \frac{1}{32CN}\left(\ln e^{i\theta}\det M - \ln e^{-i\theta}\det M^\dagger\right)^2 \\ &+ A\, tr\left(M^\dagger \hat{m}^{(0)} + M\left(\hat{m}^{(0)}\right)^\dagger\right) .\end{aligned} \tag{12.213}$$

By using the expansion (12.132) for M, or considering the nonlinear limit, when $M = \frac{F}{\sqrt{2}}\exp(2i\pi/F)$, we find that $\frac{1}{32CN} = \frac{F^2 M^2(P^0)}{16N}$ in the tree approximation. As a result, we come to the expression (12.204) for $V(M, M^\dagger)$.

Few comments are in order.

It is clear that the form (12.206) is not the only one reproducing the anomalous conservation law (12.8). For example, without modifying the main conclusions, the additional terms $\sum_m h_m G^m$ can be added to the lagrangian density (12.206). Here h_m are arbitrary functions of the invariants $tr(MM^\dagger)^n$. These additional terms will not change the equation for $\partial^\mu j^0_{5\mu}$, although the constraint (12.212) will be modified. Thus, the form (12.206) is just the simplest form reproducing the anomalous conservation law.

The constraint (12.212) expresses a dual nature of P^0 meson (η' meson at $N = 3$): it can be thought of as composed of quarks and antiquarks or of gluon fields.

The expression (12.204) clearly shows that $\bar{\theta} = \theta + \arg\det\hat{m}^{(0)}$ is a chiral invariant. In particular, at $\det\hat{m}^{(0)} = 0$, the parameter $\bar{\theta}$ is arbitrary and is not observable.

It is clear that the anomalous term in the potential $V(M, M^\dagger)$ leads to the λ_η-term in the mass matrix for pseudoscalar mesons considered in Sec.12.8.

For a detailed discussion of the effect of the anomalous term in the NJL model see Ref.[71].

12.13. Chiral Effective Lagrangian in QCD: General Structure and Comparison with the ENJL Model

In Chapter 5, we considered the general theory of nonlinear realization of spontaneous symmetry breaking in QFT. In this section, we will apply this theory to constructing the chiral effective lagrangian in QCD.

We begin by considering the field

$$\Sigma(x) = \exp(2i\pi(x)/F); \pi(x) = \sum_{\alpha=1}^{8} \frac{\lambda^\alpha}{2}\pi^\alpha(x) \tag{12.214}$$

describing 8 pseudoscalar mesons in QCD with three quark flavors : u, d and s.

We assume that the effective lagrangian for Σ admits an expansion in powers of the derivatives of Σ. We should also take into account the explicit violation of the $SU_L(3) \times SU_R(3) \times U_V(1)$ by the quark mass term $\hat{m}^{(0)}$. Since the mass term has the form

$$\bar{\psi}_L \hat{m}^{(0)} \psi_R + \bar{\psi}_R \left(\hat{m}^{(0)}\right)^\dagger \psi_L \ ,$$

it is invariant under the transformation (12.134) of the quark fields, provided the mass matrix is transformed as

$$\hat{m}^{(0)} \longrightarrow L \hat{m}^{(0)} R^\dagger \ . \tag{12.215}$$

Thus, the effective lagrangian $\mathcal{L}_{eff} = \mathcal{L}_{eff}\left(\Sigma, \partial\Sigma, \partial^2\Sigma, \cdots, \hat{m}^{(0)}\right)$ remains invariant under the transformations

$$\begin{aligned} \Sigma &\longrightarrow L \Sigma R^\dagger \\ \hat{m}^{(0)} &\longrightarrow L \hat{m}^{(0)} R^\dagger \ . \end{aligned} \tag{12.216}$$

Since $\hat{m}^{(0)}$ is proportional $M^2(\pi)$ (see Eq.(12.107), it is appropriate to atribute two powers of ∂ to $\hat{m}^{(0)}$.

The general structure of the terms in $\mathcal{L}_{eff}$ to $O(\partial^4)$ was established by Gasser and Leutwyler [78]. Since $\Sigma^\dagger\Sigma = 1$, there are no terms of $O(\partial^0) \sim O(1)$ in $\mathcal{L}_{eff}$. The $O(\partial^2)$-terms are:

$$\mathcal{L}^{(2)}_{eff} = \frac{F^2}{4} tr \left\{ \partial_\mu \Sigma^\dagger \partial^\mu \Sigma + 2B \hat{m}^{(0)} \Sigma^\dagger + 2B^* \left(\hat{m}^{(0)}\right)^\dagger \Sigma \right\} \ , \tag{12.217}$$

where the constant B will be determined in a moment.

Henceforth, we will consider the fermion basis in which

$$m^{(0)}_{ab} = m^{(0)}_a \delta_{ab}, \ m^{(0)}_a \geq 0 \ . \tag{12.218}$$

The constant B is in general complex. As follows from Eq.(12.213), the phase of B is related to the parameter θ in the θ-term $\theta\left(\frac{g^2}{64\pi^2}\right) \epsilon^{\mu\nu\delta\sigma} G^K_{\mu\nu} G^K_{\delta\sigma}$ in Eq.(12.201). The θ-term violates P and CP symmetries. The fact that the electric dipole moment of neutron is very small implies smallness of the parameter $\bar{\theta}$ defined in Eq.(12.203): $\bar{\theta} \lesssim 10^{-9}$ [79]. Therefore to a very high degree of accuracy, on can take $\bar{\theta} = 0$. This in turn implies that the constant B is real if one chooses $\theta = 0$.

The mass term in Eq.(12.217) leads to the mass for pseudoscalar mesons. Expanding $\Sigma = \exp(i\pi^\alpha\lambda^\alpha/F)$ in powers of π^α, we find that

$$M_{ab}^2 = B\left(m_a^{(0)} + m_b^{(0)}\right) + O\left(\hat{m}^{(0)^2}\right) . \tag{12.219}$$

Comparing this relation with Eq.(12.105), we find that

$$B = -F^{-2}\langle\bar{\psi}\psi\rangle + O\left(\hat{m}^{(0)}\right) , \tag{12.220}$$

i.e. the constant B is expressed through the chiral condensate.

There are the WZW terms amongst the $O(\partial^4)$ terms in $\mathcal{L}_{eff}$ (see Eq.(12.167) at $V_\mu = A_\mu = 0$). All other terms in $\mathcal{L}_{eff}$ of $O(\partial^4)$ were classified by Gasser and Leutwyler [78]:

$$\begin{aligned}\mathcal{L}_{eff}^{(4)} =& L_1\left(tr\partial_\mu\Sigma\partial^\mu\Sigma^\dagger\right)^2 + L_2 tr\left(\partial_\mu\Sigma\partial^\nu\Sigma^\dagger\right)tr\left(\partial^\mu\Sigma\partial_\nu\Sigma^\dagger\right)\\ &+ L_3 tr\left(\partial_\mu\Sigma\partial^\mu\Sigma^\dagger\partial_\nu\Sigma\partial^\nu\Sigma^\dagger\right) + L_4 tr\left(\partial_\mu\Sigma\partial^\mu\Sigma^\dagger\right)tr\left[2B\hat{m}^{(0)}(\Sigma^\dagger+\Sigma)\right]\\ &+ L_5 tr\left[\partial_\mu\Sigma\partial^\mu\Sigma^\dagger 2B\hat{m}^{(0)}(\Sigma^\dagger+\Sigma)\right] + L_6\left[tr2B\hat{m}^{(0)}(\Sigma^\dagger+\Sigma)\right]^2\\ &+ L_7\left[tr2B\hat{m}^{(0)}(\Sigma^\dagger-\Sigma)\right]^2 + L_8\left[tr4B^2\left(\hat{m}^{(0)}\right)^2(\Sigma^{\dagger^2}+\Sigma^2)\right] .\end{aligned} \tag{12.221}$$

We note that in the case of $N \geq 4$, there is an additional term in $\mathcal{L}_{eff}^{(4)}$:

$$L_0\left(tr\partial_\mu\Sigma^\dagger\partial_\nu\Sigma\partial^\mu\Sigma\partial^\nu\Sigma^\dagger\right) . \tag{12.222}$$

For $N \leq 3$, it is represented as a linear combination of the L_i-terms with $i = 1, 2, 3$. At $N = 2$, only L_i-terms with $i = 1, 2, 4, 6$ and 7 are independent.

The values of the coefficients L_i were determined by Gasser and Leutwyler from experemental data. Let us compare these values with those derived from the NJL model ($V_\mu = A_\mu = 0$).

First, we note the following. The most interesting terms are those with L_1, L_2, L_3: they are non-vanishing in the chiral limit. In the NJL model, they appear from the fermion determinant. In a sense, they are not very sensitive to the dynamics of spontaenous chiral symmetry breaking: we could calculate the fermion determinant in the constituent quark model considering Σ just as being an external field. This is what had been done in Refs.[62 - 66]. However, in the NJL model the determinant appears absolutely naturally and, moreover, Σ admits the interpretation as a composite $\bar{\psi}_L\psi_R$ field. Thus, the NJL model gives a clear physical justification for that approach.

The values of L_1, L_2 and L_3 calculated in the NJL model are [62 - 66, 40 - 42]:

$$\begin{aligned} L_2 =& 2L_1 = \frac{N_c}{16\pi^2}\frac{1}{12}\Gamma\left(2, \frac{m_{dyn}^2}{\Lambda_\chi^2}\right) \\ L_3 =& \frac{N_c}{16\pi^2}\frac{1}{6}\left[\Gamma\left(1, \frac{m_{dyn}^2}{\Lambda_\chi^2}\right) - 2\Gamma\left(2, \frac{m_{dyn}^2}{\Lambda_\chi^2}\right)\right] . \end{aligned} \tag{12.223}$$

If one lets $m_{dyn}^2/\Lambda_\chi^2 \to 0$ (what is reasonable since $m_{dyn} \sim 0.3GeV, \Lambda_\chi \gtrsim 1GeV$), then $\Gamma(n,0) = \Gamma(n) = (n-1)!$, and we get:

$$\begin{aligned} L_2 =& 2L_1 = \frac{N_c}{16\pi^2}\frac{1}{12}|_{N_c=3} = \frac{1}{64\pi^2} \simeq 1.58 \cdot 10^{-3} \\ L_3 =& -\frac{N_c}{16\pi^2}\frac{1}{6}|_{N_c=3} = -\frac{1}{32\pi^2} \simeq -3.17 \cdot 10^{-3} . \end{aligned} \tag{12.224}$$

The experimental data yield [78]:

$$\begin{aligned} L_1 =& (0.7 \pm 0.3) \cdot 10^{-3}, L_2 = (1.3 \pm 0.7) \cdot 10^{-3} , \\ L_3 =& (-4.4 \pm 2.5) \cdot 10^{-3} . \end{aligned} \tag{12.225}$$

The agreement between these two sets of the values of L_i is quite remarkable.

Let us extend this analysis to the dynamics with the V_μ and A_μ fields. Actually, for reasons which will become clear below, we will introduce, besides the collective fields S, P, V_μ and A_μ, also external fields s, p, v_μ and a_μ. In the lagrangian density, they appear in the combinations

$$\begin{aligned} \tilde{S} &= S + s , \\ \tilde{P} &= P + p , \\ \tilde{V}_\mu &= V_\mu + v_\mu , \\ \tilde{A}_\mu &= A_\mu + a_\mu . \end{aligned} \tag{12.226}$$

The external fields s, p, v_μ and a_μ can be considered as sources for Green's functions of the composite operators $\bar{\psi}\frac{\lambda}{2}^\alpha\psi, i\bar{\psi}\gamma_5\frac{\lambda}{2}^\alpha\psi, \bar{\psi}\gamma_\mu\frac{\lambda}{2}^\alpha\psi$ and $\psi\gamma_\mu\gamma_5\frac{\lambda}{2}^\alpha\psi$, respectively. On the other hand, at least some of s, p, v_μ, a_μ can be treated as fields connected with some fundamental interactions, as the $SU_L(2) \times U_Y(1)$ electroweak interactions and Higgs interactions, for example.

We include the bare mass term $\hat{m}^{(0)}$ in the external field $s(x) + ip(x)$.

Now $\mathcal{L}_{eff}^{(2)}$ takes the form:

$$\mathcal{L}_{eff}^{(2)} = \frac{F^2}{4} tr \left\{ \tilde{\nabla}_\mu \Sigma^\dagger \tilde{\nabla}^\mu \Sigma + \varphi \Sigma^\dagger + \varphi^\dagger \Sigma \right\} , \tag{12.227}$$

where $\varphi(x) \equiv 2B[s(x) + ip(x)]$ and $\tilde{\nabla}_\mu$ denotes the covariant derivative:

$$\tilde{\nabla}_\mu \Sigma = \partial_\mu \Sigma - i\tilde{A}_{L\mu}\Sigma + i\Sigma\tilde{A}_{R\mu} \tag{12.228}$$

(compare with Eq.(12.196)), where

$$\begin{aligned} \tilde{A}_{L\mu} &= \tilde{V}_\mu + \tilde{A}_\mu \\ \tilde{A}_{R\mu} &= \tilde{V}_\mu - \tilde{A}_\mu . \end{aligned} \tag{12.229}$$

The terms in $\mathcal{L}_{eff}$ of $O(\partial^4)$ now are [78]:

$$\begin{aligned} \mathcal{L}_{eff}^{(4)} =& L_1 \left(tr \tilde{\nabla}_\mu \Sigma \tilde{\nabla}^\mu \Sigma^\dagger \right)^2 + L_2 tr \left(\tilde{\nabla}_\mu \Sigma \tilde{\nabla}_\nu \Sigma^\dagger \right) tr \left(\tilde{\nabla}^\mu \Sigma \tilde{\nabla}^\nu \Sigma^\dagger \right) \\ &+ L_3 tr \left(\tilde{\nabla}_\mu \Sigma \tilde{\nabla}^\mu \Sigma^\dagger \tilde{\nabla}_\nu \Sigma \tilde{\nabla}^\nu \Sigma^\dagger \right) + L_4 tr \left(\tilde{\nabla}_\mu \Sigma \tilde{\nabla}^\mu \Sigma^\dagger \right) tr \left(\varphi \Sigma^\dagger + \varphi^\dagger \Sigma \right) \\ &+ L_5 tr \left[\tilde{\nabla}_\mu \Sigma \tilde{\nabla}^\mu \Sigma^\dagger \left(\varphi \Sigma^\dagger + \varphi^\dagger \Sigma \right) \right] + L_6 \left[tr \left(\varphi \Sigma^\dagger + \varphi^\dagger \Sigma \right) \right]^2 \\ &+ L_7 \left[tr \left(\varphi \Sigma^\dagger - \varphi^\dagger \Sigma \right) \right]^2 + L_8 tr \left(\varphi \Sigma^\dagger \varphi \Sigma^\dagger + \varphi^\dagger \Sigma \varphi^\dagger \Sigma \right) \\ &- iL_9 tr \left(\tilde{F}_R^{\mu\nu} \tilde{\nabla}_\mu \Sigma^\dagger \tilde{\nabla}_\nu \Sigma + \tilde{F}_L^{\mu\nu} \tilde{\nabla}_\mu \Sigma \tilde{\nabla}_\nu \Sigma^\dagger \right) \\ &+ L_{10} tr \left(\Sigma \tilde{F}_R^{\mu\nu} \Sigma^\dagger \tilde{F}_{L\mu\nu} \right) + H_1 tr \left(\tilde{F}_R^{\mu\nu} \tilde{F}_{R\mu\nu} + \tilde{F}_L^{\mu\nu} \tilde{F}_{L\mu\nu} \right) + H_2 tr(\varphi\varphi^\dagger) , \end{aligned} \tag{12.230}$$

where

$$\begin{aligned} F_{L\mu\nu} &= \partial_\mu \tilde{A}_{L\nu} - \partial_\nu \tilde{A}_{L\mu} - i \left[\tilde{A}_{L\mu}, \tilde{A}_{L\nu} \right] , \\ F_{R\mu\nu} &= \partial_\mu \tilde{A}_{R\nu} - \partial_\nu \tilde{A}_{R\mu} - i \left[\tilde{A}_{R\mu}, \tilde{A}_{R\nu} \right] . \end{aligned} \tag{12.231}$$

Before proceeding the analysis of these terms, we would like to indicate that, as always in quantum field theory, the parameters (coupling constants) L_i depend on a renormalization scale σ (or cutoff Λ) where they are defined. Accordingly to the Wilson theory of effective lagrangians discussed in Sec.11.2, $L_i(\sigma)$ are obtained by integrating out particles with masses of order σ and heavier, and also the contribution connected with "high" momenta, $p \gtrsim \sigma$, of light particles.

The values of L_i were determined [78] by comparison with experimental low energy information such as the D-wave $\pi\pi$ scattering lengths, the electromagnetic charge radius of the pion and the decay $\pi \to e\nu\gamma$. In that analysis, the electromagnetic and weak $W_\mu^\pm$ fields are identified with certain combinations of the fields v_μ and a_μ. The conventional scale σ for $L_i(\sigma)$ is the mass of ρ meson, $\sigma = M_\rho = 0.77GeV$. In particular, the values of L_1, L_2 and L_3 in Eq.(12.224) correspond to this scale.

The constants L_i in the effective lagrangian derived from the ENJL model should be identified with $L_i(\sigma)$ at $\sigma \sim \Lambda_\chi \sim 1 - 1.5GeV$. In this model, in the chiral limit, besides the constants L_1, L_2, L_3, there are two additional constants [62 - 66, 40 - 42],

$$\begin{aligned} L_9 =& \frac{N_c}{16\pi^2}\frac{1}{3}\Gamma\left(1, \frac{m_{dyn}^2}{\Lambda_\chi^2}\right)|_{N_c=3} \underset{\frac{m_{dyn}^2}{\Lambda_\chi^2}\to 0}{\longrightarrow} 6.33 \cdot 10^{-3} , \\ L_{10} =& -\frac{N_c}{16\pi^2}\frac{1}{6}\Gamma\left(1, \frac{m_{dyn}^2}{\Lambda_\chi^2}\right)|_{N_c=3} \underset{\frac{m_{dyn}^2}{\Lambda_\chi^2}\to 0}{\longrightarrow} -3.17 \cdot 10^{-3} . \end{aligned} \tag{12.232}$$

The experimental data yield [78, 80]:

$$\begin{aligned} L_9(M_\rho) =& (6.9 \pm 0.7) \cdot 10^{-3} , \\ L_{10}(M_\rho) =& (-5.2 \pm 0.3) \cdot 10^{-3} . \end{aligned} \tag{12.233}$$

Thus, there is an essential difference between $L_{10}(M_\rho)$ and L_{10} in Eq.(12.232). However, according to the phylosophy of the Wilson theory outlined above, we should compare $L_i(M_\rho)$ with L_i in the ENJL model calculated at the same scale. For this purpose, we must integrate out the heavier fields, as S, V_μ and A_μ, and also we must take into account the contribution of one-loop graphs generated by $\mathcal{L}_{eff}^{(2)}$, in which momenta of NG bosons change from $\sigma = M_\rho$ through $\sigma = \Lambda_\chi$.

Such a program was realized by the authors of Refs.[42, 81] by eliminating the heavier fields at the mean field level up to fourth order by means of the classical equations of motion. This leads to $\mathcal{L}_{eff}^{(2)}$ (12.227) and $\mathcal{L}_{eff}^{(4)}$ (12.230) in which $\tilde{\nabla}_\mu \to \nabla_\mu \equiv$

$\tilde{\nabla}_\mu\big|_{\substack{S=0\\ V_\mu=A_\mu=0}}$ and $\tilde{F}_R^{\mu\nu} \to F_R^{\mu\nu} \equiv \tilde{F}_R^{\mu\nu}\big|_{\substack{S=0\\ V_\mu=A_\mu=0}}$, $\tilde{F}_L^{\mu\nu} \to F_L^{\mu\nu} \equiv \tilde{F}_L^{\mu\nu}\big|_{\substack{S=0\\ V_\mu=A_\mu=0}}$. As to the parameters L_i in $\mathcal{L}_{eff}^{(4)}$ these authors found that while the values of L_1, L_2, L_3 and L_9 do not change significantly with respect to the values in Eqs.(12.223),(12.224), the value of L_{10} changes in the right direction: it becomes $L_{10} \simeq -5.8 \cdot 10^{-3}$.

Such a good agreement with experimental data indicates that the ENJL model captures at least some basic points of the dynamics of light mesons.

12.14. More about Chiral Effective Lagrangians in QCD. Naive Dimensional Analysis

As we already indicated in Sec.12.7, the ENJL model is not the only model leading to a successful description of the dynamics of light mesons. In this section, we will compare it with other models. We will also consider heuristic rules for estimating the sizes of terms in the general chiral effective theory (the so called naive dimensional analysis) [82, 83].

Let us recall that in Sec.12.7 we "derived" the ENJL model from the improved ladder approximation (see Eq.(12.87)), in which the gluon propagator $D_{\mu\nu}(p)$ was taken as $D_{\mu\nu} = -\frac{1}{\Lambda_\chi^2} g_{\mu\nu}$, and the running coupling constant $\bar{g}^2(q,p)$ was chosen as $\bar{g}^2(q,p) = g_s^2 \theta\left(\Lambda_\chi^2 - (q-p)^2\right)$. As a result, we obtained the four-fermion interaction model (12.90).

Let us consider now what will happen if the gluon propagator is chosen as

$$D_{\mu\nu}(p) = \frac{g_{\mu\nu} - \left(1 - \xi^{-1}\right) p_\mu p_\nu / p^2}{p^2 - M_g^2}, \tag{12.234}$$

where M_g is a gluon "mass", and, for simplicity, the running coupling is chosen the same as before.

We will show, following Kleinert and Shrauner [84], and others [85 - 87], that in this case, integrating out gluons, we obtain *nonlocal* four-fermion interaction system.

Let us start from the "abelianized" QCD:

$$\mathcal{L} = -\frac{1}{4}\bar{G}_{\mu\nu}^K \bar{G}^{\mu\nu K} + \bar{\psi}(i\gamma^\mu \partial_\mu - \frac{1}{2} g_s t^K G_\mu^K)\psi - \frac{\xi}{2}\left(\partial^\mu G_\mu^K\right)^2, \tag{12.235}$$

where ξ is a gauge parameter and

$$\bar{G}_{\mu\nu}^K \equiv \partial_\mu G_\nu^K - \partial_\nu G_\mu^K \tag{12.236}$$

(compare with Eq.(12.1)).

After integrating out the gluon field in the path integral of this system, the lagrangian density (12.235) transforms to

$$\mathcal{L}' = \bar{\psi}(x_1)i\gamma^\mu\partial_\mu\psi(x_1) - \frac{i}{2}\psi(y_1)\bar{\psi}(x_1)K(y_1x_1,x_2y_2)\psi(x_2)\bar{\psi}(y_2) , \tag{12.237}$$

where $K(y_1x_1, x_2y_2)$ is the BS kernel (see Eq.(8.101)) (here integration and summation over x_2, y_1, y_2 are understood). In this approximation, it is

$$K_{m_1n_1,n_2m_2}(y_1x_1,x_2y_2) = \frac{1}{(2\pi)^{12}}\int dpdqdPe^{-i(pz_1+qz_2+Pz_3)}K_{m_1n_1,n_2m_2}(q,p;P) , \tag{12.238}$$

where $K_{m_1n_1,n_2m_2}$ is given in Eq.(12.87), and

$$\begin{aligned} Z_1 &= x_1 - y_1 , \\ Z_2 &= y_2 - x_2 , \\ Z_3 &= \frac{1}{2}(x_1-x_2) + \frac{1}{2}(y_1-y_2) . \end{aligned} \tag{12.239}$$

(compare with Eq.(8.76)). We indicated explicitly the group indices in Eq.(12.238). We also note that since we consider the chiral limit, in which all quarks have the same mass, the parameters η_x and η_y in Eq.(8.76) equal $\frac{1}{2}$ now. As in the ENJL model, we will introduce the collective fields. The essential difference, however, as compared to that case, is that now they are bilocal and not local fields. For this purpose, let us add the following Gaussian term to (12.237) so as to cancel the nonlocal four-fermion interaction:

$$\begin{aligned} &\frac{1}{2i}\left[N(y_1x_1) - iK(y_1x_1,x_1'y_1')\psi(x_1')\bar{\psi}(y_1')\right] \\ &\times K^{-1}(y_1x_1,x_2y_2)\left[N(x_2y_2) - iK(x_2y_2,x_2'y_2')\psi(x_2')\bar{\psi}(y_2')\right] \\ &\equiv \frac{1}{2i}\left(N - iK\psi\bar{\psi}\right)K^{-1}\left(N - iK\psi\bar{\psi}\right) , \end{aligned} \tag{12.240}$$

where $N_{mn}(y,x)$ is the auxiliary field:

$$N_{mn}(y,x) = iK_{mn,n'm'}(yx,x'y')\psi_{n'}(x')\bar{\psi}_{m'}(y') \tag{12.241}$$

(compare with Eq.(6.4)).

By using just the same procedure as in the case of the NJL model, we find the generating functional Z for Green's functions:

$$Z(\eta,\bar{\eta}) = C\int \mathcal{D}N\mathcal{D}\psi\mathcal{D}\bar{\psi}\exp\left\{i\int d^4x\left[\left(\bar{\psi}i\partial\psi - N\right)\psi + \frac{i}{2}\left(NK^{-1}N\right) + \bar{\eta}\psi + \bar{\psi}\eta\right]\right\}$$
$$\equiv \int \mathcal{D}N\exp\left\{i\left[\frac{1}{2i}\left(NK^{-1}N\right) - i\ln\mathrm{Det}\left(i\hat{\partial} - N\right) - \bar{\eta}\left(i\hat{\partial} - N\right)\eta\right]\right\} \tag{12.242}$$

(compare with Eqs.(6.17), (12.129)).

Thus, we get the following effective action :

$$S(N) = \frac{1}{2i}\left(NK^{-1}N\right) - i\ln\mathrm{Det}\left(i\hat{\partial} - N\right) = \frac{i}{2}\left(NK^{-1}N\right) - i\mathrm{TrLn}\left(i\hat{\partial} - N\right) \tag{12.243}$$

(compare with Eqs.(6.20), (12.146)).

When does this model reduce to the ENJL model ? Let us recall that in that model, the BS kernel is

$$K(y_1x_1, x_2y_2) \sim \frac{1}{M_g^2}\cdot\delta^4(x_1-y_1)\delta^4(x_1-x_2)\delta^4(x_1-y_2) \tag{12.244}$$

(see Eq.(8.108)). Therefore the present model becomes a NJL-like model when the support of its BS kernel has such a δ-like form. In this case the bilocal field $N(y,x)$ effectively becomes a local one, $N(x,x)$. Otherwise, $N(y,x)$ describes an infinite number of local fields:

$$N(y,x) = \sum_{n_0,n_1,n_2,n_3}\frac{1}{n_0!n_1!n_2!n_3!}N^{(n_0n_1n_2n_3)}(x)\partial_{y_0}^{n_0}\partial_{y_1}^{n_1}\partial_{y_2}^{n_2}\partial_{y_3}^{n_3}\delta^4(y-x)\ , \tag{12.245}$$

where

$$N^{(n_0n_1n_2n_3)}(x) = (-1)^n\int d^4y(y_0-x_0)^{n_0}(y_1-x_1)^{n_1}(y_2-x_2)^{n_2}(y_3-x_3)^{n_3}N(y,x)\ , \tag{12.246}$$

with $n = n_0 + n_1 + n_2 + n_3$.

It is clear that this approach gives just a new language for the description of the dynamics of the improved ladder approximation model discussed in Sec.12.7. But this language is useful for the comparison of this dynamics with that of the ENJL model. In particular, since a "harmonic" $N^{(n_0n_1n_2n_3)}(x)$ appears with the coefficient proportional to $\frac{1}{M_g^2}\left(\frac{\partial}{M_g}\right)^n, n = \sum_{i=0}^{3} n_i$, in the effective action, the harmonics with higher n are suppperssed in the low energy region where $\frac{p^2}{M_g^2} \ll 1$. It is clear that these harmonics can be considered as additional matter fields in the chiral effective lagrangian.

The discussion in the previous section shows the role of heavy matter fields in chiral dynamics. Although the heavy fields, such as S, V_μ and A_μ, decouple from the low energy dynamics, they, in accord with the Wilson theory, may essentially influence the values of the parameters in the effective lagrangian of the NG field Σ, as it happens with the constant L_{10} in the ENJL model. There has been a lot of papers demonstrating a certain improvement of the description of the chiral dynamics by taking into account such heavier fields [88 - 92].*)

We note also that a rather succesful calculation of the Gasser-Leutwyler constants L_i in the framework of the so called non-local quark model (which is essentially equivalent to the model with the bilocal field considered above) was done in Ref.[93].

There is a number of chiral models based on the general theory of nonlinear realization of spontaneous symmetry breaking considered in Chapter 5. These models do not use any specific approximation for the microscopic chiral dynamics, such as the improved ladder approximation for example. We will discuss one of them.

The Georgi-Manohar (GM) model [82] is based on the hypothesis of the two scales in QCD discussed in Sec.12.6. The model includes the fields of only light particles, with masses lighter than Λ_χ: the field Σ of the eight NG bosons, the quark field χ and the gluon field G_μ^K. Under the chiral $SU_L(3) \times SU_R(3)$, the quark field χ transforms nonlinearly:

$$\chi \rightarrow U\chi \tag{12.247}$$

(see Eq.(5.57)). The field Σ transforms as usual:

$$\Sigma \rightarrow L\Sigma R^\dagger \,. \tag{12.248}$$

The effective lagrangian is written down as the most general lagrangian which is invariant under $SU_L(3) \times SU_R(3)$ and discrete transformations P, C and T. The first few terms

---

*) In those papers, the important role of isosinglet scalar ($|\sigma\rangle = \frac{1}{\sqrt{2}}(|\bar{u}u\rangle + |\bar{d}d\rangle)$) in the chiral dynamics was especially emphasized.

are:

$$\mathcal{L} = \bar{\chi}\left(i\hat{D} + \hat{V}\right)\chi + g_A^{(0)}\bar{\chi}\hat{A}\gamma_5\chi - m_{dyn}^{(0)}\bar{\chi}\chi + \frac{1}{4}F^{(0)^2}tr\partial^\mu\Sigma^\dagger\partial_\mu\Sigma - \frac{1}{4}G_{\mu\nu}^K G^{\mu\nu K} + \cdots , \tag{12.249}$$

where the quantities V_μ and A_μ were defined in Eqs.(5.30), (5.31),

$$\begin{aligned} V_\mu &= \frac{1}{2}i\left(\xi^\dagger\partial_\mu\xi + \xi\partial_\mu\xi^\dagger\right) , \\ A_\mu &= \frac{1}{2}i\left(\xi^\dagger\partial_\mu\xi - \xi\partial_\mu\xi^\dagger\right) , \end{aligned} \tag{12.250}$$

and $D_\mu = \partial_\mu + \frac{i}{2}g_s t^K G_\mu^K$ is the $SU_c(3)$ covariant derivative. The parameters $g_A^{(0)}, m_{dyn}^{(0)}$ and $F^{(0)}$ relate to the scale $\sigma = \Lambda_\chi$. For simplicity, we consider the chiral limit only; $m_{dyn}^{(0)}$ is the *dynamical* mass of quarks.

Let us compare this model with the NJL model, i.e. with the ENJL model in which $V_\mu = A_\mu = 0$. In the nonlinear limit, when $M \to \frac{v}{\sqrt{2}}\Sigma$, the NJL model leads to the form (12.249), in which $F^{(0)} = 0$ and gluon field is omitted. The equality $F^{(0)} = 0$ implies that in the NJL model, the kinetic term of Σ is induced by quark loops. $F^{(0)} \neq 0$ in the GM model would imply that a part of the kinetic term is induced by heavy fields integrated out. In particular, the higher harmonics $N^{(n_0n_1n_2n_3)}(x)$ (see Eq.(12.246)) or baryon fields might give such a contribution.

The values of the "bare" parameters $g_A^{(0)}, F^{(0)}, m_{dyn}^{(0)}$ are determined from experimental data, and, as usual in field theory, they depend on a renormalization scheme used in the calculations of loop diagrams. Although the GM model is nonrenormalizable, one can carry out the calculations by using an infinite number of counter terms (see Sec.5.5).

Phenomenological applications of the model were considered in Refs.[82, 94 - 96]. In particular, Weinberg initiated the anaysis of the model in the framework of $1/N_c$ expansion [94]. A particularly interesting result is that $g_A^{(0)} = 1$ in leading order in $1/N_c$ [94] and that $1/N_c$ loop corrections lead to $g_A(\sigma) < 1$ at $\sigma \sim m_{dyn}$ [95, 96]. The latter is necessary to explain the old result of nonrelativistic quark model concerning the nucleon coupling $g_A^{(N)}$ in β-decay. It is

$$g_A^{(N)} = \left(\frac{5}{3}\right) g_A , \tag{12.251}$$

which yields $g_A \simeq 0.75$ for the experimental value $g_A^{(N)} \simeq 1.25$.

However, all such calculations of the parameters in the framework of effective lagrangians are nesessarily based on some assumptions about the contribution of heavy fields (which are integrated out) to these parameters; for a recent discussion of this point, see Ref.[97].

We also note that, as was already pointed out in Sec.5.3, there are ambiguities in the choice of interpolating fields for the description of particles in chiral effective lagrangians. For example, instead of the quark field χ, we could use the field

$$\psi = \xi \chi_L + \xi^\dagger \chi_R \,, \tag{12.252}$$

which transforms as

$$\psi_L \to L\psi_L, \; \psi_R \to R\psi_R \tag{12.253}$$

under $SU_L(3) \times SU_R(3)$.

Although different fields lead to the same S matrix elements, a certain choice may be practically preferable. This is connected in particular with the point that in the reality, we usually can calculate only off-mass shell matrix elements, as in the case of the PCAC dynamics (See Sec.4.5). Therefore it is important to choose the "smoothest" interpolating fields, whose matrix elements are slowly varying as one goes a way off-shell.

From the practical viewpoint, this problem is particularly important for the description of vector mesons in the chiral lagrangian [80]. The success of vector-meson dominance (VMD) hypothesis [98], which is an analogue of PCAC for vector mesons, indicates that they, like pions, might play a special role in low energy hadronic dynamics. On the other hand, unlike PCAC, there is no nearby symmetric limit in which $M_V \to 0$, where M_V is a mass of vector mesons.

One interesting possibility is to use the field ρ_μ^W, introduced in Sec.5.3, as an interpolating field for vector mesons. As was indicated there, this field plays the role of a gauge field in the nonlinear realization of spontaneous chiral symmetry breaking. Recently this point has been developed further in a form of the hidden local $U_V(N)$ (or $SU_V(N)$) symmetry (for a review, see Ref.[99]). The subject is rapidly developing, and a more detailed discussion of this subject is beyond the scope of the present book.

In conclusion, we will consider, following Ref.[82, 83], heuristic rules for estimating the sizes of terms in the chiral effective lagrangian (the so called naive dimensional analysis).

There are two dimensional parameters in the chiral lagrangian: the decay constant F and the scale Λ_χ. The problem is to express the coefficients in the chiral lagrangian in terms of these two parameters.

Let us recall some basic points in the derivation of the effective lagrangian in the NJL model ($V_\mu = A_\mu = 0$). The fundamental scale in the model is Λ_χ; all other dimensional parameters are expressed through it. It is naturally to assume that Λ_χ is close to the mass of the lightest non-NG boson particle, ρ meson in QCD: $\Lambda_\chi \sim M_\rho \sim 2m_{dyn}$. As follows from the consideration in Sec.12.11, the characteristic point in the effective lagrangian derived from the NJL model is the appearance of $\frac{N_c}{16\pi^2}$ as an overall factor. It appears there because the lagrangian was derived by intergrating out quark fields, i.e. by calculating one-loop quark diagrams: the factor $\frac{1}{16\pi^2}$ comes from the measure $\frac{1}{(2\pi)^4}\int d^4q$ (see Eq.(9.66)). On the other hand, the decay constant F is

$$F^2 \simeq m_{dyn}^2 \frac{N_c}{4\pi^2} \ln\frac{\Lambda^2}{m_{dyn}^2} \sim \frac{N_c}{16\pi^2}\Lambda_\chi^2 \tag{12.254}$$

(see Eq.(12.197)). Therefore, the overall factor $\frac{N_c}{16\pi^2}$ is of the order of F^2/Λ_χ^2.

Let us look at the kinetic term (12.192). Introducing the interpolating fields

$$\begin{aligned} S_{\text{int}}^\alpha &= \left(\frac{N_c}{8\pi^2}\ln\frac{\Lambda_\chi^2}{m_{dyn}^2}\right)^{1/2} S^\alpha \sim \frac{F}{\Lambda_\chi}S^\alpha\,, \\ P_{\text{int}}^\alpha &= \left(\frac{N_c}{8\pi^2}\ln\frac{\Lambda_\chi^2}{m_{dyn}^2}\right) P^\alpha \sim \frac{F}{\Lambda_\chi}P^\alpha\,, \end{aligned} \tag{12.255}$$

which have the canonical kinetic term,

$$\mathcal{L}_k = \frac{1}{2}\partial_\mu S_{\text{int}}^\alpha \partial^\mu S_{\text{int}}^\alpha + \frac{1}{2}\partial_\mu P_{\text{int}}^\alpha \partial^\mu P_{\text{int}}^\alpha\,, \tag{12.256}$$

we come to the following rules to estimate the size of a dimensional coefficient related to any term in the effective lagrangian expressed through the interpolating fields [83]:

1. Include an overall factor F^2/Λ_χ^2;
2. Include a factor Λ_χ/F for each field;
3. Add factors of Λ_χ to get the dimension to four.

Let us discuss the physical meaning of these rules in more detail. Since Λ_χ/F is of order of the Yukawa coupling g_Y, the rule 2 reflects the point that a vertex producing a particle is proportional to g_Y. The rule 3 reflects the fact that there are no other dynamical sources (than those indicated in the rules 1 and 2) leading to the appearance of F in the lagrangian.

Looking at the kinetic term (12.192) in the ENJL model, we see that the same rules work for vector and axial-vector mesons in that case. This point reflects a universality of the interactions in the model: all coupling constants are of the order of g_Y there. Therefore the rules may be useful for the description of chiral lagrangians including also heavy particles if the universality of such a kind takes place in a theory.

If we are interested in a chiral lagrangian including NG bosons only, it is convenient to use the field $\Sigma = \exp(2i\pi/F)$ which has the kinetic term of the form $\frac{F^2}{4}\left(\partial_\mu\Sigma^\dagger\partial^\mu\Sigma\right)$. Then, one can check that the rules 1 - 3 lead to the estimate (5.137) for the coefficients in the chiral lagrangian expressed in terms of Σ.

12.15. Why are all these Models so Successful ? Green's Functions of Composite Operators in QCD.

Of course, the main reason of the success of those rather simple models considered in Sec.12.7 - 12.10 is connected with the fact that they combine the two most important features of hadron dynamics: spontaneous chiral symmetry breaking and the quark structure of hadrons. In this section, we will argue that another important reason of their success is intimately connected with the fact that hadrons are loose bound states, i.e. that their form factors rapidly decrease with increasing the momentum q at $q^2 \gtrsim \Lambda_\chi^2 > \Lambda_{QCD}^2$.

Let us start from the very beginning, that is from the question: what is the general framework for the description of bound states in local field theories ? Since the classical papers of Haag, Nishijima, and Zimmermann [100], it has been known that *any* local composite operator $O(x)$ can be chosen as an interpolating field for a bound state $|P\rangle$, as long as it satisfies the relation

$$\langle 0|O(x)|P\rangle = e^{-iPx} . \tag{12.257}$$

Then, the S matrix for bound states is expressed through Green's functions of these composite operators by means of the standard reduction formulas [100].

Thus the local composite operators can be used for the description of the dynamics of hadrons not just in the low energy region but at any energy ! The full information about the dynamics of hadrons contains in Green's functions of the local composite operators. More precisely, it contains in the on-mass shell matrix elements of the operators. The latter is an important point, especially in such theories as QCD. Let us discuss this issue in more detail.

In free theory, there is a one-to-one correspondence between particles and their fields. The situation remains essentially the same in weakly interacting theories such as QED: each particle is connected with its own interpolating field; the choice of the latter in perturbation theory is evident.

However, the situation in theories such as QCD, describing strongly interacting particles, is essentially different. Let us, for example, consider pion π^α in QCD. A possible (and natural) choice for its interpolating field is the local composite operator

$$O_\pi^\alpha(x) = C\bar{\psi}(x)i\gamma_5\frac{\lambda^\alpha}{2}\psi(x) , \tag{12.258}$$

where the normalization factor C provides the fulfillment of the condition (12.257). We can use this operator to describe the S matrix of pions. Let us now look at the propagator of this field:

$$G_\pi^{\alpha\beta}(x) = i\langle 0|TO_\pi^\alpha(x)O_\pi^\beta(0)|0\rangle = \frac{1}{(2\pi)^4}\int d^4pe^{-ipx}G_\pi^{\alpha\beta}(p) . \tag{12.259}$$

There is of course the pion pole contribution in $G_\pi^{\alpha\beta}(x)$, but there are also contributions from other dynamical sources. Actually, their contribution in $G_\pi^{\alpha\beta}(x)$ dominates. The point is that, since $O_\pi^\alpha(x)$ is a local operator, the propagator $G_\pi^{\alpha\beta}(x)$ contains contributions connected with all scales, from $p = 0$ through $p \to \infty$. At large p, because of asymptotic freedom, the dominant contribution coincides, up to some logarithmic corrections, with the contribution in free theory of quarks (see Fig.12.1). This contribution is quadratically divergent.

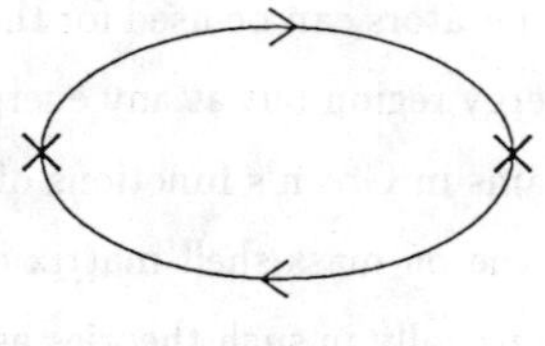

Fig. 12.1. The lowest order diagram contributing to the Green's function (12.259).

In a more formal language, this means that we cannot write the dispersion relation without subtractions for the Fourier transform $G_\pi^{\alpha\beta}(p)$; in this case, the ultraviolet behavior of its spectral function $\rho_\pi^{\alpha\beta}(s)$ requires two subtractions in the dispersion relation.*)

Although this fact is irrelevant for the on-mass shell matrix elements of pions, it is important from the practical viewpoint: the point is that usually Green's functions of composite operators can be determined only off-mass shell. This point is, for example, important for such a popular approach as QCD sum rules [52] based on dispersion relations for Green's functions of composite operators. In the practical use of the QCD sum rules, one has to adopt some prescriptions for defining a nonperturbative part of composite operators by subtracting the perturbative part connected with the short distance quark-gluon dynamics.

To make the present discussion to be more concrete, let us consider some examples.

Let us first consider the mass relation for the mass of pion. We now know two relations: the first is exact (see Eqs.(9.50), (12.48)),

$$M_{ab}^2 = -\frac{2iN_c\left(m_a^{(0)} + m_b^{(0)}\right)}{F_{ab}(2\pi)^4}\int d^4q\chi_{ab;1}^{(p)}\,, \tag{12.260}$$

the second is the famous PCAC relation

$$M_{ab}^2 = -\frac{\langle\bar{\psi}\psi\rangle}{F^2}\left(m_a^{(0)} + m_b^{(0)}\right) \tag{12.261}$$

(see Eq.(12.51)), which is valid to first order in $m_a^{(0)}$.

---

*) The present, somewhat heuristic, consideration is confirmed in a more rigorous treatment of this problem (see for example Ref.[101]).

Here $\chi^{(p)}_{ab;1}$ is the BS wave function of a pion composed of the a-th quark and the b-th antiquark, and $\langle\bar{\psi}\psi\rangle$ is the condensate which, in the chiral limit, is:

$$\langle\bar{\psi}\psi\rangle = -\lim_{x\to 0}\frac{i}{N}trG(x) = \frac{-4iN_c}{(2\pi)^4}\int d^4q\frac{B_{dyn}(q^2)}{q^2A^2(q^2)-B^2_{dyn}(q^2)} \tag{12.262}$$

(see Eq.(9.52)).

As we know, to first order in $m_a^{(0)}$, the relations (12.260) and (12.261) are equivalent. On the other hand, if we naively try to extend the relation (12.261) beyond this approximation by substituting $B_{dyn}(q^2)$ in Eq.(12.262) by the constituent quark mass function $B_a(q^2)$, the result would be disastrous since then $B_a(q^2) \sim m_a^{(0)}\left(\ln q^2/\Lambda^2_{QCD}\right)^{-c_m}$ (see Eq.(12.83)) and the condensate $\langle\bar{\psi}\psi\rangle$ would be quadratically divergent, while $m_a^{(0)} \sim \left(\ln\Lambda^2/\Lambda^2_{QCD}\right)^{-c_m}$ (see Eq.(12.19)). As a result, we would get an absurd formula for M^2_{ab} : M^2_{ab} quadratically diverges !

In order to avoid this and similar disasters, the standard prescription in the QCD sum rules approach is to substitute the composite operator $\bar{\psi}\psi$ for

$$N\left(\bar{\psi}_a\psi_b\right) = \bar{\psi}_a\psi_b - \langle 0|\bar{\psi}_a\psi_b|0\rangle_{pert}\ , \tag{12.263}$$

where $\langle 0|\bar{\psi}_a\psi_b|0\rangle_{pert} \sim O\left(m_a^{(0)}, m_b^{(0)}\right)$ is a perturbative contribution in the condensate. Unfortunately, this quantity is not well defined in the literature: although it is intuitively clear what we want to reach, there are ambiguities in separating the perturbative and nonperturbative contributions in $\langle 0|\bar{\psi}_a\psi_b|0\rangle$.

Since now we have the exact relation, Eq.(12.260), for M^2_{ab}, we can understand better the meaning of this prescription. The crucial point is that, unlike $B_a(q^2)$, the BS wave function $\chi^{(p)}_{ab;1}$ decreases rapidly with increasing q^2 at *any* value of $m_a^{(0)}$:

$$\chi^{(p)}_{ab;1} \sim (q^2)^{-2}\left(\ln q^2/\Lambda^2_{QCD}\right)^{-1+c_m} \tag{12.264}$$

(see Eq.(12.52)). Because of this and since $m_a^{(0)} \sim \left(\ln\Lambda^2/\Lambda^2_{QCD}\right)^{-c_m}$, we find that M^2_{ab} in the relation (12.260) remains finite as $\Lambda\to\infty$.

It is useful to rewrite this relation in the following form:

$$M_{ab}^2 = -\frac{\langle N\,(\bar{\psi}_a\psi_b)\rangle}{F_{ab}^2}\left(m_a^{(0)} + m_b^{(0)}\right)\ , \tag{12.265}$$

where

$$\langle N(\bar{\psi}_a\psi_b)\rangle \equiv \frac{2iN_c}{(2\pi)^4}\int d^4q\chi_{ab;1}^{(p)}\ . \tag{12.266}$$

Thus the BS wave function of pions takes care of subtracting the perturbative contribution from M_{ab}^2.

Of course, it is not easy to determine such a nonperturbative object as the BS wave function in QCD, but it is important that now a somewhat ambiguous prescription (12.263) becomes well defined.

Another example is the relation for the decay constant F_{ab}. In Sec.9.4, we derived the exact formula for F_{ab}:

$$iF_{ab}P^\mu = \frac{2N_c}{(2\pi)^4}\int d^4q\left[P^\mu\chi_{ab;2}^{(p)} + q^\mu\chi_{ab;3}^{(p)}\right]\ . \tag{12.267}$$

(see Eq.(9.71)). On the other hand, there exists the approximate Pagels-Stokar formula (12.126) for $F^2 = F_{ab}^2$ in the chiral limit:

$$F^2 = \frac{N_c}{8\pi^2}\int\limits_0^\infty duu\frac{B_{dyn}(u)\left[2B_{dyn}(u) - uB'_{dyn}(u)\right]}{\left(u + B_{dyn}^2(u)\right)^2}\ . \tag{12.268}$$

If we naively try to extend this formula beyond the chiral limit by substituting $B_{dyn}(q^2)$ by $B_a(q^2)$, then the irregular, slowly decreasing, piece in $B_a(q^2)$ (see Eq.(12.83)) would lead to a rather large contribution to F_{ab}^2. On the other hand, since the BS wave functions in the exact relation (12.267) rapidly decrease with increasing the momentum at any value of $m_a^{(0)}$, the dominant contribution to the decay constant is always connected with the infrared region [46, 102].

These two examples illustrate the general situation in QCD: since hadrons are loose bound states, all their characteristics are determined by the nonperturbative dynamics in the infrared region. This is in particular valid for the Gasser-Leutwyler constants L_i in the chiral effective lagrangian considered in Sec.12.13.

Now let us return to the question raised in the title of the present section: why are all these models so successful ? The answer to the question following from our consideration is: because all of them qualitatively correctly reproduce the property of looseness of hadrons. Let us recall what we were doing, for example, in the ENJL model. The crucial point in that model is the ultraviolet cutoff $\Lambda = \Lambda_\chi$. It cuts all form factors of hadrons at scale $q \sim \Lambda_\chi$. It is certainly a rather crude approximation to substitute the asymptotics (12.264) by a step-function, however, since hadron characteristics (at least many of them) are not very sensitive to the ultraviolet tail with $q \gtrsim \Lambda_\chi$, the approximation works reasonably well.

The models based on the improved ladder approximation take into account the ultraviolet tail, thus improving that approximation in the ENJL model. However the central point in the hadron dynamics remains the same in those models: hadrons are loose bound states.

In Chapters 14 - 15, discussing models of dynamical electroweak symmetry breaking, we will consider another type of bound states (tightly bound states).

Chapter 13

Introduction to Electroweak Symmetry Breaking

13.1. Introduction

Besides gravity, all the known fundamental interactions are described by the Standard Model (SM). The SM is a gauge theory with the gauge group $SU_c(3) \times SU_L(2) \times U_Y(1)$. The color group $SU_c(3)$ is connected with QCD considered in the previous chapter. The gauge group $SU_L(2) \times U_Y(1)$ is connected with electroweak interactions. There are four gauge bosons mediating electroweak interactions: the $W^\alpha, \alpha = 1, 2, 3$, relating to the $SU_L(2)$, and the B relating to the $U_Y(1)$. The electroweak part of the SM was elaborated by Glashow, Weinberg and Salam [1].

Unlike the color $SU_c(3)$, the $SU_L(2) \times U_Y(1)$ realizes in nature as a spontaneously broken gauge group. In particular, as a result, the physical vector bosons, $W^\pm$ and Z, and photon γ, are some combinations of the vector bosons W^α and B.

Electroweak physics is one of the central topics in modern particle physics.*) In the present chapter, we will focus on only one aspect of this subject: electroweak symmetry breaking which gives masses to the $W^\pm$ and Z, and to the quarks and leptons.

13.2. The Higgs Mechanism in the Standard Model

The SM includes three types of fields: a) twelve gauge bosons; b) three families of quarks and leptons; c) the Higgs field which is responsible for the breakdown of the electroweak group $SU_L(2) \times U_Y(1)$ to the $U_Q(1)$ of electromagnetism.

In this section, we will consider the dynamics in the sectors a) and c); the fermion sector will be considered in the next section.

Let us start from the $SU_L(2) \times SU_R(2) \times U_V(1)$ linear σ-model considered in Sec. 4.4. Its lagrangian density is:

$$\mathcal{L}_h = \frac{1}{2} tr \left[\partial_\mu M^+ \partial^\mu M\right] - \frac{\lambda}{12} tr(M^+ M M^+ M) - \frac{\mu^2}{2} tr(M^+ M), \qquad (13.1)$$

*) For a thorough consideration of this subject see the books [2,3,4]

where

$$M = \frac{1}{\sqrt{2}}(h + iw^\alpha \tau^\alpha);\, \alpha = 1, 2, 3 \tag{13.2}$$

(compare with Eqs.(4.53), (4.66) and (4.68); here we use different notations for the spinless bosons (h and w^α) to distinguish them from mesons σ and π^a in QCD.) As we know, at $\mu^2 < 0$,the chiral group $SU_L(3) \times SU_R(2) \times U_V(1)$ breaks spontaneously down to $SU_V(2) \times U_V(1)$:

$$\langle 0|M|0\rangle = \frac{1}{\sqrt{2}}\begin{pmatrix} v & 0 \\ 0 & v \end{pmatrix}, \tag{13.3}$$

where, in tree approximation, the vacuum expectation value $v = \langle 0|h|0\rangle$ is equal to

$$v = \left(\frac{-6\mu^2}{\lambda}\right)^{1/2} \tag{13.4}$$

(see Eq.(4.58)). The mass of the Higgs boson is equal to

$$M_h = \left(\frac{\lambda}{3}\right)^{1/2} v \tag{13.5}$$

(see Eq.(4.70)). The NG fields w^α are of course massless.

Under $SU_L(2) \times SU_R(2)$, M transforms as follows:

$$M \to LMR^+ \tag{13.6}$$

(see Eq. (4.52)).

Let us now gauge the subgroup $SU_L(2) \times U_{3R}(1)$ of the $SU_L(2) \times SU_R(2)$, where the $U_{3R}(1)$ transformations are given by $exp(i\varphi\tau_R^3)$. Then the gauge-field lagrangian density is

$$\mathcal{L}_g = -\frac{1}{4}W^\alpha_{\mu\nu}W^{\mu\nu;\alpha} - \frac{1}{4}B_{\mu\nu}B^{\mu\nu}, \tag{13.7}$$

where

$$\begin{aligned} W^\alpha_{\mu\nu} =& \partial_\mu W^\alpha_\nu - \partial_\nu W^\alpha_\mu - g_2\epsilon^{\alpha\beta\gamma}W^\beta_\mu W^\gamma_\nu, \\ B_{\mu\nu} =& \partial_\mu B_\nu - \partial_\nu B_\mu \end{aligned} \tag{13.8}$$

($\epsilon^{\alpha\beta\gamma}$ are the structure constants of $SU(2)$), and the lagrangian density for the Higgs field M becomes:

$$\mathcal{L}_h = \frac{1}{2} tr\left[D_\mu M^+ D^\mu M\right] - \frac{\lambda}{12} tr(M^+ M M^+ M) - \frac{\mu^2}{2} tr(M^+ M), \tag{13.9}$$

where the covariant derivative $D_\mu M$ is:

$$D_\mu M = \partial_\mu M + i g_2 W_\mu^\alpha \frac{\tau^a}{2} M - i g_1 B_\mu M \frac{\tau^3}{2}, \tag{13.10}$$

This form of $D_\mu M$ is dictated by the transformation law (13.6).

In this model, the gauge interactions break explicity $SU_L(2) \times SU_R(2)$ to $SU_L(2) \times U_{3R}(1)$. Therefore, the combined effect of spontaneous chiral symmetry breaking and explicit chiral symmetry breaking breaks $SU_L(2) \times SU_R(2)$ (and consequently $SU_L(2) \times U_{3R}(1)$) to the $U(1)$ subgroup generated by $\frac{1}{2}\left(\tau_L^3 + \tau_R^3\right)$.

Let us consider the Higgs mechanism in this model (see Sec.3.6). Substituting $\langle 0|M|0\rangle$ (13.3) into $\mathcal{L}_h$ (13.9), we find the physical mass eigenstates $W_\mu^\pm$, Z_μ and A_μ:

$$W_\mu^\pm = \frac{1}{\sqrt{2}}\left(W_\mu^1 \mp i W_\mu^2\right), \tag{13.11}$$

$$Z_\mu = \cos\theta_W W_\mu^3 - \sin\theta_W B_\mu, \tag{13.12}$$

$$A_\mu = \sin\theta_W W_\mu^3 + \cos\theta_W B_\mu, \tag{13.13}$$

where the weak mixing angle is given by

$$\cos\theta_W = \frac{g_2}{\sqrt{g_1^2 + g_2^2}}. \tag{13.14}$$

The photon A_μ is massless, the mass of the Z_μ is

$$M_Z = \frac{1}{2}\sqrt{g_1^2 + g_2^2}\, v, \tag{13.15}$$

and the mass of the $W_\mu^\pm$ is

$$M_W = \frac{1}{2} g_2 v = \frac{e}{2\sin\theta_W} v, \tag{13.16}$$

where $e = g_2 \sin\theta_W$ is the electromagnetic coupling constant. Thus we conclude that the Z-boson is heavier than the W-boson:

$$M_Z = \frac{1}{\cos\theta} M_W. \tag{13.17}$$

This equation implies that

$$\rho = M_W^2/(M_Z^2 \cos^2\theta_W) = 1 \tag{13.17a}$$

in tree approximation. As will be shown in Sec.13.5, this relation follows from the fact that at $g_1 = 0$ (or, what is the same, at $e = 0$) the lagrangian density $\mathcal{L}_h$ (13.9) is invariant with respect to the $SU_L(2) \times SU_R(2) \times U_V(1)$ which is larger than the $SU_L(2) \times U_Y(1)$.

The experimental values for M_Z, M_W, θ_W are [5]:

$$\begin{aligned} &M_Z = 91.175 \pm 0.021 GeV,\ M_W = 80.35 \pm 0.37 GeV, \\ &\sin^2\theta_W = 0.2261 \pm 0.0030 \end{aligned} \tag{13.18}$$

From here and Eq.(13.16), we find that $v \simeq 246 GeV$.

Thus, in the SM, three would-be NG bosons w^α are "eaten" by the W and Z, i.e. they transform into the longitudinal components of electroweak bosons. The only physical spinless boson in the SM is the Higgs boson h.

Sometimes it is convenient to present the M-field as

$$M = \begin{pmatrix} \varphi_1 & -\varphi_2^* \\ \varphi_2 & \varphi_1^* \end{pmatrix} \equiv (\varphi,\ \tilde{\varphi}), \tag{13.19}$$

where $\tilde{\varphi} = i\tau^2\varphi^*$ and

$$\varphi = \frac{1}{\sqrt{2}} \begin{pmatrix} h + iw^3 \\ -w^2 + iw^1 \end{pmatrix} \tag{13.20}$$

(see Eq.(13.2)).

In the higgs -gauge sector of the model, the gauge group is $SU_L(2) \times U_Y(1)$, where the generator of the $U_Y(1)$, hypercharge Y, can be identified with $\frac{1}{2}\tau_R^3$,

$$Y = \frac{1}{2}\tau_R^3, \tag{13.21}$$

and the electric charge generator Q is

$$Q = \frac{1}{2}\tau_L^3 + \frac{1}{2}\tau_R^3 = \frac{1}{2}\tau_L^3 + Y\,. \tag{13.22}$$

In this sector, the baryon and lepton numbers of all the fields equal zero. As we will see below, the definition of the hypercharge Y will change when the quark and lepton fields are introduced in the model.

13.3. Quarks and Leptons in the Standard Model

Introducing the quarks and leptons in electroweak theory is a nontrivial problem. The key point is to assign the correct quantum numbers to the fermions with respect to $SU_L(2) \times U_Y(1)$.

We know experimentally that there are at least six quarks in nature: u, d, s, c, b, and t (although the top quark has not yet been discovered, we know that there should exist an isospin partner of the b quark). The current masses of quarks are (see Sec.12.8) [5]*):

$$\begin{array}{lll} m_{cur;u} \simeq 5.6 \pm 1.1MeV & m_{cur;c} \simeq 1.35 \pm 0.05GeV & m_{cur;b} \simeq 5GeV \\ m_{cur;d} \simeq 9.9 \pm 1.1MeV & m_{cur;s} \simeq 199 \pm 33MeV & m_{cur;t} \gtrsim 91GeV\,. \end{array} \tag{13.23}$$

The masses of the four lightest quarks are evaluated at the scale $\sigma = 1GeV$; for the heavier b and t quarks, $m_{cur;b}$ and $m_{cur;t}$ are the " pole " masses.**)

There are also six leptons: electron e with $m_e \simeq 0.51MeV$, muon μ with $m_\mu \simeq 105.7MeV$, τ lepton with $m_\tau \simeq 1784MeV$, and their neutrinos, ν_e, ν_μ and ν_τ. So far,

*) At the time of finishing the writing of this book, the lower bound for $m_{cur;t}$, obtained from experiments at the Tevatron collider, has increased to $113GeV$.

**) The pole mass m is defined at the nearest singularity of the quark propagator $G(p)$ at $p^2 = m^2$ (for such heavy quarks, the confinement effects are neglected.)

only the left-handed neutrinos have been detected. The present laboratory limit on neutrino masses are as (at 95 % confidence limit) [5]:

$$m_{\nu e} < 7.3eV, m_{\nu\mu} < 270KeV, \ m_{\nu\tau} < 35MeV. \tag{13.24}$$

These fermions group in three families:

$$q^i_L = \begin{pmatrix} u^i_L \\ d^i_L \end{pmatrix}, \ q^i_R = \begin{pmatrix} u^i_R \\ d^i_R \end{pmatrix}, \ l^i_L = \begin{pmatrix} \nu^i_L \\ e^i_L \end{pmatrix}, \ l^i_R = \begin{pmatrix} \nu^i_R \\ e^i_L \end{pmatrix}, \tag{13.25}$$

where the family index i runs from 1 to 3. The first family includes

$$q^1_L = \begin{pmatrix} u_L \\ d_L \end{pmatrix}, \ q^1_R = \begin{pmatrix} u_R \\ d_R \end{pmatrix}, \ l^1_L = \begin{pmatrix} \nu_{Le} \\ e_L \end{pmatrix}, \ l^1_R = \begin{pmatrix} \nu_{Re} \\ e_R \end{pmatrix}; \tag{13.26}$$

the second includes

$$q^2_L = \begin{pmatrix} c_L \\ s_L \end{pmatrix}, \ q^2_R = \begin{pmatrix} c_R \\ s_R \end{pmatrix}, \ l^2_L = \begin{pmatrix} \nu_{L\mu} \\ \mu_L \end{pmatrix}, \ l^2_R = \begin{pmatrix} \nu_{R\mu} \\ \mu_R \end{pmatrix}, \tag{13.27}$$

and the third family consists of

$$q^3_L = \begin{pmatrix} b_L \\ t_L \end{pmatrix}, \ q^3_R = \begin{pmatrix} b_R \\ t_R \end{pmatrix}, \ l^3_L = \begin{pmatrix} \nu_{L\tau} \\ \tau_L \end{pmatrix}, \ l^3_R = \begin{pmatrix} \nu_{R\tau} \\ \tau_R \end{pmatrix}. \tag{13.28}$$

The experimental data concerning the decay of Z boson has proved that there exist only these three families with light neutrinos; other families (if they exist at all) must include heavy neutrinos with masses $m_\nu > M_Z/2 \simeq 45GeV$ [5].

We also note that although the right-handed neutrinos are not necessary for the consistency of the Standard Model, they are included here for completeness.

The three families have the same quantum numbers with respect to the $SU_c(3) \times SU_L(2) \times U_Y(1)$. In particular, the left-handed fermions q^i_L, l^i_L are doublets with respect to $SU_L(2)$, and the right-handed fermions q^i_R, l^i_R are singlets with respect to this group. The definition (13.21) of the hypercharge generator is modified now in the following way:

$$Y = T^3_R + \frac{B-L}{2}, \tag{13.29}$$

where T^3_R is the generator of $U_{3R}(1)$ (of course $T^3_R = \frac{1}{2}\tau^3_R$ in the fundamental representation of $SU_R(2)$), and B and L are baryon and lepton quantum numbers: $B = \frac{1}{3}, L = 0$

for quarks, and $B = 0, L = 1$ for leptons. We note that $\frac{B-L}{2}$ can be identified with the generator of the $U_V(1)$ in the $SU_L(2) \times SU_R(2) \times U_V(1)$. Its gauged subgroup $SU_L(2) \times U_Y(1)$ coincides with the $SU_L(2) \times U_{3R}(1)$ in the gauge-higgs sectors. The electric charge Q is

$$Q = T_L^3 + Y. \tag{13.30}$$

The quantum numbers of all the fields of the Standard Model are shown in Table 13.1.

Particle	$SU_C(3) \times SU_L(2) \times U_Y(1)$	Baryon Number	Lepton Number
G_μ^K	(8, 1, 0)	0	0
W_μ^α	(1, 3, 0)	0	0
B_μ	(1, 1, 0)	0	0
$q_L^i = \begin{pmatrix} u_L^i \\ d_L^i \end{pmatrix}$	(3, 2, 1/6)	1/3	0
$q_R^i = \begin{pmatrix} u_R^i \\ d_R^i \end{pmatrix}$	(3, 1, $T_R^3 + 1/6$)	1/3	0
$l_L^i = \begin{pmatrix} \nu_L^i \\ e_L^i \end{pmatrix}$	(1, 2, $-1/2$)	0	1
$l_R^i = \begin{pmatrix} \nu_R^i \\ e_R^i \end{pmatrix}$	(1, 1, $T_R^3 - 1/2$)	0	1
φ	(1, 2, $-1/2$)	0	0

Table 13.1. The fields of the Standard Model

Let us now consider the lagrangian density for quarks and leptons $\mathcal{L}_f$. It is convenient to distinguish two parts in $\mathcal{L}_f$:

$$\mathcal{L}_f = \mathcal{L}_{gf} + \mathcal{L}_{hf} \tag{13.31}$$

The part $\mathcal{L}_{gf}$ describes the interactions of the fermions with the $SU_L(2) \times U_Y(1)$ gauge fields (here we omit QCD dynamics considered in the previous chapter):

$$\mathcal{L}_{gf} = i\bar{q'}_L^i \gamma^\mu D_\mu {q'_L}^i + i\bar{q'}_R^i \gamma^\mu D_\mu {q'_R}^i + i\bar{l'}_L^i \gamma^\mu D_\mu {l'_L}^i + i\bar{l'}_R^i \gamma^\mu D_\mu {l'_R}^i \tag{13.32}$$

(the meaning of the primes here will be clear in a moment), where

$$D_\mu {q'_L}^i = \partial_\mu {q'_L}^i + ig_2 W_\mu^\alpha \frac{\tau^\alpha}{2} {q'_L}^i + ig_1 B_\mu Y {q'_L}^i, \tag{13.33}$$

$$D_\mu {q'_R}^i = \partial_\mu {q'_R}^i + i g_1 B_\mu Y {q'_R}^i, \tag{13.34}$$

$$D_\mu {l'_L}^i = \partial_\mu {l'_L}^i + i g_2 W_\mu^\alpha \frac{\tau^\alpha}{2} {l'_L}^i + i g_1 B_\mu Y {l'_L}^i, \tag{13.35}$$

$$D_\mu {l'_R}^i = \partial_\mu {l'_R}^i + i g_1 B_\mu Y {l'_R}^i. \tag{13.36}$$

The part $\mathcal{L}_{hf}$ describes Yukawa interactions between the fermions and the Higgs field:

$$\begin{aligned}\mathcal{L}_{hf} = & -\sqrt{2}\bar{q'}_L R_u M \frac{1+\tau_3}{2} q'_R - \sqrt{2}\bar{q'}_L R_d M \frac{1-\tau_3}{2} q'_R - \sqrt{2}\bar{l'}_L R_\nu M \frac{1+\tau_3}{2} l'_R \\ & - \sqrt{2}\bar{l'}_L R_e M \frac{1-\tau_3}{2} l'_R + H.c.,\end{aligned} \tag{13.37}$$

where R_u, R_d, R_ν, R_e are matrices in the family space, and $H.c.$ means the hermitian conjugated part.

A nontrivial point is that these family matrices can be non-diagonal. This leads to mixing between different families. Such a mixing in the case of two quark families was described by Cabibbo [6]; the mixing effect for three quark families was described by Kobayashi and Maskawa [7].

The primes in Eqs.(13.32)–(13.37) imply that these fermion fields are eigenstates of electroweak gauge couplings (a weak basis). Each of the matrices R_u, R_d, R_ν, R_e can be diagonalized by a biunitary transformation:

$$V_u^{-1} R_u W_u = \tilde{R}_u, \text{etc.}, \tag{13.38}$$

where V_u and W_u are unitary matrices, and $\tilde{R}_u$ stands for a diagonal matrix. This result can be shown by applying the polar decomposition theorem staying that any matrix K can be written as

$$K = HU, \tag{13.39}$$

where H is an hermitian matrix and U is a unitary matrix. Then, since H can be diagonalized by a unitary matrix V,

$$V^{-1} H V = \tilde{H}, \tag{13.40}$$

we obtain that

$$V^{-1} K W = \tilde{H} \tag{13.41}$$

with $W = U^{-1} V$.

The biunitary transformations (13.38) correspond to the transition to the mass eigenstate basis:

$$\begin{aligned} u'_L &= V_u u_L,\ u'_R = W_u u_R, \\ d'_L &= V_d d_L,\ d'_R = W_d d_R, \\ \nu'_L &= V_\nu \nu_L,\ \nu'_R = W_\nu \nu_R, \\ e'_L &= V_e e_L,\ e'_R = W_e e_R, \end{aligned} \tag{13.42}$$

where the fields without primes are mass eigenstates.

One can see that while these transformations do not change the form of the neutral (electromagnetic and Z_μ) currents, they change the form of the charge $SU_L(2)$ currents:

$$\begin{aligned} \jmath^+_{q\mu} \equiv& \bar{u}'_L \gamma_\mu d'_L = \bar{u}_L V_q \gamma_\mu d_L, \\ \jmath^+_{l\mu} \equiv& \bar{\nu}'_L \gamma_\mu e'_L = \bar{\nu}_L V_l \gamma_\mu e_L, \\ \jmath^-_{q\mu} \equiv& (\jmath^+_{q\mu})^\dagger, \jmath^-_{l\mu} \equiv (\jmath^+_{l\mu})^\dagger, \end{aligned} \tag{13.43}$$

where the matrices

$$V_q = V_u^\dagger V_d \tag{13.44}$$

$$V_l = V_\nu^\dagger V_e \tag{13.45}$$

are called the Cabibbo - Kobayashi - Maskawa (CKM) matrices for quarks and leptons, respectively.

We note that in the SM, among the eight transformation matrices in Eq.(13.42), only the two matrices V_q and V_l are observable. This result follows from the fact that the gauge part of the SM is invariant with respect to the following flavor symmetry

$$SU_{q_L}(n) \times SU_{u_R}(n) \times SU_{d_R}(n) \times SU_{l_L}(n) \times SU_{\nu_R}(n) \times SU_{e_R}(n), \tag{13.46}$$

where n is a number of the fermion families. This symmetry implies that without loss of generality, we can choose a weak basis in which all the W-matrices and one of V_u, V_d and one of V_ν, V_e in Eq.(13.42) are equal to the identity matrix I. In this basis, $V_q = V_u^\dagger$ or $V_q = V_d$, and $V_l = V_\nu^\dagger$, or $V_l = V_e$. In particular, if neutrinos are massless, then, without loss of generality, one can choose $V_e = V_\nu = I$ which leads to $V_l = 1$. Therefore in the SM, the lepton mixing becomes physically meaningful only if neutrinos have finite masses.

We note that the flavor symmetry (13.46) is a generalization of the Glashow - Illiopoulos - Maiani (GIM) symmetry originally introduced for the case $n = 2$ [8]. The GIM flavor symmetry guarantees the absence of flavor-changing neutral current effects in tree approximation in the SM (for more detail see the next section).

Let us describe the physical (observable) parameters in the CKM quark matrix V (henceforth we will omit a subscript q in this matrix). In the general case, a $n \times n$ matrix contains $2n^2$ real parameters. The unitary condition $V^\dagger V = I$ reduces the number of parameters to n^2. Of the remaining n^2 parameters, $2n - 1$ are non-observable phases which can be removed by redefining the phases of n U-quarks and $(n-1)$ D-quarks. Therefore the total number of observable parameters is $n^2 - (2n-1) = (n-1)^2$. Since the number of independent rotations in an n dimensional space is $n_\theta = \frac{1}{2}n(n-1)$, we find that among those $(n-1)^2$ parameters, there are $n_\theta = \frac{1}{2}n(n-1)$ angles and $n_\delta = \frac{1}{2}(n-1)(n-2)$ phases. The phase parameters are responsible for CP violation.

Thus, at $n = 2$, we find that $n_\theta = 1$ and $n_\delta = 0$, i.e. there is no CP violation for two quark families in the SM. At $n = 3$, the number of angles is $n_\theta = 3$ and the number of phases is $n_\delta = 1$.

There are several parametrizations of the 3×3 CKM matrix [5]. For example, the original parametrization of Kobayashi and Maskawa takes the following form [7]:

$$V = \begin{pmatrix} V_{ud} & V_{us} & V_{ub} \\ V_{cd} & V_{cs} & V_{cb} \\ V_{td} & V_{ts} & V_{tb} \end{pmatrix} = \begin{pmatrix} c_1 & -s_1 c_3 & -s_1 s_3 \\ s_1 c_2 & c_1 c_2 c_3 - s_2 s_3 e^{i\delta} & c_1 c_2 s_3 + s_2 c_3 e^{i\delta} \\ s_1 s_2 & c_1 s_2 c_3 + c_2 s_3 e^{i\delta} & c_1 s_2 s_3 - c_2 c_3 e^{i\delta} \end{pmatrix}, \tag{13.47}$$

where $c_i = \cos\theta_i$ and $s_i = \sin\theta_i$ for $i = 1, 2, 3$. At present, the 90 % confidence limits on the magnitude of the elements of V are [5]:

$$V = \begin{pmatrix} 0.9747 - 0.9759 & 0.218 - 0.224 & 0.001 - 0.007 \\ 0.218 - 0.224 & 0.9734 - 0.9752 & 0.030 - 0.058 \\ 0.003 - 0.019 & 0.029 - 0.058 & 0.9983 - 0.9996 \end{pmatrix}. \tag{13.48}$$

In conclusion, we would like to point out that important fact that, unlike QCD, the electroweak gauge theory is a chiral gauge theory , i.e., in accordance with the definition in Sec.9.1, the fermions, expressed in terms of left-handed fields, make up a complex representation of the $SU_L(2) \times U_Y(1)$. Indeed, let us look at the Table 13.1. Now, instead of right-handed fields q_R^i and l_R^i, we have to use the left-handed fields $q_L^{c\,i} \equiv (q_R^i)^c$ and $l_L^{c\,i} \equiv (l_R^i)^c$. One can see that each family makes up a complex representation of the $U_Y(1)$ (as is known, all representations of $SU(2)$ are real).

Among the important consequences, which this fact leads to, we would like to indicate just one connected with the condition of the absence of the anomaly connected with the $SU_L(2) \times U_R(1)$ fermion currents.*) As was first shown by Bouchiat, Iliopoulos and Meyer [9] and Gross and Jackiw [10], the ABJ anomaly spoils the renormalizability of gauge theories. Since there is no anomaly in real representations of groups [11], vector-like gauge theories, such as QCD, are automatically anomaly free. In the case of a chiral gauge theory, one should take care in constructing anomaly-free representations of the gauge group.

One can show that in the case of the $SU_L(2) \times U_Y(1)$ gauge theory with fermions being either doublets or singlets under $SU_L(2)$, the condition of the absence of the anomaly takes the following simple form [2,11]

$$\sum_{doublets} Q = \sum_{singlets} Q = 0 \tag{13.49}$$

By using the Table 13.1, one can easily check that this condition is fulfilled in each fermion family of the SM.

13.4. Flavor-Changing Neutral Currents in the Standard Model

One of the most striking experimental fact in electroweak physics is a strong suppression of flavor-changing neutral current (FCNC) effects. For example, the decay $K_L^0 \to \mu^+\mu^-$ is suppressed with respect to the "normal" weak decay $K^+ \to \mu^+\nu_\mu$ (connected with charged weak currents) by a factor of 10^{-8} [5]. The SM yields a natural explanation of this suppression based on the GIM mechanism (flavor symmetry). As we noted in the previous section, the flavor symmetry (13.46) of the gauge interactions guarantees that the neutral, electromagnetic and Z_μ, fermion currents are diagonal in the family space in tree approximation. This leads to the absence of FCNC effects in this approximation.

*) This anomaly is an analogue of the ABJ anomaly connected with the $U_5(1)$ flavor current in QCD considered in Chapter 12. The essential difference as compared to that case is that now the anomaly is connected with gauge currents.

Actually, it is easy to see that FCNC effects would completely disappear in the SM if quarks with the same electric charge were degenerate. Indeed, let us consider the unitary gauge in which the Higgs field $M = \frac{1}{\sqrt{2}}h$ (w^α are "eaten" by $W^\pm$ and Z). If quarks with the same charge are degenerate, the lagrangian density $\mathcal{L}_{hf}$ (3.37) is invariant with respect to the flavor symmetry

$$SU_u(n) \times SU_d(n).$$

Let us assume that the CKM matrix V is $V = V_u^+ V_d$ in a basis of mass eigenstates u^i and d^i. Then, we can choose another mass-eigenstate basis $u = V_u^+ \tilde{u}, d = V_d^+ \tilde{d}$, in which $V \to I$.

Therefore FCNC effects in the SM appear only because quarks from different families have different masses. Let us illustrate this by considering the $K^0 - \bar{K}^0$ mixing.

In QCD, the $SU_V(3)$ wave functions of K^0 and $\bar{K}^0$ are: $K^0 = |d\bar{s}\rangle, \bar{K}^0 = |s\bar{d}\rangle$ (see Eq.(12.106)). They are degenerated in QCD. However, weak interactions lead to mixing of these states which leads to a mass splitting: the states with definite CP parity, $K_1^0 = \frac{K^0 + \bar{K}^0}{\sqrt{2}}, K_2^0 = \frac{K^0 - \bar{K}^0}{\sqrt{2}}$, have different masses (here we neglect small CP violating effects). In the SM, the dominant contribution to this mixing is given by the box diagrams shown in Fig.13.1.

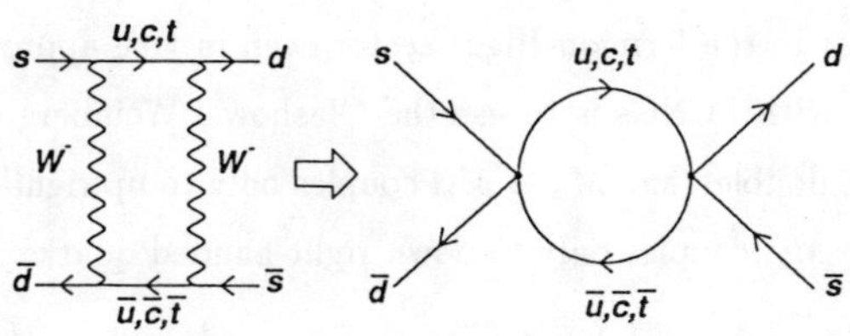

Fig. 13.1. The box diagrams leading to K^0–$\bar{K}^0$ mixing.

These diagrams were first calculated by Gaillard and Lee [12] (see also the books [2,3,4]). Neglecting the contribution of the t quark (which is small because of the very weak mixing between the first and third families (see Eq.(13.48)), one finds the following low energy effective lagrangian describing this process:

$$\mathcal{L} = G_2 \bar{s}\gamma_\mu(1+\gamma_5)d \cdot \bar{s}\gamma^\mu(1+\gamma_5)d, \tag{13.50}$$

where

$$G_2 = \frac{1}{16\pi^2} G_F^2 (m_{cur;c} - m_{cur;u})^2 \sin^2\theta_c \cos^2\theta_c. \tag{13.51}$$

Here the Fermi constant G_F is

$$G_F = \frac{\sqrt{2}g_2^2}{8M_W^2} \simeq 1.166 \cdot 10^{-5} GeV^{-2}, \tag{13.52}$$

and θ_c is the Cabibbo angle ($\sin\theta_c \simeq 0.22$; $\theta_c \simeq -\theta_1$ in Eq.(13.47)).

The lagrangian density (13.50) leads to the following mass difference between K_1^0 and K_2^0 [2-4,12]:

$$\Delta M_{12} = M_2 - M_1 \simeq \frac{4m_{cur;c}^2 \cos^2\theta_c}{3\pi m_\mu^2} \Gamma(K^+ \to \mu^+ \nu_\mu). \tag{13.53}$$

A comparison with the experimental data yields $m_{cur;c} \sim 1GeV$.

We note that the situation with flavor-changing neutral currents (FCNCs) may be essentially different beyond the SM. For example , in models with two Higgs fields,

$$M_1 = (\varphi_1, \tilde{\varphi}_1), \;\; M_2 = (\varphi_2, \tilde{\varphi}_2), \tag{13.54}$$

there are five physical spinless particles (for a review of these models see the book [13]). Three of them are neutral (h^0, H^0 and A^0), and two , $H^\pm$, are charged. In this case FCNCs may appear in the fermion-Higgs sector even in tree approximation. A way to avoid the problem with FCNCs is to use the Glashow - Weinberg (GW) scheme [14] in which one Higgs doublet, say $M_u \equiv M_1$, couples only to up right-handed quarks, and another, say $M_d \equiv M_2$, couples only to down right-handed quarks:

$$\mathcal{L}_{hf} = -\sqrt{2}\bar{Q}'_L R_u M_u \frac{1+\tau_3}{2} Q'_R - \sqrt{2}\bar{Q}'_L R_d M_d \frac{1-\tau_3}{2} Q'_R + H.c. \tag{13.55}$$

Since in this scheme the diagonalization of the mass matrices for quarks automatically leads to a diagonal coupling of quarks to the neutral Higgs bosons, FCNCs are absent here in tree approximation.

However, even in the GW scheme, there are additional sources of induced FCNCs in two-Higgs-doublet models as compared to the SM. For example, besides the box diagrams in Fig.13.1, there are also the ones with $W^\pm$ replaced by $H^\pm$. This leads to some constraints for masses and coupling constants of $H^\pm$ [15].

Although the GW scheme is not only way to suppress FCNCs in two - Higgs - doublet models [16], it is apparently the most interesting and natural. We will return to this scheme in Sec.15.6.

13.5. The Technicolor Scenario

The Higgs sector of the SM is just the $SU_L(2) \times SU_R(2) \times U_V(1)$ linear σ-model (see Eq.(13.1)). In the case of the hadronic dynamics with two lightest quark flavors (u and d), we know that the σ-model is not a fundamental theory but a low energy effective theory. The fundamental theory of the hadronic dynamics is QCD. The field $M_\sigma = \frac{1}{\sqrt{2}}(\sigma + i\pi^\alpha \tau^\alpha)$ is a composite field:

$$\sigma \sim -\bar{\psi}\psi, \ \pi^\alpha \sim i\bar{\psi}\tau^\alpha\gamma_5\psi, \tag{13.56}$$

where ψ is a quark field (compare with Eq.(4.43)). Thus, spontaneous chiral symmetry breaking in the low energy effective theory (σ-model) becomes dynamical chiral symmetry breaking in the fundamental theory (QCD).

The main idea of the technicolor scenario introduced by Weinber [17] and Susskind [18] is to use a QCD-like dynamics for the description of electroweak symmetry breaking. In this scenario the Higgs field

$$M = \frac{1}{\sqrt{2}}(h + iw^\alpha \tau^\alpha) \tag{13.57}$$

is composite:

$$h \sim -\bar{F}F, \ W^\alpha \sim i\bar{F}\tau^\alpha\gamma_5 F, \tag{13.58}$$

where F is a technifermion field. Technifermions F are new fermions participating in a hypothetical QCD-like interaction called the technicolor interaction (TC). The decay constant F_{TC} of technipions are defined in the usual way

$$\langle 0|J^\alpha_{5\mu}|P;\beta\rangle = i\delta_{\alpha\beta}F_{TC}P_\mu, \tag{13.59}$$

where $J^\alpha_{5\mu}$ are technifermion axial currents. Since a low energy effective theory of TC coincides with a $SU_L(N) \times SU_R(N) \times U_V(1)$ linear σ-model , where N is a number of the technifermion flavors, it is evident that if technipions (at least some of them) are assigned to nontrivial representations of the electroweak group $SU_L(2) \times U_Y(1)$, then $W^{\pm}$ and Z acquire masses by "eating" some technipions. To be more concrete, let us consider some examples.

The simplest TC model includes just one (colorless) $SU_L(2)$-doublet of techni - fermions (the minimal TC model):

$$F_L = \begin{pmatrix} U_L \\ D_L \end{pmatrix}, \; F_R = \begin{pmatrix} U_R \\ D_R \end{pmatrix}. \tag{13.60}$$

The hypercharge of F is (see Eq.13.30)

$$Y(F) = Q(F) - T_L^3(F). \tag{13.61}$$

The value of the electric charge $Q(F)$ is determined from the condition (13.49) that F_L, F_R make up an anomaly free representation of the $SU_L(2) \times U_V(1)$, From that condition and Eq.(13.61) we find that

$$Q(U) = +\frac{1}{2}, \; Q(D) = -\frac{1}{2}. \tag{13.62}$$

We assume that the doublet F is assigned to the fundamental representation of the gauge technicolor group $SU(N_{TC})$. The chiral group is $SU_L(2) \times SU_R(2) \times U_V(1)$ here, and it is spontaneously broken to $U_V(2)$.*) Therefore there is only one composite Higgs field in the model, and its low energy effective theory coincides with the SM.**) In particular, the decay constant F_{TC} (13.59) is

$$F_{TC} = v \simeq 246 GeV. \tag{13.63}$$

This decay constant is more than three orders of magnitude larger than the decay constant $F = F_\pi \simeq 93 MeV$ in QCD.

---

*) At $N_{TC} = 2$, because representations of $SU(2)$ are real (strictly speaking, pseudo-real), the chiral group is $SU(4)$, and it is spontaneously broken to $SP(4)$ [19]. Here, for simplicity, we will consider the case with $N_{TC} \geq 3$, although $N_{TC} = 2$ may be also viable [20].

**) As in QCD, because of the ABJ anomaly, the chiral group here is $SU_L(2) \times SU_R(2) \times U_V(1)$ and not $U_L(2) \times U_R(2)$ The fourth pseudoscalar technimeson, a TC analogue of η', is heavy due to this anomaly (see Sec.12.12)

Although the low energy effective action of this model coincides with the SM, the model predicts a lot of new particles, technihadrons, in the TeV region. The spectrum of technihadrons should resemble the usual QCD spectrum rescaled by the factor

$$\frac{F_{TC}}{F_\pi}\sqrt{\frac{N_c}{N_{TC}}}|_{N_c=3} = \frac{F_{TC}}{F_\pi}\sqrt{\frac{3}{N_{TC}}} \simeq \sqrt{\frac{3}{N_{TC}}}2600 GeV. \tag{13.64}$$

Here we have taken into account the fact that $F_{TC} \sim \sqrt{N_{TC}}$ and $F_\pi \sim \sqrt{N_c}$ (see Eq.(12.126)).

Let us discuss the origin of the relation (13.17a),

$$\rho = M_W^2/(M_Z^2 \cos^2\theta_W) = 1 + 0(\alpha), \tag{13.65}$$

in this TC model.

For this purpose, it is convenient to use an interpretation of the dynamical Higgs mechanism introduced originally by Schwinger [21] and developed further by Migdal and Polyakov, Jackiw and Johnson, and Cornwall and Norton [22]. Let us first consider the case with $g_1 = 0$. The W^α-boson propagator is:

$$D_{\mu\nu}^{\alpha\beta}(q^2) = \delta^{\alpha\beta}(g_{\mu\nu} - \frac{q_\mu q_\nu}{q^2})\frac{1}{q^2 - g_2^2\Pi(q^2)}, \tag{13.66}$$

where $\Pi(q^2)$ is the polarization operator (for simplicity, we consider the Landau gauge):

$$\Pi_{\mu\nu}^{\alpha\beta}(q) = \delta^{\alpha\beta}(g_{\mu\nu} - \frac{q_\mu q_\nu}{q^2})\Pi(q^2) = i\int d^4x e^{iqx}\langle 0|TJ_{L\mu}^\alpha(x)J_{L\nu}^\beta(0)|0\rangle, \tag{13.67}$$

where $J_{L\mu}^\alpha$ are the chiral currents of technifermions (we are interested here just in this piece of the $SU_L(2)$ currents). Because of spontaneous chiral symmetry breaking, there is a contribution of the NG bosons $|P;\alpha\rangle$ to $\Pi(q^2)$ which is (see Eq.(13.59)):*)

$$\delta^{\alpha\beta}(g_{\mu\nu} - \frac{q_\mu q_\nu}{q^2})\Pi_{NG}(q^2) = \delta^{\alpha\beta}(q^2 g_{\mu\nu} - q_\mu q_\nu)\cdot\frac{F_{TC}^2}{4q^2}. \tag{13.68}$$

The pole at $q^2 = 0$ leads to the same mass $M_W = \frac{g_2}{2}F_{TC}$ for three vector bosons W^α.

---

*) The constant term (proportional to $g_{\mu\nu}$) in Eq.(13.68) is determined from the condition of the transversality of the polarization tensor $\Pi_{\mu\nu}^{\alpha\beta}(q)$.

We note that in the general case, the spontaneous breakdown of $SU_L(2)$ to I (the trivial identity group) does not guarantee that the charged, $W^{\pm}$, and the neutral, W^3, vector bosons have the same mass: the decay constant $F_{TC}^{\pm}$ of the charged NG bosons $w^{\pm}$ may be in general different from the decay constant F_{TC}^{0} of the w^0. However, the present TC model (at $g_1 = 0$) has a larger, $SU_L(2) \times SU_R(2)$, symmetry which breaks spontaneously to the diagonal subgroup $SU_V(2)$. Since three NG bosons w^{α} are assigned to the irreducible representation of the $SU_V(2)$, they have the same decay constant. At $g_1 \neq 0$, the situation is slightly more complicated due to mixing between the W^3 and B. Considering the polarization tensor in the neutral sector, we are led to the mass matrix for the neutral gauge bosons:

$$M^2 = \begin{pmatrix} g_2^2 & g_1 g_2 \\ g_1 g_2 & g_1^2 \end{pmatrix} \frac{F_{TC}^2}{4}. \tag{13.69}$$

The eigenvalues of this matrix are

$$M_\gamma^2 = 0, \ M_Z^2 = \frac{1}{4}(g_1^2 + g_2^2) F_{TC} = \frac{M_W^2}{cos^2\theta_W}. \tag{13.70}$$

Thus, we have got the relation (13.65):

$$\rho = M_W^2/(M_Z^2 \cos^2\theta_W) = 1 + 0(\alpha) \tag{13.71}$$

(we recall that $\alpha = \frac{e^2}{4\pi} \sim g_1^2$). In the case of the absence of the isospin $SU_V(2)$ symmetry, we would get

$$\rho = (F_{TC}^{\pm}/F_{TC}^{0})^2. \tag{13.72}$$

Thus this symmetry is crucial for providing the validity of the relation (13.65). It is called a custodial symmetry in the literature [23].

Since experimental data show that $\delta\rho = \rho - 1$ is no more than a few tenth percent [24], they imply that the custodial $SU_V(2)$ is a good approximate symmetry in nature.

We also note that, unlike the SM, there is no unitary gauge (in which the composite NG bosons are excluded) in models with dynamical gauge symmetry breaking. In this case, one has to prove that S matrix is unitary in a physical subspace, where the NG bosons are absent. This fact was proved in Ref.[22].

Let us now consider another TC model: the one family TC model [25]. In this model, technifermions make up a standard family:

$$Q^k_{Lc} = \begin{pmatrix} U^k_{Lc} \\ D^k_{Lc} \end{pmatrix}, \; L^k_L = \begin{pmatrix} N^k_L \\ E^k_L \end{pmatrix}, \tag{13.73}$$

$$Q^k_{Rc} = \begin{pmatrix} U^k_{Rc} \\ D^k_{Rc} \end{pmatrix}, \; L^k_R = \begin{pmatrix} N^k_R \\ E^k_R \end{pmatrix}, \tag{13.74}$$

where $c = 1,2,3$ is a color index, and $k = 1,2,\cdots,N_{TC}$ is a technicolor index; we consider the $SU(N_{TC})$ technicolor group. Neglecting color and electroweak forces, we find that the chiral group is $SU_L(8) \times SU_R(8)$ in this model. The technicolor interactions lead to the spontaneous breakdown of this symmetry to $SU_V(8)$. As a result, 63 massless NG bosons appear. However, switching on color and electroweak interactions will lead to masses for a part of these bosons (these interactions act as electromagnetic interactions giving a contribution to the masses of $\pi^\pm$). We will return to the problem of their masses in the next section.

We will label the 63 NG bosons as $\Pi^{\tilde{\alpha}} \sim i\bar{F}\frac{\lambda^{\tilde{\alpha}}}{2}\gamma_5 F$, where

$$F = \begin{pmatrix} Q_1 \\ Q_2 \\ Q_3 \\ L \end{pmatrix}, \tag{13.75}$$

and $\lambda^{\tilde{\alpha}}$ are the traceless $SU(8)$ generators; $\tilde{\alpha} = 1,2,\cdots,63$. The NG bosons couple to the axial currents $J^{\tilde{\alpha}}_{5\mu} = \bar{F}\frac{\lambda^{\tilde{\alpha}}}{2}\gamma_\mu\gamma_5 F$:

$$\langle 0|J^{\tilde{\alpha}}_{5\mu}|\Pi^{\tilde{\beta}}\rangle = i\delta_{\tilde{\alpha}\tilde{\beta}}P_\mu F_{TC}. \tag{13.76}$$

The triplet of the NG bosons absorbed by the $W^\pm$ and Z is given by the following combination

$$\Pi^\alpha \sim (\bar{Q}\tau^\alpha\gamma_5 Q + \bar{L}\tau^\alpha\gamma_5 L); \alpha = 1,2,3. \tag{13.77}$$

Since the $SU_L(2)$ electroweak currents

$$J^\alpha_{L\mu} = \bar{F}\frac{\tau^\alpha}{2}\frac{1+\gamma_5}{2}F = 2\bar{F}\frac{\lambda^{\tilde{\alpha}}}{2}\frac{1+\gamma_5}{2}F|_{\tilde{\alpha}=\alpha}, \tag{13.78}$$

where $\frac{\lambda^{\tilde{\alpha}}}{2}$ with $\tilde{\alpha} = \alpha$ are the corresponding $SU(8)$ matrices normalized (as usual) such that $tr\left(\frac{\lambda^\alpha}{2}\right)^2 = \frac{1}{2}$, we find that

$$\langle 0|J^\alpha_{5\mu}|\Pi^\beta\rangle = 2i\delta_{\alpha\beta}P_\mu F_{TC}. \tag{13.79}$$

Comparing this equation with Eq.(13.68) in the one doublet TC model, we find that the $W^\pm$ mass is

$$M_{W^\pm} = \frac{1}{2}g_2(2F_{TC}) \tag{13.80}$$

in the present model. Therefore, the decay constant F_{TC} in this model is

$$F_{TC} = \frac{1}{2}v \simeq 123GeV. \tag{13.81}$$

In fact, it is easy to show that the relation (13.80) is a manifestation of the following general rule: if a TC model has r technifermion doublets creating NG bosons with the same decay constant F_{TC}, then F_{TC} is

$$F_{TC} = \frac{1}{\sqrt{r}}v \simeq \frac{1}{\sqrt{r}}246GeV. \tag{13.82}$$

Indeed, the flavor matrices in the axial currents $J^\alpha_{5\mu} = \bar{F}r^\alpha\gamma_5 F$ are $r^\alpha \equiv I_r\frac{\tau^\alpha}{2}$, where I_r is the $r\times r$ identity matrix. These matrices are normalized as

$$tr(r^\alpha r^\beta) = \frac{1}{2}r\delta_{\alpha\beta}. \tag{13.83}$$

Therefore if the decay constant F_{TC} for the axial currents $J^{\tilde{\alpha}}_{5\mu} = \bar{F}\frac{\lambda^{\tilde{\alpha}}}{2}\gamma_\mu\gamma_5 F$, $tr\left(\frac{\lambda^{\tilde{\alpha}}}{2}\frac{\lambda^{\tilde{\beta}}}{2}\right) = \frac{1}{2}$, is defined from Eq.(13.76), then

$$\langle 0|J^\alpha_{5\mu}|\Pi^\beta\rangle = \sqrt{r}i\delta_{\alpha\beta}P_\mu F_{TC}, \tag{13.84}$$

and we are led to the relation (13.81).

Since there are four doublets in the one family TC model, we recover the relation (13.80) in this case.

So far we have not considered the mechanism of generating masses for quarks and leptons in TC modes. We will discuss this problem in the next section.

13.6. Extended Technicolor

Since there is a composite Higgs field $M(x)$ (13.57) in all TC models, there is no problem to write a $SU_L(2) \times U_Y(1)$ invariant lagrangian density leading to masses for the ordinary fermions (quarks and leptons). It has exactly the same form as $\mathcal{L}_{hf}$ (13.37) in the SM. We now rewrite it in the following form:

$$\mathcal{L}_{hf} = -\sqrt{2}\bar{f}'_L R_{f_u} M \frac{1+\tau_3}{2} f'_R - \sqrt{2}\bar{f}'_L R_{f_d} M \frac{1-\tau_3}{2} f'_R + H.c.. \tag{13.85}$$

Here the field f includes both quarks and leptons, $f = \begin{pmatrix} q \\ l \end{pmatrix}$, and the composite Higgs field M is

$$M \sim \frac{-1}{\sqrt{2}}(\bar{F}F + \tau^\alpha \bar{F}\tau^\alpha \gamma_5 F). \tag{13.86}$$

Thus, we have obtained the nonrenormalizable four-fermion interaction responsible for generating the masses for quarks and leptons. The question is: can one write a renormalizable interaction with respect to which this four-fermion interaction will play the role of low energy effective model? One such an example is well known: the four-Fermi weak interaction plays such a role with respect to the $SU_L(2) \times U_Y(1)$ electroweak theory. In that case, the gauge symmetry is spontaneously broken.

The same idea of a spontaneously broken gauge symmetry lays in the basis of the extended technicolor (ETC) (or sideways) scenario [26]. It is assumed in this scenario that the technicolor group G_{TC} is embedded in a larger gauge group, ETC group G_{ETC}, in which fermions and technifermions are assigned to the same irreducible representation. Then it is assumed that, at some scale $\Lambda_{ETC} > F_{TC}, G_{ETC}$ spontaneously breaks down to G_{TC}. The vector bosons connected with the broken generators acquire masses of order of Λ_{ETC}; some of them mediate transition from technifermions to ordinary fermions (see Fig.13.2a).

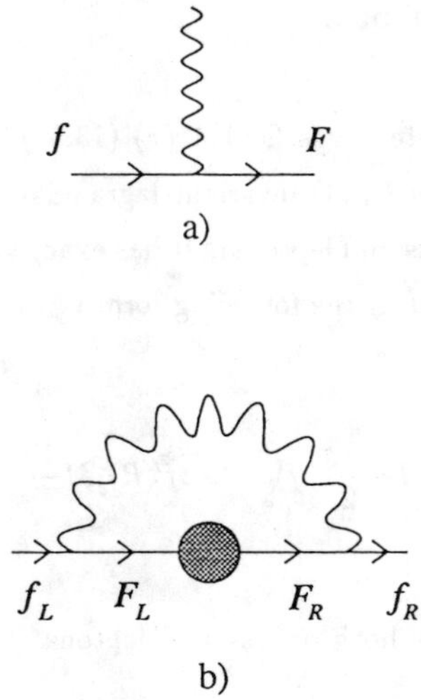

Fig. 13.2. a) The basic transition from an ordinary fermion to a technifermion, mediated by a heavy ETC gauge boson. b) Typical graph producing a mass for a quark or lepton. The shaded blob represents the dynamical mass of the technifermion.

It is clear now that diagrams like that in Fig.13.2b lead to masses for fermions. Let us discuss this point in more detail.

It is easy to write the contribution of such a diagram to the mass function $B_f(q^2)$ of a fermion. In Landau gauge, it is (compare with Eq.(12.99)):

$$B_f(q^2) = -3ig_{ETC}^2 C_{ETC} \int \frac{d^4p}{(2\pi)^4} \frac{B_F(p^2)}{p^2 - B_F^2(p^2)} \cdot \frac{1}{(q-p)^2 - \Lambda_{ETC}^2}, \tag{13.87}$$

where g_{ETC} is the ETC gauge coupling, $B_F(p^2)$ is the dynamical mass function of technifermions, and C_{ETC} is some group factor. Since TC is a QCD-like dynamics, we can use the analysis of spontaneous chiral symmetry breaking in QCD (see Sec.12.3) to describe the properties of the function $B_F(p^2)$.

From that analysis, we know that at $p^2 \lesssim \Lambda_{TC}^2$, where

$$\Lambda_{TC} \sim \frac{F_{TC}}{F_\pi} \sqrt{\frac{3}{N_{TC}}} \Lambda_{QCD} \tag{13.88}$$

is the confinement scale in TC (see Eq.(13.64)), $B_F(p^2)$ is a smooth function of momentum. At $p^2 \gg \Lambda_{TC}^2$, it rapidly decreases:

$$B_F(p^2) \sim \frac{\Lambda_{TC}^3}{p^2}\left(\ln p^2/\Lambda_{TC}^2\right)^{-1+\tilde{c}_m} \tag{13.89}$$

(compare with Eq.(12.28)), where $\tilde{c}_m$ is a TC analogue the QCD factor c_m (see Eq.(12.18)).

Now we are ready to describe the form of the fermion mass function $B_f(q^2)$. First, we note that because of Eq.(13.89), the integral in Eq.(13.87) is convergent: the contribution of the region with $p^2 \gtrsim \Lambda_{ETC}^2$ to the integral is strongly suppressed. Therefore, at $q^2 < \Lambda_{ETC}^2$ one can use the following approximate equation (in Euclidean region):

$$B_f(q^2) \simeq m_f = 3g_{ETC}^2 C_{ETC} \int^{\Lambda_{ETC}} \frac{d^4p}{(2\pi)^4}\frac{B_F(p^2)}{p^2 + B_F^2(p^2)} \cdot \frac{1}{\Lambda_{ETC}^2}. \tag{13.90}$$

In other words, in the region with $q^2 < \Lambda_{ETC}^2$, the ETC interaction reduces to a four-fermion interaction. In the present approximation, the fermion mass function B_f is just a constant in that region.

On the other hand, at $q^2 \gg \Lambda_{ETC}^2$, equation (13.87) reduces to the following asymptotic equation:

$$B_f(q^2) \simeq \frac{3g_{ETC}^2 C_{ETC}}{q^2} \int^{\Lambda_{ETC}} \frac{d^4p}{(2\pi)^4}\frac{B_F(p^2)}{p^2 + B_F^2(p^2)}. \tag{13.91}$$

Therefore the fermion mass function $B_f(q^2)$ rapidly, as $\frac{1}{q^2}$, decreases in the ultraviolet region with $q^2 \gg \Lambda_{ETC}^2$.

Thus, we conclude that in the "low energy " region, with $q^2 < \Lambda_{ETC}^2$, the ETC interaction can be, with a good accuracy, substituted by a four-fermion interaction. The latter has the following general ($SU_c(3) \times SU_L(2) \times U_Y(1)$ invariant) form [26]

$$\mathcal{L}_{ETC} = \mathcal{L}_{LR} + \mathcal{L}_{LL} + \mathcal{L}_{RR}, \tag{13.92}$$

with

$$\mathcal{L}_{LR} = -\frac{g_{ETC}^2}{\Lambda_{ETC}^2}\sum_{i,j=1}^{3}\sum_{r=1}^{n}\sum_{k=1}^{N_r}\Big\{[\bar{u}_{iL}\gamma_\mu U_{kL}^r + \bar{d}_{iL}\gamma_\mu D_{kL}^r][\Gamma_{ij,r}^u \bar{U}_{kR}^r\gamma_\mu u_{iR} + \Gamma_{ij,r}^d \bar{D}_{kR}^r\gamma_\mu d_{iR}]$$
$$+ [\bar{\nu}_{iL}\gamma_\mu U_{kL}^r + \bar{e}_{iL}\gamma_\mu D_{kL}^r][\Gamma_{ij,r}^\nu \bar{U}_{kR}^r\gamma_\mu \nu_{jR} + \Gamma_{ij,r}^e \bar{D}_{kR}^r\gamma_\mu e_{iR}] + H.c.\Big\},$$
$$\mathcal{L}_{LL} = -\frac{g_{ETC}^2}{\Lambda_{ETC}^2}\sum_{i,j=1}^{3}\sum_{r=1}^{n}\sum_{k=1}^{N_r}\Big\{[\bar{u}_{iL}\gamma_\mu U_{kL}^r + \bar{d}_{iL}\gamma_\mu D_{kL}^r][\tilde{\Gamma}_{ij,r}^u \bar{U}_{kL}^r\gamma_\mu u_{iL} + \tilde{\Gamma}_{ij,r}^d \bar{D}_{kL}^r\gamma_\mu d_{iL}]$$
$$+ [\bar{\nu}_{iL}\gamma_\mu U_{kL}^r + \bar{e}_{iL}\gamma_\mu D_{kL}^r][\tilde{\Gamma}_{ij,r}^\nu \bar{U}_{kL}^r\gamma_\mu \nu_{jL} + \tilde{\Gamma}_{ij,r}^e \bar{D}_{kL}^r\gamma_\mu e_{iL}] + H.c.\Big\}, \qquad (13.93)$$

and an analogous structure for L_{RR}. Here we consider three families of fermions, and n (N_r-dimensional) irreducible representations of the TC group. The matrices $\Gamma^{u,d,\nu,e}$, $\tilde{\Gamma}^{u,d,\nu,e}$ are determined by ETC couplings and by the mass matrix of ETC vector bosons. Here we did not indicate explicitly the color indices. As to them, there are two possibilities:

a) The ETC generators commute with the color generators (this may for example happen in the case of the one family TC model).

In this case the color indices of left-handed (right-handed) quarks are contracted with the color indices of left-handed (right-handed) techniquarks;

b) The ETC generators do not commute with the color generators (this, for example, certainly takes place in the case of the one doublet TC model in which technifermions are colorless).

In this case the color indices of left-handed quarks are contracted with the color indices of right-handed quarks.

It is useful to rewrite the expressions for $\mathcal{L}_{LR}, \mathcal{L}_{LL}$, and $\mathcal{L}_{RR}$ in the following schematic form:

$$\mathcal{L}_{LR} \sim -\left[\frac{g_{ETC}^2}{\Lambda_{ETC}^2}\right][\bar{f}_L\gamma_\mu F_L \bar{F}_R\gamma_\mu f_R + H.c.]$$
$$\sim \left[\frac{g_{ETC}^2}{\Lambda_{ETC}^2}\right][\bar{f}f\bar{F}F - \bar{f}\gamma_5 f\bar{F}\gamma_5 F], \qquad (13.94)$$

$$\begin{aligned}\mathcal{L}_{LL} &\sim -\left[\frac{g^2_{ETC}}{\Lambda^2_{ETC}}\right]\bar{f}_L\gamma_\mu F_L\bar{F}_L\gamma_\mu f_L \\ &\sim -\left[\frac{g^2_{ETC}}{\Lambda^2_{ETC}}\right]\bar{f}_L\gamma_\mu f_L\bar{F}_L\gamma_\mu F_L,\end{aligned} \tag{13.94a}$$

$$\begin{aligned}\mathcal{L}_{RR} &\sim -\left[\frac{g^2_{ETC}}{\Lambda^2_{ETC}}\right]\bar{f}_R\gamma_\mu F_R\bar{F}_R\gamma_\mu f_R \\ &\sim -\left[\frac{g^2_{ETC}}{\Lambda^2_{ETC}}\right]\bar{f}_R\gamma_\mu f_R\bar{F}_R\gamma_\mu F_R,\end{aligned} \tag{13.94b}$$

(here we used the Fierz identities for the γ_μ matrices (see Sec.B.3 in Appendix B.) These equations show that the L_{LR} - interaction is responsible for fermion masses. In particular Eq.(13.94) implies that

$$m_f \sim -\frac{g^2_{ETC}}{\Lambda^2_{ETC}}\langle\bar{F}F\rangle_{\Lambda_{ETC}}, \tag{13.95}$$

where the subscript Λ_{ETC} in the condensate $\langle\bar{F}F\rangle$ indicates that it is calculated with the cutoff Λ_{ETC}

$$\begin{aligned}\langle\bar{F}F\rangle_{\Lambda_{ETC}} &\sim \frac{-N_{TC}}{4\pi^2}\int^{\Lambda^2_{ETC}} dp^2\, B_F(p^2) \\ &\sim \frac{-N_{TC}}{4\pi^2}\Lambda^3_{TC}\left(\ln\Lambda^2_{ETC}/\Lambda^2_{TC}\right)^{\tilde{c}_m}\end{aligned} \tag{13.96}$$

(compare with Eqs.(12.23) and (12.27)). Taking into account the relation (13.88), one can express the condensate $\langle\bar{F}F\rangle \equiv \langle\bar{F}F\rangle_{\Lambda_{ETC}}$ through the chiral condensate $\langle\bar{q}q\rangle$ in QCD:

$$\langle\bar{F}F\rangle \sim \langle\bar{q}q\rangle\frac{F^3_{TC}}{F^3_\pi}\left(\frac{3}{N_{TC}}\right)^{1/2}, \tag{13.97}$$

where $\langle\bar{q}q\rangle \equiv \langle\bar{q}q\rangle_1 \sim -(207-228)^3 MeV^3$ (see Eq.(12.112)).

From Eqs.(13.95) and (13.97), we find the following estimate for m_f

$$\begin{aligned}m_f &\sim -\frac{g^2_{ETC}}{\Lambda^2_{ETC}}\frac{F^3_{TC}}{F^3_\pi}\left(\frac{3}{N_{TC}}\right)^{1/2}\langle\bar{q}q\rangle \\ &\sim 10\left(\frac{g_{ETC}F_{TC}}{\Lambda_{ETC}}\right)^2\left(\frac{3}{N_{TC}}\right)^{1/2}F_{TC}.\end{aligned} \tag{13.98}$$

To feel better this estimate, let us find $M_{ETC} \equiv \frac{\Lambda_{ETC}}{g_{ETC}} \sim \Lambda_{ETC}$ for $m_f \sim 1MeV$ and

$m_f \sim 100 GeV$ at $N_{TC} = 3$. One gets

$$M_{ETC} \sim 3r^{-3/4} \left(\frac{v}{m_f} \right)^{1/2} v, \tag{13.99}$$

and therefore

$$M_{ETC}^{(1)} \sim r^{-3/4} 400 TeV \tag{13.100}$$

at $m_f \sim 1 MeV$, and

$$M_{ETC}^{(3)} \sim r^{-3/4} 1 TeV \tag{13.101}$$

at $m_f \sim 100 GeV$ (the notations $M_{ETC}^{(1)}$ and $M_{ETC}^{(3)}$ imply that $m_f \sim 1 MeV$ and $m_f \sim 100 GeV$ are relevant for the first family and the third family (the t quark), respectively).

There estimates indicate a way how ETC can generate the complicated fermion mass spectrum realized in nature: the ETC group G_{ETC} may break down to G_{TC} through a sequence of breakings connected with different scales $\Lambda_{ETC}^{(i)}$ ($i = 1, 2, 3$ is the family index): a large hierarchy $\Lambda_{ETC}^{(1)} > \Lambda_{ETC}^{(2)} > \Lambda_{ETC}^{(3)}$ will produce a large hierarchy between fermion masses from different families.

We will however see that it is not so easy to implement this scenario in practice. We will also see that there are other possibilities in the ETC model building (see Chapter 14 and Sec.15.6).

13.7. Problems in the ETC Model Building

Up to now, no very satisfactory ETC model has been found (see, however, the recent papers [27] which will be briefly discussed below). In this section, we will discuss some problems in the ETC model building.

a) *Pseudo - NG Bosons in ETC*

In the simplest (one doublet) TC model, there are no physical NG bosons: the three would-be NG bosons are absorbed by the $W^{\pm}$ and Z vector bosons. However, this is not a typical situation in TC: we recall that, for example, in the one family model (when all forces, except technicolor, are ignored), there are the 63 NG bosons connected with the spontaneously broken $SU_L(8) \times SU_R(8)$ chiral symmetry. Let us consider them in more detail.

These 63 NG bosons are described by the composite fields

$$\Pi^{\tilde{\alpha}} \sim i\bar{F}\frac{\lambda^{\tilde{\alpha}}}{2}\gamma_5 F, \tag{13.102}$$

where $\lambda^{\tilde{\alpha}}$ are the traceless $SU(8)$ generators, and F is the technifermion field (see Eq.(13.75)). It is convenient to choose a basis for the $\lambda^{\tilde{\alpha}}$ materials in which the color, $SU_L(2)$ and charge quantum numbers are exhibited explicitly. Then the NG bosons can be enumerated as follows [25]:

$$\begin{aligned}
&\theta_K^\alpha \sim i\bar{Q}t_K\tau^\alpha\gamma_5 Q,\ K = 1,\cdots,8; \alpha = 1,2,3,\\
&\theta_K \sim i\bar{Q}t_K\gamma_5 Q,\ K = 1,\cdots,8,\\
&T_l^\alpha \sim i\bar{Q}_l\tau^\alpha\gamma_5 L,\ \bar{T}_l^\alpha \sim i\bar{L}\tau^\alpha\gamma_5 Q_l, l = 1,2,3\\
&T_l \sim i\bar{Q}_l\gamma_5 L,\ \bar{T}_l \sim i\bar{L}\gamma_5 Q_l,\\
&P^\pm \sim i\bar{Q}\tau^\pm\gamma_5 Q - 3i\bar{L}\tau^\pm\gamma_5 L,\\
&P^3 \sim i\bar{Q}\tau^3\gamma_5 Q - 3i\bar{L}\tau^3\gamma_5 L,\\
&P^0 \sim i\bar{Q}\gamma_5 Q - 3i\bar{L}\gamma_5 L,\\
&\Pi^\alpha \sim i\bar{Q}\tau^\alpha\gamma_5 Q + i\bar{L}\tau^\alpha\gamma_5 L,
\end{aligned} \tag{13.103}$$

where t_K are the $SU_c(3)$ matrices and l is the color index of techniquarks.

It is easy to check that only the three NG bosons Π^α couple to the axial currents $J_{5\mu}^\alpha$ (and therefore to the $SU_L(2)$ currents $J_{L\mu}^\alpha$); they are absorbed by the $W^\pm$ and Z bosons. The others are physical particles which acquire non-zero masses through their QCD, electroweak or ETC interactions. They are called pseudo-NG bosons (or pseudos).

The masses of pseudos can be estimated by the standard methods of current algebra [25,28]. We will just list the final results (for a detailed analysis, see Ref.[25]).

The particles $\theta_K^\alpha, \theta_K$ are color octets, and the particles $T_l^\alpha, \bar{T}_l^\alpha, T_l, \bar{T}_l$, are color triplets. They acquire masses of order $100GeV$ through their color interactions. $P^\pm$ acquire masses of order $10GeV$ through their electroweak interactions. The P^3 and P^0 are singlets of the $SU_c(3) \times SU_L(2) \times U_Y(1)$ and therefore they remain massless at this stage.

In order to generate masses for them, one has to make use of the four-fermion interactions among technifermions; these interactions are a low-energy image of spontaneously broken ETC (or possibly other) gauge interactions among technifermions. Consider, for example, the following interaction

$$\mathcal{L}_{QL} = \frac{g_{ETC}^2}{\Lambda_{ETC}^2}\left(\bar{Q}Q\bar{L}L - \bar{Q}\gamma_5\tau^\alpha Q\bar{L}\tau^\alpha\gamma_5 L\right). \tag{13.104}$$

This interaction is $SU_c(3) \times SU_L(2) \times U_Y(1) \times SU(N_{TC})$ invariant. However, it violates separate techniquark and technilepton chiral symmetries. As a result, the interaction leads to masses for P^0 and P^3 and raises the masses of $P^\pm$. Considering the interaction (13.104) to be perturbative, we can estimate these masses by using Dashen's formula [28]:

$$M_{P^a}^2 = \frac{1}{F_{TC}^2}\langle 0|[Q_5^a, [Q_5^a, \mathcal{H}_{QL}]]|0\rangle, \tag{13.105}$$

where $\mathcal{H}_{QL} = -\mathcal{L}_{QL}$ and $Q_5^a = \int d^3x J_{50}^a(x)$ with J_{50}^a being the time component of the current corresponding to P^a:

$$J_{50}^a = \frac{1}{\sqrt{48}}(\bar{Q}\gamma_0\gamma_5\tau^a Q - 3\bar{L}\gamma_0\gamma_5\tau^a L) \tag{13.106}$$

(see Eq.(13.103); the coefficient $\frac{1}{\sqrt{48}}$ ensures the correct normalization factor $\frac{1}{2}$ for the flavor matrices here).

Evaluating the commutators in Eq.(13.105), we get

$$M_{P^a}^2 = \frac{1}{F_{TC}^2}\frac{g_{ETC}^2}{\Lambda_{ETC}^2}\frac{40}{48}\langle 0|\bar{Q}Q\bar{L}L|0\rangle. \tag{13.107}$$

In order to estimate this vacuum expectation value, we will use the vacuum saturation approximation [12]:

$$\langle 0|\bar{Q}Q\bar{L}L|0\rangle = 3\langle 0|\bar{L}L\bar{L}L|0\rangle \simeq 3(\langle 0|\bar{L}L|0\rangle)^2 \equiv 3\langle\bar{F}F\rangle^2. \tag{13.108}$$

therefore $M_{P^a}^2$ is

$$M_{P^a}^2 \simeq \frac{1}{F_{TC}^2}\frac{g_{ETC}^2}{\Lambda_{ETC}^2}\frac{5}{2}\langle\bar{F}F\rangle^2. \tag{13.109}$$

By using Eq.(13.95) for the fermion mass m_f, this relation can be rewritten as

$$M_{P^a} = a\frac{M_{ETC}}{F_{TC}}m_f, \tag{13.110}$$

where $M_{ETC} = \Lambda_{ETC}/g_{ETC}$ and a is a coefficient of order one.

For the two examples, with $m_f \sim 1MeV$ and $M^{(1)}_{ETC} \sim 400TeV$, and with $m_f \sim 100GeV, M^{(3)}_{ETC} \sim 1TeV$, considered in the previous section, we get $M^{(1)}_{P^a} \sim 3aGeV$, and $M^{(3)}_{P^a} \sim 800aGeV$, respectively. Thus M_{P^a} , increases with decreasing M_{ETC}. This result can be easily understood: indeed, Eqs.(13.95), (13.96) imply that $m_f \sim \Lambda^3_{TC}/M^2_{ETC}$. Therefore

$$M_{P^a} \sim \frac{M_{ETC}}{F_{TC}} m_f \sim \frac{\Lambda^3_{TC}}{F_{TC} M_{ETC}} \sim \frac{\Lambda^2_{TC}}{M_{ETC}}. \tag{13.111}$$

Since no light spinless particles have been seen in e^+e^- colliders, we need a rather low M_{ETC}, $M_{ETC} \lesssim 10TeV$, to get reasonable values for M_{P^a} ($M_{P^a} \gtrsim 50GeV$). Is it possible? We will see below that it is not so easy to construct ETC models with such a low M_{ETC}.

b) *Flavor-Changing Neutral Currents in ETC*

In Sec.13.4, we discussed the flavor-changing neutral currents in the SM. Because of the GIM mechanism (approximate flavor symmetry (13.46)), FCNCs are strongly suppressed in the SM. What is the situation with FCNCs in ETC?

Still recently, there has been no ETC models with the GIM mechanism. However, now there is such a class of ETC models [27,29,30]. We will discuss them later. Now we will consider FCNCs in generic ETC models without the GIM mechanism.

First, let us consider the following question: why is it not possible to extend directly the GIM mechanism from the SM to ETC? Let us recall that in the SM fermions from different families couple to the same vector bosons, $W^\pm$, Z and γ. In other words, the gauge interactions in the SM preserve the flavor symmetry (13.46) and it is broken by fermion mass terms only. The situation in ETC is different. There is a lot of new vector bosons connected with broken ETC symmetries. Since fermion masses are proportional to the inverse square of the ETC gauge boson masses (see Eq.(13.95)), in generic ETC models the hierarchy between fermion masses from different families is produced by a hierarchy of ETC gauge boson masses. This yields an additional source violating the flavor symmetry, and now a unitary transformation from a weak to a mass basis introduces FCNCs unsuppressed by the GIM mechanism.

To be more concrete, let us consider $M^0 - \bar{M}^0$ mixing ($M^0 = K^0, D^0$, or B^0 mesons) in ETC describing by the box diagrams shown in Fig.13.3.

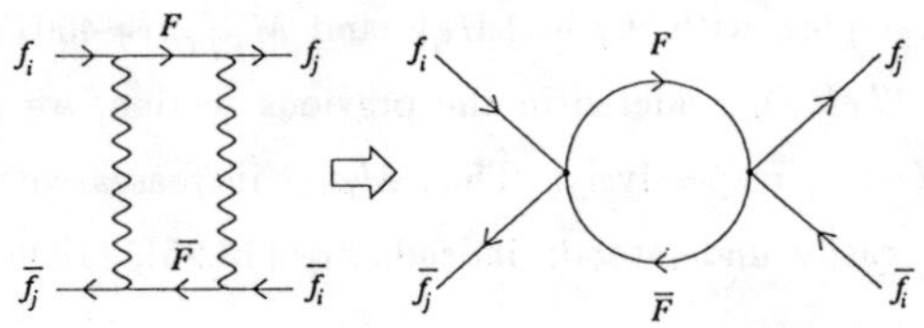

Fig. 13.3. The box diagrams leading to M^0–$\bar{M}^0$ mixing. The wavy lines correspond to ETC vector bosons.

From Eqs.(13.92)–(13.94), in the second order in g^2_{ETC}, one finds the amplitude leading to $M^0 - \bar{M}^0$ mixing:

$$A = -\frac{1}{2}\chi\left[\frac{g^2_{ETC}}{\Lambda^2_{ETC}}\right]^2 \times \langle p_1', p_2'| \int d^4x d^4y T\left\{\bar{f}(x)\gamma_a f(x)\bar{f}(y)\gamma_b f(y)\bar{F}(x)\gamma_a F(x)\bar{F}(y)\gamma_b F(y)\right\}|p_1 p_2\rangle, \tag{13.112}$$

where $a, b = S, P, V_L, V_R$ with $\gamma_S = 1, \gamma_P = i\gamma_5, \gamma_{V_L} = \gamma_\mu(1+\gamma_5)/2, \gamma_{V_R} = \gamma_\mu(1-\gamma_5)/2$. The factor χ includes Cabibbo-like mixing angles. Neglecting $SU_c(3) \times SU_L(2) \times U_Y(1)$ interactions, the amplitude A can be rewritten as

$$A = -\frac{1}{2i}\chi\left[\frac{g^2_{ETC}}{\Lambda^2_{ETC}}\right]^2 (2\pi)^{-2}\delta^4(p_1' + p_2' - p_1 - p_2) \times \bar{u}(p_1')\gamma_a v(p_2')\bar{v}(p_2)\gamma_b u(p_1)\Delta_{ab}(p_1 + p_2), \tag{13.113}$$

where

$$\Delta_{ab}(q) = i\int d^4x e^{iqx}\langle 0|T\left\{\bar{F}(x)\gamma_a F(x)\bar{F}(0)\gamma_b F(0)\right\}|0\rangle, \tag{13.114}$$

and $\bar{u}(u), v(\bar{v})$ are spinor functions of outgoing (incoming) quarks and antiquarks, respectively (see Eq.(1.31)).

The crucial point for FCNCs is that the Green's function Δ_{ab} of the composite operators $\bar{F}\gamma_a F$ is quadratically divergent in a free field theory. This implies that $\Delta_{ab} \sim \Lambda^2_{ETC}$ in ETC and therefore the amplituded A (13.113) is $A \sim 1/\Lambda^2_{ETC}$ (and not $A \sim 1/\Lambda^4_{ETC}$), which is the well-known result for the box diagram in perturbation theory. One can show [31] that TC interactions do not change essentially this result (they only introduce some additional logarithmic factors in Δ_{ab}).

In the SM, because of the flavor symmetry (13.46), the leading quadratically divergent contributions in the box diagrams in Fig.(13.4) cancel. However, in ETC models with non-degenerate ETC gauge bosons, such contributions in FCNCs do not cancel, thus leading to a new source for FCNCs.

The known experimental data for FCNCs processes give lower bounds for the mass scales $M_{ETC}^{(i)} = \Lambda_{ETC}^{(i)}/g_{ETC}^{(i)}$ ($i = 1, 2, 3$ is the family index). With the Cabibbo-type mixing angle $\theta \sim \theta_c \sim 0.2$, the data relating to fermions in the first two families lead to the bound $M_{ETC}^{(1)}, M_{ETC}^{(2)} \gtrsim 300TeV$ [32]. This bound leads to the following two problems: 1) as we showed in the previous section, with $M_{ETC} \gtrsim 300TeV$, one can get fermion masses $m_f \lesssim 0(1MeV)$ only , so that $m_{cur;s}$ and $m_{cur;c}$ are too small in these models; 2) as follows from our discussion concerning pseudo-NG bosons, such high M_{ETC} and small m_f lead to light $P^0, P^3, P^{\pm}$which has been excluded by experiment.

We will consider how one can cure this "disease" of generic ETC models (without the GIM mechanism) in Chapter 14. Now we will briefly discuss the new ETC models with the GIM mechanism [27,29,30]. The program of constructing TC models with the GIM (techni-GIM) mechanism was initiated by Dimopoulos, Georgi and Raby (DGR) [29]. The crucial point in this program is the following. As was already pointed out above, in order to implement the techni-GIM mechanism, the ETC gauge bosons have to be nearly degenerate in mass. On the other hand, this leads to difficulties with getting a non-degenerate fermion mass spectrum. To avoid this problem, DGR suggested to consider different ETC groups for left-handed fermions and technifermions and for right-handed fermions and technifermions: $G_{ETC}^{(L)}, G_{ETC}^{(U_R)}, G_{ETC}^{(D_R)}$. The ETC gauge bosons are approximately degenerate, however, the mass mixing between left- and right-handed gauge bosons is different for different families. In a sense, one can say that the fermion mass degeneracy is removed here not by different $M_{ETC}^{(i)}$ but by different (effective) coupling constants $g_{ETC}^{(i)}$ in Eq.(13.95).

This idea was realized in concrete models ("moose models") by Chivukula, Georgi and Randall [30]. The payment for the realization is introducing a lot of new fermions (which are neither ordinary fermions nor technifermions) in the models. Although the moose models are rather complicated, and there are open questions concerning their nonperturbative dynamics, the fact that the concrete ETC models with the techni-GIM mechanism have been constructed is noticeable.

It is especially noticeable that very recently, Randall and Georgi [27] have constructed an ETC model with the minimal TC model including only one doublet of technifermions (see Sec.13.5). In the model, all of the three ETC group, $G_{ETC}^{(L)}$, $G_{ETC}^{(U_R)}$, and $G_{ETC}^{(D_R)}$ are isomorphic to $SU(12+N_{TC})$ (the $(12+N_{TC})$ states include three families of quarks and leptons and one TC doublet). As we will see in Sec.13.9, the choice of the minimal TC is important to satisfy the restrictions connected with the radiative corrections to electroweak interactions.

13.8. The Appelquist-Carazzone Decoupling Theorem. Can One Detect the Electroweak Symmetry Breaking Mechanism in Low Energy Experiments?

At present, all the existing experimental facilities cannot reach the energy threshold of order $1TeV$, where the secrets of electroweak symmetry breaking mechanism should manifest themselves. The question is: can one get an information about this mechanism in low energy (with $E \ll 1TeV$) experiments? Or, in more theoretical terms, does the high energy physics exert a strong influence on the low energy physics? In the present section, we will discuss this problem.

In Sec.11.2, we considered the effect of nonrenormalizable (irrelevant) interactions on the low energy dynamics. The main conclusion was that the effect of an irrelevant interaction of dimension $n+4$, with $n > 0$, is suppressed by the factor $\left(\frac{E}{\Lambda}\right)^n$ in the low energy dynamics where E is the characteristic low energy scale and Λ is the ultraviolet cutoff.

As we indicated in that section, one can expect that nonrenormalizable interactions come from some more fundamental interaction connected with the scale Λ. A particularly interesting and important case is that one when the fundamental interaction is described by a renormalizable theory including heavy ($M \sim \Lambda$) and light ($m \ll \Lambda$) particles. Then, integrating out heavy fields, one obtains a *non-local* effective theory describing the dynamics of the light particles. Expanding this effective theory in a series in powers of the derivatives of the light fields, we obtain a tower of nonrenormalizable (irrelevant) interactions.

This is however not the whole story. The structure of the low energy effective lagrangian arising after integrating out heavy fields is described by the Appelquist-Carazzone (AC) decoupling theorem [33]. The original version of the AC theorem describes the following situation. Let us start from a renormalizable theory including both light and heavy particles. Let us also assume that the theory which is obtained from the initial one by just omitting from the lagrangian the terms including heavy fields is also interacting and renormalizable. Then the AC theorem states that the heavy particles decouple at low energies except for their contribution to renormalization effects (i.e. to the counter terms) and for a contribution suppressed by power factors $\left(\frac{E}{M}\right)^n, n > 0$. We stress that the heavy particles do not introduce counter terms of a new form; all of them are of the same type as those already present in the lagrangian without the heavy fields. As a result, the heavy particles are almost (up to small effects of order $\left(\frac{E}{M}\right)^n$) unobservable in the low energy physics.

In order to illustrate the theorem, let us consider QED with one light and one heavy fermions. Their masses are m and M, respectively. For simplicity, we assume that they have the same electric charge.

Let us consider now the photon propagator (we use the Landau gauge)

$$D_{\mu\nu}(q) = \left(g_{\mu\nu} - \frac{q_\mu q_\nu}{q^2}\right)\frac{1}{q^2 - g^{(0)2}\Pi(q^2)} \tag{13.115}$$

where $g^{(0)}$ is the bare coupling and $\Pi(q^2)$ is the polarization operator. By using the Pauli-Villars regularization, we find that in the one-loop approximation, $\Pi(q^2)$ is [34]:

$$\Pi(q^2) = -\frac{q^2}{12\pi^2}\left[\left(\ln\frac{\Lambda^2}{m^2} - \frac{5}{6}\right) + \left(\ln\frac{\Lambda^2}{M^2} - \frac{5}{6}\right)\right] + \frac{q^2}{2\pi^2}\left[I_m(q^2) + I_M(q^2)\right]. \tag{13.116}$$

The function $I_\mu(q^2), \mu = m$ or M, is:

$$I_\mu(q^2) = \int_0^1 dx x(1-x)\ln\frac{\mu^2 - q^2 x(1-x)}{\mu^2}. \tag{13.117}$$

We need to know that

$$I_\mu(q^2) = \begin{cases} -\frac{1}{30}\frac{q^2}{\mu^2} & \text{at } |q^2| \ll \mu^2 \\ \frac{1}{6}\ln\frac{|q^2|}{\mu^2} & \text{at } |q^2| \gg \mu^2 \end{cases}. \tag{13.118}$$

Therefore at $m^2 \ll q^2 \ll M^2$, we get:

$$\Pi(q^2) \simeq -q^2\left\{\frac{1}{12\pi^2}\left(\ln\frac{\Lambda^2}{m^2} - \frac{5}{6}\right) + \frac{1}{12\pi^2}\left(\ln\frac{\Lambda^2}{M^2} - \frac{5}{6}\right) - \frac{1}{12\pi^2}\ln\frac{|q^2|}{m^2} + \frac{1}{60\pi^2}\frac{q^2}{M^2}\right\}. \tag{13.119}$$

We can see that the heavy fermion contributes to the divergent part of $\Pi(q^2)$, however, its contribution to the piece of $\Pi(q^2)$ depending on q^2 is suppressed by the factor q^2/M^2.

It is instructive to introduce the following observable (through which some amplitudes are expressed):

$$g(q^2) \equiv \frac{g^{(0)^2}}{1 - g^{(0)^2} q^{-2}\Pi(q^2)}. \tag{13.120}$$

Then

$$\begin{aligned} \frac{1}{g(q^2)} &= \frac{1}{g^{(0)^2}} - \frac{1}{12\pi^2}\left(\ln\frac{\Lambda^2}{m^2} + \ln\frac{\Lambda^2}{M^2} - \frac{10}{6}\right) \\ &- \frac{1}{12\pi^2}\ln\frac{|q^2|}{m^2} + \frac{1}{60\pi^2}\frac{q^2}{M^2} \simeq \frac{1}{g^2_{ren}} - \frac{1}{12\pi^2}\ln\frac{|q^2|}{m^2}, \end{aligned} \tag{13.121}$$

where

$$\frac{1}{g^2_{ren}} = \frac{1}{g^{(0)^2}} - \frac{1}{12\pi^2}\left(\ln\frac{\Lambda^2}{m^2} + \ln\frac{\Lambda^2}{M^2} - \frac{10}{6}\right). \tag{13.122}$$

Here we use the conventional (Dyson) renormalization scheme in which the renormalized coupling constant g_{ren} is defined from the condition that $\lim_{q^2\to 0} q^{-2}\Pi_{ren}(q^2) = 0$ [34].

The expression (13.121) shows that, as a result of the renormalization, the trace of the heavy particle disappears (up to a small correction $\sim q^2/M^2$) from the low energy dynamics.

This example clearly illustrates the essence of the decoupling theorem. Heavy particles contribute to the counter terms of a renormalizable low energy effective theory and, additionally, they lead to a tower of nonrenormalizable (irrelevant) interactions. As we know from the discussion in Sec.11.2, the only effect of these irrelevant operators in higher orders of perturbation theory is again their contribution to the counter terms.

Let us consider the same example by using the dimensional regularization and the $\overline{MS}$ renormalization scheme (see Appendix C). Now $1/g^2(q^2)$ is

$$\begin{aligned} \frac{1}{g^2(q^2)} &= \frac{1}{g^{(0)^2}(\sigma)} + \frac{1}{6\pi^2}\left(\frac{1}{\epsilon} - \gamma + \ln 4\pi\right) + \\ &- \frac{1}{12\pi^2}\ln\frac{|q^2|}{\sigma^2} - \frac{1}{12\pi^2}\ln\frac{M^2}{\sigma^2} + \frac{1}{60\pi^2}\frac{q^2}{M^2}, \end{aligned} \tag{13.123}$$

where $g^{(0)}$ is the bare coupling, σ is the renormalization scale, $\gamma \simeq 0.577$ is the Euler constant and $\epsilon = \frac{4-D}{2}$. Since the observable $\bar{g}^2(q^2)$ is independent of σ, the bare coupling

$g^{(0)}(\sigma)$ depends on σ. In the $\overline{MS}$ scheme, the renormalized coupling $g(\sigma)$ is:

$$\frac{1}{g^2(\sigma)} \equiv \frac{1}{g^{(0)2}(\sigma)} + \frac{1}{6\pi^2}\left(\frac{1}{\epsilon} - \gamma + \ln 4\pi\right), \tag{13.124}$$

and therefore

$$\begin{aligned} \frac{1}{g^2(q^2)} &= \frac{1}{g^2(\sigma)} - \frac{1}{12\pi^2}\ln\frac{|q^2|}{\sigma^2} - \frac{1}{12\pi^2}\ln\frac{M^2}{\sigma^2} + \frac{q^2}{60\pi^2 M^2} \\ &\simeq \frac{1}{g^2(\sigma)} - \frac{1}{12\pi^2}\ln\frac{|q^2|}{\sigma^2} - \frac{1}{12\pi^2}\ln\frac{M^2}{\sigma^2}. \end{aligned} \tag{13.125}$$

It looks now as if the observable $g^2(q^2)$ logarithmically depends on the heavy mass M. We note however that, as in the Dyson renormalization scheme, the heavy mass M does not contribute (up to small terms $\sim q^2/M^2$) to the piece of $g^{-2}(q^2)$ depending on q^2. Let us, following Ref.[35], introduce a new renormalizable constant:

$$\tilde{g}^{-2}(\sigma) \equiv \begin{cases} g^{-2}(\sigma) & \text{at } |q^2| \geq M^2 \\ g^{-2}(\sigma) - \frac{1}{12\pi^2}\ln\frac{M^2}{\sigma^2} & \text{at } |q^2| < M. \end{cases} \tag{13.126}$$

In other words, now we use different renormalization prescriptions at $|q^2| \geq M^2$ and at $|q^2| < M^2$: at high momenta, this is just the standard $\overline{MS}$ scheme, but at low momenta, the coupling $g(\sigma)$ is modified. As a result, there is no trace of the heavy particle in low energy effective theory.

This approach is called a continuum effective field theory in the literature (for a recent review see Ref.[36]). In this approach, the effective theory is switching from one form to another when it is crossing particle thresholds, $|q^2| = M_i^2$. The important part of the continuum effective field theory approach is the so called matching condition. The matching condition provides that more fundamental theories (defined at thresholds $|q^2| = M_i^2$ with $M_i^2 > M_{i0}^2$) and the effective theory defined at the threshold $|q^2| = M_{i0}^2$, lead to the same results in the region with $|q^2| \ll M_{i0}^2$. In our simple example, the matching condition is just Eq.(13.126). It is evident that it provides such an equivalence.

So far, we have discussed the decoupling theorem in theories without spontaneous symmetry breaking. We will show below that the situation with the decoupling of heavy particles may be rather nontrivial when spontaneous symmetry breaking occurs.

Let us begin by returning to the problem of the nonlinear limit in the $SU_L(2) \times SU_R(2)$ linear σ-model considered in Sec.5.2. In this limit, the mass of σ, μ_σ, goes to infinity with the vacuum expectation value $v = \langle 0|\sigma|0\rangle$ being fixed (see Eq.(5.21)). It is clear that the dynamics of the light particles, including the NG bosons $\boldsymbol{\pi}$ only, cannot be described by a renormalizable interaction in this case: the nonlinear σ-model is nonrenormalizable. Thus, in this limit, we will get a tower of new divergences in the low energy effective theory. The regularization cutoff Λ may be identified with the σ mass μ_σ; thus, the low energy physics depends, without any suppression, on the heavy mass μ_σ (actually, it depends on μ_σ logarithmically [37]).

We also note that in this limit the σ-coupling constant $\lambda = 3\mu_\sigma^2/v$ goes to infinity. Therefore the nondecoupling effect can be understood as a result of the strong interaction between light and heavy particles.

As to the application of the AC theorem to this case, we notice the following. Removing the terms including the heavy σ field from the lagrangian of the linear σ-model, we will get a free theory corresponding to the nonlinear σ-model with the decay constant $F \to \infty$. This point reflects the two general facts that 1) there is no renormalizable chiral invatiant theory describing interactions of pions (without σ) and that 2) the interaction in the nonlinear σ-model is proportional to negative powers of F. In perturbation theory, there are nondecoupling effects at $\mu_\sigma \gg F$ in this case, however, as we will argue below, the situation may be essentially different beyond pertubation theory.

Let us now discuss another example. We consider the $SU_L(2) \times SU_R(2)$ linear σ model with a light fermion doublet ψ. The Yukawa interaction for ψ is:

$$\mathcal{L}_Y = \sqrt{2}g[\bar{\psi}_L M \psi_R + \bar{\psi}_R M^\dagger \psi_L] , \tag{13.127}$$

where the field M is

$$M = \tfrac{1}{\sqrt{2}}(\sigma + i\pi^\alpha \tau^\alpha), \ \alpha = 1, 2, 3.$$

Let us now introduce an additional (heavy) fermion doublet ψ' with the following Yukawa interaction:

$$\mathcal{L}'_Y = -\sqrt{2}\left[g_1 \bar{\psi}'_L M \psi'_R + g_3 \bar{\psi}'_L M \tau^3 \psi'_R + H.c.\right] . \tag{13.128}$$

When $g_3 \neq 0$, this interaction breaks the $SU_L(2) \times SU_R(2)$ to $SU_L(2) \times U_{3R}(1)$ (with the τ^3-generator for the $U_{3R}(1)$). Now, at $v = \langle 0|\sigma|0\rangle \neq 0$, the $SU_L(2) \times U_{3R}(2)$ breaks spontaneously down to the $U_{3V} \equiv U_{3(L+R)}$. There are still three NG bosons, $\pi^0 = \pi^3$ and $\pi^\pm$, however, now the decay constant F^0 of π^0 can in principle be different from the decay constant $F^\pm$ of $\pi^\pm$.

Let us study this effect when both Yukawa coupling constants, g_1 and g_3, (and therefore the masses of the heavy fermions) go to infinity with v and the ratio $g_3/g_1 \equiv a$ being fixed. In this model, the light particles (σ and π, and the light fermion doublet ψ) *can* be described by a renormalizable interaction. Since the Yukawa interaction (13.127) of the light doublet ψ is $SU_L(2)\times SU_R(2)$ invariant, one might think that this interaction should be the $SU_L(2) \times SU_R(2)$ symmetric linear σ-model. As we shall see this is not case.

To show this, let us consider the parameter

$$\delta\rho \equiv \frac{(F^\pm)^2 - (F^0)^2}{(F^0)^2} \tag{13.129}$$

characterizing the violation of the isospin $SU_V(2)$ symmetry in the model. The one-loop contribution of a fermion doublet into $\delta\rho$ was first calculated by Veltman [38] (see also Ref.[39]). It is:

$$\delta\rho = \frac{M_u^2}{16\pi^2 v^2}\left[\frac{2}{1+a} - \frac{1-a}{a}\ln\frac{1+a}{1-a}\right], \tag{13.130}$$

where M_u is the mass of the up fermion, $\psi'_u = \frac{1+\tau_3}{2}\psi'$, and $a = g_3/g_1$.

From the formal viewpoint, the fact that $F^\pm \neq F^0$ corresponds to the appearance of a counter term of a new type in the low energy effective lagrangian. Indeed, without the heavy doublet ψ', the theory, and therefore its counter terms, were invariant with respect to $SU_V(2)$. The fact that $F^\pm \neq F^0$ may be interpreted as the appearance of different kinetic counter terms for $\pi^\pm$ and π^0.

Thus, the nondecoupling effect takes place in the model. Again, as in the previous example, we can attribute this effect to a strong coupling between light and heavy particles ($g_1, g_3 \to \infty$).

So far, we have considered these effects in the framework of perturbation theory. Let us argue that the situation may actually be essentially different beyond it.

In the two examples considered above we assumed that some of the masses generated dynamically were infinite (or at least much larger than the vacuum expectation value v). In reality, however, it may happen that such a dynamical regime does not exist.

In Sec.11.3, we considered the $O(N)$ nonlinear σ-model and the model with four-fermion interaction in three dimensional space-time ($D = 3$). They correspond to the strong coupling regime of the linear σ-model and the Yukawa model, respectively. We emphasize, however, that in this regime, the bare coupling constants, and not renormalized ones, go to infinity. We found that while in perturbation theory the φ-particle and the σ-particle in these models (see Eq.(11.32) and Eq.(11.85)) acquire infinite masses, the real situation is different: they reappear as bound states with finite masses in the spectrum of the models, and their dynamics are governed by the nonperturbative infrared stable fixed points.

The essentially the same situation take place in these models at all D in the range $2 < D < 4$ (see Sec.11.5). As to the linear σ-model at $D = 4$, the conventional viewpoint is that the value of the infrared fixed point equals zero there (for recent reviews, see Refs.[40,41]). More precisely, this means that the coupling constants $\lambda(\sigma)$ and $g(\sigma)$, defining the interaction strength at the scale σ, are

$$\lambda(\sigma),\ g^2(\sigma) \sim \frac{1}{\ln \Lambda/\sigma}, \tag{13.131}$$

even when the bare coupling constants $\lambda^{(0)}$ and $g^{(0)}$ go to infinity. Assuming that cutoff Λ, playing here the role of the scale of a "new" physics, is at least few times larger than the heaviest mass in the model, one finds that neither fermions nor bosons can be essentially heavier than the decay constant $F \sim v$. This in turn implies that the renormalized ("physical") λ and g^2 are at most of order one.

In fact, at $D = 4$, we know a "real" theory with a strong interaction in the infrared region. This is of course QCD. We know that the decay constant F is $F = F_\pi \simeq 93MeV$ and the dynamical quark mass m_{dyn} is $m_{dyn} \sim 300MeV$ in QCD. Therefore, even in this case, the Yukawa coupling $g = m_{dyn}/F$ is of order one (see Chapter 12).

It is possible that the QCD-like dynamics yields a kind of the upper bound for the coupling constants in the infrared region for all possible chiral dynamics at $D = 4$. This will imply that at $D = 4$ dynamical masses can never be much larger than the decay constant.

From the practical viewpoint, this will imply that the contribution of the "heavy" physics into the observables of the low energy dynamics is rather small (see Eq.(13.130)). However, as we will see in the next section, even such a small contribution can be very important in precision electroweak measurements. This topic, pioneered by Veltman [38] (see also Refs.[39,42]), has become one of the central topics in electroweak physics.

13.9. Radiative Corrections to Electroweak Interactions and Technicolor Models

As is well known, precision electromagnetic measurements played the crucial role in establishing QED as the fundamental theory of electromagnetic interactions. Precision electroweak measurements play the same role with respect to the SM and with respect to all models modifying the SM (for a review see Ref.[24]).

Our goal here is to discuss only a small piece of this important topic: radiative corrections to electroweak interactions in TC models. This subject has intensively developed since the appearance of the papers [43-46] in which the basic points of the subject were established.

What we will consider is the influence of the TC dynamics, which plays here the role of a high energy physics with the characteristic scale $\Lambda_{TC} \sim \sqrt{\frac{3}{N_{TC}}} \frac{F_{TC}}{F_\pi} \Lambda_{QCD} \sim 1TeV$, on the "low" energy electroweak physics with the scale $q \sim M_W, M_Z \sim 100GeV$. The ideal tool for this purpose is the chiral effective lagrangian approach considered in the framework of QCD in Sec.12.13: since the dynamics in the TC models which we are studying in this chapter is intimately connected with the QCD dynamics, this approach can be directly applied to them. Such an analysis was first done in the works of Golden and Randall [44] and Holdom and Terning [45], and since these works, this approach has been considered in numerous papers [47].

We begin by considering the basic points of the problem.

The low energy electroweak physics is described by the following effective current-current lagrangian following from the $SU_L(2) \times U_Y(1)$ model in tree approximation:

$$\mathcal{L}_{eff}^{EW} = -\frac{4G_F}{\sqrt{2}} \left[J_\mu^+ J^{-\mu} + \rho (J_{3\mu} - s^2 J_{Q\mu})(J_3^\mu - s^2 J_Q^\mu) \right], \tag{13.132}$$

where the Fermi constant $G_F = \frac{\sqrt{2}e^2}{8s^2 M_W^2}$, $s = \sin\theta_W, c = \cos\theta_W, \rho = M_W^2/M_Z^2 c^2$, and the fermion currents are:

$$\jmath_\mu^+ = (\jmath_\mu^-)^\dagger = \sum_{i=1}^{3} \left(\bar{\nu'}^i_L \gamma_\mu e'^i_L + \bar{u'}^i_L \gamma_\mu d'^i_L \right) , \tag{13.133}$$

$$\jmath_\mu^3 = \sum_{i=1}^{3} \left(\bar{l'}^i_L \frac{\tau^3}{2} \gamma_\mu l'^i_L + \bar{q'}^i_L \frac{\tau^3}{2} \gamma_\mu q'^i_L \right) , \tag{13.134}$$

$$\begin{aligned} \jmath_Y^\mu = \sum_{i=1}^{3} \Bigg[& \left(-\frac{1}{2} \bar{l'}^i_L \gamma^\mu l'^i_L + \frac{1}{6} \bar{q'}^i_L \gamma^\mu q'^i_L \right) - \bar{e}'^i_R \gamma^\mu e'^i_R \\ & + \frac{2}{3} \bar{u'}^i_R \gamma^\mu u'^i_R - \frac{1}{3} \bar{d'}^i_R \gamma^\mu d'^i_R \Bigg] , \end{aligned} \tag{13.135}$$

$$\jmath_Q^\mu = -\bar{e'}^i \gamma^\mu e'^i + \frac{2}{3} \bar{u'}^i \gamma^\mu u'^i - \frac{1}{3} \bar{d'}^i \gamma^\mu d'^i , \tag{13.136}$$

where i is the family index and, as usual, the primes denote a weak basis. Thus, there are the following current-current interactions:

$$\mathcal{M}_{NC} = \frac{8G_F}{\sqrt{2}} \rho (\jmath_\mu^3 - s^2 \jmath_{Q\mu})(\jmath'^{3\mu} - s^2 \jmath'^\mu_Q) \tag{13.137}$$

(neutral currents),

$$\mathcal{M}_{CC} = -\frac{4G_F}{\sqrt{2}} \jmath_\mu^+ \jmath^{-\mu} \tag{13.138}$$

(charged currents). These interactions in the SM (in tree approximation) look as

$$\mathcal{M}_{NC} = \frac{e^2}{s^2 c^2} (\jmath_\mu^3 - s^2 \jmath_{Q\mu}) \left[\frac{g^{\mu\nu} - q^\mu q^\nu / M_Z^2}{q^2 - M_Z^2} \right] (\jmath'^3_\mu - s^2 \jmath'_{Q\mu}), \tag{13.139}$$

$$\mathcal{M}_{CC} = \frac{e^2}{2s^2} \jmath_\mu^+ \left[\frac{g^{\mu\nu} - q^\mu q^\nu / M_W^2}{q^2 - M_W^2} \right] \jmath_\nu^- . \tag{13.140}$$

The problem is to find the radiative corrections to these interactions. These corrections are of three types: vertex corrections, box diagrams and vacuum polarizations. It is important that among them, the vacuum polarization corrections dominate [43]. They are called the oblique corrections in the literature.

To describe them, let us introduce the vacuum polarization tensor $\Pi^{\mu\nu}_{XY}$, where $X, Y = 1, 2, 3, Q$:,

$$\begin{aligned}\Pi^{\mu\nu}_{XY}(q) =& g^{\mu\nu}\Pi_{XY}(q^2) + q^\mu q^\nu\text{–terms} \\ =& i\int d^4x e^{iqx}\langle 0|TJ^\mu_X(x)J^\nu_Y(0)|0\rangle .\end{aligned} \tag{13.141}$$

We emphasize that the currents J^μ_X are different from the currents in Eqs.(13.133)–(13.136): they include both fermion and technifermion fields.

The $q^\mu q^\nu$ terms give a small contribution to processes including light fermions (they lead to corrections of order $\frac{m_f^2}{v^2}$ which are very small). Then, the relevant part of the vector boson propagators take the following form:

$$(D^{\mu\nu}_{WW})^{-1} = g^{\mu\nu}\left(q^2 - \frac{e^2v^2}{4s^2} - \frac{e^2}{s^2}\Pi_{11}\right), \tag{13.142}$$

$$(D^{\mu\nu}_{AB})^{-1} = g^{\mu\nu}\begin{vmatrix} q^2 - e^2\Pi_{QQ} & \Pi_{Z\gamma} \\ \Pi_{Z\gamma} & q^2 - \frac{e^2v^2}{4s^2c^2} - \Pi_{ZZ}\end{vmatrix}, \tag{13.143}$$

where $A, B = \gamma, Z$ and

$$\Pi_{Z\gamma} \equiv \frac{e^2}{sc}\left(\Pi_{3Q} - s^2\Pi_{QQ}\right), \tag{13.144}$$

$$\Pi_{ZZ} \equiv \frac{e^2}{s^2c^2}\left(\Pi_{33} - 2s^2\Pi_{3Q} + s^4\Pi_{QQ}\right). \tag{13.145}$$

Now it is easy to derive from Eqs. (13.142)–(13.145) the expressions for the interactions $\mathcal{M}_{NC}$ and $\mathcal{M}_{CC}$ including the oblique corrections:

$$\begin{aligned}\mathcal{M}_{NC} =& e^2\frac{\jmath_Q\jmath'_Q}{q^2 - e^2\Pi_{QQ}} + \frac{e^2}{s^2c^2}\cdot \\ & \frac{\left[\jmath_3 - \left(s^2 - sc\frac{\Pi_{Z\gamma}}{q^2 - e^2\Pi_{QQ}}\right)\jmath_Q\right]\left[\jmath'_3 - \left(s^2 - sc\frac{\Pi_{Z\gamma}}{q^2 - e^2\Pi_{QQ}}\right)\jmath'_Q\right]}{q^2 - \frac{e^2v^2}{4s^2c^2} - \Pi_{ZZ} - \frac{(\Pi_{Z\gamma})^2}{q^2 - e^2\Pi_{QQ}}},\end{aligned} \tag{13.146}$$

$$\mathcal{M}_{CC} = \frac{e^2}{2s^2}\frac{\jmath^+\jmath^-}{q^2 - \frac{e^2v^2}{4s^2} - \frac{e^2}{s^2}\Pi_{11}}. \tag{13.147}$$

(here the amplitude $\mathcal{M}_{NC}$ includes also the electromagnetic interaction). The further analysis of this expressions can be found in Refs.[43, 48]. For our purpose, it is sufficient to know the following facts. As was shown by Peskin and Takeuchi [43], there are three parameters, S, T and U, in terms of which all the radiative corrections to low energy ($q \lesssim M_Z,\ M_W$) electroweak interactions can be described. These parameters are defined in the following way. Let us divide the functions Π_{XY} into two parts:

$$\Pi_{XY}(q^2) = \Pi_{XY}^{SM}(q^2) + \Pi_{XY}^{new}(q^2), \tag{13.148}$$

where the term Π_{XY}^{SM} is connected with the contribution in the SM and the second term $\Pi_{XY}^{new}(q^2)$ is connected with a new physics (TC in the present case). Since the term Π_{XY}^{SM} depends on still unknown masses of the t quark and Higgs boson, we must somehow choose these parameters. The reasonable values are $m_f = 150GeV$ and $M_h = 1TeV$: $m_f \sim 150GeV$ is preferred by experimental data, and $M_h = 1TeV$ is natural in the case of the TC dynamics, in which the Higgs boson is just one of technihadrons, and the scale $1TeV$ is a characteristic scale of technihadron masses.

The next step is to expand the $\Pi_{XY}^{new}(q^2)$ in a series in powers of momentum:

$$\begin{aligned}
\Pi_{QQ}^{new}(q^2) =& q^2 \Pi'^{new}_{QQ}(0) + \cdots, \\
\Pi_{3Q}^{new}(q^2) =& q^2 \Pi'^{new}_{3Q}(0) + \cdots, \\
\Pi_{33}^{new}(q^2) =& \Pi_{33}^{new}(0) + q^2 \Pi'^{new}_{33}(0) + \cdots, \\
\Pi_{11}^{new}(q^2) =& \Pi_{11}^{new}(0) + q^2 \Pi'^{new}_{11}(0) + \cdots.
\end{aligned} \tag{13.149}$$

Here we took into account the fact that because of the Ward-Takahashi identities for the electromagnetic current , $\Pi_{QQ}(0) = \Pi_{3Q}(0) = 0$. The important point is that while the characteristic TC scale is $\Lambda_{TC} \sim 1TeV$, we are interested in momenta $q \sim 100GeV$, near the mass shell of the W and Z. Therefore all other terms in the series (13.149) are suppressed by the factor $\frac{q^2}{\Lambda_{TC}^2} \sim 10^{-2}$. The Peskin-Takeuchi parameters S, T and U are defined as

$$\alpha S \equiv 4e^2 \left[\Pi'^{new}_{33}(0) - \Pi'^{new}_{3Q}(0)\right] = -4e^2 \Pi'^{new}_{3Y}(0), \tag{13.150}$$

$$\alpha T \equiv \frac{e^2}{s^2 c^2 M_Z^2} \left[\Pi_{11}^{new}(0) - \Pi_{33}^{new}(0)\right], \tag{13.151}$$

$$\alpha U \equiv 4e^2 \left[\Pi'^{new}_{11}(0) - \Pi'^{new}_{33}(0)\right]. \tag{13.152}$$

Thus the current-current interactions (13.146), (13.147) can be parametrized in the terms of these three parameters.

Let us discuss the physical meaning of these parameters.

We begin by considering the parameter T. Eqs.(13.146), (13.147) imply that

$$M_Z^2 = \frac{e}{s^2c^2}\frac{v^2}{4} + \frac{e^2}{s^2c^2}\left(\Pi_{33}(M_Z^2) - 2s^2\Pi_{3Q}(M_Z^2) + s^4\Pi_{QQ}(M_Z^2)\right), \tag{13.153}$$

$$M_W^2 = \frac{e}{s^2}\frac{v^2}{4} + \frac{e^2}{s^2}\Pi_{11}(M_W^2) \tag{13.154}$$

(we neglected the term $\frac{(\Pi_{Z\gamma})^2}{q^2-e^2\Pi_{QQ}}$ in M_Z^2: as follows from Eq.(13.144) it is suppressed by the factor α with respect to other terms in Eq.(13.153)). These relations and Eqs.(13.149), (13.151) imply that

$$\alpha T = M_W^2/M_Z^2c^2 - 1 = \rho - 1 \equiv \delta\rho. \tag{13.155}$$

Thus, this parameter, being equivalent to $\delta\rho$, describes the violation of the isospin symmetry due to a new physics (see the discussion below Eq.(13.65)).

Let us now consider the parameter S. Since the function Π_{3Y} is expressed through the Green's function (13.141) of the currents J_3 and J_Y relating to the different gauge groups, $SU_L(2)$ and $U_Y(1)$, it would be equal zero if the $SU_L(2) \times U_Y(1)$ was not spontaneously broken. Therefore, S is a measure of the electroweak condensate connected with a new physics.

Like T, the parameter U describes the isospin violation. Usually, it is small:

$$\alpha U = 4e^2\left[\Pi'^{new}_{11}(0) - \Pi'^{new}_{33}(0)\right] \sim 4e^2\left[\frac{\Pi^{new}_{11}(0) - \Pi^{new}_{33}(0)}{\Lambda^2_{TC}}\right] \sim \frac{M_Z^2}{\Lambda^2_{TC}}\alpha T. \tag{13.156}$$

Let us estimate these parameters in TC models. We begin by considering the case of TC models with the custodial $SU_V(2)$ symmetry (see Sec.13.5). In this case the isospin is a good quantum number in TC, and therefore the T and U parameters equal zero. In order to calculate S, we will use the chiral effective lagrangian

$$\mathcal{L}_{eff} = \mathcal{L}^{(2)}_{eff} + \mathcal{L}^{(4)}_{eff} \tag{13.157}$$

with $\mathcal{L}^{(2)}_{eff}$ and $\mathcal{L}^{(4)}_{eff}$ given in Eqs. (12.227) and (12.230). Now the decay constant F is

$F = v \simeq 246GeV$, and the chiral field Σ is

$$\Sigma = \frac{v}{\sqrt{2}} exp(iw^\alpha \tau^\alpha / v), \ \alpha = 1,2,3, \tag{13.158}$$

where w^α are the would-be NG bosons absorbed by the $W^\pm$ and Z. Besides possible light pseudo-NG bosons, there are no other light technihadrons, with masses of order $100GeV$ (or lighter) in TC. We will take into account the contribution from the pseudos somewhat later.

Now we will treat the gauge fields $\tilde{A}_{L\mu}$ and $\tilde{A}_{R\mu}$ in Eq.(12.229) as the gauge fields connected with the electroweak group $SU_L(2) \times U_Y(1)$:

$$\tilde{A}_{L\mu} = -\, g_2 W_\mu^\alpha \frac{\tau^\alpha}{2} = -\frac{e}{s} W_\mu^\alpha \frac{\tau^\alpha}{2}, \tag{13.159}$$

$$\tilde{A}_{R\mu} = -\, g_1 B_\mu Y = -\frac{e}{c} B_\mu Y, \tag{13.160}$$

$$\tilde{\nabla}_\mu \Sigma = D_\mu \Sigma = \partial_\mu \Sigma + i\frac{e}{s} W_\mu^\alpha \frac{\tau^\alpha}{2} \Sigma - i\frac{e}{c} B_\mu \Sigma \frac{\tau^3}{2} \tag{13.161}$$

(we recall that under the $SU_L(2) \times U_Y(1)$, Σ transforms as $\Sigma \to L\Sigma R^+$ and that $Y = \frac{\tau_R^3}{2}$ for Σ).

The S parameters (13.150) is expressed through the polarization operator Π_{3Y} It is clear that the term in $\mathcal{L}_{eff}^{(4)}$ (12.230) which contributes to Π_{3Y} is the following one:

$$L_{10} tr(\Sigma \tilde{F}_R{}^{\mu\nu} \Sigma^\dagger \tilde{F}_{L\mu\nu}) \tag{13.162}$$

with

$$\tilde{F}_R^{\mu\nu} = \frac{e}{c}(\partial^\mu B^\nu - \partial^\nu B^\mu)\frac{\tau^3}{2}, \tag{13.163}$$

$$\tilde{F}_L^{\mu\nu} = -\frac{e}{s}(\partial^\mu W^\nu - \partial^\nu W^\mu + i\frac{e}{s}[W^\mu, W^\nu]), \tag{13.164}$$

where $W_\mu \equiv W_\mu^\alpha \frac{\tau^\alpha}{2}$. Its contribution to Π_{3Y} is

$$\Pi_{3Y} = L_{10}(M_Z) q^2. \tag{13.165}$$

Here we indicated that L_{10} relates to the scale $\sigma = M_Z$.

Therefore S is:

$$S = -4\frac{e^2}{\alpha}\Pi'^{new}_{3Y}(0) = -16\pi L_{10}(M_Z). \tag{13.166}$$

Let us now estimate $L_{10}(M_Z)$ in the TC models with a QCD-like dynamics. The value of L_{10} in QCD, estimated in the framework of the ENJL model, was

$$L_{10}(\Lambda_\chi) = -\frac{N_c}{96\pi^2}\Gamma\left(1, \frac{m^2_{dyn}}{\Lambda^2_\chi}\right) \simeq -\frac{N_c}{96\pi^2} \tag{13.167}$$

(see Eq.(12.232)). This value relates to the scale $\Lambda_\chi \sim 2m_{dyn}$. In TC models with the technicolor group $SU(N_{TC})$ and with N_D technifermion doublets, this estimate looks as

$$L_{10}(\Lambda_\chi) \simeq -\frac{N_{TC}N_D}{96\pi^2}; \ \Lambda_\chi \sim 1TeV. \tag{13.168}$$

The value $L_{10}(M_Z)$ at the scale $\sigma = M_Z$ is

$$L_{10}(M_Z) = L_{10}(\Lambda_\chi) + \Delta L_{10}, \tag{13.169}$$

where, due to the standard rules of the effective field theory approach (see Sec.11.2), ΔL_{10} is the contribution connected with the light technihadron loops in the region $M^2_Z \lesssim q^2 \lesssim \Lambda^2_\chi$, i.e. ΔL_{10} is the contribution of pseudo-NG bosons.

Let us apply these relations to concrete TC models.

There are no pseudo-NG bosons in the minimal TC model with one doublet of technifermions. Therefore, in this model S is

$$S = -16\pi L_{10}(\Lambda_\chi) \simeq \frac{N_{TC}}{6\pi}. \tag{13.170}$$

The situation in the one family TC is quite different: there are 60 pseudos (see Eq.(13.103)) and four technifermion doublets. Therefore, in this case, S is

$$S = -16\pi(L_{10}(\Lambda_\chi) + \Delta L_{10}) \sim \frac{2N_{TC}}{3\pi} - 16\pi\Delta L_{10}. \tag{13.171}$$

The contribution of the 60 pseudos (4 color singlet, 32 color octet, and 24 color triplet) is [45,49]:

$$\Delta L_{10} \simeq -(f_1 + 8f_8 + 6f_3), \tag{13.172}$$

where

$$f_i = \frac{1}{96\pi^2} \ln(\Lambda_\chi / \mathrm{Max}(M_{P_i}, M_Z)) \tag{13.173}$$

(M_{P_i} are masses of the pseudos). Basically each charged pair contributes one unit of the appropriate f_i. It is important that ΔL_{10} has the same sign as $\Delta L_{10}(\Lambda_\chi)$.

Assuming "reasonable" values for the masses of the pseudos ($M_{P_1} < M_Z, M_{P_8} \sim 250GeV, M_{P_3} \sim 150GeV$) and taking $\Lambda_\chi \sim 1TeV$, Holdom and Terning got [45]

$$\Delta L_{10} \simeq -0.017. \tag{13.174}$$

Therefore the parameter S is

$$S \sim 0.2N_{TC} + 0.9 > 1.3 \tag{13.175}$$

(the minimal N_{TC} is $N_{TC} = 2$).

Peskin and Takeuchi, using another approach, based on a dispersion relation for S, found the following genelar relation [43]:

$$S \simeq 0.1N_{TC}N_{TD} \tag{13.176}$$

which yields

$$S \simeq 0.4N_{TC} > 0.8 \tag{13.177}$$

for the one family model.

Now what about the present experimental data concerning S ? This parameter can be extracted from the forward-backward asymmetries of fermions [43]. The latest results yield [24]

$$S < -0.23\ (-0.13) \tag{13.178}$$

at the 90 (95) % confidence level. These results certainly "kill" the one family TC model (which has been already killed by the problems with FCNCs !). Even the minimal TC model, with

$$S > 0.1 N_{TC} > 0.2, \tag{13.179}$$

seems to be in trouble. Thus we conclude that the experimental data imply that dynamical electroweak symmetry breaking may be realized only in models which are apparently different from QCD-like TC models. We will consider such models in Chapters 14 and 15.

In conclusion, we will discuss the T parameter describing the isospin violation. To illustrate the situation, we will just indicate the contribution of a fermion doublet $\begin{pmatrix} U \\ D \end{pmatrix}$ to $T = \delta\rho/\alpha$ in the one-loop approximation. This contribution was first calculated by Veltman [38] (see also Ref. [39]). His result is (compare with Eq.(13.130)):

$$T = \frac{1}{16\pi s^2 c^2 M_Z^2}\left[m_u^2 + m_d^2 - \frac{2m_u^2 m_d^2}{m_u^2 - m_d^2}\log\left(\frac{m_u^2}{m_d^2}\right)\right] \xrightarrow[m_u^2 \gg m_d^2]{} \frac{m_u^2}{16\pi s^2 c^2 M_Z^2} \tag{13.180}$$

The upper bound on this quantity is [24]: $T < 0.2(0.3)$ at the 90 (95)% confidence level. This implies that the custodial $SU_V(2)$ is a good approximate symmetry in nature.

We also note that in order to describe the isospin violation in the chiral effective lagrangian, one should introduce the term

$$\Delta\rho_2 \frac{v^2}{4}\left(tr(\tau_3 \Sigma^\dagger D_\mu \Sigma)\right)\left(tr(\tau_3 \Sigma^\dagger D^\mu \Sigma)\right) \tag{13.181}$$

in $\mathcal{L}_{eff}^{(2)}$ (there are of course also terms with higher powers of the derivatives violating the isospin symmetry in $\mathcal{L}_{eff}$).

The present discussion shows that precision electroweak measurements yield rather severe restrictions for the model building in electroweak symmetry breaking. In the next two chapters, we will discuss recent developments in dynamical electroweak symmetry breaking.

Chapter 14

New Technicolor Scenarios

14.1. Introduction

In Sec.13.7, we discussed problems in the ETC model building connected with FCNCs and with pseudo-NG bosons. The essence of these problems is in the following three relations:

$$m_f \sim -\frac{g_{ETC}^2}{\Lambda_{ETC}^2}\langle\bar{F}F\rangle_{\Lambda_{ETC}} \equiv -\frac{\langle\bar{F}F\rangle_{\Lambda_{ETC}}}{M_{ETC}^2} , \tag{14.1}$$

$$\langle\bar{F}F\rangle_{\Lambda_{ETC}} \simeq -\frac{N_{TC}}{4\pi^2}\int_0^{\Lambda_{ETC}^2} dp^2 B_F(p^2) \sim -\frac{N_{TC}}{4\pi^2}\Lambda_{TC}^3\left(\ln\Lambda_{ETC}^2/\Lambda_{TC}^2\right)^{\tilde{c}_m} , \tag{14.2}$$

$$M_{P^a} = a\frac{\Lambda_{ETC}}{g_{ETC}F_{TC}}m_f \equiv a\frac{M_{ETC}}{F_{TC}}m_f; a \sim 1 \tag{14.3}$$

(see Eqs.(13.95), (13.96) and (13.110). FCNCs lead to the bound $M_{ETC} \gtrsim 300TeV$ for fermions in the first two families. This, together with Eqs.(14.1) - (14.3), implies that m_f are at most of $O(1MeV)$ and M_{P^a} are at most of $O(1GeV)$. Thus, the ETC models lead to too light quarks and to too light pseudo-NG bosons (already experimentally excluded).

Is there a possibility to change the relations (14.1) - (14.3), or, at least, some of them ? The relations (14.1), (14.3) follow from rather general principles, and it is very difficult to change them. The relation (14.2) is based on the asymptotic formula (13.89) for the technifermion mass function,

$$B_F(p) \sim \frac{\Lambda_{TC}^3}{p^2}\left(\ln p^2/\Lambda_{TC}^2\right)^{-1+\tilde{c}_m} , \tag{14.4}$$

which is valid in gauge theories with a QCD-like dynamics. Actually, as was shown in Sec.12.3, Eq.(14.4) follows from the fact that in asymptotically free gauge theories, the dynamical dimension $d_{\bar{\psi}\psi}$ of the operator $\bar{\psi}\psi$ is close to its canonical (engineering) dimension $d_c = 3$.*) Indeed, by definition,

$$d_{\bar{\psi}\psi} = d_c - \gamma_m = 3 - \gamma_m \; , \tag{14.5}$$

where the anomalous dimension γ_m of $\bar{\psi}\psi$ is small near the ultraviolet fixed point $g = g_c = 0$ which governs the ultraviolet behavior of Green's functions in QCD-like dynamics. Actually, γ_m is

$$\gamma_m \simeq b\tilde{c}_m g^2 \; , \tag{14.6}$$

where b is the first coefficient of the technicolor β function (compare with Eq.(12.42)).

If γ_m were rather large, the relation (14.4) would be different. By applying the analysis based on the operator product expansion (OPE), considered in Sec.12.3, to this case, we find:

$$B_F(p) \sim \frac{(\Lambda_{TC})^{3-\gamma_m}}{(p)^{2-\gamma_m}} \tag{14.7}$$

(up to some logarithmic correction). Substituting this expression for B_F in the relation for the condensate, we get:

$$\langle \bar{F}F \rangle_{\Lambda_{ETC}} \simeq -\frac{N_{TC}}{4\pi^2} \int\limits_0^{\Lambda_{ETC}^2} dp^2 B_F(p^2) \sim -\frac{N_{TC}}{2\pi^2 \gamma_m} (\Lambda_{TC})^{3-\gamma_m} (\Lambda_{ETC})^{\gamma_m} \; . \tag{14.8}$$

Now Eqs.(14.1) and (14.3) lead to the following relations for m_f and M_{P^a}:

$$m_f(\gamma_m) \sim \left(\frac{\Lambda_{ETC}}{\Lambda_{TC}} \right)^{\gamma_m} m_f \tag{14.9}$$

$$M_{P^a}(\gamma_m) \sim \left(\frac{\Lambda_{ETC}}{\Lambda_{TC}} \right)^{\gamma_m} M_{P^a} \; , \tag{14.10}$$

where $m_f(\gamma_m)$, $M_{P^a}(\gamma_m)$ and m_f, M_{P^a} relate to TC with $\gamma_m \neq 0$ and to TC with $\gamma_m = 0$ (QCD-like dynamics), respectively. Since $\Lambda_{ETC} \gtrsim 300TeV$ is much larger than the TC scale $\Lambda_{TC} \sim 1TeV$, both fermion and pseudo-NG boson masses are now enhanced. Thus, a large anomalous dimension γ_m may solve these two problems in ETC !

---

*) For a detailed discussion of the notion of the dynamical dimension of the operators, see Chapter 11.

The idea of enhancing the fermion and pseudo-NG boson masses by using a TC dynamics with a large anomalous dimension γ_m was introduced by Holdom [1, 2]. However, the model building in this direction had not been realized until a simple toy model with a large anomalous dimension appeared in the literature: Bardeen, Leung and Love, and the present author [3, 4] interpreted the critical coupling constant $\alpha_c = \pi/3$ in the rainbow (ladder) QED as an ultraviolet fixed point at which the composite operator $\bar{\psi}\psi$ has a large anomalous dimension $\gamma_m = 1$. To understand this, let us look at the expression for the fermion mass function $B_{dyn}(p^2)$ in this model (see Eqs.(10.16), (10.17)). We see that

$$B_{dyn}(p^2) \underset{p^2 \gg m_{dyn}^2}{\sim} \frac{m_{dyn}^2}{p} \tag{14.11}$$

at $\alpha = \alpha_c = \pi/3$. Comparing this equation with Eq.(14.7), we find that $\gamma_m = 1$ in this case.

This interpretation was originally suggested without any connection with TC models. The first who recognized that this model can be relevant for TC were Yamawaki, Bando and Matumoto [5], and Akiba and Yanagida [6] who used the model in a scenario with non-asymptotically free TC.

These works also paved the way for the TC models known as walking TC (WTC) in the literature. WTC was introduced by Appelquist, Karabali and Wijewardhana [7, 8] (for earlier ideas, see Ref.[2]). Although WTC is based on asymptotically free models, its dynamics is essentially different from that of QCD: in a sense, there is an effective nontrivial fixed point which leads to a large effective $\gamma_m \sim 1$ (for a more detail, see Sec.14.3).

The next step was done by Appelquist, Einhorn, Takeuchi and Wijewardhana [9], and Miransky and Yamawaki [10] (see also Ref.[11]) who introduced a class of ETC models now known as strong ETC (SETC). Unlike previous TC scenarios, where ETC interactions were treated as perturbative, SETC contains ETC interactions with a strong coupling g_{ETC}^2.

This development inevitably led to a nonperturbative treatment of the whole ETC dynamics. This in turn revealed some new physical effects which are interesting in themselves. Here we would like to indicate one of them: nondecoupling of the high energy physics from the low energy one in the near-critical dynamical regime with large anomalous dimensions. The importance of this effect for the description of symmetry breaking in the TC model building was first pointed out by Kogut, Dagotto and Kocić [12] and Chivukula [13].

In the description of such a nonperturbative dynamics, the theory of critical phenomena is extremely useful. As was shown in Chapter 11, the problem is equivalent to the description of the dynamics of the second order phase transition in some statistical models. At present, the techniques available for its study are very limited. One such approach which offers a great deal of promise involves lattice computer simulations. The studies of the dynamics in lattice models relevant for WTC and SETC have been realized by various groups [14 - 19].

In the present chapter, we will discuss these, rather recent, developments in TC.

14.2. The Gauged Nambu-Jona-Lasinio Model as a Basic Model for New Dynamical Electroweak Symmetry Breaking Scenarios

In this section, we will describe the gauged NJL model [4] which provides a basic example for different scenarios of dynamical electroweak symmetry breaking.

The gauged NJL model is described by the following lagrangian density:

$$\mathcal{L} = -\frac{1}{4}F_{\mu\nu}F^{\mu\nu} + \bar{\psi}\left(i\gamma^{\mu}D_{\mu} - m^{(0)}\right)\psi + \frac{G^{(0)}}{2}\left[(\bar{\psi}\psi)^2 + (\bar{\psi}i\gamma_5\psi)^2\right], \tag{14.12}$$

where the gauge interaction ($D_\mu = \partial_\mu + igA_\mu$) is treated in the ladder (rainbow) approximation and the four-fermion interactions are treated in the mean-field theory approximation (see Chapter 6). Since our main purpose here is just to illustrate the main features of this dynamics, we, for simplicity, will consider chiral $U_L(1)\times U_R(1)$ symmetry and a gauge $U(1)$ symmetry (the extension to non-Abelian case will be discussed at the end of this section).

In this approximation, the Schwinger-Dyson (SD) equation for the fermion propagator

$$G(p) = [\hat{p}A(p^2) - B(p^2)]^{-1} \tag{14.13}$$

can be derived in the same way as it was derived in the NJL model and QED (see Sec.8.4). In Landau gauge, we get:

$$A(p^2) = 1 \tag{14.14}$$

$$B(p^2) = m^{(0)} + \frac{\kappa}{\Lambda^2}\int_0^{\Lambda^2} dq^2 \frac{q^2 B(q^2)}{q^2 + B^2(q^2)} + \frac{3\alpha}{4\pi}\int_0^{\Lambda^2} dq^2 \frac{q^2 B(q^2)}{q^2 + B^2(q^2)}\left[\frac{1}{p^2}\theta(p^2 - q^2) + \frac{1}{q^2}\theta(q^2 - p^2)\right] \tag{14.15}$$

(compare with Eqs.(8.54), (8.55) and Eqs.(8.62), (8.63)). Here $\kappa \equiv G^{(0)}\Lambda^2/4\pi^2$ and $\alpha = \frac{g^2}{4\pi}$; the equations are written in Euclidean region.

It is easy to check that the solutions to this equation satisfy the differential equation

$$\frac{d}{dp^2}\left[p^2 \frac{dB(p^2)}{dp^2}\right] + \frac{3\alpha}{4\pi}\frac{B(p^2)}{(p^2 + B(p^2))} = 0 \tag{14.16}$$

and two boundary conditions:

$$\lim_{p^2 \to 0}\left((p^2)^2 \frac{dB(p^2)}{dp^2}\right) = 0 \, , \tag{14.17}$$

$$\lim_{p^2 \to \Lambda^2}\left[\left(1 + \frac{4\pi\kappa}{3\alpha}\right) p^2 \frac{dB}{dp^2} + B(p^2)\right] = m^{(0)} \, . \tag{14.18}$$

Eqs.(14.16) and (14.17) coincide with Eqs.(10.9) and (10.10) in QED. The four-fermion coupling κ appears in the ultraviolet boundary condition (14.18) only (compare with Eq.(10.11)).

The analysis of Eqs.(14.16) - (14.18) will just follow the analysis of Eqs.(10.9) - (10.11) in QED. In particular, we will use the linearized version of Eq.(14.16):

$$\frac{d}{dp^2}\left[p^2 \frac{dB(p^2)}{dp^2}\right] + \frac{3\alpha}{4\pi}\frac{B(p^2)}{(p^2 + m^2)} = 0 \, , \tag{14.19}$$

where $m \equiv B(0)$.

As in Sec.10.2, we find that the solution to Eqs.(14.19), (14.17) is

$$B(p^2) = mF\left(\frac{1+\omega}{2}, \frac{1-\omega}{2}, 2; -p^2/m^2\right) \, , \tag{14.20}$$

where F is the hypergeometric function and $\omega = \left(1 - \frac{3\alpha}{\pi}\right)^{1/2}$. The second boundary condition (14.18) yields a relation between $m, m^{(0)}$ and Λ.

Let us now consider the chiral limit with $m^{(0)} = 0$. Substituting the asymptotic expressions (10.15) - (10.17) for $B_{dyn}(p^2)$ (with $m = m_{dyn}; m^{(0)} = 0$) into Eq.(14.18), we find that a nontrivial solution with $m_{dyn} \neq 0$ exists in the following region in the (α, κ) plane: at $\alpha < \alpha_c = \pi/3$ and $\kappa > \kappa_c(\alpha)$, where

$$\kappa_c(\alpha) = \frac{1}{4}(1+\omega)^2 , \tag{14.21}$$

and at $\alpha > \alpha_c = \pi/3$. The critical line, separating the symmetric and asymmetric phases in the model is shown in Fig.14.1. This critical line in the gauged NJL model was obtained by Kondo, Mino and Yamawaki [20] and by Appelquist, Soldate, Takeuchi and Wijewardhana [21].

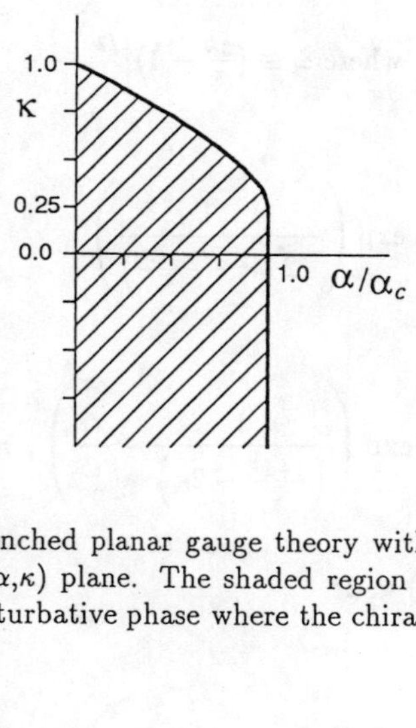

Fig. 14.1. Critical line of quenched planar gauge theory with four-fermion interaction is depicted by the solid line in (α,κ) plane. The shaded region bounded by the critical line and the vertical axis is the perturbative phase where the chiral symmetry is spontaneously unbroken.

One can see that as $\alpha \to 0$, the critical value $\kappa_c(\alpha) \to 1$ (the pure NJL model; see Chapter 6). As $\kappa \to 0$, we finds that $\alpha_c(\kappa) \to \alpha_c = \pi/3$ (the rainbow QED; see Chapter 10). Thus, this critical line interpolates between the critical points in the NJL model and rainbow QED.

The explicit form of the dynamical mass function m_{dyn} as a function of Λ, κ and α near the critical line (i.e. the scaling law in the model) is [20 - 24]:

1) $0 < \alpha < \alpha_c = \frac{\pi}{3}$:

$$m_{dyn} \simeq \Lambda \left[\frac{\Gamma(1+\omega)\Gamma^2\left(\frac{3-\omega}{2}\right)}{\Gamma(1-\omega)\Gamma^2\left(\frac{3+\omega}{2}\right)} \cdot \frac{\kappa - \left(\frac{\omega+1}{2}\right)^2}{\kappa - \left(\frac{1-\omega}{2}\right)^2} \right]^{1/2\omega} \tag{14.22}$$

in the near-critical ($\kappa - \frac{1}{4}(1+\omega)^2 \ll 1$) region;

2) $\alpha = \alpha_c = \frac{\pi}{3}, \kappa > \frac{1}{4}$:

$$m_{dyn} \simeq 4\Lambda \exp\left(-\frac{1}{2(\kappa - \frac{1}{4})}\right) \tag{14.23}$$

at $\kappa - \frac{1}{4} \ll 1$;

3) $\alpha > \alpha_c = \frac{\pi}{3}$:

a) $\alpha > \frac{\pi}{3}, \kappa > \frac{1}{4}(1-\tilde{\omega}^2)$, where $\tilde{\omega} \equiv \left(\frac{3\alpha}{\pi} - 1\right)^{1/2}$:

$$m_{dyn}^{(0)} \simeq 4\Lambda \exp\left(-\frac{1}{2\left(\kappa - \frac{1+\tilde{\omega}^2}{4}\right)}\right), \tag{14.24}$$

$$m_{dyn}^{(n)} \simeq 4\Lambda \exp\left(\frac{1}{\left(\frac{1+\tilde{\omega}^2}{2} - 2\kappa\right)} - \frac{\pi n}{\tilde{\omega}}\right), \quad n = 1, 2, \cdots, \tag{14.25}$$

at $\tilde{\omega} \ll 1$ and $\kappa - \frac{1}{4}(1-\tilde{\omega}^2) \ll 1$;

b) $\alpha > \frac{\pi}{3}, \kappa < \frac{1}{4}(1-\tilde{\omega}^2)$:

$$m_{dyn}^{(n)} \simeq \frac{4}{e}\Lambda \exp\left(\frac{1+4\kappa}{1-4\kappa} - \frac{\pi n}{\tilde{\omega}}\right), \quad n = 1, 2, \cdots, \tag{14.26}$$

at $\tilde{\omega} \ll 1$.

Few comments concerning this scaling law for m_{dyn} are in order:

a) if we applied the relation (14.22) to the pure NJL model, when $\alpha \to 0 (\omega \to 1)$, we would obtain the relation

$$m_{dyn} \simeq \Lambda \lim_{\omega \to 1} (1-\omega)(\kappa - 1)^{1/2} , \tag{14.27}$$

instead the correct one (see Eq.(6.14))

$$m_{dyn} \simeq \Lambda \left(\ln \frac{\Lambda^2}{m_{dyn}^2} \right)^{-1/2} (\kappa - 1)^{1/2} . \tag{14.28}$$

This discrepancy will be removed if the next-to-next-to-leading term is taken into account in the asymptotic expression (10.15) for $B_{dyn}(p^2)$ [24];

b) as follows from the discussion in Sec.10.3, the value of the dynamical mass defining the stable vacuum coincides with the largest value of $m_{dyn}^{(n)}$, which is $m_{dyn}^{(0)}$ in Eq.(14.24), or $m_{dyn}^{(1)}$ in Eq.(14.26), at $\alpha > \pi/3$. The presence of other solutions for m_{dyn} implies that there are radial excitations of the NG bosons (see Sec.10.3). Eqs.(14.22) - (14.26) imply the following structure of the spectrum of these excitations in the model [22]. In the region where $\alpha < \alpha_c = \pi/3$, there are no such radial excitations. In the region with the supercritical gauge coupling $\alpha > \alpha_c = \pi/3$, there are an infinite number of the pseudoscalar excitations. Moreover, in the region with $\alpha > \frac{\pi}{3}$ and $\kappa > \frac{1-\tilde{\omega}^2}{4}$, there is an additional pseudoscalar composite in comparison with pure QED.

This result admits the following simple interpretation. In the region with $\alpha < \alpha_c = \pi/3$, the *short-range* four-fermion interaction is mostly responsible for the formation of fermion-antifermion bound states. But in the region with $\alpha > \alpha_c = \pi/3$ supercritical *long-range* Coulomb-like forces lead to the appearance of an infinite number of additional composites. When the four-fermion coupling constant decreases and becomes smaller than $\frac{1-\tilde{\omega}^2}{4}$, one composite disappears. In this region Coulomb-like forces are mostly responsible for the formation of the bound states. The four-fermion interaction just enhances the dynamical fermion mass (see Eq.(14.26)).

One can also show that a similar picture takes place for the radial excitations in the scalar channel [22].

Let us now calculate the chiral condensate along the critical line in this model.

Differentiating equation (14.15) in p^2, we obtain the expression for the condensate:

$$\langle 0|\bar{\psi}\psi|0\rangle_\Lambda = -\frac{1}{4\pi^2}\int_0^{\Lambda^2} dp^2 \frac{p^2 B(p^2)}{p^2 + B^2(p^2)} = \frac{1}{3\pi\alpha}\left((p^2)^2 \frac{dB}{dp^2}\right)|_{p^2=\Lambda^2} . \tag{14.29}$$

By using Eqs.(10.15) - (10.17) for the ultraviolet asymptotics of $B_{dyn}(p^2)$ in the near-critical region, we find from Eq.(14.29) that

$$\langle 0|\bar{\psi}\psi|0\rangle_\Lambda \sim -\Lambda^{(\omega+1)} m_{dyn}^{(2-\omega)} \tag{14.30}$$

near the critical lime. Comparing this relation with the general relation for the condensate (14.8) derived from OPE, we find that the anomalous dimension γ_m of the operator $\bar{\psi}\psi$ is [10]:

$$\gamma_m = 1 + \omega = 1 + \left(1 - \frac{3\alpha}{\pi}\right)^{1/2} . \tag{14.31}$$

Thus, γ_m is large, $\gamma_m \geq 1$, along the whole critical line.

One can check the consistency of this conclusion by calculating γ_m from its standard definition:

$$\gamma_m = \frac{\partial \ln Z_m^{(\sigma)}}{\partial \ln \sigma} = -\frac{\partial \ln Z_m^{(\sigma)}}{\partial \ln \Lambda} \tag{14.32}$$

(compare with Eq.(12.42)), where $Z_m^{(\sigma)}$ is the renormalization constant of the fermion mass:

$$m_{\text{cur}}^{(\sigma)} = Z_m^{(\sigma)^{-1}} m^{(0)} \tag{14.33}$$

(see Eq.(12.19)). Taking now $m^{(0)} \neq 0$ in Eq.(14.18), we find the following relation for this renormalization constant at the critical line:

$$Z_m \equiv Z_m^{(\sigma)}|_{\sigma=0} \simeq Z_m^{(\sigma)}|_{\sigma=m} \sim \left(\frac{m}{\Lambda}\right)^{(1+\omega)} , \tag{14.34}$$

where here the total mass m is $m = m_{\text{cur}}$.*)

*) Indeed, $m_{dyn} = 0$ at the critical line

Taking Λ to be equal to Λ_{ETC} and m_{dyn} to be of order of Λ_{TC}, we rewrite the relation (14.30) as

$$\langle 0|\bar{\psi}\psi|0\rangle_{\Lambda_{ETC}} \sim -\Lambda_{ETC}^{\gamma_m}\Lambda_{TC}^{3-\gamma_m} \tag{14.35}$$

with $\gamma_m = 1 + \left(1 - \frac{3\alpha}{\pi}\right)^{1/2} \geq 1$. Thus, there is a huge enhancemet of the chiral condensate in this model as compared to QCD-like theories (see Eq.(14.2)). Let us discuss this point in more detail.

In asymptotically free theories, as QCD, the critical value α_c equals zero. Therefore exactly at the critical point (with a finite cutoff Λ), these theories are free. In particular, the relation between $m^{(0)}$ and m becomes trivial at the critical point:

$$m^{(0)} = m\ . \tag{14.36}$$

This in turn implies that $\gamma_m = 0$. When $\alpha^{(0)}(\Lambda)$ is small but nonzero, there are logarithmic corrections to this relation (see Eqs.(12.19), (12.20)).

A small γ_m implies that the dynamical mass function of fermions $B_{dyn}(p^2)$ rapidly, as $B_{dyn}(p^2) \sim \frac{1}{p^2}$ (see Eq.(14.4)), decreases with momentum. As we discussed in detail in Sec.12.4, this in turn implies that BS wave functions of spinless fermion-antifermion bound states (i.e. their form factors) also rapidly decrease with momentum; these bound states are loose ones.

In the gauged NJL model, the situation is essentially different. The interaction at small distances, governed by the values of the coupling constants κ and α at the critical line, is now strong. The dynamical mass function $B_{dyn}(p^2)$ decreases with p^2 much slower than in the QCD-like dynamics:

$$B_{dyn}(p^2) \sim m_{dyn}\left(p/m_{dyn}\right)^{(\gamma_m-2)}\ , \tag{14.37}$$

$\gamma_m = 1 + \omega > 1$. This in turn implies that form factors of fermion-antifermion bound states decrease with momentum also much slower than in QCD. These states are tightly bound states. This fact, first pointed out in Ref.[10], will be discussed in more detail in Sec.14.5.

Let us extend this analysis to the case when QED is substituted by an asymptotically free gauge theory [25 - 28]. We will use the improved ladder approximation for the gauge interaction, already utilized in QCD in Secs.12.5 and 12.7.

Now the coupling constant α is replaced by the running one (in the one-loop approximation). We will use the following ansätz for it:

$$\bar{\alpha}(p,q) \equiv \bar{g}^2(p,q)/4\pi = \theta(p^2-q^2)\bar{\alpha}(p^2) + \theta(q^2-p^2)\bar{\alpha}(q^2) \tag{14.38}$$

with

$$\bar{\alpha}(p^2) = \frac{1}{2\pi b}\frac{1}{\ln\left((p^2+p_c^2)/\Lambda_{TC}^2\right)} \tag{14.39}$$

(compare with Eqs.(12.96) and (12.100)). The SD equation for the mass functions $B(p^2)$ takes the following form in Landau gauge:

$$\begin{aligned} B(p^2) =& m^{(0)} + \frac{\kappa}{\Lambda^2}\int_0^{\Lambda^2} dq^2 \frac{q^2 B(q^2)}{q^2+B^2(q^2)} \\ &+ \frac{3C_2(F)}{4\pi}\int_0^{\Lambda^2} dq^2 \frac{q^2 B(q^2)}{q^2+B^2(q^2)}\left[\frac{\bar{\alpha}(p^2)}{p^2}\theta(p^2-q^2) + \frac{\bar{\alpha}(q^2)}{q^2}\theta(q^2-p^2)\right] \end{aligned} \tag{14.40}$$

(compare with Eq.(14.15) and with Eqs.(12.71), (12.99)). Here $C_2(F)$ is the quadratic Casimir invariant of the fermion representation of the gauge group ($C_2(F) = \frac{N_c^2-1}{2N_c}$ in QCD).

Our goal is to determine the influence of the four-fermion coupling κ on the ultraviolet asymptotics of $B_{dyn}(p^2)$ in the vicinity of the critical point where, by definition, m_{dyn} is small. For this purpose we can use the linearized form of Eq.(14.40) corresponding to the bifurcation approximation considered in Sec.10.9. In this approximation, $\frac{B(q^2)}{q^2+B^2(q^2)} \longrightarrow \frac{B(q^2)}{q^2}$, and the infrared cutoff $q_0^2 = m^2$ is introduced in the equation:

$$\begin{aligned} B(p^2) =& m^{(0)} + \frac{\kappa}{\Lambda^2}\int_{m^2}^{\Lambda^2} dq^2 B(q^2) \\ &+ \frac{3C_2(F)}{4\pi}\int_{m^2}^{\Lambda^2} dq^2 B(q^2)\left[\frac{\bar{\alpha}(p^2)}{p^2}\theta(p^2-q^2) + \frac{\bar{\alpha}(q^2)}{q^2}\theta(q^2-p^2)\right] . \end{aligned} \tag{14.41}$$

At $\kappa = 0$, this equation coincides with the QCD equation (12.71) in which $q_0^2 = m^2$ and $\frac{N_c^2-1}{2N_c} = C_2(F)$. We recall that the main conclusion from the analysis of that equation was that, in the chiral limit, the term with the irregular asymptotics was suppressed by the factor $\frac{1}{\Lambda^2}$ (see Eq.(12.77)). We will show that when $\kappa \simeq 1$, the situation changes drastically: the term with the irregular asymptotics dominates in this case [25 - 28].

The analysis of Eq.(14.41) is just a copy of the analysis of Eq.(12.71). We find that the solutions to Eq.(14.41) satisfy the differential equation

$$\frac{4\pi}{3C_2(F)}\frac{d}{dp^2}\left\{\left[\frac{d}{dp^2}\left(\frac{\bar{\alpha}(p^2)}{p^2}\right)\right]^{-1}\frac{dB(p^2)}{dp^2}\right\} = B(p^2) \tag{14.42}$$

and the ultraviolet boundary condition

$$\left[B(p^2) - \frac{(4\pi/3C_2(F))\kappa + \bar{\alpha}(p^2)}{p^2\frac{d}{dp^2}\left(\frac{\bar{\alpha}(p^2)}{p^2}\right)}\frac{dB(p^2)}{dp^2}\right]|_{p^2=\Lambda^2} = m^{(0)} \tag{14.43}$$

(compare with Eqs.(12.72), (12.73)). They also satisfy the infrared boundary condition:

$$\frac{dB(p^2)}{dp^2}|_{p^2=m^2} = 0\,. \tag{14.44}$$

Since the bare coupling $\alpha^{(0)} \equiv \bar{\alpha}(\Lambda^2)$ goes to zero when $\Lambda \to \infty$, the analysis in the preceding model, with a "standing" coupling α, suggests that the dynamics with a large anomalous dimension $\gamma_m \simeq 2$ should be realized around the value $\kappa = 1$ in the present model. As we will see, this is indeed the case.

The general solution of the linear differential equation (14.42) is

$$B(p^2) = C_1B_1(p^2) + C_2B_2(p^2)\,, \tag{14.45}$$

where $B_1(p^2)$ and $B_2(p^2)$ are the irregular and regular solutions, respectively (compare with Eq.(12.74)), i.e.

$$B_1(p^2) \sim z^{-\tilde{c}_m} + \sum_{n=1}\frac{I_n}{z^n}\,, \tag{14.46}$$

$$B_2(p^2) \sim e^{-z}z^{(\tilde{c}_m-1)}\left(1 + \sum_{n=1}\frac{R_n}{z^n}\right)\,, \tag{14.47}$$

where $z \equiv \ln p^2/\Lambda_{TC}^2$, $\tilde{c}_m = \frac{3C_2(F)}{8\pi^2 b}$ and I_n, R_n are coefficients which can be calculated by substituting $B(p^2)$ (14.45) into Eq.(14.42) (the explicit form of these coefficients can be found in Ref.[27]).

Now the crucial quantity for us is the ratio C_1/C_2 in the chiral limit, which can be calculated from the boundary condition (14.43) with $m^{(0)} = 0$. In the case of QCD, the ratio was $C_1/C_2 \sim 1/\Lambda^2$ (see Eq.(12.77)). But now it is different [26 - 28]:

$$\frac{C_2}{C_1} \simeq -\frac{\left[\kappa - \left(\frac{3C_2(F)}{4\pi}\bar{\alpha}(\Lambda)\right)^2\right] B_1'(\Lambda^2)}{\left[\kappa - \left(1 - \frac{3C_2(F)}{2\pi}\bar{\alpha}(\Lambda)\right)\right] B_2'(\Lambda^2)} , \tag{14.48}$$

where $B_i'(p^2) \equiv \frac{dB_i}{dp^2}$. This relation shows that when

$$\kappa - \kappa_c^{(eff)} \lesssim \frac{\Lambda_{TC}^2}{\Lambda^2} , \tag{14.49}$$

where

$$\kappa_c^{(eff)} \equiv 1 - \frac{3C_2(F)}{2\pi}\bar{\alpha}(\Lambda) , \tag{14.50}$$

there is no more suppression of the irregular solution and it dominates in the ultraviolet asymptotics.

Let us discuss the meaning of the quantity $\kappa_c^{eff} = 1 - \frac{3C_2(F)}{2\pi}\bar{\alpha}(\Lambda)$ in more detail. In the gauged NJL model with the ladder QED chosen as the gauge interaction, the critical line has the following form at $\alpha \ll \alpha_c = \pi/3$:

$$\kappa_c(\alpha) = \frac{1}{4}(1+\omega)^2 \underset{\alpha \ll \alpha_c}{\simeq} 1 - \frac{3}{2\pi}\alpha . \tag{14.51}$$

(see Eq.(14.21)). Comparing this relation with Eq.(14.50) and taking into account that $C_2(F) = 1$ in QED, we conclude that in the present model, κ_c^{eff} plays the role of $\kappa_c(\alpha)$, and the bare coupling $\alpha^{(0)} \equiv \bar{\alpha}(\Lambda)$ plays the role of α.

However, one should be careful with this interpretation. The point is that the critical line in the present model (with asymptotically free gauge interaction) has a simple form shown in Fig.14.2. Indeed, at all $\alpha^{(0)} \equiv \bar{\alpha}(\Lambda) > 0$ the phase with spontaneous chiral symmetry breaking is realized due to gauge interaction already. At $\alpha^{(0)} = 0$ the dynamical fermion mass is zero at $\kappa \leq \kappa_c = 1$, while at $\kappa > \kappa_c$ spontaneous chiral symmetry breaking takes place even at $\alpha^{(0)} = 0$, due to four-fermion interaction.

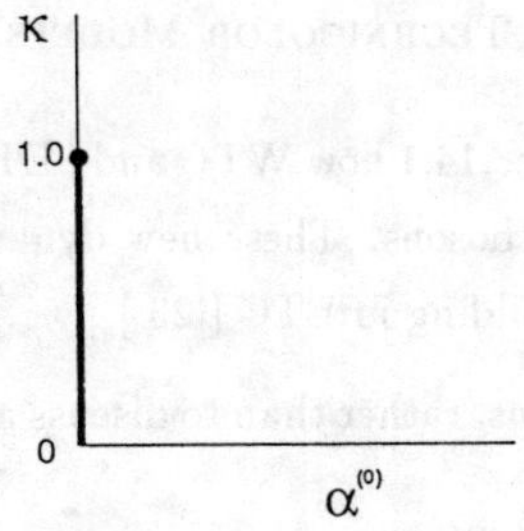

Fig. 14.2. Critical line of asymptotically free gauge theory with four-fermion interaction is depicted by the solid line in $(\alpha^{(0)},\kappa)$ plane.

As the present analysis shows, the end point $\kappa = \kappa_c = 1$ of this critical line is rather special. Eq.(14.49) implies that near this point, the character of the solution for $B_{dyn}(p^2)$ changes dramatically. Its ultraviolet asymptotics becomes

$$B_{dyn}(p^2) \sim \Lambda_{TC} \left(\ln p^2/\Lambda_{TC}^2\right)^{-\tilde{c}_m} , \tag{14.52}$$

and the condensate $\langle\bar{\psi}\psi\rangle = \frac{1}{N}\langle 0|\bar{\psi}\psi|0\rangle$ gets a huge enhancement as compared to QCD:

$$\langle\bar{\psi}\psi\rangle_\Lambda \simeq -\frac{N_{TC}}{4\pi^2}\int_0^{\Lambda^2} dp^2 B_{dyn}(p^2) \sim -\frac{N_{TC}}{4\pi^2}\Lambda_{TC}\Lambda^2 . \tag{14.53}$$

This relation corresponds to the anomalous dimension $\gamma_m = 2$ of the composite operator $\bar{\psi}\psi$.

One can check the consistency of this conclusion by calculating γ_m from the definition

$$\gamma_m = -\frac{\partial \ln Z_m^{(\sigma)}}{\partial \ln \Lambda};\; m_{\text{cur}}^{(\sigma)} = Z_m^{(\sigma)^{-1}} m^{(0)} \tag{14.54}$$

(compare with Eq.(14.32)). It is rather easy to find from Eq.(14.43) that $\gamma_m = 2$ for κ satisfying the restriction (14.49).

Thus, near the *effective* critical coupling $\kappa_c^{eff}(\alpha(\Lambda)) = 1 - \frac{3C_2(F)}{2\pi}\bar{\alpha}(\Lambda)$, this dynamics leads to a strong enhancement of the chiral condensate. Notice that gauge interaction lowers the critical value of κ.

14.3. More About New Technicolor Models

We already described in Sec.14.1 how WTC and SETC solve the problems connected with FCNCs and pseudo-NG bosons. These new dynamical possibilities have led to a recent resurgence of model building in ETC [29].

In this and the next sections, rather than to discuss all these models, we will describe their characteristic features.

As we have already learned, the chiral condensate $\langle\bar{F}F\rangle_{\Lambda_{ETC}}$ is very sensitive to the ultraviolet behavior of the technifermion mass function $B_F(p^2)$. This of course follows from the fact that the integral in the expression

$$\langle\bar{F}F\rangle_{\Lambda_{ETC}} \simeq -\frac{N_{TC}}{4\pi^2}\int\limits_0^{\Lambda_{ETC}^2} dp^2 B_F(p^2) \tag{14.55}$$

is badly divergent unless $B_F(p^2)$ decreases with p^2 more rapidly than $1/p^2$. This fact is essence of the enhancement mechanism in WTC and SETC.

On the other hand, the decay constant F_{TC} is much less sensitive to the ultraviolet behavior of $B_F(p^2)$ as compared to the condensate. Indeed, let us look at the Pagels-Stokar relation for F_{TC}:

$$F_{TC}^2 = \frac{N_{TC}}{8\pi^2}\int\limits_0^{\infty} dp^2 p^2 \frac{B_F(p^2)\left[2B_F(p^2) - p^2 B_F'(p^2)\right]}{\left(p^2 + B_F^2(p^2)\right)^2} \tag{14.56}$$

(compare with Eq.(12.126)). One can see that even for $B_F(p^2) = constant$ this integral diverges only logarithmically.

This point is in fact crucial for the model building. While a strong coupling dynamics enhances $\langle\bar{F}F\rangle_{\Lambda_{ETC}}$ thus enhancing masses of fermions and pseudo-NG bosons by the factor $(\Lambda_{ETC}/\Lambda_{TC})^{\gamma_m}$, the decay constant F_{TC}, which is fixed by the relation $F_{TC} = r^{-1/2}\cdot 246GeV$, with r being the number of technifermion doublets (see Eq.(13.82)), is not enhanced essentially by this mechanism.

It is convenient to divide all physical quantities into two classes: the class of ultraviolet sensitive (US) quantities and the class of ultraviolet insensitive (UIS) ones. In particular, it is important to show that the observables connected with FCNC processes are UIS ones. In other words, one should show that a large γ_m is responsible for an enhancement of the ordinary fermion masses without a simultaneous enhancement of FCNCs. Although there were some arguments in the support of this in the literature, a conclusive proof of this fact has been given only recently [30] (fore more detail, see Sec.14.4).

Let us now discuss different classes of TC models in which one can avoid the problems with FCNCs and light pseudos.

1) *TC with a nontrivial fixed point.*

In these models, one assumes that TC is a non-asymptotically free gauge interaction with a nontrivial ultraviolet stable fixed point at which the anomalous dimension γ_m is rather large [5, 6].

2) *Walking TC.*

In this class of models [7, 8], TC is an asymptotically free gauge theory, however, it is essentially different from the QCD-like dynamics.

Let us recall that in QCD, the running coupling $\bar{\alpha}(p^2)$ is

$$\bar{\alpha}(p^2) = \frac{1}{2\pi b \ln\left(p^2/\Lambda_{QCD}^2\right)}\,, \tag{14.57}$$

and the β function is

$$\beta_\alpha(\alpha) \equiv \frac{2g}{4\pi}\beta(g) = -4\pi b\alpha^2,\ b = \frac{11N_c - 2N}{24\pi^2}\,. \tag{14.58}$$

(see Eq.(12.11)). Let us introduce the quantity

$$\frac{\bar{\alpha}(p^2)}{\alpha_c} = \frac{2A}{\ln\left(p^2/\Lambda_{QCD}^2\right)},\ A \equiv \frac{1}{4\pi b\alpha_c},\ \alpha_c \equiv \frac{\pi}{3C_2(F)}\,. \tag{14.59}$$

In QCD with $N_c = 3,\ N = 3,\ C_2(F) = \frac{4}{3}$, we find

$$A \simeq 1.4. \tag{14.60}$$

The parameter A reflects the character of changing the running coupling $\bar{\alpha}(p^2)$ at $\bar{\alpha} \sim \alpha_c$: larger A implies slower running. Indeed, since $\frac{\beta_\alpha(\alpha_c)}{\alpha_c} = -\frac{1}{A}$, a large A corresponds to a small β function.

In QCD, the value A is of order one. The QCD dynamics is a dynamics with a fast running of $\bar{\alpha}(p^2)$. As a result, the final asymptotic form of $B_{dyn}(p^2)$,

$$B_{dyn}(p^2) \sim \frac{m_{dyn}^3}{p^2} \left(\ln p^2/\Lambda_{QCD}^2\right)^{1-c_m} , \tag{14.61}$$

sets in almost immediately beyond $p \sim 2m_{dyn} \sim \Lambda_\chi$.

In the walking TC, it is assumed that the parameter A is $A \simeq 5 - 15$. Such values of A are not exotic: it is easy to get them by choosing a rather large number of fermions in models ($A \sim 1/b \sim 1/(11N_{TC} - 2N)$) [8]. Then, $\bar{\alpha}(p^2)$ is slowly running (walking). This leads to the relatively slow initial fall of $B_{dyn}(p^2)$ before it takes on its final asymptotic form. One can say that in the region $\left(2m_{dyn}\right)^2 < p^2 \ll \Lambda_{ETC}^2$, the dynamics in WTC is governed by an effective ultraviolet stable fixed point $\alpha_c \sim \frac{\pi}{3C_2(F)}$ at which the anomalous dimension γ_m is $\gamma_m \lesssim 1$.

The numerical analysis of the SD equation in WTC with $A = 5 - 15$ shows that the relative size of the condensate here is 10 - 50 as compared with the "normal" running ($A \sim 1$) [8, 31].

Thus WTC can generate masses for the three lightest quarks, although generating the mass for the c quark (and, of course, for the t and b) looks very problematic here.

3) *Strong ETC.*

We have already described the essence of the dynamics of the SETC models in the previous section. The asymptotics of $B_F(p^2)$ takes following form there:

$$B_F(p^2) \sim C_1 \left(\ln p^2/\Lambda_{TC}^2\right)^{-\tilde{c}_m} + C_2 \frac{1}{p^2} \left(\ln p^2/\Lambda_{TC}^2\right)^{\tilde{c}_m - 1} , \tag{14.62}$$

where the ratio C_2/C_1 is yielded in Eq.(14.48). Near the effective critical point κ_c^{eff}, defined in Eq.(14.50), one can get a huge enhancement of the condensate. In principle, by tuning κ to κ_c^{eff}, one may generate masses even for the t and b.

The real problem for all these models is to generate a realistic spectrum of fermions without the violation of the restrictions connected with the S, T and U parameters (see Sec.13.9). Particularly important problems are the following:

a) How can one generate a large splitting between the masses of t and b without encountering a problem with the severe restriction for T ?

b) Is it possible to get a small value for S in TC ?

Although we are still far from having a realistic ETC model, some very recent works suggest that the near-critical strong coupling dynamics can help in solving these problems. We will return to this point in the next section.

In the next two sections we will consider two important aspects of the near-critical dynamics: a) the non-decoupling effects and b) the appearance of spinless "tight" bound states.

14.4. Nondecoupling Effects in New Technicolor Scenarios

The effect of the enhancement of the condensate in new TC scenarios is based on the fact that the anomalous dimension γ_m of the composite operator $\bar{\psi}\psi$ is large there.*)

In Chapter 11, we discussed the effect of nonrenormalizable (irrelevant) interactions in the low energy dynamics. As was shown there, their effect is strongly suppressed in perturbative theory. This fact is intimately connected with the Appelquist-Carazzone (AC) decoupling theorem discussed in Sec.13.8.

On the other hand, in Secs.11.3 and 11.5, considering some models in D dimensional space-time, with $2 < D < 4$, we saw that the situation may change dramatically if some composite operators acquire large anomalous dimensions. Therefore one should expect that nondecoupling effects occur in new TC scenarios. Actually, already that fact that the anomalous dimension γ_m depends on the four-fermion coupling constant κ along the critical line (14.21) in the gauged NJL model proves that the four-fermion interaction leads to nondecoupling effects in the model.

Let us beging by considering the WTC scenario with $\gamma_m \simeq 1$. We assume now that ETC interactions are rather weak and that they explicitly violates the isospin symmetry:

$$\mathcal{L}_{ETC} = \frac{g_U}{\Lambda_{ETC}^2}\bar{Q}_L U_R \bar{U}_R Q_L + \frac{g_D}{\Lambda_{ETC}^2}\bar{Q}_L D_L \bar{D}_R Q_L \tag{14.63}$$

with $g_U \neq g_D$ and $Q_L = \begin{pmatrix} U_R \\ D_R \end{pmatrix}, Q_R = \begin{pmatrix} U_L \\ D_L \end{pmatrix}$ being technifermions (here we consider the part of ETC connected with self-interaction of technifermions). In perturbative theory, in accordance with the decoupling theorem, the effect of the isospin violation would be suppressed by two powers of the ETC scale Λ_{ETC} in the low energy region (see Sec.13.8). But in WTC the situation is different. It is easy to check, slightly

*) An alternative interpretation of this mechanism will be considered in Sec.14.5.

modifying the analysis of the SD equation in Sec.14.2, that in the gauged NJL model with $\alpha \simeq \alpha_c = \pi/3$, the isospin violation formed at the high scale Λ_{ETC} manifest itself without any suppression at low momenta ! The essence of this effect is that at $\gamma_m = 1$, the four-fermion interaction becomes marginal (i.e. renormalizable) operator.*) Indeed:

$$\begin{aligned} \frac{g_U}{\Lambda_{ETC}^2} \bar{Q}_L U_R \bar{U}_R Q_L \simeq \frac{g_U}{\Lambda_{ETC}^2} Z_m^{(\sigma)^{-2}} \left(\bar{Q}_L U_R \bar{U}_R Q_L\right)_{\rm ren}^{(\sigma)} \\ \sim \frac{g_U}{\sigma^2} \left(\bar{Q}_L U_R \bar{U}_R Q_L\right)_{ren}^{(\sigma)} ; \; \sigma \sim \Lambda_{TC} \; , \end{aligned} \tag{14.64}$$

where the renormalization constant $Z_m^{(\sigma)}$ is $Z_m^{(\sigma)} \sim \sigma/\Lambda_{ETC}$ at $\alpha \simeq \alpha_c = \pi/3$.

The fact that four-fermion interaction becomes marginal at $\alpha = \alpha_c$ was pointed out by Bardeen, Leung and Love [4]. The first who recognized that this effect can be used for the description of symmetry breaking in model building were Kogut, Dagotto and Kocić [12]. The present example with the isospin violation in WTC was first considered by Chivukula [13].

This example clearly illustrates the general situation in new TC scenarios: their dynamics includes (without a strong suppression) high momentum degrees of freedom with $\Lambda_{TC} \ll p \lesssim \Lambda_{ETC}$. This is a consequence of the fact that in these models, at least one of the two coupling constants $\left(\bar{\alpha}_{TC}(p^2) \text{ or } g_{ETC}\right)$ is rather strong. This is a new situation as compared to the old TC scenarios in which both g_{ETC} and $\bar{\alpha}_{TC}(p^2)|_{p^2 \sim \Lambda_{ETC}^2}$ are small.

This feature of the new scenarios can be important for the description of the mass spectrum of the ordinary fermions (quarks and leptons). We described this spectrum in Sec.13.3. One can see that the spectrum is rather complicated: while the isospin violation is rather moderate in the first family of quarks, it is rather strong in the heavier families (and especially in the third one); the mass of the b quark is 3 orders of magnitude larger than the mass of the d; the t quark is especially heavy. One can speculate that such a bizarre mass spectrum reflects a very specific underlying dynamics. If this dynamics is sensitive to formally irrelevant interactions (as that one leading to the isospin violation in the example above), it may produce such a complicated mass spectrum.

Actually, one should be careful in implementing this feature of the dynamics. For example, the mechanism of the isospin violation considered above can be used in gen-

---

*) For the definition of relevant, irrelevant and marginal operators, see Sec.11.2.

erating a large m_c/m_s or m_t/m_b ratios, however, it also may give a large contribution to the T parameter whose value is severely restricted (see Sec.13.9) [13].

Even the situation with FCNCs looks now more complicated than in the regular TC. Why is there no enhancement of FCNCs in WTC and SFTC ? The usual consideration of FCNCs in TC is based on the analysis of the box diagrams leading to M^0-$\bar{M}^0$ mixing ($M^0 = K^0, D^0$, or B^0) (see Fig.13.3). However, these diagrams correspond only to the lowest order in perturbation theory in technicolor coupling. While this approximation is justified in the case of QCD-like dynamics with a small running coupling $\bar{\alpha}_{TC}(p^2)|_{p^2\sim\Lambda^2_{ETC}} \ll 1$, it is unclear why it is also appropriate for WTC with the walking coupling $\bar{\alpha}_{TC}(p^2)|_{p^2\sim\Lambda^2_{ETC}}$ being rather strong: $\bar{\alpha}_{TC}(p^2)|_{p^2=\Lambda^2_{ETC}} \lesssim \alpha_c \sim 1$. In the case of strong ETC, the TC coupling $\bar{\alpha}_{TC}(p^2)|_{p^2=\Lambda^2_{ETC}}$ may be small, however, g_{ETC} is strong in that case, and therefore the analysis of FCNCs based on the box diagrams is again not justified.

The careful analysis of FCNCs in WTC and SETC confirms that FCNCs can indeed be suppressed there if ETC interactions are chosen in an appropriate way [30]. However, this example again shows that one should be careful in using such a powerful "weapon" as the new dynamics in WTC and SETC.

Another interesting possibility is the enhancement of the effect of the ordinary, $SU_c(3) \times SU_L(2) \times U_Y(1)$, interactions in WTC and SETC. This effect was first pointed out by Holdom in WTC [32]. But it is even more dramatic in SETC [9, 10]. Let us describe the essence of this effect.

In the pure NJL model, the relation for the dynamical mass m_{dyn} (i.e. scaling law) has the following form:

$$m_{dyn} \simeq \Lambda \left(\ln \frac{\Lambda^2}{m^2_{dyn}} \right)^{-1/2} (\kappa - 1)^{1/2} . \tag{14.65}$$

As was shown in Sec.14.2, in the gauged NJL model, with an asymptotically free gauge interaction, this relation is substituted by the following one:

$$m_{dyn} \sim \Lambda(\kappa - \kappa_c^{(eff)})^{1/2} , \tag{14.66}$$

where $\kappa_c^{(eff)} \simeq 1 - \frac{3C_2(F)}{2\pi}\bar{\alpha}(\Lambda)$ (see Eq.(14.50)). Let us use this relation in SETC. In this case Λ is Λ_{ETC}, and m_{dyn} is the mass of technifermions: $m_{dyn} = m_F \sim \Lambda_{TC} \ll \Lambda_{ETC}$.

The gauge interaction includes both TC interaction and ordinary gauge interactions. In particular, in the one family TC model, for example, QCD will contribute to the dynamics of techniquarks Q but not to the dynamics of technileptons L. Then, we have

$$\kappa_{cL}^{(eff)} \simeq 1 - \frac{3C_2(F)}{2\pi}\bar{\alpha}_{TC}(\Lambda_{ETC}) , \tag{14.67}$$

$$\kappa_{cQ}^{(eff)} \simeq 1 - \frac{3C_2(F)}{2\pi}\bar{\alpha}_{TC}(\Lambda_{ETC}) - \frac{2}{\pi}\bar{\alpha}_s(\Lambda_{ETC}) , \tag{14.68}$$

where $\bar{\alpha}_s$ is the QCD running coupling. Assuming that g_{ETC} is nearly the same for Q and L, we obtain the following mass relations:

$$m_L \sim \Lambda_{ETC}\left(g_{ETC} - \kappa_{cL}^{(eff)}\right)^{1/2} , \tag{14.69}$$

$$m_Q \sim \Lambda_{ETC}\left(g_{ETC} - \kappa_{cQ}^{(eff)}\right)^{1/2} . \tag{14.70}$$

Since m_L and m_Q are much less than Λ_{ETC}, $\Delta g_Q^{(eff)} \equiv \left(g_{ETC} - \kappa_{cQ}^{(eff)}\right)^{1/2}$ and $\Delta g_L^{(eff)} \equiv \left(g_{ETC} - \kappa_{cL}^{(eff)}\right)^{1/2}$ should be rather small. But what is more interesting, the difference

$$m_Q^2 - m_L^2 = (m_Q - m_L)\,(m_Q + m_L) \sim \bar{\alpha}_s\Lambda_{ETC}^2 \tag{14.71}$$

can be large. As a result, we find that QCD may yield an essential enhancement of techniquark masses as compared to technilepton masses.

It is instructive to consider this effect in the gauged NJL model with a standing gauge coupling α. Let us assume that the gauge coupling constant of techniquarks α_Q is $\alpha_Q = \alpha_L + \delta\alpha$, where α_L is the gauge coupling of technileptons. Then, the scaling law (14.22) implies that

$$m_Q - m_L \sim \Lambda\left[\left(g_{ETC} - \kappa_c(\alpha_Q)\right)^{1/2\omega_Q} - \left(g_{ETC} - \kappa_c(\alpha_L)\right)^{1/2\omega_L}\right] , \tag{14.72}$$

where

$$\begin{aligned} \kappa_c(\alpha_Q) &= \frac{1}{4}\left(1+\omega_Q\right)^2 , \\ \kappa_c(\alpha_L) &= \frac{1}{4}\left(1+\omega_L\right)^2 , \end{aligned} \tag{14.73}$$

and $\omega_Q = \left(1 - \frac{3\alpha_Q}{\pi}\right)^{1/2}$, $\omega_L = \left(1 - \frac{3\alpha_L}{\pi}\right)^{1/2}$.

Since the parameters $\omega_Q^{-1} \simeq \omega_L^{-1}$ grow with α_Q and α_L (and they are infinite at $\alpha_Q,\ \alpha_L = \pi/3$), one can see that the splitting effect is largest near the pure NJL model, where α_Q, α_L are small and $\gamma_m = 2$. It reduces essentially near the critical value $\alpha = \alpha_c = \pi/3$ where $\gamma_m = 1$. A more careful analysis [32] shows that the effect still remains even at $\gamma_m = 1$ but the qualitutive picture we have discussed above is correct.

All of these examples demonstrate rich possibilities of this dynamics. In the language of the theory of critical phenomena (see Chapter 11), the general dynamical situation in these scenarios can be described in the following way. To get a large anomalous dimension γ_m, one needs in strong high energy ($p \gg \Lambda_{TC}$) interaction. To arrange a hierarchy of the scales $m_F \sim \Lambda_{TC}$ and Λ_{ETC}, $\Lambda_{TC} \ll \Lambda_{ETC}$, this interaction should be rather specific: it must be realized in the critical (or near-critical) regime. The latter means that TC and ETC (effective) coupling constants are close to a critical point where the ratio m_F/Λ_{ETC} is rather small (at the critical point, corresponding to a second order phase transition, this ratio is exactly zero). A toy model of this scenario is the gauged NJL model considered in Sec.14.2.

This dynamical picture implies a certain ammount of fine tuning of the coupling constants, but it certainly yields new possibilities in model building.

As the last illustration of these possibilities, let us consider a way to get small (or even negative) values for the S parameter, considered in Sec.13.9, in TC models. This possibility has been pointed recently by Appelquist and Terning [33].

Let us again consider the one family model (see Sec.13.5). But this time we will use SETC. We will assume that, due to QCD, techniquark are essentially heavier than technileptons. We assume further that while the U and D techniquarks are nearly degenerate in mass, the technineutrino N is lighter than the technielectron E (due to ETC interaction). For concreteness, we take

$$m_E = 150GeV,\ m_N = 50GeV\ , m_Q \equiv m_U = m_D = 1TeV\ . \tag{14.74}$$

In order to illustrate the essence of the idea, we will calculate S and T in the one-loop approximation, neglecting TC and ETC interactions. Then, the relations for S and T are:

$$S = \frac{N_{TC}}{6\pi}\left[\ln\left(\frac{m_N^2}{m_E^2}\right) + 1\right] + \frac{3N_{TC}}{6\pi}\ , \tag{14.75}$$

$$T = \frac{N_{TC}}{16\pi s^2 c^2 M_Z^2} \left[m_N^2 + m_E^2 - \frac{2m_N^2 m_E^2}{m_N^2 - m_E^2} \ln \left(\frac{m_N^2}{m_E^2} \right) \right] \tag{14.76}$$

(compare with Eqs.(13.170) and (13.180)). Notice that the relation for the technilepton contribution to S,

$$S_L = \frac{N_{TC}}{6\pi} \left[\ln \left(\frac{m_N^2}{m_E^2} \right) + 1 \right] , \tag{14.77}$$

generalizes the relation (13.170) for the case of unequal masses $m_N^2 \neq m_E^2$ [33]. With the values for the masses in Eq.(14.74) and $N_{TC} = 2$, we find:

$$S_L \simeq -0.2, \; S_Q \simeq 0.3, \; S = S_L + S_Q \simeq 0.1 \; , \tag{14.78}$$

$$\alpha T_L \simeq 0.0026 = 0.26\%, \; \alpha T_Q = 0, \; \alpha T \simeq 0.26\% \; , \tag{14.79}$$

while in the QCD-like scenario we would get $S \simeq 0.46$.

Thus, while, because of a big ratio $m_E^2/m_N^2 = 9$, technileptons give a large negative contribution to S, their contribution to T is small because their masses are much less then the masses of techniquarks.

All these examples, although only illustrative, clearly reveal interesting features of new TC scenarios.

14.5. Tight Bound States in New Technicolor Scenarios. Bosonic Technicolor

In this section, we will show that the dynamical effects which we discussed in the previous section can be understood as a result of the presence of tight bound states in a system (we will give a more precise definition of such states below).

In Secs.11.3 and 11.5, we described the dynamics with a large anomalous dimension γ_m of the composite operator $\bar{\psi}\psi$ in models with four-fermion interaction in D dimensional space-time ($3 \leq D < 4$). A nice feature of those models was that they are interacting models in the strict continuum limit, although their dynamics is not asymptotically free. They will serve as a very useful laboratory in our discussion.

As follows from the discussion in Secs.11.3, 11.5, the essence of the nondecoupling effect in these models was connected with transmutation of the dynamical dimension

$d_{\bar{\psi}\psi}$ of the composite operator $\bar{\psi}\psi$. At the critical point $\kappa = \kappa_c$ in the models, it becomes

$$d_{\bar{\psi}\psi} = d_{c;\bar{\psi}\psi} - \gamma_{\bar{\psi}\psi} = 1 \,, \tag{14.80}$$

where the canonical (engineering) dimension $d_{c;\bar{\psi}\psi} = D - 1$ and the $\gamma_{\bar{\psi}\psi} \equiv \gamma_m = D - 2$ is the anomalous dimension of this operator.

As was also shown there, at the critical point (i.e. in the strict continuum limit), the models become equivalent to the Yukawa models in the strong coupling regime. In particular, there is a fermion-antifermion scalar bound state in the spectrum of the models described by the interpolating field $\bar{\psi}\psi$.

On the other hand, as was pointed out at the end of Sec.11.3, if a finite cutoff Λ remains in the models and is treated as a scale of a "new" physics, the Yukawa models can be considered as low energy effective theories of the "fundamental" theories with four-fermion interaction. These effective theories describe the interactions at low ($p \ll \Lambda$) momenta.

The description of nondecoupling effects becomes very clear in the language of effective theories. The crucial point is the presence of a *light* fermion-antifermion scalar bound state; "light" means here that its mass is much lighter than the cutoff Λ. At low momenta, the Yukawa coupling of this scalar to fermions is given by an infrared fixed point (see Secs.11.3, 11.5). Its value is finite and *is not suppressed by powers of* Λ. Thus, near the critical point, the four fermion interaction becomes relevant and produces the light bound state which is the main dynamical source of the interaction in the low energy region. Of course, as was stressed in Secs.11.1, 11.3, the existence of such light states is possible only near the critical point defining a second order phase transition.

In fact, almost the same situation takes place in the NJL model at in $D = 4$ at leading order in $1/N_c$. The low energy effective action of this model was derived in Sec.12.11. The Yukawa coupling $g_Y^{(\sigma)}$, describing the interaction at the scale $p \sim \sigma$, is now

$$g_Y^{(\sigma)^2} \sim \frac{1}{N_c} \frac{1}{\ln \Lambda^2/\sigma^2} \tag{14.81}$$

(see the discussion below Eq.(12.197)). It goes to zero as $\Lambda/\sigma \to \infty$, but only logarithmically; again, there is no suppression of the low-momenta interaction by powers of Λ.

In a more formal language, this means that at $D = 4$, the infrared fixed point equals zero, and the model becomes free in the strict continuum limit. But the important point is that as $\Lambda \to \infty$, the coupling constants in the model go to zero only logarithmically, and therefore it may serve as a perfect effective theory at finite Λ.

From the viewpoint of the effective theories, the large anomalous dimension $\gamma_{\bar{\psi}\psi} \equiv \gamma_m$ of the composite operator $\bar{\psi}\psi$ can be interpreted as a result of the mixing of the operator $\bar{\psi}\psi$ and the scalar field φ (which explicitly presents in the effective lagrangian). Look at Fig.(11.4) describing the propagator of $\bar{\psi}\psi$. The first diagram yields the same contribution as that in free theory. The second diagram corresponds to the contribution connected with the $\bar{\psi}\psi - \varphi$ mixing. As was shown in Sec.11.3, this contribution dominates near the critical point where $m/\Lambda \ll 1$. Actually, in the strict continuum limit (i.e. exactly at the critical point), the second diagram yields the full contribution to the renormalizable propagator; the renormalizable fields $\varphi^{(\sigma)}$ and $(\bar{\psi}\psi)^{(\sigma)}$ become identical (see the discussion at the end of Sec.11.3).

The language of effective theories is appropriate for SETC models where the dynamics is formed near the effective critical point $\kappa_C^{eff}(\bar{\alpha}(\Lambda)) = 1 - \frac{3C_2(F)}{2\pi}\bar{\alpha}(\Lambda)$ (see Eq.(14.50)). The existence of light (with masses $M \ll \Lambda_{ETC}$) spinless bound states, composed of technifermions and antitechnifermions, in such a dynamics was first pointed out in Ref.[10] (see also Refs.[34, 35]). Chivukula, Cohen and Lane [36] pointed out how nondecoupling effects in SETC can be described in the language of effective theories. In particular the enhancement of the technifermion condensate in SETC is now described as just a unsuppressed (or very mildly suppressed) Yukawa coupling of technifermions and light bound states. The effective theory includes a piece with a Higgs-like dynamics.

As was already indicated in Sec.14.2, the Bethe-Salpeter wave function (i.e. form factor) of such bound states in SETC decreases with increasing momentum p very slowly, up to $p \sim \Lambda_{ETC}$. So, while masses of these bound states are of order $\Lambda_{TC} \sim 1TeV$, their compositeness scale is of order Λ_{ETC} which is much larger than Λ_{TC}. In other words, they look as point-like particles at momenta $p < \Lambda_{ETC}$. Therefore the contribution to the chiral condensate, and to Green's functions of composite operators, in general, can be divided into two parts: the first part is connected with the "soft" TC dynamics, and the second ("hard") part is provided by the dynamics of the tight, point-like, bound states. As in 3 dimensional four-fermion model considered in Sec.11.3, the second contribution dominates in the near-critical regime, thus providing a huge enhancement of the condensate.

The existence of such bound states is interesting from the phenomenological viewpoint. They play the role of (composite) Higgs bosons here and their detection should be similar to the detection of usual Higgs particles. Therefore it is certainly interesting to derive the low energy effective lagrangian describing these particles. Such a program, iniciated by Bardeen and Love [37, 38], was completed in the gauged NJL model with a standing gauge coupling in Refs.[39 - 43].

The appearance of an additional gauge interaction in the model makes the derivation of the effective action here essentially more complicated than that in the pure NJL model. Indeed, if in the pure NJL model, the derivation reduces to the study of the fermion determinant in the external (auxiliary) σ and π fields (see Sec.12.11), in the gauged NJL model, we have the fermion determinant in the external σ and π fields *and* in *quantum* gauge fields, which also must be integrated out in the path integral. This essentially complicates the problem.

Gusynin and the present author [39, 40] have elaborated a regular approach to the derivation of the low energy effective action in gauge theories based on the formalism of Green's functions of composite operators on the nonperturbative vacuum [44]. They applied the method to the derivation of the effective action in the gauged NJL model and derived it as a series in powers of the derivatives of the σ and π fields.

Appelquist, Terning and Wijewardhana [41] applied the effective action in the gauged NJL model to the description of the properties of composite Higgs bosons in SETC. In particular, they estimated their masses and their widths.

We will return to composite Higgs bosons in the next chapter.

In conclusion, we will discuss the connection between SETC and a TC scenario known as the bosonic TC in the literature.

The history of using additional Higgs-like fields in TC traces back to the paper of Georgi and Glashow [45] in which a Higgs field (instead of ETC) was used to generate masses of ordinary fermions. The concrete TC models with additional Higgs fields were considered by Georgi and Kaplan [46], Simmons [47], and Samuel and Kagan [48]. Following Samuel [48], we will refer all TC models with additional Higgs fields as bosonic TC.

The connection between bosonic TC and SETC is evident. SETC may be considered as a dynamical realization of bosonic TC. It provides the dynamics which naturally leads to additional light composite bosons in TC.

14.6. New Technicolor Scenarios and Lattice Computer Simulations

As we have already stated several times, the crucial point for the realization of new TC scenarios is a near-critical dynamics with a strong bare coupling providing the existence of light tight bound state. The analytical methods available for its study are very limited. This is why the studies of such a dynamics in lattice computer simulations are so important.

In describing a critical dynamics (second order phase transition) in lattice computer simulations, the central point is calculating various critical exponents which were defined in Sec.11.4 (see Eqs.(11.89) - (11.93)).

As a useful laboratory for TC scenarios, noncompact lattice QED was studied in Refs.[14 - 19]. The main conclusion of these studies is the existence of a second order chiral phase transition in QED with a relatively small number of fermion flavors ($N \leq 8$). Unlike QCD, the critical coupling $\alpha_c^{(0)}$ is nonzero and separates symmetric and asymmetric phases.

Although, formally, the action of noncompact lattice QED coincides with the standard QED action (as the lattice spacing a goes to zero), the quantum dynamics of the lattice theory at rather large values of the bare coupling may be essentially different from that in conventional QED. The point is that the action of lattice QED is nonlocal. Expanding it in a series in powers of $a \equiv 1/\Lambda$, we will get, besides the conventional terms, a tower of nonrenormalizable (formally irrelevant) operators suppressed by powers of a. However, as we have already learned in Chapter 11 (and in this chapter), at large values of coupling constants these *perturbatively* irrelevant interactions may become marginal and relevant ones.

This is that crucial point which distinguishes non-asymptotically free theories on lattice (as QED) from asymptotically free ones (as QCD). The actions of the latter also include nonrenormalizable operators, but, as $a \to 0$, because of weakness of the interaction near the critical point $\alpha^{(0)} = \alpha_c = 0$ in asymptotically free theories, they are really irrelevant in the continuum limit.

The gauged NJL model (QED plus (formally irrelevant) four-fermion interaction) considered in Sec.14.2, gives a simple example of the transmutation of irrelevant interactions in relevant ones. It is instructive to calculate the critical exponents in this model [15, 38].

The critical exponents $\alpha, \beta, \gamma, \nu$ and δ were defined in Eqs.(11.89) - (11.93). They can be calculated in the gauged NJL model in the same way as in the 3 dimensional model with four-fermion interaction in Sec.11.4. For this purpose, we need in the expression for the effective potential $V(\sigma, \pi)$ describing the collective fields σ and π in the gauged NJL model. This potential was derived in Refs.[37, 38] It is

$$\begin{aligned} V(\sigma,\pi) = \left(m^{(0)^2}/2G^{(0)}\right) - \left(m^{(0)}/G^{(0)}\right)\sigma + \frac{1}{2}\left[1/G^{(0)} - (\Lambda^2/\pi^2)(1+\omega)^{-2}\right](\sigma^2+\pi^2) \\ + \left(\tilde{A}^2/16\pi^2\right)(\alpha_c/\alpha)(2/\omega - 1)\left[(\tilde{A}/4)(1+1/\omega)e^{\omega\cdot\delta}\Lambda^{(-1+\omega)}\right]^{-4/(2-\omega)} \\ \times\left(\sigma^2+\pi^2\right)^{2/(2-\omega)}, \end{aligned} \tag{14.82}$$

where $\tilde{A}(\alpha)$ and $\delta(\alpha)$ are certain functions of α [37, 38]. The values of the critical exponents, calculated from this expression, are [15, 38]:

$$\begin{aligned} \tilde{\alpha} &= 2\left(\sqrt{1-\alpha/\alpha_c} - 1\right)/\sqrt{1-\alpha/\alpha_c}\,, \\ \beta &= \left(2-\sqrt{1-\alpha/\alpha_c}\right)/2\sqrt{1-\alpha/\alpha_c}\,, \\ \gamma &= 1\,, \\ \nu &= 1/2\sqrt{1-\alpha/\alpha_c}\,, \\ \delta &= \left(2+\sqrt{1-\alpha/\alpha_c}\right)/\left(2-\sqrt{1-\alpha/\alpha_c}\right)\,, \end{aligned} \tag{14.83}$$

where $\alpha_c = \pi/3$, and here we use $\tilde{\alpha}$ instead of α to avoid confusion with the gauge coupling constant.

The characteristic feature of these exponents is that only at $\alpha = 0$ (the pure NJL model) they coincide with the mean-field exponents (see Eq.(11.102)); at all nonzero values of α they are different from them. As $\alpha \to \alpha_c$, the exponents $\tilde{\alpha}$, β and ν become infinite. These facts reflect the point that while the universality class of the pure NJL model is that of a free theory, there is an essential singularity in the scaling law for m_{dyn} at $\alpha = \alpha_c$ (see Eqs.(14.25), (14.26)).

These results can be understood in somewhat different way. The ultraviolet asymptotics of the dynamical mass function in this model is:

$$B_{dyn}(p^2) \sim m_{dyn}\left(p/m_{dyn}\right)^{(-1+\sqrt{1-\alpha/\alpha_c})} \tag{14.84}$$

(see Eq.(14.20)). By using the Pagels-Stokar formula (14.56) for calculating the decay constant F, we find that while at $\alpha = 0$ F^2 logarithmically diverges $\left(F^2 \sim m_{dyn}^2\right.$

$\cdot \ln \Lambda^2/m_{dyn}^2)$, it is finite at all non-vanishing values of α. Because of the Goldberger-Treiman relation (4.80), this implies that, in the continuum limit Yukawa coupling

$$g_Y \sim m_{dyn}/F$$

is zero at $\alpha = 0$ and is non-vanishing at $\alpha \neq 0$ in the continuum limit.

We recall that the contribution of fermion loops are neglected in the present approximation (quenched approximation).

Let us now turn to the data in the computer simulations of lattice QED. As we have already indicated above, the data clearly show the existence of the second order chiral phase transition in lattice noncompact QCD with a relatively small number N of fermion flavors. At $N = 0$ (quenched QED), the parameters δ and β are $\delta \simeq 2.2$ and $\beta \simeq 0.8$ [15], i.e. while they deviate from the mean-filed values ($\delta = 3$, $\beta = 0.5$), they are also essentially deviate from values corresponding to a critical point with an essential singularity in the scaling law ($\delta = 1$, $\beta = \infty$). These data also yields $\gamma = 1$ and $\nu = 0.68$. All these values are in good agreement with Eq.(14.83) at the point $(\alpha, \kappa) = (0.44\alpha_c, 0.76)$.

Thus, these data indicate that even in quenched lattice QED, the formally irrelevant four-fermion interactions survive in the continuum limit.

The situation concerning the data in full lattice QED (with $N = 2$ and $N = 4$) is more controversial and complicated. While the fit of the data in Refs.[16, 19] indicate a significant deviation of critical exponents from the mean-field values, the fit of the data in Refs.[17, 18] corresponds to a mean-field theory. Also, as the authors of Ref.[16] argue, there may be an essential effect due to monopole percolation in lattice QED.

As was already emphasized in Sec.11.2, we are interested in finding simple models which may play the role of effective theories for new ETC scenarios. Lattice computer simulations of non-asymptotically free gauge theories have proved to be extremely valuable for this purpose.

Chapter 15

Top Mode Standard Model and Beyond

15.1. Introduction

So far, ordinary fermions (quarks and leptons) played a rather passive role in all of the scenarios of electroweak symmetry breaking which have been discussed in this book. For example, in the standard electroweak theory, discussed in Chapter 13, the fermions acquire their masses from the electroweak condensate generated by the self-interaction of a Higgs doublet. Thus, electroweak symmetry breaking in the standard model (SM) would occur even without the fermions; they are mostly spectators in this scenario.

The same is true for the technicolor (TC) scenario. In this case, the electroweak condensate appears due to TC dynamics connected with new degrees of freedom (technifermions and technigluons); again, ordinary fermions are just spectators in this scenario.

In this chapter we will consider another class of models of electroweak symmetry breaking in which some of ordinary fermions (for the most part, fermions from the third, or a possible forth, family) play the active role in triggering electroweak symmetry breaking and/or in generating the structure of the fermion mass spectrum [1 – 4].

The basic idea underlying these models was introduced by Tanabashi, Yamawaki and the present author (MTY) [1] and Nambu [2]. The idea is that, besides the standard color and electroweak couplings, quarks and leptons have new very short-range interactions. Among them only the t quark (and possibly other (b, τ, ν_τ) fermions from the third family) has a sufficiently large coupling to produce the $\bar{t}t$-condensate and its own mass. The condensate leads to dynamical electroweak symmetry breaking. There is one composite Higgs doublet in this model:

$$\varphi_t = \begin{pmatrix} \varphi_t^0 \\ \varphi_t^- \end{pmatrix} \sim \Lambda^{-2} \begin{pmatrix} \bar{t}(1+\gamma_5)t \\ \bar{t}(1+\gamma_5)b \end{pmatrix}, \tag{15.1}$$

where $\Lambda \gg F_{EW} \simeq 246 GeV$ is the scale of the new short-range interactions. Three composite Nambu-Goldstone (NG) bosons, $w^- \sim \bar{t}(1+\gamma_5)b$, $w^+ \sim \bar{b}(1-\gamma_5)t$, and $w^0 \sim i\bar{t}\gamma_5 t$, are absorbed into the $W^\pm$ and Z bosons, which acquire masses, and one composite (Higgs) boson, $h \sim \bar{t}t$, remains in the spectrum of the model.

Thus, in this scenario, the quark doublet $\begin{pmatrix} t \\ b \end{pmatrix}$ plays the same role as a technifermion doublet in TC. However, a very nontrivial new point in the present scenario is that since the t is much heavier than the b, there is a strong violation of the isospin symmetry in this doublet. One may think that this will destroy the relation $|\delta\rho| \lesssim 0.2\%$, where $\delta\rho = \rho - 1$, $\rho = M_W^2/M_Z^2 cos^2\theta_W$ (see Secs.13.5 and 13.9). However, as we shall see this is not the case.

We shall also see that the scenario naturally leads to a large value of the t quark mass.

The motivations for the Nambu approach and for the MTY approach were somewhat different. Nambu started from the bootstrap idea in the SM [5]. The basic requirement in his approach (which traces back to the work of Vetman [6]) is the requirement of the cancellation of quadratic divergences in the electroweak theory. This leads to sum rules restricting the values of the masses of the t quark and the Higgs boson.

The motivation of the MTY approach was the result of Ref.[7] (already discussed in Sec.14.2) that in the near-critical regime, four-fermion interaction leads to a slowly decreasing with momentum (irregular) asymptotics for the *dynamical* fermion mass. As was discussed in Sec.12.3, the *current* mass functions of quarks in QCD have just such an asymptotics. Therefore the four-fermion near-critical interaction provides a dynamical mechanism for generating current masses of fermions.

MTY constructed the concrete model which they called the "top mode" standard model (TMSM) in contrast to the usual one, the "Higgs mode" standard model. In the framework of the TMSM, they estimated the masses of the t quark and Higgs boson.

The next conceptually important step was done by Bardeen, Hill and Lindner (BHL) [4]. They clarified some basic points in the dynamics of electroweak symmetry breaking in the TMSM. Their analysis essentially simplified the model building in this class of models.

The idea of the $\bar{t}t$ condensate was also considered by Marciano [3] who, following earlier works [8], postulated the irregular asymptotics for the dynamical mass of the t quark without specifying the underlying dynamics.

Since these papers, the idea of the $\bar{t}t$ condensate has been extensively used in model building. In this chapter, rather than to discuss all of these models, we will consider the basic points in their dynamics.

As we will see, this class of models addresses the following questions:

1) Why is the t quark so heavy ?*)

2) Can Higgs bosons appear in other, than doublet, representations of $SU_L(2)$?

3) How many light (with masses of O(100GeV)) Higgs bosons are in electroweak theory ?

4) What is the spectrum of other (than ordinary fermions, W and Z, and Higgs bosons) particles in electroweak theory ?

15.2. Top Mode Standard Model

Let us consider the model which consists of the standard three families of quarks and leptons with the $SU_c(3) \times SU_L(2) \times U_Y(1)$ gauge interactions but without Higgs bosons. Instead of the standard Higgs sector, we introduce the $SU_c(3)\times SU_L(2)\times U_Y(1)$ - invariant four-fermion interactions among quarks and leptons, the origin of which is not specified at this moment. For simplicity we shall disregard leptons for the moment. The general form of such four-fermion interactions among quarks may be written as [1]

$$\mathcal{L}_{4f} = \frac{4\pi^2}{N_c\Lambda^2}\Big[\kappa^{(1)}_{ii';jj'}\left(\bar{\psi}^{ia}_L\psi^{i'b}_R\right)\left(\bar{\psi}^{jb}_R\psi^{j'a}_L\right) + \kappa^{(2)}_{ii';jj'}\left(\bar{\psi}^{ia}_L\psi^{i'b}_R\right)(i\tau_2)^{ac}(i\tau_2)^{bd}\left(\bar{\psi}^{jc}_L\psi^{j'd}_R\right) + \kappa^{(3)}_{ii';jj'}\left(\bar{\psi}^{ia}_L\psi^{i'b}_R\right)(\tau_3)^{bc}\left(\bar{\psi}^{jc}_R\psi^{j'a}_L\right)\Big] + H.c., \tag{15.2}$$

where N_c is the number of quark colors ($N_c = 3$ in reality), $\kappa^{(1)}_{ii';jj'}, \kappa^{(2)}_{ii';jj'}$, and $\kappa^{(3)}_{ii';jj'}$ are the dimensionless four-fermion couplings, (i, i', j, j') and (a, b, c, d) are the family and the isospin indices, respectively, and Λ is a scale of the new interaction. Out of the terms in Eq.(15.2), we are particularly interested in those corresponding to the self-interaction of quarks of the third family $\psi \equiv \delta_{i3}\psi^i, \kappa^{(r)} \equiv \kappa^{(r)}_{33;33}$ with $(r = 1, 2, 3)$;

$$\mathcal{L}_{4f} = \frac{4\pi^2}{N_c\Lambda^2}\Big[\kappa^{(1)}\left(\bar{\psi}^a_L\psi^b_R\right)\left(\bar{\psi}^b_R\psi^a_L\right) + \kappa^{(2)}\left(\bar{\psi}^a_L\psi^b_R\right)(i\tau^2)^{ac}(i\tau^2)^{bd}\left(\bar{\psi}^c_L\psi^d_R\right) + \kappa^{(3)}\left(\bar{\psi}^a_L\psi^b_R\right)(\tau_3)^{bc}\left(\bar{\psi}^c_R\psi^a_L\right)\Big] + H.c.\,. \tag{15.3}$$

*) By the time of finishing of the writing of this book, the lower bound for the mass of the t quark (from experiments at the Tevatron collider) has become $113 GeV$. The central value for m_t, from neutrino experiments and from experiments at the CERN e^+e^- collider LEP, has become $164 GeV$.

MTY considered two scenarios [1]:

1) the scenario with the only one coupling constant $\kappa_t \equiv \kappa^{(1)}+\kappa^{(3)}$ being supercritical ($\kappa_t \gtrsim \kappa_c \simeq 1$) and with two other couplings, $\kappa^{(2)}$ and $\kappa_b \equiv \kappa^{(1)} - \kappa^{(3)}$, being small. In a simplified version with $\kappa^{(2)} = 0$, we get:

$$\mathcal{L}_{4f} = \frac{8\pi^2}{N_c\Lambda^2}\kappa_t \left(\bar{\psi}_L^a t_R\right) \left(\bar{t}_R \psi_L^a\right) + \frac{8\pi^2}{N_c\Lambda^2}\kappa_b \left(\bar{\psi}_L^a b_R\right) \left(\bar{b}_R \psi_L^a\right) . \tag{15.4}$$

When $\kappa_b \to 0$, the right-handed b becomes sterile with respect to this new interaction;

2) the scenario with both coupling constants κ_t and κ_b being supercritical. In this case both masses m_t and m_b are dynamical.

Let us discuss these two scenarios in more detail.

Let us study the Schwinger-Dyson (SD) equations for the mass functions $B_t(p^2)$ and $B_b(p^2)$. We will use the improved ladder approximation to describe QCD interactions and four-fermion interactions (15.4) (in Landau gauge). The equations are:

$$B_t(p^2) = \frac{\kappa_t}{\Lambda^2}\int_0^{\Lambda^2} dq^2 \frac{q^2 B_t(q^2)}{q^2 + B_t^2(q^2)} + \int_0^{\Lambda^2} dq^2 K(p^2,q^2)\frac{q^2 B_t(q^2)}{q^2 + B_t^2(q^2)} , \tag{15.5}$$

$$B_b(p^2) = \frac{\kappa_b}{\Lambda^2}\int_0^{\Lambda^2} dq^2 \frac{q^2 B_b(q^2)}{q^2 + B_b^2(q^2)} + \int_0^{\Lambda^2} dq^2 K(p^2,q^2)\frac{q^2 B_b(q^2)}{q^2 + B_b^2(q^2)} , \tag{15.6}$$

where

$$K(p^2,q^2) = \frac{3(N_c^2-1)}{8\pi N_c}\left[\frac{\bar{\alpha}(p^2)}{p^2}\theta(p^2-q^2) + \frac{\bar{\alpha}(q^2)}{q^2}\theta(q^2-p^2)\right] \tag{15.7}$$

with $\bar{\alpha}(p^2)$ being the QCD running coupling (see Eq.14.40)). At large $p^2, \bar{\alpha}(p^2)$ is:

$$\bar{\alpha}(p^2) \longrightarrow \frac{1}{2\pi b \ln\left(p^2/\Lambda_{QCD}^2\right)} , \quad b = \frac{11N_c - 2N}{24\pi^2} . \tag{15.8}$$

As was shown in Sec.14.2, for the near-critical κ, $1-\kappa \sim \bar{\alpha}(\Lambda^2)$, the solutions of the

SD equations (15.5), (15.6) have the irregular asymptotics:

$$B_{t(b)}(p^2) \longrightarrow m_{t(b)} \left[\frac{\bar{\alpha}(p^2)}{\bar{\alpha}\left(m^2_{t(b)}\right)} \right]^{c_m} , \quad c_m = \frac{3(N_c^2 - 1)}{16\pi^2 b N_c} . \tag{15.9}$$

Let us now consider the first scenario, in which κ_b is small and κ_t is near-critical. Then, the SD equation (15.5) has a solution with the irregular asymptotics (15.9). Let us estimate the mass of the t quark. For this purpose, we will calculate the decay constants F^0 and $F^{\pm}$ of the NG bosons w^0 and $w^{\pm}$ in the Higgs doublet (15.1). In Sec.12.9, we derived the Pagels-Stoker (PS) formula in QCD. It is straightforward to generalize this formula to the present case. In the approximation in which m_b is neglected,*) we get:

$$(F^0)^2 = \frac{N_c}{8\pi^2} \int_0^{\Lambda^2} dq^2 q^2 \frac{B_t^2 - \frac{q^2}{4}\frac{d}{dq^2} B_t^2}{(q^2 + B_t^2)^2} , \tag{15.10}$$

$$(F^{\pm})^2 = \frac{N_c}{8\pi^2} \int_0^{\Lambda^2} dq^2 \left[\frac{B_t^2 - \frac{q^2}{4}\frac{d}{dq^2} B_t^2}{(q^2 + B_t^2)} + \frac{q^2}{2} \frac{B_t^2}{(q^2 + B_t^2)} \cdot \frac{d}{dq^2} \ln\left(\frac{q^2 + B_t^2}{q^2}\right) \right] .$$

The mass of the top quark is estimated from Eq.(15.10) with $F^0 \simeq F^{\pm} \simeq 246 GeV$ by using the following parameterization for the solution of Eq.(15.5):

$$B_t(p^2) = \begin{cases} m_t \left(\frac{\ln(m_t^2/\Lambda_5^2)}{\ln(\Lambda_{IR}^2/\Lambda_5^2)} \right)^{12/23} & \text{at } p^2 < \Lambda_{IR}^2 , \\ m_t \left(\frac{\ln(m_t^2/\Lambda_5^2)}{\ln(p^2/\Lambda_5^2)} \right)^{12/23} & \text{at } \Lambda_{IR}^2 < p^2 < m_t^2 , \\ m_t \left(\frac{23 \ln(m_t^2/\Lambda_5^2)}{2\ln(m_t^2/\Lambda_5^2) + 21 \ln(p^2/\Lambda_5^2)} \right)^{4/7} & \text{at } m_t^2 < p^2 , \end{cases} \tag{15.11}$$

where the auxiliary infrared cutoff $\Lambda_{IR} > \Lambda_5$ is introduced to regularize the singularity at $p^2 = \Lambda_5^2$. Here the parameter Λ_5 is Λ_{QCD} with $N = 5$ ($\Lambda_5 \simeq 110 MeV$ in the $\overline{MS}$ renormalization scheme), and $c_m = \frac{4}{7}$ and $c_m = \frac{12}{23}$ correspond to QCD with $N = 6$ and $N = 5$, respectively.

Eqs.(15.10), (15.11) lead to the following estimate for m_t [1]:

---

*) Even in the real situation, with $m_b \simeq 5 GeV$, the b quark contribution to the decay constants is very small because m_b is much less than $m_t > 113 GeV$.

Λ (GeV)	10^{10}	10^{11}	10^{12}	10^{13}	10^{14}	10^{15}	10^{16}	10^{17}	10^{18}	10^{19}
m_t (GeV)	303	292	284	277	271	266	261	257	253	250

Table 15.1

The following comments are in order.

Why should we trust this analysis ? It is clear that the ansatz (15.11) is rather crude in the infrared region, where QCD forces dominate; also the parameter Λ_{IR} is free in Eq.(15.11), and the result should depend on its choice. The answer to this question is that, because of the irregular asymptotics (15.9), the dominant contribution in the integrals in Eq.(15.10) comes from the ultraviolet region where the ansatz (15.11) is certainly reliable. This fact also explains the results of numerical study which show that the value of m_t does not change significantly for wide range of Λ_{IR}. We also notice that the results for m_t in Table 15.1 were confirmed by the analysis using a numerical solution of the SD equation [9].

The dominance of the ultraviolet region in the integrals (15.10) suggests the following approximate form for $F^0 \simeq F^{\pm}$:

$$(F^0)^2 \simeq \frac{N_c}{8\pi^2}\int_{m_t^2}^{\Lambda^2} dq^2 \frac{B_t^2(q^2)}{q^2} \simeq \frac{N_c}{2\pi^2} m_t^2 \frac{\left(\bar{\alpha}(m_t^2)\right)^{1/7} - \left(\bar{\alpha}(\Lambda^2)\right)^{1/7}}{\left(\bar{\alpha}(m_t^2)\right)^{8/7}} . \tag{15.12}$$

Therefore

$$m_t^2 \simeq \frac{2\pi^2}{N_c} \frac{\left(\bar{\alpha}(m_t^2)\right)^{8/7}}{\left(\bar{\alpha}(m_t^2)\right)^{1/7} - \left(\bar{\alpha}(\Lambda^2)\right)^{1/7}} (246GeV)^2 . \tag{15.13}$$

This form of the estimate for m_t^2 was first obtained by Marciano [3].

Let us now estimate the parameter $\delta\rho = \rho - 1$. Eq.(15.10) implies that

$$\begin{aligned}\delta\rho = \frac{(F^{\pm})^2 - (F^0)^2}{(F^{\pm})^2} &= \frac{\frac{N_c}{16\pi^2}\int_0^{\Lambda^2} dq^2 \frac{B_t^2[B_t^2 + \frac{1}{2}q^2(B_t^2)']}{(q^2+B_t^2)^2}}{(246GeV)^2} \\ &\simeq \frac{N_c}{16\pi^2}\frac{\int_0^{m_t^2} dq^2}{(246GeV)^2} = \frac{N_c}{16\pi^2}\left(\frac{m_t}{246GeV}\right)^2 .\end{aligned} \tag{15.14}$$

This estimate is essentially the same as the one-loop Veltman result in the SM [10]. Therefore $\delta\rho$ leads to the usual upper bound for m_t in the TMSM.

Table 15.1 shows that m_t decreases with increasing Λ. The reason of this is evident from the relation (15.13).

It is instructive to compare these results for m_t with those in the NJL model, where m_t is just a constant. In this case, we find from Eq.(15.12) that

$$(F^0)^2 \simeq \frac{N_c}{8\pi^2} \ln \frac{\Lambda^2}{m_t^2}. \tag{15.15}$$

This relation leads to $m_t \simeq 144 GeV$ for $\Lambda = 10^{19} GeV$. Thus, QCD interaction enhances the value of m_t. The reason of this is clear: QCD interaction leads to the mass function $B_t(p^2)$ which decreases with increasing the momentum. Therefore to obtain the same value for F^0 as in the NJL model, m_t has to be increased in this case.

To generate the current masses of quarks other than the t quark, we will use the interaction (15.2). Notice that three down quarks (d, s and b) may acquire their masses from the $\bar{t}t$ condensate through the perturbative $\kappa^{(2)}$ interaction. It is also evident that the interaction (15.2) can produce a realistic Cabibbo - Kobayashi - Maskawa mass matrix for quarks (see Sec.13.3). In the model, besides the mass of t, all other quark masses are free parameters.

Also, it is clear that charge leptons acquire masses in a similar way, through the $\kappa^{(2)}$ interaction with the $\bar{t}t$ condensate.

Let us turn now to the second scenario in which both κ_t and κ_b are supercritical (see the second paper in Ref.[1]). Now both t and b acquire dynamical masses m_t and m_b. Let us first consider the case when $\kappa^{(2)}$ interactions are absent. We come to an elegant scheme in which the $\bar{t}t$ condensate provides masses for up quarks, and the $\bar{b}b$ condensate provides masses for down quarks and leptons. Notice that when $\kappa^{(2)}_{ii';jj'} = 0$, the lagrangian density (15.2) possesses an additional (global) $U_5(1)$ symmetry. However, as we know from Chapter 12, the $U_5(1)$ symmetry is explicitly broken by the strong Adler-Bell-Jackiw anomaly (see Eq.(12.8)). This symmetry plays here the role of the Peccei-Quinn (PQ) symmetry [11] introduced to explain the smallness of the $\bar{\theta}$ parameter ($\bar{\theta} < 10^{-9}$) in the SM (see Secs.12.12, 12.13). In the absence of the strong anomaly, the PQ symmetry would be only spontaneously broken by the $\bar{t}t$ and $\bar{b}b$ condensates. The NG boson associated with this spontaneous breakdown is known as axion. However, because of the strong anomaly, axion acquires a mass. By using the standard methods, one estimates the decay constant F_{PQ} of axion and the axion mass in the present model as [1, 12]

$$F_{PQ} = 2F^0 \frac{m_b}{m_t}, \; m_a \simeq 75 \frac{m_t}{m_b} keV. \tag{15.16}$$

Unfortunately, such an visible axion is phenomenologically unacceptable [13].

To escape from this difficulty, we are forced to reintroduce the $\kappa^{(2)}$-interaction in the model. This interaction explicity violates the $U_5(1)$ and may easily enhance the axion mass to $O(100GeV)$.

So far, we have discussed the mass spectrum of fermions in the TMSM. What is the spectrum of composite Higgs bosons in this model ? In the first scenario, with only one, κ_t, supercritical four-fermion coupling, there is only one composite Higgs doublet (15.1). In the original papers [1,2], the mass of the Higgs boson $h \sim \bar{t}t$ was estimated, by using the NJL formula (9.126), as $M_h \simeq 2m_t$. However, later, Nambu [14] and Shuto, Tanabashi and Yamawaki [15] showed that QCD interaction reduces M_h to $M_h \simeq \sqrt{2}m_t$.

In the second scenario, with supecritical κ_t and κ_b, in addition to the Higgs fields φ_t (15.1), there is another Higgs field

$$\varphi_b = \begin{pmatrix} \varphi_b^0 \\ \varphi_b^- \end{pmatrix} \sim \Lambda^{-2} \begin{pmatrix} -\bar{b}(1-\gamma_5)b \\ \bar{t}(1-\gamma_5)b \end{pmatrix}. \tag{15.17}$$

The spectrum of φ_t and φ_b was studied by Suzuki [16] and Luty [17]. For more detail, we refer the readers to their papers.

15.3. Renormalization Group Method in the Top Mode Standard Model

In this section, we will discuss a renormalization group approach to the TMSM elaborated by Bardeen, Hill and Lindner (BHL) [4].

In Chapters 6,11 – 12, we used the method of auxiliary (collective) fields in the analysis of models with four-fermion interactions. This is also the initial step in the BHL approach. Let us consider the interaction (15.4) in which, for simplicity, $\kappa_b = 0$.

Then, we can rewrite this interaction in the following form:

$$\mathcal{L}_{4f} = -\bar{\psi}_L \varphi_t t_R - \bar{t}_R \varphi_t^\dagger \psi_L - \frac{1}{G_t^{(0)}} \varphi_t^\dagger \varphi_t, \tag{15.18}$$

where

$$G_t^{(0)} \equiv \frac{8\pi^2}{N_c \Lambda^2} \kappa_t \tag{15.19}$$

(compare with Eq.(6.2)). Therefore, ignoring, for simplicity, gauge interactions, we get the lagrangian density in the TMSM in the form:

$$\mathcal{L} = i\bar{\psi}\gamma^\mu \partial_\mu \psi + \mathcal{L}_{4f} = i\bar{\psi}\gamma^\mu \partial_\mu \psi - \bar{\psi}_L \varphi_t t_R - \bar{t}_R \varphi_t^\dagger \psi_L - \frac{1}{G_t^{(0)}} \varphi_t^\dagger \varphi_t. \tag{15.20}$$

There is no kinetic term for the Higgs field φ_t.

Let us compare this lagrangian density with that in the SM (again, gauge interactions are neglected):

$$\begin{aligned}\mathcal{L}_{SM} =& i\bar{\psi}^{(0)}\gamma^\mu \partial_\mu \psi^{(0)} + \partial^\mu \varphi^{(0)\dagger} \partial_\mu \varphi^{(0)} - \sqrt{2} g_t^{(0)} \bar{\psi}_L^{(0)} \varphi^{(0)} t_R^{(0)} - \sqrt{2} g_t^{(0)} \bar{t}_R^{(0)} \varphi^{(0)\dagger} \psi_L^{(0)} \\ & - \mu^{(0)2} \varphi^{(0)\dagger} \varphi^{(0)} - \frac{\lambda^{(0)}}{3!} (\varphi^{(0)\dagger} \varphi^{(0)}))^2.\end{aligned} \tag{15.21}$$

Here all quantities are bare (unrenormalized). Rescaling the Higgs fields $\varphi^{(0)}$ as $\varphi^{(0)} \rightarrow \varphi = \sqrt{2} g_t^{(0)} \varphi^{(0)}$, we rewrite $\mathcal{L}_{SM}$ as

$$\begin{aligned}\mathcal{L}_{SM} =& i\bar{\psi}^{(0)}\gamma^\mu \partial_\mu \psi^{(0)} + \left(\frac{1}{2 g_t^{(0)2}} \right) \partial^\mu \varphi^\dagger \partial_\mu \varphi - \bar{\psi}_L \varphi t_R - \bar{t}_R \varphi^\dagger \psi_L - \frac{\mu^{(0)2}}{2 g_t^{(0)2}} \varphi^\dagger \varphi \\ & - \frac{\lambda^{(0)}}{4! g^{(0)4}} (\varphi^\dagger \varphi)^2.\end{aligned} \tag{15.22}$$

Comparing this equation with Eq.(15.20), we find that the TMSM corresponds to the following special limit of the SM:

$$g_t^{(0)} \rightarrow \infty, \tag{15.23}$$

$$\frac{\lambda^{(0)}}{g_t^{(0)^4}} \to 0, \tag{15.24}$$

$$\frac{\mu^{(0)^2}}{2g_t^{(0)^2}} \to \frac{1}{G_t^{(0)}}. \tag{15.25}$$

We recall that the condition that the bare Yukawa coupling goes to infinity was also used in 3 dimensional four-fermion model in Chapter 11.

Since the running couplings approach the corresponding bare couplings as the renormalization scale σ goes to Λ, the relations (15.23), (15.24) can be considered as boundary conditions for the renormalization-group (RG) equations.

Let us consider the one-loop RG equation for the Yukawa coupling constant $\bar{g}_t(\sigma)$. It is [4]:

$$\frac{d\bar{g}_t}{d\ln\sigma} = \frac{\bar{g}_t}{16\pi^2}\left[(2N_c+3)\bar{g}_t^2 - \left(8\bar{g}_3^2 + \frac{9}{4}\bar{g}_2^2 - \frac{17}{12}\bar{g}_1^2\right)\right], \tag{15.26}$$

where $\bar{g}_1, \bar{g}_2$ and $\bar{g}_3$ are the running couplings of $U_Y(1), SU_L(2)$ and $SU_c(3)$, respectively. The one-loop RG equations for them take the usual form:

$$\frac{d\bar{g}_i}{d\ln\sigma} = -\frac{c_i}{16\pi^2}\bar{g}_i^3 \tag{15.27}$$

with

$$c_1 = -\frac{1}{6} - \frac{20}{9}N_f,\ c_2 = -\frac{43}{6} - \frac{4}{3}N_f,\ c_3 = 11 - \frac{4}{3}N_f, \tag{15.28}$$

where N_f is the number of families.

It is instructive to compare the RG equation (5.26) with the MTY analysis discussed in the previous section. In that analysis of the SD equations, the gauge $U_Y(1)$ and $SU_L(2)$ interactions and the contribution connected with the exchange of virtual Higgs bosons were neglected. The RG equation for g_t takes the following form in this approximation:

$$\frac{d\bar{g}_t}{d\ln\sigma} = \frac{\bar{g}_t}{16\pi^2}\left[2N_c\bar{g}_t^2 - 8\bar{g}_3^2\right]. \tag{15.29}$$

One can easily check that with the boundary condition $\bar{g}_t(\sigma)\xrightarrow[\sigma\to\Lambda]{}\infty$, this equation and equation (15.27) for $\bar{g}_3$ lead to the relation (15.13) for m_t. Therefore, in this approximation, the MTY and the BHL analysises are perfectly consistent.

Numerical study of Eqs.(15.26), (15.27) yields the following values for $m_t \simeq \bar{g}_t(m_t)\times F_{EW}$ [4]:

$$\begin{aligned} m_t &\simeq 229 GeV \text{at } \Lambda = 10^{15} GeV \\ m_t &\simeq 218 GeV \text{at } \Lambda = 10^{19} GeV \end{aligned} \tag{15.30}$$

These values are somewhat smaller than the corresponding values obtained by MTY (see Table 15.1). This effect appears mostly due to the additional contribution of virtual Higgs bosons in Eq.(15.26) which reflects itself in the term 3 in the first parenthesis on the right-hand side of Eq.(15.26). Let us discuss this effect in more detail.

As was noted at the end of the previous section, QCD interaction enhances the mass of the t quark. Since the Yukawa interaction gives the contribution to Eq.(15.26) with the opposite sign as compared to the QCD contribution, it tends to decrease m_t. It is clear that this effect is stronger in equation (15.26) than in Eq.(15.29)*).

This effect can be also clarified in the following way. The structure of the right-hand side in equation (15.26) suggests that its solutions are attracted toward the "effective low-energy fixed point" [18,19]

$$\bar{g}_t^2(\sigma) \simeq (2N_c + 3)^{-1} 8\bar{g}_3^2(\sigma) \underset{N_c=3}{=} \frac{8}{9}\bar{g}_3^2(\sigma). \tag{15.31}$$

In equation,(15.29), such a fixed point is at

$$\bar{g}_t^2(\sigma) = \frac{4}{3}\bar{g}_3^2(\sigma), \tag{15.32}$$

which is larger than the value (15.31). This in turn leads to somewhat larger values of m_t.

The presence of this effective fixed point is also important for the following. It is clear that the boundary condition

$$\bar{g}_t^2(\sigma) \underset{\sigma\to\Lambda}{\longrightarrow} \infty \tag{15.33}$$

indicates strong coupling at the high scale $\sigma \sim \Lambda$ which is generally an indication of a new-physics threshold. Using perturbative RG equation at $\sigma \sim \Lambda$ can be dangerous.

*) This discussion implies that the appearance of additional Higgs fields may reduce the value of m_t (see Sec.15.7).

Therefore it is important to check how stable is the prediction of m_t in the model under the variation of the boundary condition for $\bar{g}_t(\sigma)$ at $\sigma = \Lambda$. BHL showed that at high Λ ($\Lambda \geq 10^{15} GeV$), changing the value of $\bar{g}_t(\Lambda)$ to unity instead of infinity results in a less than 2 % changing the value of m_t. Such a remarkable stability of the prediction for m_t is a consequence of the presence of the effective fixed point (15.31).

Let us now turn to the prediction of the mass of the Higgs boson. This mass is determined by using the tree-level formula (see Eq.13.5)

$$M_h^2 = \frac{\bar{\lambda}^2(\sigma)}{3} F_{EW}^2, \tag{15.34}$$

with the coupling $\bar{\lambda}(\sigma)$ defined at a scale $\sigma \simeq M_h$. The RG equation for $\bar{\lambda}(\sigma)$ is solved with the boundary condition (15.24). The result is that at $\Lambda > 10^{15} GeV$ M_h is [4]

$$M_h \simeq 1.1 m_t. \tag{15.35}$$

This is again smaller than the result

$$M_h \simeq \sqrt{2} m_t \tag{15.36}$$

corresponding to neglecting the contribution of virtual Higgs bosons (see the discussion at the end of the previous section).

Now, how about the consistency of these predictions of the TMSM with experimental data ? As was already indicated at the end of Sec.15.1, the present lower bound for m_t is $113 GeV$. The combinated data from neutrino experiments and from experiments at CERN e^+e^- collider LEP yields for m_t the following estimate: $m_t = 164 \pm 27 GeV$. Thus, at present, the predicted $m_t \simeq 218 GeV$ at $\Lambda = 10^{19} GeV$ in the TMSM seems too large (although not completely excluded). However, as we will see below, there are some modifications of the TMSM leading to lower values for m_t.

15.4. What does the Top Mode Standard Model Explain in Electroweak Dynamics ? Is it Equivalent to the Standard Model ?

At the end of Sec.15.1, we raised four questions concerning electroweak dynamics. What answers to these questions can be obtained in the TMSM ?

The first question was: why is the t quark so heavy ? The TMSM yields a very natural explanation for this fact: the t quark is a very special particle providing the dynamics of electroweak symmetry breaking. Therefore it is not surprising that its mass is of order $F_{EW} \simeq 246GeV$.

The second question was: can Higgs bosons be in other, then doublet, representations of $SU_L(2)$? In the TMSM the representation for the Higgs boson is defined by the representations of quarks (see Eq.(15.1)). It is a doublet representation because just this representation is selected by the four-fermion interaction (15.3). Of course, one can look for higher representations connected with multi-quark bound states. But as we will argue below, no such bound states are expected in the TMSM.

Let us turn to the third question: how many light (with masses of $O(100GeV)$) Higgs bosons are in electroweak theory ? If we assume that a strong short-range interaction is attributed only to one family of quarks, we will immediately conclude that there may be at most two light Higgs doublets. In the first scenario in the TMSM, there is one light composite Higgs field (see Eq.(15.1)). In the second scenario, in which a strong short-range interaction is attributed to both t and b, there are two composite Higgs fields (see Eqs.(15.1), (15.17)). Thus the TMSM gives a concrete answer to this question either.

The last question was: what is the spectrum of other, than ordinary fermions, W and Z, and Higgs bosons, particles in electroweak theory?

As we shown in Chapter 5, all theories with spontaneous breakdown of the same chiral symmetry have the same low energy effective theories, σ-models. The only difference between them is in the values of their parameters (coupling constants). For example, in the σ-model corresponding to QCD, considered in Sec.12.14, the Yukawa coupling quark is

$$g_Y^{(q)} \sim m_{dyn}/F_\pi \simeq 3 \qquad (15.37)$$

with $m_{dyn} \sim 300MeV$ and $F_\pi \sim 93MeV$. Nearly the same value should be valid for Yukawa couplings of technifermions in TC.

This fact correlates with the point that, as we saw in Chapter 12, the NJL model with cutoff $\Lambda_\chi \sim 1GeV$ is a reasonable effective theory of the chiral dynamics in QCD. Indeed, let us look at the relation (9.133) for the decay constant F_π in the NJL model:

$$F_\pi^2 \sim m_{dyn}^2 \frac{N_c}{4\pi^2} \ln \Lambda_\chi^2 / m_{dyn}^2. \tag{15.38}$$

This relation with $N_c = 3$ implies that in order to get $g_Y^{(q)} \sim m_{dyn}/F_\pi \simeq 3$, the ratio Λ_χ/m_{dyn} must be $\Lambda_\chi/m_{dyn} \sim 3$, i.e., $\Lambda_\chi \sim 1GeV$.

Such a relatively low Λ_χ implies that in this case a new-physics threshold is also low. Of course, in QCD, we are perfectly familiar with this threshold: it is nothing else but the production threshold of such "non-chiral" particles as vector mesons, etc.. In Chapter 12, in order to include them, the NJL model was extended to the ENJL model.

Let us compare this situation with that in electroweak dynamics. Of course in the case of TC the situation will be similar to that in QCD, at least in that there will be a lot of technihadrons. Let us however assume that the heaviest (elementary) fermion participating in electroweak interactions is the t quark with the mass $m_t \lesssim F_{EW}$. Then, its Yukawa coupling is $g_t \sim 1$, and the characteristic parameter $g_t^2/4\pi^2$ in chiral perturbative theory is now at least 10 times smaller than in QCD. What does this imply for a new-physics threshold Λ in electroweak dynamics? Let us assume that m_t is a dynamical mass and that the t quark is mostly responsible for electroweak symmetry breaking. Then, the analogue of the relation (15.38) is now:

$$F_{EW}^2 \sim m_t^2 \frac{N_c}{8\pi^2} \ln \Lambda^2 / m_t^2 \tag{15.39}$$

(compare with Eq.(15.15)). Then, we find that

$$\Lambda/m_t \gtrsim 10^8 \Lambda_\chi/m_{dyn}, \tag{15.40}$$

i.e., in agreement with the analysis in TMSM, we find that the new-physics threshould in such a scenario has to be very high. Thus, the rather minor factor 3 in the ratio $g_Y^{(q)}/g_t$ transforms, because of the logarithmic factors in the relations (15.38), (15.39), in a huge enhancement of the ratio Λ/m_t in electroweak dynamics as compared with the ratio Λ_χ/m_{dyn} in QCD.

This of course implies a very strong fine-tuning of the four-fermion coupling κ_t to the critical value $\kappa_c = 1$: indeed

$$m_t/\Lambda \sim (\kappa_t - 1)^{1/2} \tag{15.41}$$

(compare with Eq.(6.14)).

Since the new-physics threshold Λ is much higher than F_{EW}, no light non-chiral bound states are expected in this case; all these particles are very heavy (with $M \sim \Lambda$) and very unstable. In particular, returning to the question number 2 above, we conclude that no light Higgs bosons in non-doublet representations of $SU_L(2)$ exist in such a scenario.

This conclusion can also be interpreted in somewhat different way. What we have encountered above is that the effective theory works in a wide range of momenta, $F_{EW} \lesssim p \lesssim \Lambda \gtrsim 10^8 F_{EW}$. The only possibility in this case is that such a "perfect" effective theory is a *renormalizable* theory. Indeed, the renormalizability is synonymous with the fact that a theory is well defined and can be used in a wide range of momenta.

Of course, in the present example, where the t quark is responsible for electroweak symmetry breaking, this perfect effective theory is nothing else but the Standard Model! One can consider the present consideration as a "derivation" of the microscopic dynamics underlying the SM. In a sense, the TMSM is the SM in which two parameters (m_t and M_h) are fixed, or, at least, estimated. Let us discuss this point in more detail.

In Sec.11.3, 11.5 we considered a four-fermion model in D dimensional space-time with $D < 4$. We concluded that, at least for a large N_c, this model relates to the same class of universality as the (renormalizable) Yukawa model. In the continuum limit, the dynamics in both these models is governed by the same infrared stable fixed point.

The conclusion remains valid (in this approximation) at $D = 4$ either, but in this case the fixed point equals zero, i.e. both these theories become non-interacting theories in the continuum limit (compare with the discussion at the end of Sec.13.8).

Thus, one may expect that a) the TMSM and the SM relate to the same class of universality and b) they are non-interacting theories in the continuum limit. The latter implies that both these theories are actually effective : both of them must be considered with an ultraviolet cutoff Λ (a new-physics threshold).

In connection with this the following questions arise: Is the TMSM more predictable than SM ? Or is the predictive power of the TMSM illusive ?

These questions have been recently intensively discussed in the literature [20 – 23].

The central point in the discussion is the following. One can expect that the four-fermion interaction comes from some more fundamental interaction connected with the scale Λ (for example, gauge interactions with heavy ($M_V \sim \Lambda$) vector bosons). How unique are predictions for low energy dynamics in a theory with the fundamental interactions replaced by the four-fermion ones ?

In the language of the effective theory, what happens is the following. Integrating out fields of heavy particles connected with the scale Λ (as those vector bosons with $M_V \sim \Lambda$), one obtains a tower of non-renormalizable interactions. These interactions are organized in terms of a derivative expansion, in increasing powers of derivatives and numbers of fields. The term with the nonderivative four-fermion interaction is just the leading one in this expansion. Why can we neglect all others ?

It is clear that without knowing the characteristics of the fundamental dynamics, one cannot justify this approximation. However, as we will show below, for a rather wide class of the fundamental dynamics (including, in particular, asymptotically free interactions for heavy vector bosons with $M_V \sim \Lambda$), this is a good approximation for the description of the low energy dynamics.

Let us begin by recalling the main arguments and conclusions of Ref.[22]. These authors, following Suzuki [20], considered the so called generalized NJL model, with four-fermion interaction of the form

$$\mathcal{L}_{4f} = \frac{G^{(0)}}{2}\left[(\bar{\psi}\psi) + S^{(0)}(\partial^\mu\bar{\psi}\partial_\mu\psi)\right]^2 + \frac{G^{(0)}}{2}\left[(\bar{\psi}i\gamma_5\psi) + S^{(0)}(\partial^\mu\bar{\psi}i\gamma_5\partial_\mu\psi)\right]^2. \quad (15.42)$$

Introducing the auxiliary fields σ and π one gets the following lagrangian density:

$$\mathcal{L}_{4f} \to \mathcal{L} = -(\bar{\psi}\psi + S^{(0)}\partial^\mu\bar{\psi}\partial_\mu\psi)\sigma - (\bar{\psi}i\gamma_5\psi + S^{(0)}\partial^\mu\bar{\psi}i\gamma_5\partial_\mu\psi)\pi - \frac{1}{2G^{(0)}}(\sigma^2 + \pi^2). \quad (15.43)$$

The model was analyzed at the leading order of the $1/N_c$ expansion. The main conclusions are the following[22]:

a) the model relates to the same universality class as the linear σ-model. In particular, varying the parameter $S^{(0)}$ in Eq.(15.42), one can change the dynamical mass m_{dyn} of fermions as a function of the decay constant F;

b) to leading order in $1/N_c$, the scalar mass M_σ is

$$M_\sigma = 2m_{dyn} + O(1/(\ln \frac{\Lambda^2}{m_{dyn}^2})) \qquad (15.44)$$

in both generalized NJL model and linear σ-model.

The conclusion of the authors of Ref.[22] is that without specifying a fundamental dynamics at $p^2 \gtrsim \Lambda^2$, determining in particular the parameter $S^{(0)}$ in Eq.(15.42), one cannot fix the fermion mass m_{dyn}.

Although we agree with this general statement, let us show that on the one hand, for a rather wide class of models of the fundamental dynamics one gets the predictions for m_{dyn} close to those in the NJL model without the $S^{(0)}$-term, and on the other hand, a large deviation from this value of m_{dyn} implies a rather specific character of that fundamental dynamics.

The central point is that to leading order in $1/N_c$, the solution of the SD equation for $\Sigma(p)$ in the generalized NJL model takes the form [20,22]:

$$B_{dyn}(p) = m_{dyn} - m_{dyn}s(p^2/\Lambda^2), \qquad (15.45)$$

where $s \equiv \Lambda^2 S^{(0)}$. Using the Pagels-Stoker formula (12.126) for the decay constant F,

$$F^2 = \frac{N_c}{8\pi^2}\int_0^{\Lambda^2} du u \frac{B_{dyn}(u)\left[2B_{dyn}(u) - uB'_{dyn}(u)\right]}{(u + B_{dyn}^2(u))^2}, \qquad (15.46)$$

we find

$$(F/m_{dyn})^2 = (N_c/4\pi^2)(\ln(\Lambda^2/m_{dyn}^2) + 4s^2 - 6s - 1). \qquad (15.47)$$

This relation implies that the $S^{(0)}$-term in the generalized NJL model becomes essential only if $s^2 \gtrsim \ln(\Lambda^2/m^2)$. What does such an enhancement of the derivative interaction term in Eq.(15.42) mean from the physical viewpoint ?

To answer this question, let us note that the additional term in $\Sigma(p)$ (15.45) connected with the derivative interactions looks rather unusually: it becomes relevant only at very high momenta $p \sim \Lambda$, and, moreover, at $p \gtrsim \Lambda$, it *increases* with momentum. Such a pathological behavior suggests that the only appropriate interpretation of this term is as a term in the power expansion of a piece $B^h_{dyn}(p)$ in $B_{dyn}(p)$ induced by the high energy dynamics (with $p \gtrsim \Lambda$). What kind of the dynamics might lead to $B^h_{dyn}(p)$ which essentially changes the relation between F and m_{dyn} in Eq.(15.47) ? Let us first consider a rather conservative scenario with the short range dynamics described by asymptotically free gauge interactions with heavy ($M_V \sim \Lambda$) vector bosons. Then $B_{dyn}(p)$ is

$$B_{dyn}(p) = B^l_{dyn}(p) + B^h_{dyn}(p) \tag{15.48}$$

with $B^l_{dyn}(p) \sim \theta(\Lambda - p)m_{dyn}$, $B^h_{dyn}(p) \sim \theta(p - \Lambda) \cdot m_{dyn}\Lambda^2/p^2$. As was shown in Sec.12.3, because of asymptotic freedom, $B^h_{dyn}(p)$ rapidly decreases with increasing the momentum (the coefficient $m_{dyn}\Lambda^2$ in $B^h_{dyn}(p)$ is determined by continuity of $B_{dyn}(p)$ at $p = \Lambda$). Since the contribution of $B^l_{dyn}(p)$ in F^2 is (see Eqs (15.46) and (15.47))

$$(F^2)_l \sim \frac{N_c m^2_{dyn}}{4\pi^2} \ln(\Lambda^2/m^2_{dyn}) \tag{15.49}$$

and that of $B^h_{dyn}(p)$ is

$$(F^2)_h \sim \frac{N_c m^2_{dyn}}{4\pi^2}, \tag{15.50}$$

this latter is irrelevant if $\ln(\Lambda^2/m^2_{dyn}) \gg 1$. We recall that this condition (with $\Lambda \sim 10^{15} \div 10^{19} GeV, m_{dyn} \sim 100 GeV$) is fulfilled in the TMSM scenario.

In fact it is not even necessary to accept that the dynamics beyond cutoff Λ is an asymptotically free dynamics with rapidly decreasing (like in QCD) running coupling. For example, in walking TC (see Chapter 14), where $B^h_{dyn}(p) \sim m_{dyn}\Lambda/p$ the logarithmic term (15.49) will again dominate in F^2 at $\ln \Lambda^2/m^2_{dyn} \gg 1$.

It is clear that in the general case of the high energy dynamics with the asymptotics

$$B^h_{dyn}(p) \sim m_{dyn}(\Lambda/p)^\delta, \tag{15.51}$$

$\delta \geq 0$, the high energy piece $(F^2)_h$ can be relevant only if δ is very small, i.e. if there is no practically any form factor connected with the scale Λ. One can say that the cutoff Λ itself becomes rather illusive in this case.

Introducing the $SU_c(3) \times SU_L(2) \times U_Y(1)$ gauge interactions does not change essentially this conclusion.

Thus one expects that there exists a wide class of models with high energy dynamics leading to the same predictions as the TMSM. One such a model will be considered in the next section.

15.5. Dynamical Origin of the Top Mode Standard Model

As was noted in the previous section, one should expect that the TMSM comes from a more fundamental interaction. Attempts to derive the TMSM from renormalizable interactions were undertaken in Refs.[24,25]. Here we will consider one such a model (the topcolor model) introduced by Hill [24].

Let us begin by considering the TMSM lagrangian density (15.4) with $\kappa_b = 0$:

$$\mathcal{L}_{4f} = G_t^{(0)} (\bar{\psi}_L^a t_R)(\bar{t}_R \psi_L^a), \tag{15.52}$$

where $G_t^{(0)} = \frac{8\pi^2}{N_c \Lambda^2}\kappa_t$. By using the Fierz transformations, considered in Appendix B, this interaction can be rewritten as

$$G_t^{(0)} (\bar{\psi}_L^a t_R)(\bar{t}_R \psi_L^a) = -G_t^{(0)} (\bar{\psi}_L^a \gamma_\mu \cdot \frac{1}{2}\tilde{t}^K \psi_L^a)(\bar{t}_R \gamma^\mu \cdot \frac{1}{2}\tilde{t}^K t_R) + O(1/N_c), \tag{15.53}$$

where $\tilde{t}^K$ are the matrices from the fundamental representation of $SU(N_c)$. This suggests that the TMSM can be derived from a (broken) gauge interaction mediated by heavy ($M_V \sim \Lambda$) vector bosons (this resembles the ETC scenario (see Chapter 13)).

The basic point in the topcolor model is an extension of the Standard Model to the model with the gauge interaction ($N_c = 3$)

$$SU_1(3) \times SU_2(3) \times SU_L(2) \times U_Y(1). \tag{15.54}$$

Thus, in the model, instead of the color interaction $SU_c(3)$, we have the gauge interaction $SU_1(3) \times SU_2(3)$.

In order to distinguish the top quark from other fermions, one assigns light quarks and leptons to the following representations under ($SU_1(3), SU_2(3), SU_L(2)$):

$$\begin{gathered}\begin{pmatrix} u \\ d \end{pmatrix}_L, \begin{pmatrix} c \\ s \end{pmatrix}_L \to (3,1,2), \\ \begin{pmatrix} \nu_e \\ e \end{pmatrix}_L, \begin{pmatrix} \nu_\mu \\ \mu \end{pmatrix}_L, \begin{pmatrix} \nu_\tau \\ \tau \end{pmatrix}_L \to (1,1,2), \\ u_R, d_R, c_R, s_R, b_R \to (3,1,1), \\ e_R, \mu_R, \tau_R, (\nu_{iR}) \to (1,1,1),\end{gathered} \tag{15.55}$$

while the t quark is assigned

$$\begin{pmatrix} t \\ b \end{pmatrix}_L \to (1,3,2),\ t_R \to (1,3,1). \tag{15.56}$$

One can see that the interactions connected with the $SU_1(3)$ and $SU_2(3)$ groups are not anomaly free (compare with a discussion in Sec.13.3). In order to cancel these anomalies, one introduces the following electroweak singlet quark:

$$Q_R \to (1,3,1),\ Q_L \to (3,1,1). \tag{15.57}$$

Both Q_R and Q_L have electric charge $Q = -\frac{1}{3}$, hence hypercharge $Y = -\frac{1}{3}$.

The crucial dynamical assumption in the topcolor model is that at the scale Λ the symmetry $SU_1(3)\times SU_2(3)$ breaks down to the color group $SU_c(3)$. For this purpose, one uses a scalar field $\phi^l_{m'}$ which transforms as $(3,\bar{3},1)$ and whose effective potential leads to a vacuum expectation value of the form $\langle 0|\phi|0\rangle = diag(\Lambda,\Lambda,\Lambda)$. Thus, the usual (unbroken) color $SU_c(3)$ is the diagonal subgroup of the $SU_1(3) \times SU_2(3)$. There are eight massless gluon fields G^K_μ connected with the $SU_c(3)$; also, there are eight massive vector bosons ("colorons") $\tilde{G}^K_\mu$ connected with the generators of the coset $[SU_1(3) \times SU_2(3)]/SU_c(3)$.

To avoid additional light fermions, Q must be given a large Dirac mass:

$$-\sqrt{2}g^Q_Y\bar{Q}^{m'}_L\phi^l_{m'}Q_{eR} + H.c. \tag{15.58}$$

In order to arrange a strong interaction of the t with colorons, one assumes that the gauge coupling h_2 of the $SU_2(3)$ is much larger than the gauge coupling h_1 of the

$SU_1(3)$. Then, the gluon and coloron fields are defined by

$$G^K_{1\mu} = \cos\theta G^K_\mu - \sin\theta \tilde{G}^K_\mu,$$
$$G^K_{2\mu} = \sin\theta G^K_\mu + \cos\theta \tilde{G}^K_\mu, \tag{15.59}$$

where the mixing angle θ is

$$\theta = \arctan\frac{h_1}{h_2}. \tag{15.60}$$

The gauge coupling g_3 of the color interaction is

$$g_3 = \frac{h_1 h_2}{\sqrt{h_1^2 + h_2^2}}. \tag{15.61}$$

The mass of the degenerate octet of colorons is

$$M_V = \Lambda\sqrt{h_1^2 + h_2^2} = \Lambda(g_3/\sin\theta\cos\theta). \tag{15.62}$$

The $SU_c(3)$ current is the usual QCD current while the current interacting with colorons (multiplied by its coupling strength) is:

$$\tilde{j}^K_\mu = g_3 \cot\theta(\bar{t}\gamma_\mu \cdot \frac{1}{2}\tilde{t}^K t + \bar{b}_L\gamma_\mu \cdot \frac{1}{2}\tilde{t}^K b_L + \bar{Q}_R\gamma_\mu \cdot \frac{1}{2}\tilde{t}^K Q_R) + g_3\tan\theta(\bar{b}_R\gamma_\mu \cdot \frac{1}{2}\tilde{t}^K b_R + \sum_{i=1}^{2}\bar{q}^i\gamma_\mu \cdot \frac{1}{2}\tilde{t}^K q^i). \tag{15.63}$$

Since $h_2 \gg h_1$, the ratio

$$g_3\cot\theta/g_3\tan\theta = \frac{h_2^2}{h_1^2} \tag{15.64}$$

is very large, i.e. the t quark interacts with colorons much stronger than other quarks.

Eqs.(15.53) and (15.63) imply that

$$G_t^{(0)} = \frac{8\pi^2}{N_c\Lambda^2}\kappa_t = g_3^2\cot^2\theta/M_V^2. \tag{15.65}$$

Thus, the TMSM is derived from the renormalizable interaction.

Few comments are in order. In the previous section, we discussed the possibility of changing the predictions of the TMSM due to the appearance of higher dimension operators in the model. The topcolor model allows computation of the strength of these operators (as the coefficient $S^{(0)}$ in Eq.(15.42)). The result is that their strength is small [24]. This agrees with the general conclusion of the previous section.

In the TMSM, the coupling constant κ_t must be tuned very precisely to the critical value κ_c in order to arrange a huge hierarchy between the scales $m_t \sim F_{EW}$ and Λ. The same fine-tuning for the gauge coupling $g_3^2 \cot^2\theta$ has to be arranged in the topcolor model. The best-known way to overcome this problem is supersymmetric extension of the dynamics [26]. However, there may exist other (dynamical) possibilities to deal with this problem (or even to use the fine-tuning; see the next section). In fact, it is possible that this problem reflects our poor understanding of short distance dynamics connected with the scale Λ. There is a famous analogue of this in cosmology: the change of the standard cosmological scenario at early times (introducing the inflation scenario) solved the fine-tuning problem connected with the initial density of the matter in the universe [27].

Also the recent development of the wormhole approach in quantum gravity has shown that the problem of the fixation of values of coupling constants can be dynamical [28]. In this context, it is worth noting that critical values of coupling constants connected with second order phase transitions (conformal points; see Chapter 11) are very special.

For some speculations in this direction see Ref.[29,30]. We also would like to indicate recent interesting speculations concerning solving the fine-tuning problem in the framework of self-organizing criticality approach [31].

The topcolor model, in its original form, does not address the problem of generating masses of the fermions other than the t quark. As in the TMSM, to generate these masses from the t quark condensate, one must introduce additional interactions. A particularly nontrivial problem is generating the masses of down fermions from this condensate in models in which (as in the topcolor model) the fundamental interactions are gauge interactions. The point is the following [32,33]. To generate a mass of a down fermion, say the b quark, from the $\bar{t}t$-condensate, one must use the $\kappa^{(2)}$-interaction:

$$\kappa^{(2)}(\bar{\psi}_L^a \psi_R^b)(i\tau^2)^{ac}(i\tau^2)^{bd}(\bar{\psi}_L^c \psi_R^d), \tag{15.66}$$

where the field ψ describes the t and the b quarks (see Eq.(15.3)). As we already pointed out in Sec.15.2, this interaction explicitly violates the axial (Peccei-Quinn [11]) $U_5(1)$ symmetry. Of course, there is nothing in principle that is wrong with this its feature, but the point is that it is rather difficult to obtain a relatively strong $\kappa^{(2)}$-interaction (which may produce the mass of the b) in gauge theories with *short-range* interactions. Let us discuss this point in more detail.

In Sec.12.12, we considered the effective action in QCD obeying the anomalous conservation law for the $U_5(1)$. The microscopic dynamics which breaks the $U_5(1)$ is essentially nonperturbative and is connected with instanton-like configurations in the QCD path integral [34,35]. The reason of this is that although the ABJ anomaly (12.8) violates the conservation of the $U_5(1)$ *current* even in perturbation theory, the $U_5(1)$ *charge* is perturbatively conserved (for a review see Ref.[35]). As a result, the $U_5(1)$ remains a good symmetry in perturbative theory. Moreover, since the contribution of an instanton-like configuration is proportional to the factor $exp\left(-\frac{const}{\bar{\alpha}(\rho^{-1})}\right)$, where $\bar{\alpha}(\rho^{-1})$ is the running coupling calculated at the scale ρ^{-1} with ρ being the size of the configuration, the dominant contribution is provided by "large" instantons, with $\rho \gtrsim \Lambda_{QCD}^{-1}$, for which $\bar{\alpha}(\rho^{-1})$ is large.

In gauge theories with broken gauge groups, there are only "small" instanton-like configurations, with $\rho \lesssim M_V^{-1}$, where M_V is a mass of gauge bosons. In this case, $\bar{\alpha}(\rho^{-1})$ is small, and the contribution of these configurations is strongly suppressed (for example, in the $SU_L(2) \times U_Y(1)$ electroweak theory, $\bar{\alpha} \sim 10^{-2}$). As a result the coefficient $\kappa^{(2)}$ in the interaction (15.66) is very small, and one cannot generate realistic masses for down fermions.

In the next section we will consider a model which solves this problem.

15.6. Why is the t quark Much Heavier than the b quark ?

In the original TMSM, besides the mass of the t, all masses of other fermions remain undefined. In the present section we will consider a model which shares some features with the TMSM and which suggests a dynamical explanation of the origin of the fermion mass spectrum. The model was elaborated by Mendel and the present author [36,37].

The spectrum of the ordinary fermions was considered in Sec.13.3. The quark spectrum is characterized by the following striking features: 1) There is a large hierarchy between quark masses from different families. In particular, the mass of the b quark is 3 orders of magnitude larger than the mass of the d; 2) The isospin violation is very different in different families. Assuming $164GeV$ for the mass of the t quark (the central value from neutrino experiments and from experiments at the CERN e^+e^- collider LEP), one finds that while $m_u/m_d \sim 1$, the ratios m_c/m_s and m_t/m_b are $m_c/m_s \sim 7$ and $m_t/m_b \sim 30$ *). These experimental facts point out that the isospin violation becomes stronger with increasing the quark masses.

As we will see, the model [36,37] suggests a dynamical explanation of a correlation between these two features of the quark spectrum. The basic idea in the model is to separate the dynamics primarily responsible for the generation of the W and Z masses from that responsible for triggering the complicated structure of the fermion mass matrix. As a first step, we assume that the dynamics leading to the W and Z masses generates the same (or nearly the same) mass $m^{(0)} \sim 1MeV$ for all six quarks ($SU_f(6)$ symmetric mass spectrum) **). What kind of the dynamics can do this job ? This, for example, can be a dynamical Higgs mechanism with a Higgs doublet composed of t' and b' quarks of a possible fourth family. The Higgs mechanism will lead to the masses of the W and Z and to the $SU_f(6)$ symmetric quark mass spectrum ***). Another possibility is that a regular (and not "strong", considered in Sec.14.3) extended technicolor (ETC) does this. As was shown in Sec.13.7, to suppress flavor-changing neutral currents (FCNCs), one must use a large, $M_{ETC} \equiv \Lambda_{ETC}/g_{ETC} \gtrsim 100TeV$, scale for ETC. As a result, only light fermion masses, $O(1MeV)$, can be generated. If all fermions were so light, this would be a great triumph of ETC ! However, the actual situation is quite different: the masses of quarks from the second and third families are much larger than $1MeV$. This was a serious problem for ETC. However, in the present model, at the first stage, all of the quark masses are of $O(1MeV)$. Therefore we may use ETC to generate such masses. Moreover, we can consider this as a natural explanation of a mystery why the u and d quarks are so light: we assume that the scale $1MeV$ occurs in ETC.

---

*) Here all the masses are current masses of fermions.

**) We will concentrate on the quark spectrum, mentioning leptons briefly at the end.

***) Of course, even the conventional Higgs mechanism with an elementary Higgs field can do this.

Actually, the scenario with the fourth family and the TC scenario can be very similar. For example, recently, Martin has shown how "self-breaking" TC may produce the fourth family responsible for electroweak symmetry breaking[38].

Henceforth, for concreteness, we will consider a scenario with ETC, although one should keep in mind the possibility with the fourth family.

To generate the W and Z masses, the minimal anomaly-free TC model with one doublet $\begin{pmatrix} U \\ D \end{pmatrix}$ of technifermions, having charges $Q = \pm\frac{1}{2}$, is used (see Sec.13.5). Choosing the fermion mass $m^{(0)}$, generated by ETC, as $m^{(0)} \simeq 4MeV$ and using rescaling relations to get the TC condensate from the QCD one, one finds $\Lambda_{ETC}/g_{ETC} \sim 3 \cdot 10^2 TeV$ for $SU_{TC}(N_{TC})$ with $N_{TC} \leq 4$.

To generate the realistic masses for quarks from $m^{(0)} \simeq 4MeV$, one introduces an additional horizontal interaction acting in the family space and commuting with the $SU_L(2) \times SU_R(2)$. It is broken in such a way that it is essentially stronger for the third family than for the two first ones: $G_3^h \equiv 4\pi^2\kappa_3^h/N_c(\Lambda_3^h)^2 \gg G_1^h \equiv 4\pi^2\kappa_1^h/N_c(\Lambda_1^h)^2, G_2^h \equiv 4\pi^2\kappa_2^h/N_c(\Lambda_2^h)^2$, where Λ_i^h, $i = 1, 2, 3$, are masses of corresponding vector bosons. FCNCs are suppressed by choosing the scales Λ_i^h large enough. We will choose $\Lambda_3^h \sim \Lambda_{ETC} \equiv \Lambda$ and $\Lambda_1^h, \Lambda_2^h \geq \Lambda_3^h$. In the low-energy region, this short-range interaction can be approximated (as ETC) by the four-fermion interaction.

Since the horizontal interaction commutes with the $SU_L(2) \times SU_R(2)$, it cannot lead to splitting the masses of up and down quarks. We will return to this point below.

Let us show how the horizontal interaction enhances the fermion masses. First, for illustration, let us neglect the SM interactions. Then in the mean-field approximation, the Schwinger-Dyson equation for the mass functions $B_t(p^2)$ and $B_b(p^2)$ have the same form:

$$B_b(p^2) = \frac{g_{ETC}^2}{\Lambda^2} \int_0^{\Lambda^2} dq^2 \frac{q^2 B_D(q^2)}{q^2 + B_D^2(q^2)} + \frac{\kappa_3^h}{\Lambda^2} \int_0^{\Lambda^2} dq^2 \frac{q^2 B_b(q^2)}{q^2 + B_b^2(q^2)}, \tag{15.67}$$

where $B_D(q^2)$ is the technifermion mass function (for concreteness, we consider the SD equation for the mass function of the b). The first term on the left-hand side of Eq.(15.67) is connected with ETC interaction:

$$\frac{g_{ETC}^2}{\Lambda^2}\int_0^{\Lambda^2} dq^2 \frac{q^2 B_D(q^2)}{q^2+B_D^2(q^2)} = m^{(0)} \simeq 4MeV. \qquad (15.68)$$

Equation (15.67) is equivalent to the following algebraic equation:

$$m_b(1-\kappa_3^h) = m^{(0)} - \kappa_3^h \frac{m_b^3}{\Lambda^2}\ln\left(\frac{\Lambda^2+m_b^2}{m_b^2}\right). \qquad (15.69)$$

with $m_b \equiv B_b$ (compare with Eq.(9.84)). This equation implies that the ratio $m_b/m^{(0)}$ changes from one to $O(\Lambda^2/m_b^2)$ when the coupling κ_3^h increases from zero to the critical value $\kappa_c = 1$ (compare with Eq.(11.126)). In the region with

$$m_b^2/\Lambda^2 \ll 1-\kappa_3^h \qquad (15.70)$$

one finds [36,39,40]:

$$m_b/m^{(0)} \simeq \frac{1}{1-\kappa_3^h}. \qquad (15.71)$$

This relation has the form of a resonance solution with a pole at $\kappa_3^h = \kappa_c = 1$ Taking $1-\kappa_3^h \sim 10^{-3}$, one gets $m_b \sim 5GeV$.

We emphasize that the central point in this enhancement mechanism is the fact that a fermion mass function generated by ETC is a slowly changing function of momentum in the region with $q \lesssim \Lambda_{ETC} \equiv \Lambda \gg \Lambda_{TC}$ (see Sec.13.6). Indeed, if $B_b(q^2)$ decreased rapidly at $q^2 \lesssim \Lambda_{TC}^2$, then the second term on the right-hand side of Eq.(15.67) would be very small,

$$\frac{\kappa_3^h}{\Lambda^2}\int_0^{\Lambda_{TC}^2} dq^2 \frac{q^2 B_b(q^2)}{q^2+B_b^2(q^2)} \sim \frac{\kappa_3^h \Lambda_{TC}^2}{\Lambda^2} \ll 1, \qquad (15.72)$$

and the enhancement effect would disappear.

Another important point is that, in the near-critical regime, a resonance solution is unstable under small perturbations. We will use this feature below to generate a large ratio m_t/m_b due to the hypercharge $U_Y(1)$ interaction.

It is clear that the present model shares some features with the TMSM. However, there is an essential difference between them. While in the TMSM, a self-interaction of the top is responsible for both electroweak symmetry breaking and the generation of fermion masses, in the present model, the self-interaction of the t and b essentially enhances their masses generated by ETC and thus triggers the structure of the fermion mass matrix. However, it is not responsible for electroweak symmetry breaking. As a result, the problem of strong fine-tuning of the TMSM can be avoided: while $\kappa_t - \kappa_c \sim 10^{-30} - 10^{-38}$ in the TMSM (see Eq.(15.41)), $\kappa_c - \kappa_3^h \sim 10^{-3}$ in the present model. Moreover, since the mass $m^{(0)}$ plays the role of the bare mass with respect to the horizontal interaction, the coupling κ_3^h is (although near-critical) subcritical.

Let us now show how the hypercharge $U_Y(1)$ interaction may lead to a large ratio m_t/m_b in the near-critical regime.

When the $SU_c(3) \times SU_L(2) \times U_Y(1)$ interactions are taken into account, the SD equation (15.67), and a similar equation for $B_t(p^2)$, changes. We will now use the improved ladder approximation for these equations (see Sec.14.2). First of all, we note that in the present case of the four-fermion interaction with additional gauge ones, the relation (15.71) must be slightly modified. As directly follows from the analysis in Sec.14.2, this relation now takes the following form for m_b and m_t:

$$m_b = \frac{m^{(0)}}{\Delta\kappa_b}, \tag{15.73}$$

$$m_t = \frac{m^{(0)}}{\Delta\kappa_t}, \tag{15.74}$$

where

$$\Delta\kappa_b = 1 - \frac{3\bar{\alpha}_b(\Lambda)}{2\pi} - \kappa_3^h, \tag{15.75}$$

$$\Delta\kappa_t = 1 - \frac{3\bar{\alpha}_t(\Lambda)}{2\pi} - \kappa_3^h, \tag{15.76}$$

with

$$\begin{aligned} \bar{\alpha}_b(\Lambda) &= \frac{N_c^2 - 1}{2N_c}\bar{\alpha}_3(\Lambda) - \frac{1}{18}\bar{\alpha}_1(\Lambda) \\ \bar{\alpha}_t(\Lambda) &= \frac{N_c^2 - 1}{2N_c}\bar{\alpha}_3(\Lambda) + \frac{1}{9}\bar{\alpha}_1(\Lambda). \end{aligned} \tag{15.77}$$

Here $\bar{\alpha}_3(\sigma)$ and $\bar{\alpha}_1(\sigma)$ are the running coupling of the color and the hypercharge interactions, respectively. The different contributions of the $U_Y(1)$ interaction to m_b and m_t

occur because while $Y(b_R) = -\frac{1}{3}$, $Y(t_R)$ is $Y(t_R) = \frac{2}{3}$*).

At $\Lambda = 100TeV$, the color coupling $\bar{\alpha}_3(\Lambda)$ is $\bar{\alpha}_3(\Lambda) \simeq 0.055$ and the hypercharge one is $\bar{\alpha}_1(\Lambda) \simeq 0.01$. We now present a toy numerical example. Taking $m^{(0)} = 4MeV$ ("bare" mass generated by ETC) and $\Delta\kappa_t = 2.7 \cdot 10^{-5}$, one finds $\Delta\kappa_b = \Delta\kappa_t + \frac{1}{4\pi}\bar{\alpha}_1(\Lambda) \simeq 8.7 \cdot 10^{-4}$. Then, through enhamcement (15.73), (15.74), we obtain $m_t \simeq 150GeV$ and $m_b = 4.5GeV$!

Thus, in the near-critical regime, the perturbative hypercharge interaction can be the origin of a large $m_t - m_b$ splitting. An interesting point in this model is that there is a correlation between the fact that the b quark is 3 orders of magnitude larger than the mass of the d and that fact that there is a very strong isospin violation in the third family. Indeed, since we started from the mass $m^{(0)} = 4MeV$, which is of the order of the magnitude of the masses of quarks from the first family, we should use the near-critical enhancement mechanism with $\Delta\kappa_b \sim 10^{-3}$, to get m_b which is 3 orders of magnitude larger than $m^{(0)} \sim m_u, m_d$. But $\Delta\kappa_b \sim 10^{-3}$ corresponds to a near-critical regime in which the hypercharge interaction with $\frac{1}{4\pi}\bar{\alpha}_1(\Lambda) \sim 10^{-3}$ just becomes relevant!

Although the present calculations can be considered only as illustrative, they clearly reveal interesting features of the near-critical dynamics.

As was already pointed out in Sec.14.5, the nerar-critical dynamics leads to tightly bound spinless states. Since in the present model the coupling κ_3^h is (although near-critical) subcritical, there are four nearly degenerate composite states,

$$\varphi_t = \begin{pmatrix} \varphi_t^0 \\ \varphi_t^- \end{pmatrix} \sim \Lambda^{-2} \begin{pmatrix} \bar{t}(1+\gamma_5)t \\ \bar{t}(1+\gamma_5)b \end{pmatrix}, \tag{15.78}$$

and four other nearly degenerate ones,

$$\varphi_b = \begin{pmatrix} \varphi_b^0 \\ \varphi_b^- \end{pmatrix} \sim \Lambda^{-2} \begin{pmatrix} -\bar{b}(1-\gamma_5)b \\ \bar{t}(1-\gamma_5)b \end{pmatrix} \tag{15.79}$$

(two composite Higgs doublets). Of course, there is also one composite techni-Higgs

*) Since the $SU_L(2)$ gauge bosons couple only to left-handed fermions, they do not contribute to the SD equations in the present approximation.

doublet,

$$\varphi_T = \begin{pmatrix} \varphi_T^0 \\ \varphi_T^- \end{pmatrix} \sim F_{EW}^{-2} \begin{pmatrix} \bar{F}F + \bar{F}\gamma_5\tau^3 F \\ 2\bar{F}\gamma_5\tau^+ F \end{pmatrix}. \tag{15.80}$$

A low energy effective action in this model for $\varphi_t, \varphi_b, \varphi_T$ and ordinary fermions was derived in Ref.[37]. Here we will indicate main features of this action.

The key point in the derivation of the effective action is weakness of the ETC interaction. In fact, rewriting it as the Yukawa interaction.

$$\mathcal{L}_Y^{ETC} = -\sqrt{2}g_T \sum_{i=1}^{3} \left[\bar{q'}_L^i \varphi_T u_R'^i + \bar{q'}_L^i \tilde{\varphi}_T d_R'^i \right] + H.c. \tag{15.81}$$

in the low energy ($q \ll \Lambda$) region, one finds $g_T = m^{(0)}/v_T \simeq 1.5 \cdot 10^{-5}$ with $m^{(0)} = 4MeV$, $v_T = \sqrt{2}\langle 0|\varphi_T^0|0\rangle \simeq F_{EW}$ (here $\tilde{\varphi}_T = i\tau^2\varphi_T^*$, and the primes indicate that a weak basis for fermions is used). One can show [37] that, due to smallness of g_T, the only relevant effect of ETC in the action is the appearance of the mixing term

$$V_m = -\mu_m^2(\varphi_T^\dagger \varphi_t + \varphi_T^\dagger \varphi_b + H.c.) \tag{15.82}$$

in the potential, with $\mu_m \sim F_{EW}$. All other terms in the action connected with ETC are strongly suppressed (by $g_T \sim 10^{-5}$). We would like to note that the interaction (15.82) (softly) violates the PQ, $U_5(1)$, symmetry. The microscopic origin of this violation is in the TC instantons. Since all other terms connected with ETC are strongly suppressed, this interaction is the only source of the violation of the PQ symmetry in the model.

As a result of the mixing φ_T with φ_t and φ_b, the neutral components φ_t^0 and φ_b^0 acquire non-vanishing vacuum expectation values $\varphi_t^0 = \frac{v_t}{\sqrt{2}}$ and $\varphi_b^0 = \frac{v_b}{\sqrt{2}}$. Because of the hypercharge interaction, the value v_t is much larger than v_b : $v_t \simeq \frac{v_T}{2}, v_b \simeq 2GeV$, and $v_T^2 + v_t^2 + v_b^2 = F_{EW}^2$ [37].

The φ_t and φ_b interact with the t and b quarks:

$$\mathcal{L}_Y^{(3)} = -\sqrt{2}g_t \bar{\psi}'_L \varphi_t t_R'^i - \sqrt{2}g_b \bar{\psi}'_L \tilde{\varphi}_b b_R'^i + H.c., \tag{15.83}$$

where the Yukawa coupling g_t and g_b are nearly the same: $g_t \simeq g_b \simeq 1.4$ [37]. Therefore, since $v_t \gg v_b$, the t quark is much heavier than the b.

Of course, the φ_t and φ_b also interact with quarks from the first and second families. Let us show that in the model the Glashow-Weinberg (GW) scheme (discussed in Sec.13.4) naturally occurs. We recall that in the GW scheme, the doublet φ_t generates masses for up quarks and φ_b generates masses for down quarks. In this case, the Yukawa interactions preserve the PQ, $U_5(1)$, symmetry, But this is exactly what we have encountered in the present model: as was pointed above, only the interaction (15.82) violates the PQ symmetry in the model; all other terms, including the Yukawa interaction terms, preserve (up to small corrections of $O(g_T) \sim 10^{-5}$) this symmetry.

Thus, the model yields a dynamical realization for the GW scenario. The interaction (15.82), violating the PQ symmetry, plays the important role in the model: it gives a large ($M \sim O(100GeV)$) mass to axion, thus escaping the difficulties with visible light axion (see Sec.15.2) [37].

We notice that in the modification of this model, in which the fourth family substitutes TC, it is important to have a strong (although broken) gauge interaction for the fourth family fermions in order to use the instanton dynamics for the (rather strong) violation of the PQ symmetry. For a detailed discussion of this point, see Ref.[33].

One can in principle use φ_t and φ_b to generate also neutrino and $l^\pm$ masses, respectively. It looks quite natural in the case of grand unified theory scenarios. There is an alternative [36]: using two additional lepton-antilepton composite doublets φ_{ν_τ} and φ_τ to do this.

The most interesting prediction of this model is that there is a clear signature for the resonances φ_t (with the mass $M \sim 600GeV$) to be seen at large hadron colliders, or at a very high energy e^+e^- collider [37].

We conclude this section by saying that the near-critical dynamics can naturally produce the main features of the observed fermion spectrum.

15.7. More About Modifications of the Top Mode Standard Model

In this section, we will briefly discuss some modifications of the simplest TMSM.

1) *Forth Family Models*

Modifications of the TMSM by using a fourth family of fermions for electroweak symmetry breaking were already considered in Refs.[1-4]. MTY (see the second paper in Ref.[1]) and Marciano[3] discussed the modification in which the contribution to the electroweak condensate of an almost degenerate fourth family of quarks is comparable to that of the t quark. In this scenario, m_t is roughly reduced by factor $1/\sqrt{2}$ as compared to the TMSM.The low energy effective theory in this case corresponds to the two Higgs doublets SM [3].

BHL [4] considered another modification in which the top quark would not play an essential role and its mass would not be predicted.

Since experimental data have proved that there are only three light neutrino species interacting with the Z boson [41], in these modifications, a mechanism providing a mass $m_{\nu_{\tau'}} > M_Z/2$ for a forth family neutrino must be introduced. Such a mechanism was studied by Hill, Paschos and Luty [42].

The dynamical origin of fourth family models was studied by King and Elliot [33,43].

2) *Supersymmetric Models*

The supersymmetric extension of the TMSM [26] contains two composite Higgs multiplets. The predicted mass of the t quark is in the range $140GeV < m_t < 190GeV$.

3) *Models with additional Higgs bosons*

As was already pointed out in Sec.15.3, the appearance of additional Higgs fields may reduce the value of m_t. Babu and Mohapatra [44] considered a model in which, besides the composite Higgs doublet φ_t, there also are two isosinglet color-triplet composite Higgs fields. As a result, the top-quark mass was predicted to be $m_t = 160 - 200GeV$.

4) *Grand Unified Theories*

Since the composite scale Λ is necessarily very large in the TMSM, this suggests that this scenario may be naturally extended to grand unified theories (GUT) [1,4]. A new point in such GUT scenarios is that their Higgs fields are composed of the ordinary fermions.

A concrete example of the extension of the TMSM to GUT has been considered by Bando, Kugo and Suehiro [40] and by Achiman and Lukas [45]. They considered E_6 GUT in which all Higgs fields are fermion-fermion composites. The model predicts $m_t \gtrsim 100GeV$ and describes the masses and mixing angles of the light neutrinos [45].

Thus, the idea of strongly self-interacting (ordinary) fermions has yielded new rich possibilities in model building.

Appendix A

Dirac Equation

The Dirac equation takes the form

$$i\gamma^{\mu}\partial_{\mu}\psi - m\psi = 0 \ , \tag{A.1}$$

where the matrices γ^{μ} satisfy the commutation relations

$$\gamma^{\mu}\gamma^{\nu} + \gamma^{\nu}\gamma^{\mu} = 2g^{\mu\nu} \tag{A.2}$$

By using these commutation relations, one can show that all products of γ matrices are reduced to one of the following sixteen elements:

$$\begin{aligned} & I, \\ & \gamma^0,\ \gamma^1,\ \gamma^2,\ \gamma^3, \\ & \gamma^0\gamma^1, \gamma^0\gamma^2,\ \gamma^0\gamma^3,\ \gamma^1\gamma^2,\ \gamma^1\gamma^3,\ \gamma^2\gamma^3\ , \\ & \gamma^0\gamma^2\gamma^3,\ \gamma^0\gamma^1\gamma^3,\ \gamma^0\gamma^1\gamma^2,\ \gamma^1\gamma^2\gamma^3\ , \\ & \gamma^5\ \equiv\ -i\gamma^0\gamma^1\gamma^2\gamma^3\ . \end{aligned} \tag{A.3}$$

A set of matrices satisfying the commutation relations (A.2) is called a Clifford algebra. We need to know the following facts [1]. The sixteen elements Γ_i in Eq.(A.3) are linearly independent. Apart from equivalences, there exists only one irreducible representation of this algebra.

The most frequently used representations of the Dirac matrices are the following two ones: the spinor representation with (see Eq. (1.24))

$$\gamma^0 = \begin{pmatrix} 0 & \sigma_0 \\ \sigma_0 & 0 \end{pmatrix}, \boldsymbol{\gamma} = \begin{pmatrix} 0 & -\boldsymbol{\sigma} \\ \boldsymbol{\sigma} & 0 \end{pmatrix}, \tag{A.4}$$

and the standard representation with

$$\gamma^0 = \begin{pmatrix} \sigma_0 & 0 \\ 0 & -\sigma_0 \end{pmatrix}, \boldsymbol{\gamma} = \begin{pmatrix} 0 & \boldsymbol{\sigma} \\ -\boldsymbol{\sigma} & 0 \end{pmatrix}. \tag{A.5}$$

A bispinor $\psi = \begin{pmatrix} \varphi \\ \chi \end{pmatrix}$ in the standard representation is expressed through the bispinor

$\psi = \begin{pmatrix} \xi \\ \eta \end{pmatrix}$ in the spinor representation as

$$\begin{aligned} \varphi &= \frac{1}{\sqrt{2}}(\xi + \eta) \\ \chi &= \frac{1}{\sqrt{2}}(\xi - \eta) \ . \end{aligned} \tag{A.6}$$

The general solution of the Dirac equation takes the form

$$\psi(x) = \frac{1}{(2\pi)^{3/2}} \sum_{s=\pm1} \int d^3\mathbf{p} \left[b(p,s)u(p,s)e^{-ipx} + d^\dagger(p,s)v(p,s)e^{ipx} \right] , \tag{A.6}$$

$$\bar{\psi}(x) \equiv \psi^\dagger \gamma_0 = \frac{1}{(2\pi)^{3/2}} \sum_{s=\pm1} \int d^3\mathbf{p} \left[b^\dagger(p,s)\bar{u}(p,s)e^{ipx} + d(p,s)\bar{u}(p,s)e^{-ipx} \right] , \tag{A.7}$$

where $p^0 = \sqrt{\mathbf{p}^2 + m^2}$ and we take the spinors $u(p,s)$ and $v(p,s)$ to be helicity eigenstates, that is

$$\frac{\mathbf{\Sigma} \cdot \mathbf{p}}{|\mathbf{p}|} u(p,s) = s u(p,s), \frac{\mathbf{\Sigma} \cdot \mathbf{p}}{|\mathbf{p}|} v(p,s) = -s v(p,s) \text{ for } s = \pm 1 \ , \tag{A.8}$$

where

$$\mathbf{\Sigma} = \begin{pmatrix} \sigma & 0 \\ 0 & \sigma \end{pmatrix} . \tag{A.9}$$

They satisfy the Dirac equations

$$\begin{aligned} (\gamma^\mu p_\mu - m)\, u(p,s) &= 0 \ , \\ (\gamma^\mu p_\mu + m)\, v(p,s) &= 0 \ , \end{aligned} \tag{A.10}$$

which in terms of the adjoint are

$$\begin{aligned} \bar{u}(p,s)\,(\gamma^\mu p_\mu - m) &= 0 \\ \bar{v}(p,s)\,(\gamma^\mu p_\mu + m) &= 0 \ . \end{aligned} \tag{A.11}$$

We will use the following normalization for them

$$u^\dagger(p,s')u(p,s) = v^\dagger(p,s')v(p,s) = \delta_{ss'} \ . \tag{A.12}$$

The explicit form of these bispinors in the standard representation is:

$$
\begin{aligned}
u(p,+1) &= \left(\frac{p^0+m}{2p^0}\right)^{1/2}\begin{bmatrix} w_+ \\ \frac{\mathbf{p}}{p^0+m} w_+ \end{bmatrix}, \\
u(p,-1) &= \left(\frac{p^0+m}{2p^0}\right)^{1/2}\begin{bmatrix} w_- \\ \frac{-p}{p^0+m} w_- \end{bmatrix}, \\
v(p,+1) &= \left(\frac{p^0+m}{2p^0}\right)^{1/2}\begin{bmatrix} \frac{p}{p^0+m} \tilde{w}_+ \\ -\tilde{w}_+ \end{bmatrix}, \\
v(p,-1) &= \left(\frac{p^0+m}{2p^0}\right)^{1/2}\begin{bmatrix} \frac{p}{p^0+m} \tilde{w}_- \\ \tilde{w}_- \end{bmatrix},
\end{aligned}
\tag{A.13}
$$

where

$$
\begin{aligned}
w_+ &= \frac{1}{\sqrt{2(n^3+1)}}\begin{pmatrix} n^3+1 \\ n^1+in^2 \end{pmatrix}, \\
w_- &= \frac{1}{\sqrt{2(n^3+1)}}\begin{pmatrix} -n^1+in^2 \\ n^3+1 \end{pmatrix}, \\
&\tilde{w}_+(\mathbf{n}) \equiv w_+(-\mathbf{n}),\ \tilde{w}_-(\mathbf{n}) \equiv w_-(-\mathbf{n})
\end{aligned}
\tag{A.14}
$$

with $\mathbf{n} = \mathbf{p}/p,\ p = |\mathbf{p}|$. One can check that $\sigma \mathbf{n} w_\pm = \pm w_\pm$.

In the spinor representation, using Eq.(A.6), we find

$$
\begin{aligned}
u(p,+1) &= \left(\frac{p^0+m}{4p^0}\right)^{1/2}\begin{bmatrix} \frac{p^0+m+p}{p^0+m} w_+ \\ \frac{p^0+m-p}{p^0+m} w_+ \end{bmatrix}, \\
u(p,-1) &= \left(\frac{p^0+m}{4p^0}\right)^{1/2}\begin{bmatrix} \frac{p^0+m-p}{p^0+m} w_- \\ \frac{p^0+m+p}{p^0+m} w_- \end{bmatrix}, \\
v(p,+1) &= \left(\frac{p^0+m}{4p^0}\right)^{1/2}\begin{bmatrix} \frac{p-p^0-m}{p^0+m} \tilde{w}_+ \\ \frac{p^0+m+p}{p^0+m} \tilde{w}_+ \end{bmatrix}, \\
v(p,-1) &= \left(\frac{p^0+m}{4p^0}\right)^{1/2}\begin{bmatrix} \frac{p^0+m+p}{p^0+m} \tilde{w}_- \\ \frac{p-p^0-m}{p^0+m} \tilde{w}_- \end{bmatrix}.
\end{aligned}
\tag{A.15}
$$

For a massless fermion ($m = 0$), these expressions essentially simplify:

$$
\begin{aligned}
u(p,+1) &= \begin{pmatrix} w_+ \\ 0 \end{pmatrix} , \\
u(p,-1) &= \begin{pmatrix} 0 \\ w_- \end{pmatrix} , \\
v(p,+1) &= \begin{pmatrix} 0 \\ \tilde{w}_+ \end{pmatrix} , \\
v(p,-1) &= \begin{pmatrix} \tilde{w}_- \\ 0 \end{pmatrix} .
\end{aligned} \tag{A.16}
$$

Therefore at $m = 0$, the helicity eigenstates are assigned to the irreducible representations $\left(\frac{1}{2}, 0\right)$ or $\left(0, \frac{1}{2}\right)$ of the Lorentz group: right-handed fermions with $s = +1$ are described by a spinor $\xi^{\alpha} \epsilon \left(\frac{1}{2}, 0\right)$, and left-handed ones with $s = -1$ are described by a dotted spinor $\eta_{\dot{\alpha}} \epsilon \left(0, \frac{1}{2}\right)$.

The γ_5 matrix $\gamma_5 = -i\gamma^0\gamma^1\gamma^2\gamma^3$ is

$$
\gamma_5 = \begin{pmatrix} -\sigma_0 & 0 \\ 0 & \sigma_0 \end{pmatrix} \text{ and } \gamma_5 = \begin{pmatrix} 0 & -\sigma_0 \\ -\sigma_0 & 0 \end{pmatrix} \tag{A.17}
$$

in the spinor and standard representations, respectively. Therefore the matrices

$$
P_{L(R)} = \frac{1 + (-)\gamma_5}{2} \tag{A.18}
$$

are projection operators for dotted and undotted spinors, respectively; at $m = 0, P_L$ is the projection operator for states with $s = -1$ (left-handed fermions), and P_R is the projection operator for states with $s = +1$ (right-handed fermions).

One can check that the spinors $u(p, s), v(p, s)$ satisfy the following useful relations [2]:

$$
\bar{v}(p, s)u(p, s') = 0 = v^{\dagger}(p, s)u(-p, s') , \tag{A.19}
$$

where $u(-p,s) \equiv u(p_0, -\mathbf{p}, s)$, and

$$\sum_{s=\pm 1} \left[u_\alpha(p,s)\bar{u}_\beta(p,s) - v_\alpha(p,s)\bar{v}_\beta(p,s)\right] = \frac{m}{p^0}\delta_{\alpha\beta} ,$$

$$\sum_{s=\pm 1} u_\alpha(p,s)\bar{u}_\beta(p,s) = \left(\frac{p_\mu\gamma^\mu + m}{2p^0}\right)_{\alpha\beta} , \tag{A.20}$$

$$\sum_{s=\pm 1} v_\alpha(p,s)\bar{v}_\beta(p,s) \left(\frac{p_\mu\gamma^\mu - m}{2p^0}\right)_{\alpha\beta} .$$

The field $\psi(x)$ satisfies the canonical equal time anticommutation relations

$$\begin{aligned} \{\psi(x), \bar{\psi}(y)\} \,|_{x_0=y_0} &= \gamma_0 \delta^3(\mathbf{x}-\mathbf{y}) \\ \{\psi(x), \psi(y)\} \,|_{x_0=y_0} &= \{\bar{\psi}(x), \bar{\psi}(y)\} \,|_{x_0=y_0} = 0 \ . \end{aligned} \tag{A.21}$$

With the normalization (A.12) for the spinors u and v, these relations imply the following anticommutation relations for b and d:

$$\left\{b(p,s), b^\dagger(p',s')\right\} = \left\{d(p,s), d^\dagger(p',s')\right\} = \delta_{ss'}\delta^3(\mathbf{p}-\mathbf{p}') , \tag{A.22}$$

with all other anticommutators being zero.

Let us consider the operations of charge conjugation, and space and time inversions.

The charge conjugation transformation is

$$\psi \longrightarrow \psi_c = C\bar{\psi}^\top , \tag{A.23}$$

where $\top$ means the transposition operation, and the matrix C satisfies [1, 2]

$$CC^\dagger = 1, C^\top = -C, C\gamma_\mu^\top = -\gamma_\mu C . \tag{A.24}$$

One can see the physical meaning of this transformation by considering Dirac's equation in an external electromagnetic field A_μ:

$$i\gamma^\mu\left(\partial_\mu - eA_\mu\right)\psi - m\psi = 0 . \tag{A.25}$$

Then taking the complex conjugate of this equation, we find from Eqs. (A.23) and (A.24) that

$$i\left(\gamma^\mu\partial_\mu + eA_\mu\right)\psi_c - m\psi_c = 0 . \tag{A.26}$$

This equation describes antifermions with the charge $-e$.

In the both spinor and standard representations, the conditions (A.24) are satisfied by the matrix $C = \eta_c \gamma^2 \gamma^0$, the constant η_c being arbitrary. Assuming that the charge conjugation operation $\hat{C}$ satisfies the condition $\hat{C}^2 = 1$, we find that $|\eta_c|^2 = 1$. A convenient choice for η_c is $\eta_c = 1$:

$$C = \gamma^2 \gamma^0 \, . \tag{A.27}$$

Eq.(A.23) implies that in any representation of Dirac's matrices, $\gamma'^\mu = U\gamma^\mu U^{-1}$ with U being unitary, the matrix C' is

$$C' = UCU^\top \, . \tag{A.28}$$

In the spinor representation, we find

$$\psi_c = \gamma^2 \psi^* \, , \tag{A.29}$$

and therefore:

$$\begin{aligned} \psi^\alpha_{cL} &\equiv \xi^\alpha_c = -i\epsilon^{\alpha\beta}\left(\eta_{\dot\beta}\right)^* \equiv -i\epsilon^{\alpha\beta}\psi^*_{R\beta} \\ \psi^\alpha_{cR} &\equiv \eta_{c\dot\alpha} = -i\epsilon_{\dot\alpha\dot\beta}\left(\xi^\beta\right)^* \equiv -i\epsilon_{\alpha\beta}\psi^*_{L\beta} \, . \end{aligned} \tag{A.30}$$

Thus charge conjugation transforms left-handed fields into right-handed ones, and vice versa.

Sometimes, it is very convenient to use the Majorana representations for γ-matrices. It is connected with the standard representation by the unitary matrix

$$U_M = \frac{i}{\sqrt{2}}\left(\gamma^0\gamma^2 + \gamma^0\right) \, . \tag{A.31}$$

The matrices $\gamma_M = U_M \gamma U_M^{-1}$ are

$$\begin{aligned} \gamma^0_M &= \begin{pmatrix} 0 & \sigma_2 \\ \sigma_2 & 0 \end{pmatrix}, \quad \gamma^1_M = \begin{pmatrix} i\sigma_3 & 0 \\ 0 & i\sigma_3 \end{pmatrix}, \\ \gamma^2_M &= \begin{pmatrix} 0 & -\sigma_2 \\ \sigma_2 & 0 \end{pmatrix}, \quad \gamma^3_M = \begin{pmatrix} -i\sigma_1 & 0 \\ 0 & -i\sigma_1 \end{pmatrix}. \end{aligned} \tag{A.32}$$

A remarkable property of this representation is that all γ^μ_M are imaginary. It implies that as a basis for the general solution of Dirac's equation in this representation, one can choose a set of real functions. Using Eqs.(A.28) and (A.29) we find that a charge conjugation spinor ψ_c is $\psi_c = \psi^*$ in this representation $(C_M = \gamma_M^0)$. Real fermion fields

$$\begin{aligned} \psi_1 &= (\psi + \psi_c)/\sqrt{2} = (\psi + \psi^*)/\sqrt{2}\,, \\ \psi_2 &= i(\psi_c - \psi)/\sqrt{2} = i(\psi^* - \psi)/\sqrt{2}\,. \end{aligned} \tag{A.33}$$

are called the Majorana fields.

The space inversion transformation is chosen as

$$\psi(x) \longrightarrow \psi_p(x) = i\gamma^0\psi(i_s x) \tag{A.34}$$

with $i_s x = (x^0, -\mathbf{x})$. One can easily check that $\psi_p(x)$ satisfies the Dirac equation. The choice of the factor i on the right hand of Eq.(A.34) leads to the same inversion transformation for charge conjugation spinors:

$$\psi_c(x) \longrightarrow i\gamma^0\psi_c(i_s x)\,. \tag{A.35}$$

In the spinor representation, the transformation (A.34) is

$$\begin{aligned} \psi_{pL}^{\alpha}(x) &\equiv \xi_p^{\alpha}(x) = i\eta_{\dot\alpha}(i_s x) \equiv i\psi_R^{\alpha}(i_s x)\,, \\ \psi_{pR}^{\alpha}(x) &\equiv \eta_{p\dot\alpha}(x) = i\xi^{\alpha}(i_s x) \equiv i\psi_L^{\alpha}(i_s x)\,. \end{aligned} \tag{A.36}$$

Thus space inversion transforms left-handed (right-handed) fields into right-handed (left-handed) ones.

Eqs.(A.29) and (A.34) imply that CP transformation is

$$\psi_{cp}(x) = -iC\psi^*(i_s x)\,. \tag{A.37}$$

In the spinor representation, we find from this equation that

$$\begin{aligned} \psi_{cpL}^{\alpha}(x) &\equiv \xi_{cp}^{\alpha}(x) = \epsilon_{\alpha\beta}\xi^{\beta *}(i_s x) = \epsilon_{\alpha\beta}\psi^{\beta *}{}_L(i_s x)\,, \\ \psi_{cpR}^{\alpha}(x) &\equiv \eta_{(cp)\dot\alpha}(x) = \epsilon^{\dot\alpha\dot\beta}\eta_{\dot\beta}^*(i_s x) = -\epsilon_{\alpha\beta}\psi^{\beta *}{}_R(i_s x)\,. \end{aligned} \tag{A.38}$$

We see that, unlike C and P transformations, the CP operation can be defined on two-component (Weyl) spinors.

In second quantized theory, C and P transformations can be defined as the action of unitary operators in Fock space:

$$\begin{aligned} \psi_c(x) &= U_c^{-1}\psi(x)U_c \, , \\ \psi_p(x) &= U_p^{-1}\psi(x)U_p \, . \end{aligned} \tag{A.39}$$

However, as it was shown by Wigner [3], the time inversion transformation can be defined only by antiunitary operator in Fock space (a comprehensive discussion of this point is given in Ref.[1]). It means that the operator U_t satisfies the following condition:

$$\begin{aligned} U_t\alpha|\psi\rangle &= \alpha^* U_t|\psi\rangle \, , \\ U_t\left(|\psi_1\rangle + |\psi_2\rangle\right) &= U_t|\psi_1\rangle + U_t|\psi_2\rangle \, , \\ (\psi_1, \, \psi_2) &= (U_t\psi_2, \, U_t\psi_1) \, . \end{aligned} \tag{A.40}$$

The time inversion transformation is defined as [1]

$$\psi_t(x) = U_t^{-1}\psi_t(x)U_t = \eta_t \left(C^\dagger\gamma_5\right)\psi\left(i_t x\right) \tag{A.41}$$

with $|\eta_t|^2 = 1$ and $(i_t x) = \left(-x^0, \mathbf{x}\right)$. The antiunitary character of U_t manifests in that $\psi_t(x)$ satisfies the equation

$$\left(-i\gamma^{\mu *}\partial_\mu - m\right)\psi_t(x) = 0 \tag{A.42}$$

with $\gamma^{\mu *}$ being complex conjugate of γ^μ.

Like CP, the time inversion transformation is well defined on two-component Weyl spinors.

We also remind that any local relativistic theory is invariant under PCT transformation (the PCT theorem) [4].

Appendix B

Some Useful Formulas and Relations

B.1. The Groups $U(N)$

$U(N)$ is a group of $N \times N$ unitary matrises. The unimodular matrices satisfying the condition $\det U = 1$ form its subgroup $SU(N)$.

The fundamental representation of the $U(N)$ group is given by the matrices

$$U = \exp\left(i\omega_\alpha \frac{1}{2}\lambda^\alpha\right), \quad \alpha = 0, 1, \cdots, N^2 - 1\,, \tag{B.1}$$

where $\frac{\lambda^\alpha}{2}$ make up a basis of the $U(N)$ Lie algebra ($tr\lambda^\alpha\lambda^\beta = 2\delta_{\alpha\beta}$). At $N = 2$, the Pauli matrices τ^0, $\boldsymbol{\tau}$ can be chosen as λ^α:

$$\begin{aligned} \tau^0 &= \begin{pmatrix} 1 & 0 \\ 0 & 1 \end{pmatrix}, \quad \tau^1 = \begin{pmatrix} 0 & 1 \\ 1 & 0 \end{pmatrix}, \\ \tau^2 &= \begin{pmatrix} 0 & -i \\ i & 0 \end{pmatrix}, \quad \tau^3 = \begin{pmatrix} 1 & 0 \\ 0 & -1 \end{pmatrix}. \end{aligned} \tag{B.2}$$

At $N = 3$, a convenient choice for λ^α is the Gell-Mann matrices:

$$\begin{aligned} \lambda^0 &= \sqrt{\tfrac{2}{3}}\begin{pmatrix} 1 & 0 & 0 \\ 0 & 1 & 0 \\ 0 & 0 & 1 \end{pmatrix}, \quad \lambda^1 = \begin{pmatrix} 0 & 1 & 0 \\ 1 & 0 & 0 \\ 0 & 0 & 0 \end{pmatrix}, \quad \lambda^2 = \begin{pmatrix} 0 & -i & 0 \\ i & 0 & 0 \\ 0 & 0 & 0 \end{pmatrix}, \\ \lambda^3 &= \begin{pmatrix} 1 & 0 & 0 \\ 0 & -1 & 0 \\ 0 & 0 & 0 \end{pmatrix}, \quad \lambda^4 = \begin{pmatrix} 0 & 0 & 1 \\ 0 & 0 & 0 \\ 1 & 0 & 0 \end{pmatrix}, \quad \lambda^5 = \begin{pmatrix} 0 & 0 & -i \\ 0 & 0 & 0 \\ i & 0 & 0 \end{pmatrix}, \\ \lambda^6 &= \begin{pmatrix} 0 & 0 & 0 \\ 0 & 0 & 1 \\ 0 & 1 & 0 \end{pmatrix}, \quad \lambda^7 = \begin{pmatrix} 0 & 0 & 0 \\ 0 & 0 & -i \\ 0 & i & 0 \end{pmatrix}, \quad \lambda^8 = \frac{1}{\sqrt{3}}\begin{pmatrix} 1 & 0 & 0 \\ 0 & 1 & 0 \\ 0 & 0 & -2 \end{pmatrix}. \end{aligned} \tag{B.3}$$

B.2. Second-Order Casimir Invariant

An important characteristics of a representation R of a Lie group G is the second-order (quadratic) Casimir invariant $C_2(R)$:

$$\delta_{ij} C_2(R) = \sum_\alpha (T^\alpha T^\alpha)_{ij}\,, \tag{B.4}$$

where T^α are the Lie algebra matrices in this representation. Another important characteristics is the quantity $T(R)$, which is defined as

$$\delta_{\alpha\beta}T(R) = tr(T^\alpha T^\beta) \; . \tag{B.5}$$

Eqs.(B.4), (B.5) imply the following relation between $C_2(R)$ and $T(R)$:

$$T(R) = \frac{d(R)}{r} C_2(R) \; , \tag{B.6}$$

where $d(R)$ is the dimension of the representation R and r is the dimension (number of generators) of the group.

In the case of $SU(N)$ we have for the fundamental representation $\left(\text{with } T^\alpha = \frac{\lambda^\alpha}{2} \right)$

$$\begin{aligned} d(f) =& N, \; r = N^2 - 1 \; , \\ T(f) =& \frac{1}{2} \; , \\ C_2(f) =& \frac{N^2 - 1}{2N} \; . \end{aligned} \tag{B.7}$$

The value of C_2 for the adjoint representation of $SU(N)$ is

$$C_2(a) = N \; . \tag{B.8}$$

The values of $C_2(R)$ for representations of different groups can be found in the review [1].

B.3. Fierz Identities

Let us consider some orthogonal matrix basis B^A_{ij} in the space of $n \times n$ matrices M in which the scalar product of two matrices M_1 and M_2 is defined as $tr M_1 M_2$. Thus we have

$$tr(B^A B^C) = \Delta^A \delta_{AC} \; . \tag{B.9}$$

Then, the completeness of the matrices B^A implies that

$$\sum_A \left(\Delta^A \right)^{-1} B^A_{ml} B^A_{ik} = \delta_{mk} \delta_{li} \; . \tag{B.10}$$

This equation in turn leads to the following relation for any pair of matrices M and L:

$$M_{mk}L_{il} = \sum_A \left(\Delta^A\right)^{-1} \left(MB^A L\right)_{ml} B^A_{ik} \, . \tag{B.11}$$

This is the general Fierz identity. Let us apply it to 16 Dirac's matrices γ^A (A.3):

$$\begin{aligned} A =&1 : I; \\ A =&2,3,4,5 : \gamma^0, \gamma^1, \gamma^2, \gamma^3; \\ A =&6,7,8,9,10,11 : \frac{1}{i}\sigma^{10},\ \frac{1}{i}\sigma^{20},\ \frac{1}{i}\sigma^{30},\ \frac{1}{i}\sigma^{12},\ \frac{1}{i}\sigma^{23},\ \frac{1}{i}\sigma^{31}; \\ A =&12,13,14,15 : \gamma^5\gamma^0, \gamma^5\gamma^1, \gamma^5\gamma^2, \gamma^5\gamma^3; \\ A =&16 : \gamma_5, \end{aligned} \tag{B.12}$$

where $\sigma^{\mu\nu} = \frac{i}{2}[\gamma^\mu, \gamma^\nu]$.

One can easily check that in this case the parameters Δ^A are :

$$\begin{aligned} \Delta_A =&4 \text{ for } A = 1,2,6,7,8,13,14,15,16, \\ \Delta_A =&-4 \text{ for } A = 3,4,5,9,10,11,12. \end{aligned} \tag{B.13}$$

The Fierz identities are especially useful in transforming Lorentz invariant four-fermion interactions :

$$\left(\bar\psi_a(x)O_i\psi_b(x)\right)\left(\bar\psi_c(x)O^i\psi_d(x)\right) = \sum_k C_{ik}\left(\bar\psi_a(x)O_k\psi_d(x)\right)\left(\bar\psi_c(x)O^k\psi_b(x)\right) \, , \tag{B.14}$$

where $O_S = I, O_V = \gamma^\mu, O_T = \sigma^{\mu\nu}/i\sqrt{2}, O_A = \gamma_5\gamma^\mu, O_P = \gamma_5$. Taking into account Eqs.(B.11), (B.13), we find the values for coefficients C_{ik} given in the following Fierz table.

	S	V	T	A	P
S	$-\frac{1}{4}$	$-\frac{1}{4}$	$\frac{1}{4}$	$\frac{1}{4}$	$-\frac{1}{4}$
V	-1	$\frac{1}{2}$	0	$\frac{1}{2}$	1
T	$\frac{3}{2}$	0	$\frac{1}{2}$	0	$\frac{3}{2}$
A	1	$\frac{1}{2}$	0	$\frac{1}{2}$	-1
P	$-\frac{1}{4}$	$\frac{1}{4}$	$\frac{1}{4}$	$-\frac{1}{4}$	$-\frac{1}{4}$

(B.15)

(anticommutativity of spinor fields is taken into account here).

Let us consider now the Fierz identities for λ^α matrices of $U(N)$ groups.

Since $tr\lambda^\alpha\lambda^\beta = 2\delta_{\alpha\beta}$, the parameters Δ_A are equal to 2 in this case. Therefore the completeness relation takes the following form:

$$\frac{1}{N}\delta_{ab}\delta_{cd} + \frac{1}{2}\boldsymbol{\lambda}_{ab}\boldsymbol{\lambda}_{cd} = \delta_{ad}\delta_{cb} \ , \tag{B.16}$$

where $\boldsymbol{\lambda}\boldsymbol{\lambda} \equiv \sum_{\alpha=1}^{N^2-1} \lambda^\alpha\lambda^\alpha$.

This equaation and the equation

$$\frac{1}{N}\delta_{ad}\delta_{cb} + \frac{1}{2}\boldsymbol{\lambda}_{ad}\boldsymbol{\lambda}_{cb} = \delta_{ab}\delta_{cd} \tag{B.17}$$

lead to the following relation :

$$\boldsymbol{\lambda}_{ab}\boldsymbol{\lambda}_{cd} = \frac{2(N^2-1)}{N^2}\delta_{ad}\delta_{cb} - \frac{1}{N}\boldsymbol{\lambda}_{ad}\boldsymbol{\lambda}_{cb} \ . \tag{B.18}$$

B.4. Angle Integrations in 4-Dimensional Euclidean Space

We collect here some useful formulas for the angle integrations in 4-dimensional Euclidean space (see, for example, Ref.[2]).

Let k_μ and p_μ be two momenta, and $pk = |p||k|\cos\theta; |p| = \left(p_0^2 + \mathbf{p}^2\right)^{1/2}$. Then

$$\begin{aligned} \int d^4kF\left((k-p)^2\right) =& \pi^2 \int dk^2 d\Omega_k k^2 F(k^2 - 2|k||p|\cos\theta + p^2) \ , \\ d\Omega_k =& \frac{2}{\pi}\sin^2\theta d\theta; \int d\Omega_k = 1 \ . \end{aligned} \tag{B.19}$$

For our purposes, the central formula is :

$$\int d\Omega_k \frac{1}{(k-p)^2 + m^2} = \frac{2}{k^2 + p^2 + m^2 + (k^2 + p^2 + m^2)^2 - 4k^2p^2} \ . \tag{B.20}$$

The general structure of the integrals we use takes the following form:

$$I_n\left((pk)^m\right) \equiv \int d\Omega_k \frac{(pk)^m}{(k-p)^{2n}} \ , \tag{B.21}$$

$$I_n\left(k_\mu(pk)^m\right) \equiv \int d\Omega_k \frac{k_\mu(pk)^m}{(k-p)^{2n}} \ , \tag{B.22}$$

$$I_n\left(k_\mu k_\nu(pk)^m\right) \equiv \int d\Omega_k \frac{k_\mu k_\nu(pk)^m}{(k-p)^{2n}} \ . \tag{B.23}$$

Making $\partial/\partial m^2$ act on both sides of Eq.(B.20) $n-1$ times and setting $m^2 = 0$, we obtain

$$\begin{aligned} I_0 =& 1, I_{-1} = k^2 + p^2, I_1 = \frac{1}{\max(k^2,p^2)} , \\ I_2 =& \frac{1}{|k^2 - p^2| \max(k^2,p^2)} , \text{etc.} , \end{aligned} \tag{B.24}$$

where

$$I_n \equiv \int d\Omega_k \frac{1}{(k-p)^{2n}} . \tag{B.25}$$

Using the identity $pk = \frac{1}{2}(k^2 + p^2 - (k-p)^2)$, we obtain the following relation :

$$I_n\left((pk)^m\right) = \frac{1}{2^m} \sum_{l=0}^{m} (-1)^l \begin{pmatrix} m \\ l \end{pmatrix} (k^2+p^2)^{m-l} I_{n-l} . \tag{B.26}$$

To calculate the integrals (B.22), (B.23), we use the relations following from covariance:

$$I_n\left(k_\mu (pk)^m\right) = \frac{p_\mu}{p^2} I_n\left((pk)^{m+1}\right) , \tag{B.27}$$

$$I_n\left(k_\mu k_\nu (pk)^m\right) = \delta_{\mu\nu} X_n^m + \frac{p_\mu p_\nu}{p^2} Y_n^m , \tag{B.28}$$

where

$$\begin{aligned} 4X_n^m + Y_n^m =& k^2 I_n\left((pk)^m\right) , \\ X_n^m + Y_n^m =& p^{-2} I_n\left((pk)^{m+2}\right) . \end{aligned} \tag{B.29}$$

Some useful explicit expressions are $(x \equiv p^2, y \equiv k^2)$:

$$I_1((pk)^1) = \frac{1}{2} \frac{\min(x,y)}{\max(x,y)} , \tag{B.30}$$

$$I_1((pk)^2) = \frac{1}{4} \frac{(x+y)\min(x,y)}{\max(x,y)} , \tag{B.31}$$

$$I_2((pk)^1) = \frac{\min(x,y)}{|x-y| \max(x,y)} , \tag{B.32}$$

$$I_2((pk)^2) = \frac{xy + 3\min(x^2,y^2)}{4|x-y| \max(x,y)} , \tag{B.33}$$

$$I_2((pk)^3) = \frac{(x+y)\min(x^2,y^2)}{2|x-y|\max(x,y)} \,, \tag{B.34}$$

$$X_1^0 = \frac{1}{12} \times \begin{cases} (3xy - y^2)/x^2 & x \geq y \\ (3y-x)/y & y \geq x \,, \end{cases} \tag{B.35}$$

$$Y_1^0 = \frac{1}{3} \times \begin{cases} y^2/x^2 & x \geq y \\ x/y & y \geq x \,, \end{cases} \tag{B.36}$$

$$X_2^0 = \frac{1}{4x} \times \begin{cases} y/x & x \geq y \\ x/y & y \geq x \,, \end{cases} \tag{B.37}$$

$$Y_2^0 = \begin{cases} y^2/x^2(x-y) & x \geq y \\ x/y(y-x) & y \geq x \,, \end{cases} \tag{B.38}$$

$$X_2^1 = \frac{1}{6} \times \begin{cases} y^2/x^2 & x \geq y \\ x/y & y \geq x \,, \end{cases} \tag{B.39}$$

$$Y_2^1 = \frac{1}{3} \times \begin{cases} \frac{y^2(x+2y)}{x^2(x-y)} & x \geq y \\ \frac{x(y+2x)}{y(y-x)} & y \geq x \,. \end{cases} \tag{B.40}$$

Appendix C

Renormalization Schemes and Path Integral Representation for Green's Functions

C.1. Dimensional Regularization

The dimensional regularization [1] is discussed in many text-books (see, for example, Ref.[2]). Here we indicate some useful, for our purposes, information.

The idea of the dimensional regularization is to replace the 4-dimensional integral d^4k by the n-dimensional one d^nk. The point is that the ultraviolet behavior of Feynman integrals improves with decreasing n, and they become finite at sufficiently small n. The expressions at $n \to 4$ can be obtained by using the analytical continuation in the complex n-plane.

The conventions for the dimensional regularization are :

1) The space-time index μ runs from 0 to $n-1$. The metric tensor $g^{\mu\nu} = (+, -, \cdots, -)$; $g^\mu_\mu = n$. The Dirac algebra is unchanged :

$$\{\gamma^\mu, \gamma^\nu\} = 2g^{\mu\nu} \,, \tag{C.1}$$

and

$$\begin{aligned} \gamma_\mu\gamma^\mu &= n \,, \\ \gamma_\mu\gamma_\nu\gamma^\mu &= (2-n)\gamma_\nu \,. \end{aligned} \tag{C.2}$$

The trace of the γ-matrices is normalized to

$$tr(\gamma^\mu\gamma^\nu) = 4g^{\mu\nu} \,. \tag{C.3}$$

2) The integral measure is $(2\pi)^{-n}\int d^nk$.

3) γ_5 matrix is defined as an object which satisfies $\{\gamma_5, \gamma^\mu\} = 0$.

Let us indicate explicit expressions for some n-dimensional integrals in Minkowski space ($\eta \equiv n/2$) [3]:

$$\int \frac{d^nk}{i(2\pi)^n} \ln(m^2 + 2k\cdot p - k^2) = -\frac{\Gamma(-\eta)}{(4\pi)^\eta}(m^2+p^2)^\eta \,, \tag{C.4}$$

$$\int \frac{d^n k}{i(2\pi)^n} \frac{1}{(m^2 + 2k \cdot p - k^2)^\alpha} = \frac{\Gamma(\alpha - \eta)}{(4\pi)^\eta \Gamma(\alpha)} \frac{1}{(m^2 + p^2)^{\alpha - \eta}} , \tag{C.5}$$

$$\int \frac{d^n k}{i(2\pi)^n} \frac{1}{(m^2 - p^2)p^2} = -\frac{2\Gamma(2 - \frac{n}{2})}{(4\pi)^{n/2}(n-2)} m^{n-4} , \tag{C.6}$$

$$\int \frac{d^n k}{i(2\pi)^n} \frac{k^\mu}{(m^2 + 2k \cdot p - k^2)^\alpha} = \frac{\Gamma(\alpha - \eta)}{(4\pi)^\eta \Gamma(\alpha)} \frac{p^\mu}{(m^2 + p^2)^{a - \eta}} , \tag{C.7}$$

$$\begin{aligned} &\int \frac{d^n k}{i(2\pi)^n} \frac{k^\mu k^\nu}{(m^2 + 2k \cdot p - k^2)^\alpha} \\ &= \frac{1}{(4\pi)^\eta \Gamma(\alpha)} \left[\Gamma(\alpha - \eta) \frac{p^\mu p^\nu}{(m^2 + p^2)^{\alpha - \eta}} - \Gamma(\alpha - 1 - \eta) \frac{(1/2) g^{\mu\nu}}{(m^2 + p^2)^{\alpha - 1 - \eta}} \right] , \end{aligned} \tag{C.8}$$

$$\begin{aligned} &\int \frac{d^n k}{i(2\pi)^n} \frac{k^\mu k^\nu k^\rho}{(m^2 + 2k \cdot p - k^2)^\alpha} \\ &= \frac{1}{(4\pi)^\eta \Gamma(\alpha)} \left[\Gamma(\alpha - \eta) \frac{p^\mu p^\nu p^\rho}{(m^2 + p^2)^{\alpha - \eta}} - \Gamma(\alpha - 1 - \eta) \frac{(1/2) g^{(\mu\nu} p^{\rho)}}{(m^2 + p^2)^{\alpha - 1 - \eta}} \right] , \end{aligned} \tag{C.9}$$

$$\begin{aligned} &\int \frac{d^n k}{i(2\pi)^n} \frac{k^\mu k^\nu k^\rho k^\sigma}{(m^2 + 2k \cdot p - k^2)^\alpha} \\ &= \frac{1}{(4\pi)^\eta \Gamma(\alpha)} \left[\Gamma(\alpha - \eta) \frac{p^\mu p^\nu p^\rho p^\sigma}{(m^2 + p^2)^{\alpha - \eta}} - \Gamma(\alpha - 1 - \eta) \frac{(1/2) g^{(\mu\nu} p^\rho p^{\sigma)}}{(m^2 + p^2)^{\alpha - 1 - \eta}} \right. \\ &\qquad \left. + \Gamma(\alpha - 2 - \eta) \frac{(1/4) g^{(\mu\nu} g^{\rho\sigma)}}{(m^2 + p^2)^{\alpha - 2 - \eta}} \right] , \end{aligned} \tag{C.10}$$

where the symbols $g^{(\mu\nu}p^{\rho)}, g^{(\mu\nu}p^\rho p^{\sigma)}, g^{(\mu\nu}g^{\rho\sigma)}$ imply the symmetrization of the indices: $g^{(\mu\nu}p^{\rho)} = g^{\mu\nu}p^\rho + g^{\nu\rho}p^\mu + g^{\rho\mu}p^\nu$ etc. At $\epsilon \equiv \frac{(4-n)}{2} \ll 1$, the expressions for the gamma-functions are :

$$\begin{aligned} \Gamma(\epsilon) &= \frac{1}{\epsilon} - \gamma + \frac{1}{2}\left(\gamma^2 + \frac{\pi^2}{6}\right)\epsilon + O(\epsilon^2) \\ \Gamma(\epsilon - 1) &= -\frac{1}{\epsilon} + (\gamma - 1) - \frac{1}{2}\left(\gamma^2 - 2\gamma + \frac{\pi^2}{6}\right)\epsilon + O(\epsilon^2) \\ \Gamma(\epsilon - n) &= \frac{(-1)^n}{n!}\left[\frac{1}{\epsilon} + \left(\sum_{k=1}^{n} \frac{1}{k}\right) - \gamma\right] + O(\epsilon) , \end{aligned} \tag{C.11}$$

where $\gamma \simeq 0.577$ is the Euler constant.

We note that there are ambiguities in the definition of some quantities in the dimensional regularization. For example, one may use the following alternative definitions of the effective potential V :

$$V = \lim_{n\to 4} (\sigma^2)^\epsilon V_n \ , \tag{C.12a}$$

$$V = \lim_{n\to 4} V_n \ , \tag{C.12b}$$

where σ is the renormalization group parameter, and V_n is calculated in n-dimensional space. We use the definition (C.12a) ; the combination $(\sigma^2)^\epsilon V_n$ has dimension 4 even at $n \neq 4$.

C.2. Minimal Subtraction and Modified Minimal Subtraction Renormalization Schemes

In the dimensional regularization, Feynman integrals include expressions of the kind shown in Eqs.(C.4) - (C.10). Eq.(C.11) implies that these expressions have poles at $\epsilon \to 0$. These pole singularities replace here logarithmically divergent terms appearing in other regularization schemes, as the Pauli-Villars scheme for example.

In the minimal subtraction (MS) scheme [4], only the pole term $1/\epsilon$ in expressions for Green's functions is subtracted. For example in the case of the expression (C.4), we have :

$$\begin{aligned} -\frac{\Gamma(-\eta)}{(4\pi)^\eta}(m^2+p^2)^\eta &\underset{\epsilon\ll 1}{\simeq} -\frac{1}{2(4\pi)^2}(m^2+p^2)^2 \\ &\cdot\left[\frac{1}{\epsilon}-\gamma+\ln 4\pi+\frac{3}{2}-\ln(m^2+p^2)\right] \overrightarrow{MS} \\ &-\frac{1}{2(4\pi)^2}(m^2+p^2)^2\left[-\gamma+\ln 4\pi+\frac{3}{2}-\ln(m^2+p^2)\right] \ . \end{aligned} \tag{C.13}$$

Of course, in renormalizable theories, such subtractions lead to multiplicative renormalizations of Green's functions.

As this example illustrates, in the dimensional regularization, the pole term is usually accompanied by γ and $\ln 4\pi$ in the combination

$$\frac{1}{\epsilon} - \gamma + \ln 4\pi \ . \tag{C.14}$$

Therefore in order to get a simpler expression for renormalizable Green's functions, it is convenient to subtract the whole combination (C.14) instead of subtracting only the pole term $\frac{1}{\epsilon}$. This is the modified minimal subtraction ($\overline{MS}$) scheme [5]. For example, in the $\overline{MS}$ scheme, instead of Eq.(C.13), we have

$$-\frac{\Gamma(-\eta)}{(4\pi)^\eta}(m^2+p^2)^\eta \underset{\overline{MS}}{\longrightarrow} -\frac{1}{2(4\pi)^2}(m^2+p^2)^2\left[\frac{3}{2}-\ln(m^2+p^2)\right] . \qquad (C.15)$$

In renormalizable theories the dimensionful parameter σ appears in the MS and $\overline{MS}$ schemes because dimensionless coupling constants becomes dimensionful at $n \neq 4$. For example:

$$g = \tilde{g}\sigma^\epsilon \underset{\epsilon\ll 1}{\simeq} \tilde{g}(1+\epsilon\ln\sigma) , \qquad (C.16)$$

where $\tilde{g}$ is a dimensionless gauge coupling constant. The product of the pole term $\frac{1}{\epsilon}$ and $\tilde{g}\epsilon\ln\sigma$ leads to a term depending logarithmically on σ.

C.3. Path Integrals

Path integrals are discussed in many text-books (for example, see Ref.[2, 6, 7, 8]). Here we collect some useful formulae needed for our purposes.

Let us consider a real scalar field $\varphi(x)$. The Feynman formula for the transition amplitude $\langle\varphi_f(\mathbf{x}),t_f|\varphi_i(\mathbf{x}),t_i\rangle$, where $\varphi_i(\mathbf{x})$ corresponds to a field configuration at the initial time t_i, and $\varphi_f(\mathbf{x})$ corresponds to a field configuration at the final time t_f, takes the following form :

$$\langle\varphi_f(\mathbf{x}),t_f|\varphi_i(\mathbf{x}),t_i\rangle = \int\limits_{\varphi(t_i,\mathbf{x})=\varphi_i(\mathbf{x})}^{\varphi(t_f,\mathbf{x})=\varphi_f(\mathbf{x})} \mathcal{D}(\varphi)\exp\left[i\int d^4x\mathcal{L}(x)\right] . \qquad (C.17)$$

Here $\mathcal{L}(x)$ is the lagrangian density of the theory, and the φ integration is functional ; the path integral is evaluated along all trajectories from $\varphi_i(\mathbf{x})$ to $\varphi_f(\mathbf{x})$ in the functional space.

The central mathematical problem with the path integral (C.17) is a proper definition of the measure $\mathcal{D}(\varphi)$. Usually it is considered as a continuum limit of the measure on a discrete space-time lattice with a size of a lattice link being equal a; the continuum limit corresponds to $a \to 0$.

However, fortunately, for many purposes, one may avoid the problem with the proper definition of $\mathcal{D}(\varphi)$. The point is that usually we are interested not in the calculation of the transition amplitude (C.17) but in the calculation of Green's functions. For this purpose, we introduce the generating functional

$$Z(J) = \lim_{\substack{t_f \to \infty \\ t_i \to -\infty}} N \int_{\varphi(t_i,\mathbf{x})=\varphi_i(\mathbf{x})}^{\varphi(t_f,\mathbf{x})=\varphi_f(\mathbf{x})} \mathcal{D}(\varphi) \exp\left[i \int d^4x \{\mathcal{L}(x) + J(x)\varphi(x)\}\right] , \qquad (C.18)$$

where the normalization factor N provides $Z(0) = 1$, i.e.

$$N^{-1} = \lim_{\substack{t_f \to \infty \\ t_i \to -\infty}} \int_{\varphi(t_i,\mathbf{x})=\varphi_i(\mathbf{x})}^{\varphi(t_f,\mathbf{x})=\varphi_f(\mathbf{x})} \mathcal{D}(\varphi) \exp\left[i \int d^4x \mathcal{L}(x)\right] . \qquad (C.19)$$

The condition $Z(0) = 1$ ensures that $Z(0)$ can be identified with the vacuum transition amplitude $\langle 0|0\rangle = 1$ at $J = 0$.

Green's function

$$\langle 0|T\varphi(x_1)\cdots\varphi(x_n)|0\rangle \qquad (C.20)$$

is

$$\langle 0|T\varphi(x_1)\cdots\varphi(x_n)|0\rangle = (-i)^n \frac{\delta^n Z}{\delta J(x_1)\cdots\delta J(x_n)} \qquad (C.21)$$

(see Sec.7.2).

In the practical calculations, the Gaussian integrals play the primary important role. The general form of the Gaussian integral is

$$\int \mathcal{D}(\varphi) \exp\left[-\int d^4x \{\frac{1}{2} \int d^4y \varphi(x)\hat{A}_{xy}\varphi(y) - J(x)\varphi(x)\}\right] , \qquad (C.22)$$

where $\hat{A}_{xy}$ is an integral or differential operator in the coordinate representation.

The path integral (C.22) is a functional analogue of the usual n-dimensional Gaussian integral

$$(2\pi)^{-n/2} \int dx_1 \cdots dx_n \exp\left[-\frac{1}{2}(\mathbf{x}, A\mathbf{x}) + (\mathbf{b}, \mathbf{x})\right] = \exp\left[\frac{1}{2}(\mathbf{b}, A^{-1}\mathbf{b})\right] (\det A)^{-\frac{1}{2}} , \qquad (C.23)$$

where the scalar product $(\mathbf{x}, A\mathbf{x}) = x_i A_{ij} x_j$, and the integral is well-defined if A is a real symmetric positive-definite matrix. By analytic continuation, this relation is also valid if A is a complex symmetric matrix with a positive-definite real part.

The analogue of Eq.(C.23) for the path integral (C.22) is

$$\begin{aligned} &N\int \mathcal{D}(\varphi)\exp\left[-\int d^4x\left\{\int d^4y\frac{1}{2}\varphi(x)\hat{A}_{xy}\varphi(y)-J(x)\varphi(x)\right\}\right] \\ &=N(\mathrm{Det}\hat{A})^{-\frac{1}{2}}\exp\left[\frac{1}{2}\int d^4xd^4yJ(x)\hat{A}^{-1}_{xy}J(y)\right] . \end{aligned} \tag{C.24}$$

If the normalization factor is chosen to ensure that this expression equals 1 at $J=0$, then $N=(\mathrm{Det}\hat{A})^{\frac{1}{2}}$ and the integral is equal to

$$\exp\left[\frac{1}{2}d^4xd^4yJ(x)\hat{A}^{-1}_{xy}J(y)\right] . \tag{C.25}$$

Since the integrand in Eq.(C.18) is an oscillatory function, there are difficulties in defining this integral in Minkowski space. The conventional way to avoid the difficulties is to transform the integral into Euclidean space : $d^4x \longrightarrow -id^4x$. Then

$$i\int d^4x\left\{\mathcal{L}(x)+J(x)\varphi(x)\right\} \longrightarrow -\int d^4x\mathcal{L}_E(x)+\int d^4xJ(x)\varphi(x) , \tag{C.26}$$

where $\mathcal{L}_E(x)$ is the lagrangian density in Euclidean space. For example, in free theory with

$$\mathcal{L}(x)=\frac{1}{2}\partial_\mu\varphi(x)\partial^\mu\varphi(x)-\frac{1}{2}\mu^2\varphi^2(x) , \tag{C.27}$$

$\mathcal{L}_E$ is

$$\mathcal{L}_E(x)=\frac{1}{2}(\partial_4\varphi)^2+\frac{1}{2}(\nabla\varphi)^2+\frac{1}{2}\mu^2\varphi^2 . \tag{C.28}$$

As the result, the path integral representation for $Z(J)$ becomes more manageable in Euclidean space :

$$Z(J)=\lim_{\substack{x_{4f}\to\infty \\ x_{4i}\to-\infty}} N\int_{\varphi(x_{4i},\mathbf{x})=\varphi_i(\mathbf{x})}^{\varphi(x_{4f},\mathbf{x})=\varphi_f(\mathbf{x})}\mathcal{D}(\varphi)\exp\left[-\int d^4x\{\mathcal{L}_E(x)-J(x)\varphi(x)\}\right] . \tag{C.29}$$

For the free theory with the lagrangian density $\mathcal{L}(x)$ (C.27), we get

$$Z(J)=\exp\left[\frac{1}{2}\int d^4xd^4yJ(x)D_E(x-y)J(y)\right] , \tag{C.30}$$

where

$$D_E(x)=\frac{1}{(2\pi)^4}\int d^4ke^{-ikx}(k^2+\mu^2)^{-1} \tag{C.31}$$

with $k^2=k_4^2+\mathbf{k}^2$.

The relation (C.30) is the basic one in deriving the covariant perturbative expansion in relativistic field theories [2, 7, 8].

We note that in the case of complex scalar field, in Eq.(C.24), $\mathcal{D}(\varphi) \longrightarrow \mathcal{D}(\varphi)\mathcal{D}(\varphi^*)$, $\frac{1}{2}\varphi(x)\hat{A}_{xy}\varphi(y) \longrightarrow \varphi^*(x)\hat{A}_{xy}\varphi(y)$ and $(\mathrm{Det}\hat{A})^{-\frac{1}{2}} \longrightarrow (\mathrm{Det}\hat{A})^{-1}$.

The conception of the path integral for fermion fields was elaborated by Berezin [9] (see also Ref.[8]). For our purposes, it is sufficient to know that

$$\int \mathcal{D}(\bar{\psi})\mathcal{D}(\psi)\exp\left[\int d^4x d^4y \bar{\psi}(x)\hat{A}_{xy}\psi(y)\right] = C(\mathrm{Det}\hat{A}) \,, \tag{C.32}$$

where C is a normalization factor. The appearance of $(\mathrm{Det}\hat{A})$ here instead of $(\mathrm{Det}\hat{A})^{-1}$ in the case of complex scalar field reflects the presence of the additional factor (-1) for fermion loops.

Appendix D

The Dirac Equation with the Supercritical Coulomb Potential

Here we analyse the Dirac equation with the potential

$$V(r) = -\frac{Ze^2}{4\pi}\left[\frac{\theta(r-r_0)}{r} + \frac{\theta(r_0-r)}{r_0}\right] . \qquad (D.1)$$

The Dirac equation with a spherically symmetric potential can be reduced to the form [1]

$$\begin{cases} \frac{d}{dr}(rg) + \frac{\chi}{r}(rg) - (E+m-V)rf = 0 , \\ \frac{d}{dr}(rf) - \frac{\chi}{r}(rf) + (E-m-V)rg = 0 , \end{cases} \qquad (D.2)$$

where $g(r)$ and $f(r)$ are the radial functions of the upper and the lower bispinor components in the standard representation of the Dirac matrices (see Eq.(A.5)), $\chi = \pm\left(\jmath + \frac{1}{2}\right)$ ($\jmath$ is the angular momentum) and E is the fermion energy. For the $nS_{\frac{1}{2}}$-states (to which, in particular, the lowest level belongs) we have $\chi = -1$. Performing the standard substitution

$$\begin{cases} rg(r) = r^{-\frac{1}{2}}\sqrt{m+E}(F_1 + F_2) , \\ rf(r) = r^{-\frac{1}{2}}\sqrt{m-E}(F_1 - F_2) , \end{cases} \qquad (D.3)$$

we obtain the following equation for the $nS_{\frac{1}{2}}$-states in the external region $r > r_0$:

$$\frac{d^2F}{d\varrho^2} + k_s F_s = 0, \; s = 1, 2 , \qquad (D.4)$$

where

$$\begin{cases} \varrho = 2\nu r, \; \nu = \sqrt{m^2 - E^2} , \\ k_{1,2} = -\frac{1}{4} + \frac{E\alpha\nu^{-1} \mp \frac{1}{2}}{\varrho} + \frac{\alpha^2 - \frac{3}{4}}{\varrho^2} . \end{cases} \qquad (D.5)$$

Equation (D.4) is the Whittaker equation [2]. Two independent solutions of this equation are two Whittaker functions $W_{\kappa_s, i\mu}(\varrho)$ and $M_{\kappa_s, i\mu}(\varrho)$, where

$$\kappa_{1,2} = \frac{E\alpha}{\nu} \mp \frac{1}{2}, \; \mu = \sqrt{\alpha^2 - 1} .$$

We will look for the solutions corresponding to the quasi-stationary states which define outgoing antifermion waves with the energy

$$E_{an} = \text{Re}E_{an} + i\text{Im}E_{an},\ \text{Re}E_{an} > m,\ \text{Im}E_{an} < 0,\ |\text{Im}E_{an}| \ll |\text{Re}E_{an}|.$$

Due to the conventional interpretation of the antifermion states in the Dirac equation, these states are described by the solutions with the energy

$$E = -E_{an},\ \text{Re}E < -m,\ \text{Im}E > 0\ . \tag{D.6}$$

For $\nu > 0$ and $r \to \infty$ the behavior of the Whittaker functions is as follows [2]:

$$W_{\kappa_s,i\mu}(\varrho) \underset{r\to\infty}{\simeq} \exp[-\nu r](2\nu r)^{\kappa_s}\ , \tag{D.7}$$

$$M_{\kappa_s,i\mu}(\varrho) \underset{r\to\infty}{\simeq} \frac{\Gamma(1+2i\mu)}{\Gamma\left(\frac{1}{2}+i\mu-\kappa_s\right)} \exp[\nu r](2\nu r)^{-\kappa_s}\ . \tag{D.8}$$

Therefore, stationary bound states, with $E^2 < m^2$, are described by the function $W_{\kappa_s,i\mu}(\varrho)$. The same function also describes the quasi-stationary states with the energy (D.6): for them $\text{Re}(E^2 - m^2) > 0$, $\text{Im}(E^2 - m^2) < 0$, and therefore, on the second unphysical sheet in the complex place of the variable $K = E^2 - m^2$, where the poles corresponding to these states lie, the function $W_{\kappa_s,i\mu}$ describes an outgoing wave.

In the region $r < r_0$, the solution of Eq.(D.2), regular at $r = 0$, is expressed through the elementary functions

$$F_1 = r^{\frac{1}{2}}\left[\frac{1}{\sqrt{m+E}}\sin kr - \frac{1}{\sqrt{m-E}}\frac{kr_0}{\alpha+(m+E)r_0}\left(\frac{\sin kr}{kr} - \cos kr\right)\right]\ , \tag{D.9}$$

$$F_2 = r^{\frac{1}{2}}\left[\frac{1}{\sqrt{m+E}}\sin kr + \frac{1}{\sqrt{m-E}}\frac{kr_0}{\alpha+(m+E)r_0}\left(\frac{\sin kr}{kr} - \cos kr\right)\right]\ , \tag{D.10}$$

where $k = \sqrt{(E+\alpha/r_0)^2 - m^2}$. The energy of the levels is determined by equating at $r = r_0$ the logarithmic derivatives of the solutions in the inner and the external regions. Since one of the functions F_s can be expressed through the other, it is sufficient to consider only one of them, for example F_2:

$$x\frac{W'_{\kappa_2,i\mu}(x)}{W_{\kappa_2,i\mu}(x)} = \frac{1}{2} + \frac{z\ \text{ctg}\ z + \gamma(z - z^{-1} + \text{ctg}\ z)}{1+\gamma(z^{-1} - \text{ctg}\ z)} \equiv f\ , \tag{D.11}$$

where

$$x = 2\nu r_0, \ z = r_0\sqrt{\left(E + \frac{\alpha}{r_0}\right)^2 - m^2}, \ \gamma = \frac{m+E}{\nu}.$$

We are interested in the case with $|E| \gg m$ and, in particular, in the case of massless electron ($m = 0$). The analytical answer can be obtained for the near-critical values α when $\alpha - 1 \ll 1$. Let us accept that in this case $|x| \ll 1$ (this supposition will be justified below). At $|x| \ll 1$, the Whittaker function is [2]:

$$W_{\kappa_2, i\mu} \underset{|x| \ll 1}{\simeq} \frac{\Gamma(-2i\mu)}{\Gamma\left(\frac{1}{2} - \kappa_2 - i\mu\right)} x^{\frac{1}{2}+i\mu} + \frac{\Gamma(2i\mu)}{\Gamma\left(\frac{1}{2} - \kappa_2 + i\mu\right)} x^{\frac{1}{2}-i\mu} \,. \tag{D.12}$$

Equations (D.11) and (D.12) lead to the following relation:

$$\ln 2\nu r_0 + \psi\left(\frac{E\alpha}{\nu}\right) - 2\psi(1) - \frac{2}{2f - 1} = \frac{\pi n}{\sqrt{\alpha^2 - 1}}, \ n = 1, 2, \cdots, \tag{D.13}$$

where

$$\psi(x) = \frac{d}{dx} \ln \Gamma(x).$$

Let us first consider the case of massless electron. At $m = 0$ and $\alpha - 1 \ll 1$, the parameters in Eq.(D.13) are

$$\gamma = -i, \ \nu = iE, \ \frac{2}{2f - 1} \simeq \text{tg}1 \cdot \frac{1 - i(1 - \text{ctg}1)}{1 - i} \tag{D.14}$$

(the sign of $\nu = \sqrt{-E^2}$ is determined from the condition that the function $W_{\kappa_2, i\mu}$ describes an outgoing wave corresponding to the state with $\text{Re}E < 0$). Therefore, Eq.(D.13) takes the following form:

$$\frac{1}{2}\ln(-E^2) + \ln 2r_0 + \psi(i) - \text{tg}1 + \frac{i}{i - 1} - 2\psi(1) = \frac{\pi n}{\sqrt{\alpha^2 - 1}}. \tag{D.15}$$

Since $\psi(1) = -\gamma_E \simeq -0.577$ (γ_E is the Euler constant) and

$$\text{Im}\psi(i) = \frac{\pi}{2}\text{ctgh}\pi + \frac{1}{2} \,,$$

we find from Eq.(D.15) that

$$E^{(n)} = a r_0^{-1}(\sin\beta - i\cos\beta)\exp\left[\frac{-\pi n}{\sqrt{\alpha^2 - 1}}\right], \ n = 1, 2, \cdots, \tag{D.16}$$

where

$$\beta = -\tfrac{\pi}{2}\mathrm{ctgh}\pi \simeq -\tfrac{\pi}{2}\cdot 1.004\,,$$

$$a = \tfrac{1}{2}\exp\left[\mathrm{tg}1 - 2\left(\gamma_E + \tfrac{1}{4}\right) - \mathrm{Re}\psi(i)\right] \simeq 0.4.$$

According to the conventional interpretation, the levels (D.16) determine the antifermion states with the energy

$$E_{an} = -E, \mathrm{Re}E_{an} > 0, \mathrm{Im}E_{an} < 0.$$

Note that $|\mathrm{Im}E_{an}| \ll |\mathrm{Re}E_{an}|$.

Let us consider now the solutions of Eq.(D.13) in the case of $0 < m \ll |E|$. We shall solve this equation by the iteration method, expanding its left-hand side in the double series in $E^{(n)}$ and m, where

$$E^{(n)} = E_0^{(n)} + \sum_{s=1} E_s^{(n)},\ E_0^{(n)} \equiv E^{(n)}|_{m=0},\ E_s^{(n)} \sim m^s\,. \tag{D.17}$$

Up to the second-order terms, we find

$$\ln 2\nu r_0 = \ln 2r_0 + \frac{1}{2}\ln\left(-E_0^{(n)^2}\right) + \frac{E_1^{(n)}}{E_0^{(n)}} - \frac{E_1^{(n)^2}}{2E_0^{(n)^2}} - \frac{m^2}{2E_0^{(n)^2}}\,, \tag{D.18}$$

$$\psi\left(\frac{-E^{(n)}}{\sqrt{m^2 - E^{(n)^2}}}\right) = \psi(i) + i\psi'(i)\frac{m^2}{2E_0^{(n)^2}} \simeq \psi(i) - (0.54 + 0.8i)\frac{im^2}{2E_0^{(n)^2}}\,, \tag{D.19}$$

$$\varphi \equiv \frac{-2}{2f-1} \simeq -\mathrm{tg}1 + \frac{i}{i-1} + \frac{m}{2E_0^{(n)}} - \frac{mE_1^{(n)}}{2E_0^{(n)^2}} + i\frac{m^2}{4E_0^{(n)^2}}\,. \tag{D.20}$$

Substituting these expressions in Eq.(D.13), we obtain the following relation:

$$E^{(n)} = E_0^{(n)} - \frac{m}{2} + \frac{m^2}{40|E_0^{(n)}|} - i\frac{m^2}{50|E_0^{(n)}|}\,. \tag{D.21}$$

Since $\mathrm{Im}E_0^{(n)} > 0$, the appearance of mass results in decreasing $\mathrm{Im}E^{(n)}$, i.e. in increasing the stability of the system.

Appendix E

The Calculation of the Determinant of the Dirac Operator

In this appendix, we outline the calculation of the logarithm of the modulus of the determinant of the Dirac operator (see Eq.(12.155))

$$i\mathfrak{D} = -i\gamma_\mu D_\mu - M\frac{1-\gamma_5}{2} - M^\dagger\frac{1+\gamma_5}{2} , \tag{E.1}$$

where the covariant derivative D_μ is

$$D_\mu = \partial_\mu - iV_\mu - i\gamma_5 A_\mu . \tag{E.2}$$

The logarithm of the modulus of the determinant of the operator $i\mathfrak{D}$ in Euclidean space is

$$\ln|\mathrm{Det}\, i\mathfrak{D}| = -\frac{1}{2}\int\limits_{1/\Lambda^2}^{\infty}\frac{dt}{t}\mathrm{Tr}\; e^{-t\mathfrak{D}^\dagger\mathfrak{D}} \tag{E.3}$$

(we use the proper-time regularization scheme). In calculating this quantity, we will use the method of the derivative expansion in which $\ln|\mathrm{Det}\, i\mathfrak{D}|$ is derived as a series in powers of the (covariant) derivatives of the fields [1, 2].

We will use the method considered in Ref.[2]. Let us introduce the resolvent $(A-\lambda)^{-1}$ of the operator $A \equiv \mathfrak{D}^\dagger\mathfrak{D}$. Then we get:

$$\mathrm{Tr}\; e^{-tA} = \mathrm{Tr}\int\limits_C \frac{id\lambda}{2\pi}e^{-t\lambda}\frac{1}{A-\lambda} , \tag{E.4}$$

where the contour C encloses all singularities of the resolvent in the complex λ plane. We define

$$G(x,x';\lambda) \equiv \langle x|\frac{1}{A-\lambda}|x'\rangle = \int\frac{d^4k}{(2\pi)^4}\exp\left(-ik_\mu\cdot(x-x')_\mu\right)\eta(x,k;\lambda) . \tag{E.5}$$

Then

$$\mathrm{Tr}\; e^{-tA} = \int d^4x \int\limits_C \frac{id\lambda}{2\pi}e^{-t\lambda}\int\frac{d^4k}{(2\pi)^4}tr\;\eta(x,k;\lambda) , \tag{E.6}$$

where tr denotes the trace over the Dirac and inner group indices.

Eq.(E.5) implies that $G(x, x'; \lambda)$ satisfies the equation:

$$\int \langle x|A - \lambda|x''\rangle G(x'', x'; \lambda) d^4x'' = \delta^4(x - x') \ . \tag{E.7}$$

For the operator (E.1), we find:

$$\begin{aligned} A = \mathfrak{D}^\dagger \mathfrak{D} = -D^2 - \frac{1+\gamma_5}{2}\left[i(\hat{\partial} M) + [\hat{V}, M] + \{\gamma_5 \hat{A}, M\}\right] \\ - \frac{1-\gamma_5}{2}\left[i(\hat{\partial} M^\dagger) + [\hat{V}, M^\dagger] + \{\gamma_5 \hat{A}, M^\dagger\}\right] + \frac{1+\gamma_5}{2} M M^\dagger + \frac{1-\gamma_5}{2} M^\dagger M ; \\ \hat{D}, \hat{V}, \hat{A} \equiv \gamma_\mu D_\mu, \gamma_\mu V_\mu, \gamma_\mu A_\mu \ . \end{aligned} \tag{E.8}$$

Therefore Eq.(E.7) acquires the form:

$$\begin{aligned} \Big[k^2 - 2ik_\mu D_\mu - D^2 - \frac{1+\gamma_5}{2}\left[i(\hat{\partial} M) + [\hat{V}, M] + \{\gamma_5 \hat{A}, M\}\right] \\ - \frac{1-\gamma_5}{2}\left[i(\hat{\partial} M^\dagger) + [\hat{V}, M^\dagger] + \{\gamma_5 \hat{A}, M^\dagger\}\right] \\ + \frac{1+\gamma_5}{2} M M^\dagger + \frac{1-\gamma_5}{2} M^\dagger M - \lambda\Big] \eta(x, k; \lambda) = 1 \ . \end{aligned} \tag{E.9}$$

Introducing the formal expansion parameter ϵ by

$$\eta(x, k; \lambda) = \sum_{m=0}^{\infty} \epsilon^m \eta_m(x, k; \lambda) \tag{E.10}$$

and assuming that the derivative ∂_μ and the gauge fields correspond to one power of ϵ, we obtain the following equation:

$$\begin{aligned} \Big[k^2 - 2i\epsilon k_\mu D_\mu - \epsilon^2 D^2 - \epsilon\frac{1+\gamma_5}{2}\left[i(\hat{\partial} M) + [\hat{V}, M] + \{\gamma_5 \hat{A}, M\}\right] \\ - \epsilon\frac{1-\gamma_5}{2}\left[i(\hat{\partial} M^\dagger) + [\hat{V}, M^\dagger] + \{\gamma_5 \hat{A}, M\}\right] + \frac{1+\gamma_5}{2} M M^\dagger + \frac{1-\gamma_5}{2} M^\dagger M - \lambda\Big] \\ \times \sum_{m=0}^{\infty} \epsilon^m \eta_m(x, k; \lambda) = 1 \end{aligned} \tag{E.11}$$

(at the end of the calculations $\epsilon = 1$). This equation leads to recurrence relations for $\sigma_m(x, k; \lambda)$.

Although the presence of gauge fields do not lead to principal difficulties, they essentially complicate the calculation. In order to simplify the illustration of the method, henceforth we will consider the case without gauge fields.

From Eq.(E.11) we obtain the following recurrence relations for $\eta_m(x,k;\lambda)$:

$$\begin{aligned}
&\left(k^2+\frac{1-\gamma_5}{2}M^\dagger M+\frac{1+\gamma_5}{2}MM^\dagger-\lambda\right)\eta_0=1\,,\\
&\left(k^2+\frac{1-\gamma_5}{2}M^\dagger M+\frac{1+\gamma_5}{2}MM^\dagger-\lambda\right)\eta_1\\
&\qquad-\left(2ik_\mu\partial_\mu+\frac{1+\gamma_5}{2}i\hat{\partial}M+\frac{1-\gamma_5}{2}i\hat{\partial}M^\dagger\right)\eta_0=0\,,\\
&\left(k^2+\frac{1-\gamma_5}{2}M^\dagger M+\frac{1+\gamma_5}{2}MM^\dagger-\lambda\right)\eta_2\\
&\qquad-\left(2ik_\mu\partial_\mu+\frac{1+\gamma_5}{2}i\hat{\partial}M+\frac{1-\gamma_5}{2}i\hat{\partial}M^\dagger\right)\eta_1-\partial^2\eta_0=0\,.
\end{aligned}\tag{E.12}$$

From (E.12) we find that

$$\eta_0=\left(k^2+\frac{1-\gamma_5}{2}M^\dagger M+\frac{1+\gamma_5}{2}MM^\dagger-\lambda\right)^{-1}\,.\tag{E.13}$$

For simplicity, we will consider the case with the $U_L(2)\times U_R(2)$ chiral symmetry. In this case, η_0 can be written as

$$\begin{aligned}
\eta_0=&\frac{1}{2}\left\{\frac{1}{\sqrt{\mathbf{q}^2}+(k^2+q^0-\lambda)}-\frac{1}{\sqrt{\mathbf{q}^2}-(k^2+q^0-\lambda)}\right\}\tau^0\\
&+\frac{1}{2\sqrt{\mathbf{q}^2}}\left\{\frac{1}{\sqrt{\mathbf{q}^2}+(k^2+q^0-\lambda)}+\frac{1}{\sqrt{\mathbf{q}^2}-(k^2+q^0-\lambda)}\right\}\\
&\times\left(\frac{1+\gamma_5}{2}e^{i\frac{\pi\tau}{2}}\mathbf{q}\boldsymbol{\tau}e^{-i\frac{\pi\tau}{2}}+\frac{1-\gamma_5}{2}e^{-i\frac{\pi\tau}{2}}\mathbf{q}\boldsymbol{\tau}\,e^{i\frac{\pi\tau}{2}}\right)\,,
\end{aligned}\tag{E.14}$$

where

$$\begin{aligned}
&M=\xi Q\xi\equiv\xi\sigma^\alpha\tau^\alpha\xi\,,\\
&q^0\equiv\sigma^\alpha\sigma^\alpha,\mathbf{q}\equiv2\sigma^0\boldsymbol{\sigma}
\end{aligned}\tag{E.15}$$

($\tau^\alpha\equiv(\tau^0,\boldsymbol{\tau})$ are Pauli matrices).

Using Eqs.(E.3), (E.4) and Eq.(12.154), we can calculate the effective potential V:

$$\begin{aligned}\int d^4x V = \frac{N_c}{2}\int d^4x \int_{1/\Lambda^2}^{\infty} \frac{dt}{t}\int_C \frac{id\lambda}{2\pi}e^{-t\lambda}\int \frac{d^4k}{(2\pi)^4} tr\eta_0 \\ + \int d^4x \left[\frac{1}{2G_s^{(0)}} tr\left(\left(M - \hat{m}^{(0)}\right)^\dagger \left(M - \hat{m}^{(0)}\right)\right)\right]\end{aligned} \tag{E.16}$$

(here we use x_μ for the coordinates in Euclidean space).

Carrying out the trivial integrations over $d\lambda$ and d^4k, and using the definition

$$\Gamma(-n, x) \equiv \int_x^\infty \frac{dt}{t^{n+1}}e^{-t} \tag{E.17}$$

for the incomplete gamma function $\Gamma(-n, x)$, we obtain

$$\begin{aligned}V =& \frac{1}{2G_s^{(0)}} tr\left(\left(M - \hat{m}^{(0)}\right)^\dagger \left(M - \hat{m}^{(0)}\right)\right) \\ &+ \frac{N_c}{8\pi^2}\left\{C_1^2\Gamma\left(-2, C_1/\Lambda^2\right) + C_2^2\Gamma\left(-2, C_2/\Lambda^2\right)\right\} ,\end{aligned} \tag{E.18}$$

where

$$\begin{aligned}C_1 &= \frac{1}{2}\sigma^\alpha\sigma^\alpha + \sigma^0(\sigma^2)^{1/2} \\ C_2 &= \frac{1}{2}\sigma^\alpha\sigma^\alpha - \sigma^0(\sigma^2)^{1/2} .\end{aligned} \tag{E.19}$$

The expression (E.18) coincides with that in Eq.(12.180) at $V_\mu = A_\mu = 0$.

It is easy to check that η_1 does not contribute to the effective action. This corresponds to the fact that there are no terms in the action with a single derivative.

The function η_2, determined from Eq.(E.12), contributes to the kinetic term. After straightforward calculations, retaining only the terms quadratic in the fields π^α and σ^α (which are divergent as $\Lambda \to \infty$), we find Eq.(12.192) at $V_\mu = A_\mu = 0$.

REFERENCES

Chapter 1

1. I. M. Gel'fand, R. A. Minlos and Z. Ya. Shapiro, *Representation of the Rotation and Lorentz Groups and their Applications* (Pergamon Press, Oxford, 1963).

2. R. Streater and A. Wightman, *PCT, Spin and Statistics, and All That* (W. A. Benjamin Inc., New York, 1964).

Chapter 2

1. L. D. Landau, *J. Phys. USSR* **5** (1941) 71.

2. E. M. Lifshitz and L. P. Pitaevskii, *Statistical Physics*, Part 2 (Pergamon Press, Oxford, 1980).

3. N. N. Bogolyubov, *J. Phys. USSR* **11** (1947) 23.

4. J. Bardeen, L. N. Cooper and J. R. Schrieffer, *Phys. Rev.* **108** (1957) 1175.

5. V. L. Ginzburg and L. D. Landau, *Sov. Phys. JETP* **20** (1950) 1064.

6. L. P. Gor'kov, *Sov. Phys. JETP* **36** (1959) 1918.

7. P. W. Anderson, *Phys. Rev.* **112** (1958) 1900.

8. Y. Nambu, *Phys. Rev.* **117** (1960) 648.

Chapter 3

1. Y. Nambu, *Phys. Rev. Lett.* **4** (1960) 380.

2. Y. Nambu and G. Jona-Lasinio, *Phys. Rev.* **112** (1961) 345.

3. J. Goldstone, *Nuovo Cimento* **19** (1961) 154.

4. V. G. Vaks and A. I. Larkin, *Sov. Phys. JETP* **13** (1961) 192.

5. S. Coleman, *Aspects of Symmetry* (Cambridge University Press, Cambridge, 1985).

6. E. Witten, *Nucl. Phys.* **B188** (1981) 513.

7. S. Coleman, *J. Math. Phys.* **7** (1966) 787.

8. J. A. Swieca, *Phys. Rev. Lett.* **17** (1966) 974; G. F. Dell'Antonio, *Nuovo Cimento* **47A** (1967) 1.

9. P. Federbush and K. Johnson, *Phys. Rev.* **120** (1960) 1926.

10. R. Streater and A. Wightman, *PCT, Spin and Statistics, and All That* (W. A. Benjamin, Inc., New York, 1964).

11. J. Goldstone, A. Salam and S. Weinberg, *Phys. Rev.* **127** (1962) 962.

12. N. N. Bogolyubov, *Physica* **26** (1960) 1.

13. F. Englert and R. Brout, *Phys. Rev. Lett.* **13** (1964) 321; P. Higgs, *Phys. Lett.* **12** (1964) 132; G. Guralnik, C. Hagen and T. Kibble, *Phys. Rev. Lett.* **13** (1964) 585.

14. J. Schwinger, *Phys. Rev.* **125** (1962) 397; P. W. Anderson, *Phys. Rev.* **130** (1963) 439.

15. T. Kibble, *Phys. Rev.* **155** (1967) 1554.

16. L. D. Faddeev and A. A. Slavnov, *Gauge Fields: Introduction to Quantum Theory* (Redwood City, Calif., Addison-Wisley Pub. Co., 1991).

17. G. Costa and M. Tonin, *Rivista del Nuovo Cimento* **5** (1975) 29.

18. T. Kugo, in *Dynamical Symmetry Breaking*, ed. K. Yamawaki (World Scientific Pub. Co., Singapore, 1992).

Chapter 4

1. M. Gell-Mann and M. Levy, *Nuovo Cimento* **16** (1960) 705.

2. S. L. Adler and R. F. Dashen, *Current Algebras* (W. A. Benjamin Inc., New York, 1968).

3. V. De Alfaro, S. Fubini, G. Furlan and C. Rossetti, *Currents in Hadron Physics* (North-Holland Publishing Co., Amsterdam, 1973).

4. H. Georgi, *Weak Interactions and Modern Particle Theory* (The Benjamin / Cummings Publishing Co., Menlo Park, California, 1984).

5. O. Hara, Y. Fujii and Y. Ohnuki, *Prog. Theor. Phys.* **19** (1958) 129; B. Touschek, *Nuovo Cimento* **8** (1958) 181.

6. K. Nishijima, *Nuovo Cimento* **11** (1959) 910.

7. F. Gürsey, *Nuovo Cimento* **16** (1960) 230.

8. M. L. Goldberger and S. B. Treiman, *Phys. Rev.* **110** (1958) 1178.

9. M. Levy, *Nuovo Cimento* **52** (1967) 23.

10. W. A. Bardeen and B. W. Lee, *Phys. Rev.* **177** (1969) 2389; J. Schechter and Y. Ueda, *Phys. Rev.* **D3** (1971) 168.

11. Y. Nambu, *Phys. Rev. Lett.* **4** (1960) 380.

Chapter 5

1. W. A. Bardeen and B. W. Lee, *Phys. Rev.* **177** (1969) 2389.

2. S. Coleman, J. Wess and B. Zumino, *Phys. Rev.* **177** (1969) 2239; C. Callan, S. Coleman, J. Wess and B. Zumino, *Phys. Rev.* **177** (1969) 2247.

3. S. Weinberg, *Phys. Rev.* **166** (1968) 1568.

4. D. V. Volkov, Preprint of the Institute for Theoretical Physics ITF-69-75, Kiev, 1969; *Sov. J. Part. Nucl.* **4** (1973) 1.

5. M. Bando, T. Kugo and K. Yamawaki, *Phys. Rep.* **164** (1988) 217.

6. G. Boulware and L. S. Brown, *Ann. Phys. (N.Y.)* **138** (1982) 392.

7. S. Weinberg, *Physica* **96A** (1979) 327.

8. S. Weinberg, *Phys. Rev. Lett.* **17** (1966) 616.

9. T. Muta, *Foundations of Quantum Chromodynamics* (World Scientific Publishing Co., Singapore, 1987).

10. H. Lehmann, *Phys. Lett.* **41B** (1972) 529.

11. A. Manohar and H. Georgi, *Nucl. Phys.* **B234** (1984) 189.

12. S. L. Adler and R. F. Dashen, *Current Algebras* (W. A. Benjamin Inc., New York, 1968); V. De Alfaro, S. Fubini, G. Furlan and C. Rossetti, *Currents in Hadron Physics* (North-Holland Publishing Co., Amsterdam, 1973).

Chapter 6

1. Y. Nambu and G. Jona-Lasinio, *Phys. Rev.* **122** (1961) 345.

2. V. G. Vaks and A. I. Larkin, *Sov. Phys. JETP* **13** (1961) 192.

Chapter 7

1. J. Goldstone, A. Salam and S. Weinberg, *Phys. Rev.* **127** (1962) 962.

2. G. Jona-Lasinio, *Nuovo Cimento* **34** (1964) 1790.

3. S. Coleman and E. Weinberg, *Phys. Rev.* **D7** (1973) 1888.

4. C. Itzykson and J. Zuber, *Quantum Field Theory* (McGraw-Hill Book Co., New York, 1980).

5. K. Symanzik, *Comm. Math. Phys.* **16** (1970) 48.

6. S. Coleman, *Aspects of Symmetry* (Cambridge University Press, Cambridge, 1985).

7. Y. Nambu, *Phys. Lett.* **26B** (1966) 626.

8. R. Jackiw, *Phys. Rev.* **D9** (1973) 1686.

9. A. Linde, *Particle Physics and Inflation Cosmology* (Harwood Academic Publishers, Chur, 1990).

10. A. Linde, *Sov. Phys. JETP Lett.* **23** (1976) 73.

11. S. Weinberg, *Phys. Rev. Lett.* **36** (1976) 294.

12. I. V. Krive and A. Linde, *Nucl. Phys.* **B117** (1976) 265.

13. M. Lindner, M. Sher and H. W. Zaglauer, *Phys. Lett.* **B228** (1988) 527.

14. L. D. Landau and I. M. Khalatnikov, *Sov. Phys. JETP* **2** (1956) 69.

15. G. 't Hooft and M. Veltman, *Nucl. Phys.* **B50** (1972) 318.

16. N. N. Bogolyubov and D. V. Shirkov, *Introduciton to the Theory of Quantized Fields* (3rd ed.) (John Wiley Inc., New York, 1980).

17. B. Kastening, *Phys. Lett.* **B283** (1992) 287.

18. M. Bando, T. Kugo, N. Maekawa and H. Nakano, *Phys. Lett.* **B301** (1993) 83.

19. C. Ford, D. R. T. Jones, P. W. Stephenson and M. B. Einhorn, *Nucl. Phys.* **B395** (1993) 17.

20. E. Weinberg and A. Wu, *Phys. Rev.* **D36** (1987) 2474.

21. L. D. Landau and E. M. Lifshitz, *Quantum Mechanics* (Pergamon Press, Oxford, 1977).

22. A. H. Guth and S. Y. Pi, *Phys. Rev.* **D32** (1985) 1899.

Chapter 8

1. T. D. Lee and C. N. Yang, *Phys. Rev.* **117** (1960) 22.

2. J. M. Luttinger and J. C. Ward, *Phys. Rev.* **118** (1960) 1417.

3. P. Martin and C. De Dominics, *J. Math. Phys.* **5** (1964) 5.

4. H. D. Dahmen and G. Jona-Lasinio, *Nuovo Cimento* **52A** (1967) 807.

5. A. N. Vasil'ev and A. K. Kazanskii, *Theor. Math. Phsy.* **12** (1972) 875.

6. J. M. Cornwall, R. Jackiw and E. Tomboulis, *Phys. Rev.* **D10** (1974) 2428.

7. J. Schwinger, *Proc. Nat. Acad. Sc.* **37** (1951) 452.

8. F. J. Dyson, *Phys. Rev.* **75** (1949) 1736.

9. E. E. Salpeter and H. A. Bethe, *Phys. Rev.* **84** (1951) 1232.

10. R. Fukuda, *Prog. Theor. Phys.* **78** (1987) 1487.

11. R. Fukuda, M. Komachiya and M. Ukita, *Phys. Rev.* **D38** (1988) 3747.

12. D. W. McKay and H. J. Munczek, *Phys. Rev.* **D40** (1989) 4151.

13. N. Nakanishi, *Suppl. Prog. Theor. Phys.* **43** (1969) 1.

14. K. Nishijima, *Fields and Particles* (W. A. Benjamin Inc., New York, 1969).

15. S. Mandelstam, *Proc. R. Soc. London* **A233** (1955) 248.

16. T. Banks and S. Raby, *Phys. Rev.* **D14** (1976) 2182.

17. R. W. Haymaker and J. Perez-Mercader, *Phys. Rev.* **D27** (1983) 1353.

18. R. W. Haymaker and T. Matsuki *Phys. Rev.* **D33** (1986) 1137.

19. M. Inoue, H. Katata, T. Muta and K. Shimizu, *Prog. Theor. Phys.* **79** (1988) 519.

20. R. W. Haymaker, *Rivista del Nuovo Cimento* **14** No 8 (1991) 1.

21. R. Casalbuoni, S. De Curtis, D. Dominici and R. Gatto, *Phys. Lett.* **B140** (1984) 357.

Chapter 9

1. S. L. Adler, *Phys. Rev.* **177** (1969) 2426.

2. J. S. Bell and R. Jackiw, *Nuovo Cimento* **60A** (1969) 47.

3. *Curren Algebra and Anomalies*, Eds. S. B. Treiman, R. Jackiw, B. Zumino and E. Witten (World Scientific Publishing Co., Singapore, 1985).

4. K. Fujikawa, *Phys. Rev.* **D21** (1980) 2848.

5. M. Gell-Mann, R. J. Oakes and B. Renner, *Phys. Rev.* **175** (1968) 2195.

6. T. Muta, *Foundations of Quantum Chromodynamics* (World Scientific Publishing Co., Singapore, 1987).

7. R. Jackiw and K. Johnson, *Phys. Rev.* **D8** (1973) 2386.

8. K.-I. Aoki, M. Bando, T. Kugo, M. G. Mitchard and H. Nakatani, *Prog. Theor. Phys.* **84** (1990) 683.

9. H. Bateman and A. Erdélyi, *Higher Transcendental Functions*, vol.2 (McGraw-Hill Book Co., Inc., New York, 1953).

Chapter 10

1. L. D. Landau, A. A. Abrikosov and I. M. Khalatnikov, *Nuovo Cimento Suppl.* **3** (1956) 80.

2. K. Johnson, M. Baker and R. Willey, *Phys. Rev. Lett.* **11** (1963) 518; *Phys. Rev.* **136** (1967) B1111.

3. M. Baker, K. Johnson and B. W. Lee, *Phys. Rev.* **163** (1967) 1699.

4. S. L. Adler, *Phys. Rev.* **D5** (1972) 3021.

5. S. L. Adler and W. A. Bardeen, *Phys. Rev.* **D4** (1971) 3045.

6. H. Pagels, *Phys. Rev.* **D7** (1973) 3689.

7. P. D. Mannheim, *Phys. Rev.* **D12** (1975) 1772.

8. M. Gell-Mann and F. E. Low, *Phys. Rev.* **95** (1954) 1300.

9. P. I. Fomin and V. I. Truten, *Theor. Math. Phys.* **5** (1970) 1101.

10. T. Maskawa and H. Nakajima, *Prog. Theor. Phys.* **52** (1974) 1326.

11. P. I. Fomin and V. A. Miransky, *Phys. Lett.* **B64** (1976) 166.

12. V. P. Gusynin and V. A. Miransky, *Nucl. Phys.* **B109** (1976) 526; P. I. Fomin, V. P. Gusynin, V. A. Miransky and Yu. A. Sitenko, *Nucl. Phys.* **B110** (1976) 445.

13. R. Fukuda and T. Kugo, *Nucl. Phys.* **B117** (1976) 250.

14. P. I. Fomin, V. P. Gusynin and V. A. Miransky *Phys. Lett.* **B78** (1978) 136.

15. V. A. Miransky and P. I. Fomin, *Phys. Lett.* **B105** (1981) 387.

16. V. A. Miransky, *Phys. Lett.* **B91** (1980) 421.

17. V. A. Miransky, *Nuovo Cimento* **90A** (1985) 149.

18. P. I. Fomin, V. P. Gusynin, V. A. Miransky and Yu. A. Sitenko, *Rivista del Nucovo Cimento* **6** No.5 (1983) 1; V. A. Miransky and P. I. Fomin, *Sov. J. Part. Nucl.* **16** (1985) 203.

19. W. A. Bardeen, C. N. Leung and S. T. Love, *Phys. Rev. Lett.* **56** (1986) 1230; C. N. Leung, S. T. Love and W. A. Bardeen, *Nucl. Phys.* **B273** (1986) 649.

20. J. Bartholomew, J. B. Kogut, S. H. Shenker, J. Sloan, M. Stone, H. W. Wyld, J. Shigemitsu and D. K. Sinclair, *Nucl. Phys.* **B230** (1984) 222.

21. E. Dagotto, A. Kocić and J. B. Kogut, *Phys. Rev. Lett.* **60** (1988) 772; *Nucl. Phys.* **B317** (1989) 253; *Nucl. Phys.* **B317** (1989) 271; A. Kocić, S. Hands, J. B. Kogut and E. Dagotto, *Nucl. Phys.* **B347** (1990) 551; S. Hands, J. B. Kogut, R. Renken, A. Kocić, D. K. Sinclair and K. C. Wang, *Phys. Lett.* **B261** (1991) 294.

22. A. M. Horowitz, *Nucl. Phys. B(Proc. Suppl.)* **9** (1989) 403; A. M. Horowitz, *Phys. Rev.* **D43** (1991) 43.

23. S. P. Booth, R. D. Kenway and B. J. Pendleton, *Phys. Lett.* **B228** (1989) 115.

24. M. Göckeler, R. Horsley, E. Laermann, P. Rakow, G. Schierholz, R. J. Sommer and U. J. Wiese, *Nucl. Phys.* **B334** (1990) 527; M. Göckeler, R. Horsley, P. Rakow, G. Schierholz and R. Sommer, *Nucl. Phys.* **B371** (1992) 713.

25. H. Batéman and A. Erdélyi, *Higher Transcendental Functions* vol.1 (McGraw-Hill Book Co., Inc., New York, 1953).

26. K. G. Wilson and J. B. Kogut, *Phys. Rep.* **12C** (1974) 77; C. Itzykson and J.-M. Drouffe, *Statistical Field Theory* (Cambridge University Press, Cambridge, 1989); J. Zinn-Justin, Quantum Field Theory and Critical Phenomena (Clarendon Press, Oxford, 1989).

27. L. D. Landau and E. M. Lifshitz, *Quantum Mechanics* (Pergamon Press, Oxford, 1977).

28. H. Bateman and A. Erdélyi, *Higher Transcendental Functions*, vol.2 (McGraw-Hill Book Co., New York, 1953).

29. M. Suwa and H. So, *Prog. Theor. Phys.* **83** (1990) 274.

30. A. M. Perelomov and V. S. Popov, *Theor. Math. Phys.* **4** (1970) 664.

31. V. S. Popov, *Sov. J. Nucl. Phys.* **12** (1970) 235; Ya. B. Zeldovich and V. S. Popov, *Sov. Phys. Uspekhi* **14** (1972) 673.

32. L. D. Landau and I. M. Khalatnikov, *Sov. Phys. JETP* **2** (1956) 69.

33. G. 't Hooft and M. Veltman, *Nucl. Phys.* **B50** (1972) 318.

34. T. Appelquist, K. Lane and U. Mahanta, *Phys. Rev. Lett.* **61** (1988) 1553.

35. D. Atkinson, P. W. Johnson and K. Stam, *Phys. Lett.* **201** (1988) 105.

36. P. Rembiesa, *Phys. Rev.* **D41** (1990) 2009.

37. B. Haeri, Jr., *Phys. Rev.* **D43** (1991) 2701.

38. K. -I. Kondo, *Int. J. Mod. Phys.* **A7** (1992) 7239.

39. D. C. Curtis and M. R. Pennington, *Phys. Rev.* **D42** (1990) 4165; *Phys. Rev.* **D44** (1991) 536.

40. D. C. Curtis and M. R. Pennington, Preprint of the University of Durham DTP-93/20, 1993.

41. D. Atkinson, V. P. Gusynin and P. Maris, *Phys. Lett.* **303** (1993) 157.

42. B. Holdom, *Phys. Rev. Lett.* **62** (1989) 997; U. Mahanta, *Phys. Lett.* **B225** (1989) 187.

43. T. Muta, *Foundations of Quantum Chromodynamics* (World Scientific Publishing Co., Singapore, 1987).

44. V. A. Miransky, *Phys. Lett.* **B165** (1985) 401; in *Proceedings of the VI Protvino International Seminar on High Energy Physics and Quantum Field Theory* (Protvino, 1983).

45. S. Goldstein, *Phys. Rev.* **91** (1953) 1516.

46. R. Jackiw, in *Current Algebra and Anomalies*, Eds. S. B. Treiman, R. Jackiw, B. Zumino and E. Witten (World Scientific Publishing Co., Singapore, 1985).

47. S. L. Adler, J. C. Collins and A. Duncan, *Phys. Rev.* **D15** (1977) 1712; J. C. Collins, A. Duncan and S. D. Joglekar, *Phys. Rev.* **D16** (1977) 438; N. K. Nielsen, *Nucl. Phys.* **B120** (1977) 212.

48. M. A. Shifman, A. I. Vainstein and V. I. Zakharov, *Nucl. Phys.* **B147** (1979) 385.

49. F. David, *Nucl. Phys.* **B234** (1984) 237.

50. V. P. Gusynin and V. A. Miransky, *Phys. Lett.* **B198** (1987) 79; *Phys. Lett.* **B198** (1987) 362; V. A. Miransky and V. P. Gusynin, *Prog. Theor. Phys.* **81** (1989) 426.

51. P. Griffin, *Phys. Rev.* **D47** (1993) 3530.

52. T. Appelquist and J. Carazzone, *Phys. Rev.* **D11** (1975) 2856.

53. K. Kondo, Y. Kikukawa and H. Mino, *Phys. Lett.* **B220** (1989) 270; K. Kondo and H. Nakatani, *Nucl. Phys.* **B351** (1990) 236.

54. J. Oliensis and P. W. Johnson, *Phys. Rev.* **D42** (1990) 656.

55. V. P. Gusynin, *Mod. Phys. Lett.* **A5** (1990) 133.

56. P. E. L. Rakow, *Nucl. Phys.* **B356** (1991) 27.

57. K. Kondo, H. Mino and H. Nakatani, *Mod. Phys. Lett.* **A7** (1992) 1509; D. Atkinson, H. J. de Groot and P. W. Johnson, *Int. J. Mod. Phys.* **A7** (1992) 7629.

58. D. Atkinson, *J. Math. Phys.* **28** (1987) 2494; in *Springer Tracts in Modern Physics*, **v.119** (1990) 159; D. Atkinson and P. W. Johnson, *Phys. Rev.* **D37** (1988) 2290.

59. B. G. Dragović, D. P. Mavlo and A. T. Filippov, *Fizika (Zagreb)* **10** (1978) 51.

60. D. Atkinson and D. W. E. Blatt, *Nucl. Phys.* **B151** (1979) 342.

61. A. I. Alekseev, B. A. Arbuzov and A. Ya. Rodionov, *Theor. Math. Phys.* **42** (1980) 191.

62. N. Nakanishi, in *Dynamical Symmetry Breaking*, ed. K. Yamawaki (World Scientific Publishing Co., Singapore, 1992).

63. K. Higashijima, KEK Preprint, 1993.

64. J. R. Finger and J. E. Mandula, *Nucl. Phys.* **B199** (1982) 168.

65. A. Amer, A. Le Yaouanc, L. Oliver, O. Péne and J. -C. Raynal, *Phys. Rev.* **D28** (1983) 1530; *Z. Pys.* **C17** (1983) 61.

66. S. L. Adler and A. C. Davis, *Nucl. Phys.* **B244** (1984) 496.

67. R. E. Cutkosky and K. C. Wang, *Phys. Rev.* **D37** (1988) 3024; **42** (1990) 1260; S. Capstick, R. E. Cutkosky, M. A. Joensen and K. C. Wang, *Phys. Rev.* **D42** (1990) 1270.

Chapter 11

1. K. G. Wilson, *Phys. Rev.* **B4** (1971) 3174.
2. K. G. Wilson and J. B. Kogut, *Phys. Rep.* **12** (1974) 75.
3. C. Itzykson and J. -M. Drouffe, *Statistical Field Theory* (Cambridge University Press, Cambridge, 1989); J. Zinn-Justin, *Quantum Field Theory and Critical Phenomena* (Clarendon Press, Oxford, 1989).
4. N. N. Bogolyubov and D. V. Shirkov, *Introduciton to the Theory of Quantized Fields* (3rd ed.) (John Wiley Inc., New York, 1980).
5. M. Gell-Mann and F. E. Low, *Phys. Rev.* **95** (1954) 1300.
6. L. D. Landau and E. M. Lifshitz, *Statistical Physics*, Part 1 (Pergamon Press, Oxford, 1980).
7. K. G. Wilson, *Phys. Rev.* **D7** (1973) 2911.
8. D. J. Gross and A. Neveu, *Phys. Rev.* **D10** (1974) 3235.
9. G. Parisi, *Nucl. Phys.* **B100** (1975) 368.
10. B. Rosenstein, B. J. Warr and S. H. Park, *Phys. Rev. Lett.* **62** (1989) 1433.
11. B. Rosenstein, B. J. Warr and S. H. Park, *Phys. Rep.* **205** (1991) 59.
12. T. Appelquist and J. Carazzone, *Phys. Rev.* **D11** (1975) 2856.
13. J. Polchinski, *Nucl. Phys.* **B231** (1984) 269.
14. E. Brézin and J. Zinn-Justin, *Phys. Rev.* **B14** (1976) 3110.
15. A. Kovner and B. Rosenstein, *Phys. Lett.* **B261** (1991) 97; G. Gat, A. Kovner and B. Rosenstein, *Nucl. Phys.* **B385** (1992) 76.
16. J. Zinn-Justin, *Nucl. Phys.* **B367** (1991) 105.
17. T. Appelquist, M. Bowick, D. Karabali and L. C. R. Wijewardhana, *Phys. Rev.* **D33** (1986) 3704.
18. I. Ya. Aref'eva, *Ann. Phys. (N.Y.)* **117** (1979) 393; I. Ya. Aref'eva, E. R. Nissimov and S. J. Pacheva, *Commun. Math. Phys.* **71** (1980) 213.

19. G. W. Semenoff and L. C. R. Wijewardhana, *Phys. Rev. Lett.* **63** (1989) 2633; *Phys. Rev.* **D45** (1992) 1342.

20. D. Amit, *Field Theory, the Renormalization Group and Critical Phenomena* (McGraw-Hill Book Co., New York, 1978).

21. S. Hands, A. Kocić and J. B. Kogut, *Ann. Phys. (N.Y.)* **224** (1993) 29.

22. L. P. Kadanoff, *Physics* **2** (1966) 263.

23. S. Dimopoulos, S. Raby and F. Wilczek, *Phys. Today* **44** No.10 (1991) 25.

24. Y. Kikukawa and K. Yamawaki, *Phys. Lett.* **B234** (1990) 497.

25. N. D. Mermin and H. Wagner, *Phys. Rev. Lett.* **17** (1966) 1133.

26. S. Coleman, *Commun. Math. Phys.* **31** (1973) 259.

27. B. A. Arbuzov, A. N. Tavkhelidze and R. N. Faustov, *Dokl. Akad. Nauk SSSR* **139** (1961) 345.

28. E. Witten, *Nucl. Phys.* **B145** (1978) 110.

29. V. L. Berezinski, *Sov. Phys. JETP* **32** (1970) 493; J. M. Kosterlitz and D. J. Thouless, *J. Phys.* **C6** (1973) 1181.

Chapter 12

1. Y. Nambu, *Phys. Rev. Lett.* **4** (1960) 380.

2. Y. Nambu and Jona-Lasinio, *Phys. Rev.* **122** (1961) 345.

3. J. Goldstone, *Nuovo Cimento* **19** (1961) 154.

4. S. L. Adler, *Phys. Rev.* **177** (1969) 2426.

5. J. S. Bell and R. Jackiw, *Nuovo Cimento* **60A** (1969) 47.

6. T. Muta, *Fundations of Quantum Chromodynamics* (World Scientific Publishing Co., Singapore, 1987).

7. D. Gross and F. Wilczek, *Phys. Rev. Lett.* **30** (1973) 1343; H. D. Politzer, *Phys. Rev. Lett.* **30** (1973) 1346.

8. W. J. Marciano and H. Pagels, *Phys. Rep.* **36C** (1978) 137.

9. V. A. Miransky, *Sov. J. Nucl. Phys.* **38** (1983) 280.

10. V. A. Miransky, *Phys. Lett.* **B165** (1985) 401.

11. K. Lane, *Phys. Rev.* **D10** (1974) 2605.

12. H. D. Politzer, *Nucl. Phys.* **B117** (1976) 397.

13. K. G. Wilson, *Phys. Rev.* **179** (1969) 1499.

14. V. A. Miransky, *Mod. Phys. Lett.* **A5** (1990) 1979.

15. K. Higashijima, *Phys. Rev.* **D18** (1978) 2128.

16. K. Higashijima, *Phys. Rev.* **D29** (1984) 1228.

17. V. A. Miransky and P. I. Fomin, *Phys. Lett.* **B105** (1981) 387; V. A. Miransky, V. P. Gusynin and Yu. A. Sitenko, *Phys. Lett.* **B100** (1981) 157.

18. V. A. Miransky and P. I. Fomin, *Sov. J. Nucl. Phys.* **35** (1982) 913.

19. P. I. Fomin, V. P. Gusynin, V. A. Miransky and Yu. A. Sitenko, *Rivista del Nuovo Cimento* **6** No.5 (1983) 1.

20. T. Goldman and R. W. Haymaker, *Phys. Lett.* **B100** (1981) 276.

21. T. Goldman and R. W. Haymaker, *Phys. Rev.* **D24** (1981) 724.

22. J. Finger and J. E. Mandula, *Nucl. Phys.* **B199** (1982) 168.

23. J. M. Cornwall, *Phys. Rev.* **D26** (1982) 1453.

24. R. Fukuda, *Phys. Lett.* **B73** (1978) 33.

25. V. P. Gusynin and V. A. Miransky, *Phys. Lett.* **B76** (1978) 585.

26. K. Higashijima, *Suppl. Prog. Theor. Phys.* **104** (1991) 1.

27. D. Atkinson and P. W. Johnson, *Phys. Rev.* **D37** (1988) 2296.

28. A. Barducci, R. Casalbuoni, S. De Curtis, D. Domninci and R. Gatto, *Phys. Rev.* **D38** (1988) 238.

29. C. G. Callan, R. F. Dashen and D. J. Gross, *Phys. Rev.* **D17** (1978) 2717.

30. D. G. Caldi, *Phys. Rev. Lett.* **39** (1977) 121; R. D. Carlitz and D. B. Creamer, *Ann. Phys. (N.Y.)* **118** (1979) 429.

31. E. V. Shuryak, *Nucl. Phys.* **B203** (1982) 93.

32. E. V. Shuryak, *The QCD Vacuum, Hadrons and the Superdense Matter* (World Scientific Publishing Co., Singapore, 1988).

33. D. I. Dyakonov and V. Yu. Petrov, *Nucl. Phys.* **B245** (1984) 259.

34. D. I. Dyakonov and V. Yu. Petrov, *Sov. Phys. JETP* **62** (1985) 204.

35. J. M. Cornwall, *Phys. Rev.* **D22** (1980) 1452.

36. A. I. Alekseev, B. A. Arbuzov and V. A. Baykov, *Sov. J. Nucl. Phys.* **34** (1981) 763; B. A. Arbuzov, *Sov. J. Part. Nucl.* **19** (1988) 1.

37. A. Amer, A. Le Yaouanc, L. Oliver, O. Péne and J. C. Raynal, *Phys. Rev. Lett.* **50** (1983) 87.

38. M. Baker, J. S. Ball and F. Zachariasen, *Phys. Rev.* **D38** (1988) 1926; *Phys. Rep.* **209** (1991) 73.

39. D. Ebert and M. K. Volkov, *Z. Phys.* **C16** (1983) 205; M. K. Volkov, *Ann. Phys. (N.Y.)* **157** (1984) 282.

40. A. Dhar, R. Shanker and S. R. Wadia, *Phys. Rev.* **D31** (1985) 199.

41. D. Ebert and H. Reinhardt, *Nucl. Phys.* **B271** (1986) 188.

42. J. Bijnens, C. Bruno and E. de Rafael, *Nucl. Phys.* **B396** (1993) 81.

43. L. D. Faddeev and A. A. Slavnov, *Gauge Fields: Introduction to Quantum Theory* (Redwood City, Calif., Addison-Wisley Pub. Co., 1991).

44. A. G. Williams, G. Krein and C. D. Roberts, *Ann. Phys. (N.Y.)* **210** (1991) 464.

45. P. Jain and H. J. Munczek, *Phys. Rev.* **D44** (1991) 1873.

46. P. Jain and H. J. Munczek, *Phys. Rev.* **D46** (1992) 438.

47. K. -I. Aoki, M. Bando, T. Kugo, M. G. Mitchard and H. Nakatani, *Prog. Theor. Phys.* **84** (1990) 683.

48. K. -I. Aoki, T. Kugo and M. G. Mitchard, *Phys. Lett.* **85** (1991) 467.

49. T. Kugo and M. G. Mitchard, *Phys. Lett.* **B282** (1992) 162; **B286** (1992) 355.

50. P. Jain and H. J. Munczek, *Phys. Lett.* **B282** (1992) 157.

51. H. Georgi, E. H. Simmons and A. G. Cohen, *Phys. Lett.* **B236** (1990) 183.

52. M. A. Shifman, A. I. Vainstein and V. I. Zakharov, *Nucl. Phys.* **B147** (1978) 385.

53. R. F. Dashen, *Phys. Rev.* **183** (1969) 1245.

54. J. Gasser and H. Leutwyler, *Phys. Rep.* **87** (1982) 79.

55. Particle Data Group, K. Hikasa et al., *Phys. Rev.* **D45** Part II (1992) S1.

56. G. 't Hooft, *Phys. Rev. Lett.* **37** (1976) 8.

57. E. Witten, *Nucl. Phys.* **B156** (1979) 269.

58. G. Veneziano, *Nucl. Phys.* **B159** (1979) 213.

59. A. De Rújula, H. Georgi and S. L. Glashow, *Phys. Rev.* **D12** (1975) 147.

60. H. Pagels and S. Stokar, *Phys. Rev.* **D20** (1979) 2947.

61. D. I. Dyakonov and M. I. Eides, *JETP Letters* **38** (1983) 433.

62. A. A. Andrianov and L. Bonora, *Nucl. Phys.* **B233** (1984) 232; A. A. Andrianov and Yu. V. Novozhilov, *Phys. Lett.* **B153** (1985) 422; A. A. Andrianov, V. A. Andrianov, V. Yu. Novozhilov and Yu. V. Novozhilov, *Lett. Math. Phys.* **11** (1986) 217.

63. R. I. Nepomechie, *Ann. Phys. (N.Y.)* **158** (1984) 67.

64. I. J. R. Aitchison and C. M. Fraser, *Phys. Lett.* **B146** (1984) 63; L. H. Chan, *Phys. Rev. Lett.* **54** (1985) 1222; V. P. Gusynin and V. A. Kushnir, *Sov. J. Nucl. Phys.* **51** (1990) 373.

65. J. Balog, *Phys. Lett.* **B149** (1984) 197.

66. D. W. McKay and H. J. Munczek, *Phys. Rev.* **D32** (1985) 266.

67. W. A. Bardeen, *Phys. Rev.* **184** (1969) 1848.

68. J. Wess and B. Zumino, *Phys. Lett.* **B37** (1971) 95.

69. E. Witten, *Nucl. Phys.* **B156** (1979) 269.

70. H. Bateman and A. Erdélyi, *Higher Transcendental Fuctions*, vol.1 (McGraw-Hill Book Co., New York, 1953).

71. V. Bernard, R. L. Jaffe and U. -G. Meissner, *Nucl. Phys.* **B308** (1988) 753.

72. V. Elias, Preprint of the University of Western Ontario, 1993.

73. C. Callan, R. Dashen and D. Gross, *Phys. Lett.* **63B** (1976) 334; R. Jackiw and C. Rebbi, *Phys. Rev. Lett.* **37** (1976) 172.

74. S. Coleman, *Aspects of Symmetry* (Cambridge University Press, Cambridge, 1985).

75. C. Rosenzweig, J. Schechter and C. G. Trahern, *Phys. Rev.* **D21** (1980) 3388.

76. P. Di Vecchia and G. Veneziano, *Nucl. Phys.* **B171** (1980) 253.

77. P. Nath and R. Arnowitt, *Nucl. Phys.* **B209** (1982) 234.

78. J. Gasser and H Leutwyler, *Ann Phys. (N.Y.)* **158** (1984) 142; *Nucl. Phys.* **B250** (1985) 465.

79. V. Baluni, *Phys. Rev.* **D19** (1979) 2227; R. Crewther, P. Di Vecchia, G. Veneziano and E. Witten, *Phys. Lett.* **88B** (1979) 123.

80. G. Ecker, J. Gasser, H. Leutwyler, A. Pich and E. de Rafael, *Phys. Lett.* **223B** (1989) 425.

81. E. Ruiz Arriola, *Phys. Lett.* **B253** (1991) 430.

82. A. Manohar and H. Georgi, *Nucl. Phys.* **B234** (1984) 189.

83. H. Georgi, *Phys. Lett.* **B298** (1993) 187.

84. H. Kleinert, *Phys. Lett.* **62B** (1976) 429; E. Schrauner, *Phys. Rev.* **D16** (1977) 1877.

85. T. Kugo, *Phys. Lett.* **76B** (1978) 625.

86. R. T. Cahill and C. D. Roberts, *Phys. Rev.* **D32** (1985) 2419; J. Praschifka, C. D. Roberts and R. T. Cahill, *Phys. Rev.* **D36** (1987) 209.

87. T. Morozumi and H. So, *Prog. Theor. Phys.* **77** (1987) 1434.

88. M. C. Birse, M. K. Banerjee, *Phys. Lett.* **136B** (1984) 284.

89. V. Elias and M. D. Scadron, *Phys. Rev. Lett.* **53** (1984) 1129; *Phys. Rev.* **D30** (1984) 647.

90. M. K. Volkov, M. Nagy, A. N. Ivanov and N. I. Troitskaja, *Czech. J. Phys.* **B39** (1988) 1256.

91. E. P. Shabalin, *Sov. J. Nucl. Phys.* **48** (1989) 172.

92. T. Morozumi, C. S. Lim and A. I. Sanda, *Phys. Rev. Lett.* **65** (1990) 404.

93. B. Holdom, J. Terning and K. Verbeek, *Phys. Lett.* **B245** (1990) 612.

94. S. Weinberg, *Phys. Rev. Lett.* **65** (1990) 1181.

95. S. Peris, *Phys. Lett.* **B268** (1991) 415; *Phys. Rev.* **D46** (1992) 1202.

96. S. Weinberg, *Phys. Rev. Lett.* **67** (1991) 3473; D. A. Dicus, D. Minic, U. van Kolck and R. Vega, *Phys. Lett.* **B284** (1992) 384.

97. S. Peris and E. de Rafael, *Phys. Lett.* **B309** (1993) 389.

98. J. J. Sakurai, *Currents and Mesons* (University of Chicago Press, Chicago, 1969).

99. M. Bando, T. Kugo and K. Yamawaki, *Phys. Rep.* **164** (1988) 217.

100. R. Haag, *Phys. Rev.* **112** (1958) 669; K. Nishijima, *Phys. Rev.* **111** (1958) 995; W. Zimmermann, *Nuovo Cimento* **10** (1958) 587.

101. V. A. Miransky, S. Peris and S. Raby, *Phys. Rev.* **D47** (1993) 2058.

102. D. Atkinson, H. J. de Groot and P. W. Johnson, *Phys. Rev.* **D43** (1991) 218.

Chapter 13

1. S. L. Glashow, *Rev. Mod. Phys.* **52** (1980) 539; A. Salam, *Rev. Mod. Phys.* **52** (1980) 525; S. Weinberg, *Rev. Mod. Phys.* **52** (1980) 515.

2. T. -P. Cheng and L. -F. Li, *Gauge Theory of Elementary Particle Physics* (Clarendon Press, Oxford, 1984).

3. H. Georgi, *Weak Interactions and Modern Particle Theory* (The Benjamin / Cummings Publishing Co., Menlo Park, California, 1984).

4. L. B. Okun, *Leptons and Quarks* (North-Holland Publishing Co., Amsterdam, 1982).

5. Particle Data Group, K. Hikasa et al., *Phys. Rev.* **D45** Part II (1992) S1.

6. N. Cabibbo, *Phys. Rev. Lett.* **10** (1963) 531.

7. M. Kobayashi and M. Maskawa, *Prog. Theor. Phys.* **49** (1973) 652.

8. S. L. Glashow, J. Illiopoulos and L. Maiani, *Phys. Rev.* **D2** (1970) 1285.

9. C. Bouchiat, J. Illiopoulos and Ph. Meyer, *Phys. Lett.* **38B** (1972) 519.

10. D. J. Gross and R. Jackiw, *Phys. Rev.* **D6** (1972) 477.

11. H. Georgi and S. L. Glashow, *Phys. Rev.* **D6** (1972) 429.

12. M. K. Gaillard and B. W. Lee, *Phys. Rev.* **D10** (1974) 897.

13. J. F. Gunion, H. E. Haber, G. Kane and S. Dawson, *The Higgs Hunter's Guide* (Addison-Wesley, Redwood City, California, 1989).

14. S. L. Glashow and S. Weinberg *Phys. Rev.* **D15** (1977) 1958.

15. L. F. Abbot, P. Sikivie and M. B. Wise, *Phys. Rev.* **D21** (1980) 1393; G. G. Athanasiu, P. J. Franzini and F. J. Gilman, *Phys. Rev.* **D32** (1985) 3010.

16. H. Georgi and D. V. Nanopoulos, *Phys. Lett.* **82B** (1979) 95.

17. S. Weinberg, *Phys. Rev.* **D13** (1976) 974; *Phys. Rev.* **D19** (1979) 1277.

18. L. Susskind, *Phys. Rev.* **D20** (1979) 2619.

19. M. Peskin, *Nucl. Phys.* **B175** (1980) 197; J. Preskill, *Nucl. Phys.* **B177** (1981) 21.

20. R. Sundrum, *Nucl. Phys.* **B395** (1993) 60.

21. J. Schwinger, *Phys. Rev.* **125** (1962) 397.

22. A. A. Migdal and A. M. Polyakov, *Sov. Phys. JETP* **24** (1967) 91; R. Jackiw and K. Johnson, *Phys. Rev.* **D8** (1973) 2386; J. M. Cornwall and R. E. Norton, *Phys. Rev.* **D8** (1973) 3338.

23. P. Sikivie, L. Susskind, M. B. Voloshin and V. I. Zakharov, *Nucl. Phys.* **B173** (1980) 189.

24. G. Altarelli, CERN Preprint CERN-TH-6867-93, 1993; D. C. Kennedy, Fermilab Preprint Fermilab-conf-93/023-T, 1993; J. Rosner, Enrico Fermi Institute Preprint EFI 92-58, 1992.

25. E. Farhi and L. Susskind, *Phys. Rep.* **74** (1981) 277.

26. S. Dimopoulos and L. Susskind, *Nucl. Phys.* **B155** (1979) 237; E. Eichten and K. Lane, *Phys. Lett.* **90B** (1980) 125.

27. L. Randall, *Nucl. Phys.* **B403** (1993) 122; H. Georgi, Harvard Preprint HUTP-92/A037, 1992.

28. R. F. Dashen, *Phys. Rev.* **183** (1969) 1245.

29. S. Dimopoulos, H. Georgi and S Raby, *Phys. Lett.* **B127** (1983) 101.

30. R. S. Chivukula and H. Georgi, *Phys. Lett.* **B188** (1987) 99; R. S. Chivukula, H. Georgi and L. Randall, Nucl. Phys. **B292** (1987) 93; H. Georgi, in *Proceedings of the 1990 International Workshop on Strong Coupling Gauge Theories and Beyond,*

eds. T. Muta and K. Yamawaki (World Scientific Publishing Co., Singapore, 1991).

31. V. A. Miransky, S. Peris and S. Raby, *Phys. Rev.* **D47** (1993) 2058.

32. S. Dimopoulos and J. Ellis, *Nucl. Phys.* **B182** (1981) 505.

33. T. Appelquist and J. Carazzone, *Phys. Rev.* **D11** (1975) 2856.

34. A. I. Akhiezer and V. B. Berestetsky, *Quantum Electrodynamics* (Interscience, New York, 1965).

35. S. Weinberg, *Phys. Lett.* **91B** (1980) 51; B. Ovrut and H. J. Schnitzer, *Phys. Lett.* **100B** (1981) 2518; L. Hall, *Nucl. Phys.* **B178** (1981) 75.

36. H. Georgi, *Nucl. Phys. B* (Proc. Suppl.) **29B,C** (1992) 1.

37. T. Appelquist and C. Bernard, *Phys. Rev.* **D22** (1980) 200.

38. M. Veltman, *Nucl. Phys.* **B123** (1977) 89.

39. M. B. Einhorn, D. R. T. Jones and M. Veltman, *Nucl. Phys.* **B191** (1981) 146.

40. J. Kuti, in *Electroweak Symmetry Breaking*, ed. W. A. Bardeen, J. Kodaira and T. Muta (World Scientific Publishing Co., Singapore, 1992).

41. R. Shrock, in *Quantum Fields on the Computer*, ed. M. Creutz (World Scientific Publishing Co., Singapore, 1992).

42. T. Inami and C. S. Lim, *Prog. Theor. Phys.* **65** (1981) 297.

43. M. Peskin and T. Takeuchi, *Phys. Rev. Lett.* **65** (1990) 964; *Phys. Rev.* **D46** (1992) 381.

44. M. Golden and L. Randall, *Nucl. Phys.* **B361** (1991) 3.

45. B. Holdom and J. Terning, *Phys. Lett.* **B247** (1990) 88.

46. G. Altarelli and R. Barbieri, *Phys. Lett.* **B253** (1990) 161; R. Johnson, B. -L. Young and D. W. Mckay, *Phys. Rev.* **D43** (1991) R17; W. Marciano and J. Rosner, *Phys. Rev. Lett.* **65** (1990) 2963; D. C. Kennedy and P. Langacker, *Phys. Rev. Lett.* **65** (1990) 2967.

47. H. Georgi, *Nucl. Phys.* **B363** (1991) 301; A. Dobado, D. Espriu and M. Herrero, *Phys. Lett.* **B253** (1991) 161; S. Dawson and G. Valencia, *Nucl. Phys.* **B352** (1991) 27; J. Bagger, in *Perspectives in the Standard model*, eds. R. K. Ellis, C. T.

Hill and J. D. Lykken (World Scientific Publishing Co., Singapore, 1992); M. B. Einhorn and J. Wudka, in *Electroweak Symmetry Breaking*, eds. W. A. Bardeen, J. Kodaira and T. Muta (World Scientific Publishing Co., Singapore, 1992).

48. D. C. Kennedy and B. W. Lynn, *Nucl. Phys.* **B322** (1989) 1.

49. R. Renkin and M. Peskin, *Nucl. Phys.* **B211** (1983) 93.

Chapter 14

1. B. Holdom, *Phys. Rev.* **D24** (1981) 1441.

2. B. Holdom, *Phys. Lett.* **B150** (1985) 301.

3. V. A. Miransky *Nuovo Cimento* **90A** (1985) 149.

4. W. A. Bardeen, C. N. Leung and S. T. Love, *Phys. Rev. Lett.* **56** (1986) 1335; C. N. Leung, S. T. Love and W. A. Bardeen, *Nucl. Phys.* **B273** (1986) 649.

5. K. Yamawaki, M. Bando and K. Matumoto, *Phys. Rev. Lett.* **56** (1986) 1335; M. Bando, K. Matumoto and K. Yamawaki, *Phys. Lett.* **B178** (1986) 308.

6. T. Akiba and T. Yanagida, *Phys. Lett.* **B169** (1986) 432.

7. T. Appelquist, D. Karabali and L. C. R. Wijewardhana, *Phys. Rev. Lett.* **57** (1986) 957.

8. T. Appelquist and L. C. R. Wijewardhana *Phys. Rev.* **D36** (1987) 568; T. Appelquist, D. Carrier, L. C. R. Wijewardhana and W. Zheng, *Phys. Rev. Lett.* **60** (1988) 1114.

9. T. Appelquist, M. B. Einhorn, T. Takeuchi and L. C. R. Wijewardhana *Phys. Lett.* **B220** (1989) 223; *Phys. Lett.* **B232** (1989) 211.

10. V. A. Miransky and K. Yamawaki, *Mod. Phys. Lett.* **A4** (1989) 29.

11. K. Matumoto, *Prog. Theor. Phys.* **81** (1989) 277.

12. J. Kogut, E. Dagotto and A. Kocić, *Phys. Rev. Lett.* **62** (1989) 1001.

13. S. R. Chivukula, *Phys. Rev. Lett.* **61** (1988) 2657.

14. E. Dagotto, A. Kocić and J. B. Kogut, *Phys. Rev. Lett.* **60** (1988) 1250.

15. A. Kocić, S. Hands, J. B. Kogut and E. Dagotto, *Nucl. Phys.* **B347** (1990) 217; A. Kocić J. B. Kogut, M. -P. Lombardo and K. C. Wang, *Nucl. Phys.* **B397** (1993) 451.

16. A. Kocić, J. B. Kogut and K. C. Wang, *Nucl. Phys.* **B398** (1993) 405; S. J. Hands, A. Kocić and J. B. Kogut, *Phys. Lett.* **B289** (1992) 400.

17. M. Göckeler, R. Horsley, E. Laermann, P. Rakow, G. Schierholz, R. J. Sommer and U. J. Weise, *Nucl. Phys.* **B334** (1990) 527.

18. M. Göckeler, R. Horsley, P. Rakow, G. Schierholz and R. J. Sommer, *Nucl. Phys.* **B371** (1992) 713.

19. A. Azcoiti, G. Di Carlo and A. F. Grillo, *Mod. Phys. Lett.* **A7** (1992) 3561.

20. K. -I. Kondo, H. Mino and K. Yamawaki, *Phys. Rev.* **D39** (1989) 2430.

21. T. Appelquist, M. Soldate, T. Takeuchi and L. C. R. Wijewardhana, in *Proceedings of John Hopkins Workshop on Current Problems in Particle Theory*, eds. G. Domokos and S. Kovesi-Domokos (World Scientific Publishing Co., Singapore, 1988).

22. V. P. Gusynin, V. A. Kushnir and V. A. Miransky, *Phys. Lett.* **B220** (1989) 635.

23. M. Inoue, T. Muta and T. Ochiumi, *Mod. Phys. Lett.* **A4** (1989) 605.

24. T. Nonoyama, T. Suzuki and K. Yamawaki, *Prog. Theor. Phys.* **81** (1989) 1238.

25. V. A. Miransky, T. Nonoyama and K. Yamawaki, *Mod. Phys. Lett.* **A4** (1989) 1409.

26. T. Takeuchi, *Phys. Rev.* **D40** (1989) 2697.

27. K. -I. Kondo, S. Shuto and K. Yamawaki, *Mod. Phys. Lett.* **A6** (1991) 3385.

28. M. Bando, T. Kugo and K. Suehiro, *Prog. Theor. Phys.* **85** (1991) 1299.

29. B. Holdom, *Phys. Lett.* **B246** (1990) 169; D. Kaplan, *Nucl. Phys.* **B365** (1991) 259; G. F. Giudice and S. Raby, *Nucl. Phys.* **B368** (1992) 221; S. King and S. Mannan, *Nucl. Phys* **B369** (1992) 119; M. B. Einhorn and D. Nash, *Nucl. Phys* **B371** (1992) 32; T. Appelquist and J. Terning, In *Electroweak Symmetry Breaking*, eds. W. A. Bardeen, J. Kodaira and T. Muta (World Scientific Publishing Co., Singapore, 1988); A. Sundrum, *Nucl. Phys.* **B395** (1993) 60; C. T. Hill, D. C. Kennedy, T. Onogi and Hoi-Lai Yu, *Phys. Rev.* **D47** (1993) 2940; M A. Luty, *Phys. Rev.* **D48** (1993) 1295; S. Martin, *Nucl. Phys.* **B398** (1993) 359.

30. V. A. Miransky, S. Peris and S. Raby, *Phys. Rev.* **D47** (1993) 2058.

31. M. Bando, T. Morozumi, H. So and K. Yamawaki, *Phys. Rev. Lett.* **59** (1987) 389.

32. B. Holdom, *Phys. Rev. Lett.* **60** (1988) 1233.

33. T. Appelquist and J. Terning, *Phys. Lett.* **B315** (1993) 139.

34. Y. Kikukawa and K. Yamawaki, *Phys. Lett.* **B234** (1990) 497.

35. V. A. Miransky *Phys. Lett.* **B248** (1990) 151; *Int. J. Mod. Phys.* **A6** (1991) 1641.

36. R. S. Chivukula, A. Cohen and K. Lane, *Nucl. Phys.* **B343** (1990) 554.

37. W. A. Bardeen and S. T. Love, 1990 (unpublished;) *Phys. Rev.* **D45** (1992) 4672.

38. W. A. Bardeen, S. T. Love and V. A. Miransky, *Phys. Rev.* **D42** (1990) 3514.

39. V. P. Gusynin and V. A. Miransky, *Mod. Phys. Lett.* **A6** (1991) 2443; *Sov. Phys. JETP* **74** (1992) 216.

40. V. A. Miransky, *Int. J. Mod. Phys.* **A8** (1993) 135.

41. T. Appelquist, J. Terning and L. C. R. Wijewardhana, *Phys. Rev.* **D44** (1991) 871.

42. M. Carena and C. E. M. Wagner, *Phys. Lett.* **B285** (1992) 277.

43. K. -I. Kondo, M. Tanabashi and K. Yamawaki, *Prog. Theor. Phys.* **89** (1993) 1249.

44. V. P. Gusynin, V. A. Kushnir and V. A. Miransky, *Phys. Rev.* **D39** (1989) 2355.

45. H. Georgi and S. L. Glashow, *Phys. Rev. Lett.* **47** (1981) 1511.

46. H. Georgi and D. Kaplan, *Phys. Lett.* **136B** (1984) 183.

47. E. H. Simmons, *Nucl. Phys.* **B312** (1989) 253; C. D. Carone and E. H. Simmons, *Nucl. Phys.* **B397** (1993) 591.

48. S. Samuel, *Nucl. Phys.* **B347** (1990) 625; A. Kagan and S. Samuel, *Phys. Lett.* **B252** (1990) 605.

Chapter 15

1. V. A. Miransky, M. Tanabashi and K. Yamawaki; *Phys. Lett.* **B221** (1989) 177; *Mod. Phys. Lett.* **A4** (1989) 1043.

2. Y. Nambu, Chicago Preprint EFI 89-08, 1989.

3. W. J. Marciano, *Phys. Rev. Lett.* **62** (1989) 2793; *Phys. Rev.* **D41** (1990) 219.

4. W. A. Bardeen, C. T. Hill and M. Lindner, *Phys. Rev.* **D41** (1990) 1647.

5. Y. Nambu, in *Proceedings of the 1988 Kazimierz Workshop*, eds. Z. Ajduk et al. (World Scientific Publishing Co., Singapore, 1989); in *Proceedings of 1988 International Workshop on New Trends in Strong Coupling Gauge Theories*, eds. M. Bando, T. Muta and K. Yamawaki (World Scientific Publishing Co., Singapore, 1989).

6. M. Veltman, *Acta Phys. Pol.* **B12** (1981) 437.

7. V. A. Miransky and K. Yamawaki, *Mod. Phys. Lett.* **A4** (1989) 129.

8. S. Englert and R. Brout, *Phys. Lett.* **B49** (1973) 77; A. Carter and H. Pagels, *Phys. Rev. Lett.* **43** (1979) 1845; K. Huang and R. R. Mendel, MIT CTP Report No.876, 1980; J. D. Carpenter, R. E. Norton and A. Soni, *Phys. Lett.* **B212** (1988) 63.

9. S. F. King and S. H. Mannan, *Phys. Lett.* **B241** (1990) 249; F. Barrios and U. Mahanta, *Phys. Rev.* **D43** (1991) 284.

10. M. Veltman, *Nucl. Phys.* **B123** (1977) 89.

11. R. Peccei and H. Quinn, *Phys. Rev. Lett.* **38** (1977) 1440.

12. M. Tanabashi, in *Dynamical Symmetry Breaking*, ed. K. Yamawaki (World Scientific Publishing Co., Singapore, 1992).

13. R. Peccei, in *CP violation*, ed. C. Jarlskog (World Scientific Publishing Co., Singapore, 1989).

14. Y. Nambu, in *Proceedings of the 1989 Workshop on Dynamical Symmetry Breaking*, eds. T. Muta and K. Yamawaki (Nagoya University, Nagoya, 1990).

15. S. Shuto, M. Tanabashi and K. Yamawaki, in *Proceedings of the 1989 Workshop on Dynamical Symmetry Breaking*, eds. T. Muta and K. Yamawaki (Nagoya University, Nagoya, 1990).

16. M. Suzuki, *Phys. Rev.* **D41** (1990) 3457.

17. M. Luty, *Phys. Rev.* **D41** (1990) 2893.

18. B. Pendleton and G. G. Ross, *Phys. Lett.* **B98** (1981) 291.

19. C. T. Hill, *Phys. Rev.* **D24** (1981) 691.

20. M. Suzuki, *Mod. Phys. Lett.* **A5** (1990) 1205.

21. W. A. Bardeen, Preprint FERMILAB-CONF-90/269-T, 1990.

22. A. Hasenfratz, P. Hasenfratz, K. Jansen, J. Kuti and Y. Shen, *Nucl. Phys.* **B365** (1991) 79.

23. J. Zinn-Justin, *Nucl. Phys.* **B367** (1991) 105.

24. C. T. Hill, *Phys. Lett.* **B269** (1991) 419.

25. K. -I. Kondo, M. Tanabashi and K. Yamawaki, in *Proceeding of the 1989 Workshop on Dynamical Symmetry Breaking*, eds. T. Muta and K. Yamawaki (Nagoya University, Nagoya, 1990); R. Bönish, *Phys. Lett.* **B268** (1991) 394; M. Lindner and D. Ross, *Nucl. Phys.* **B370** (1992) 30; S. P. Martin, *Phys. Rev.* **D45** (1992) 4283; *Phys. Rev.* **D46** (1992) 2197; *Phys. Rev.* **D46** (1992) 2769.

26. T. E. Clark, S. T. Love and W. A. Bardeen, *Phys. Lett.* **B237** (1990) 235; K. Sasaki, M. Carena, C. E. M. Wagner, T. E. Clark and W. A. Bardeen, *Nucl. Phys.* **B356** (1991) 46.

27. A. Linde, *Particle Physics and Inflation* Cosmology (Harvood Academic Publishers, Chur, 1990).

28. S. Coleman, *Nucl. Phys.* **B310** (1988) 643.

29. V. A. Miransky *Int. J. Mod. Phys.* **A6** (1991) 1641.

30. C. T. Hill and D. Salopek, *Ann. Phys. (N.Y.)* **213** (1992) 21.

31. S. Bornholdt and C. Wetterich, *Phys. Lett.* **B282** (1992) 399.

32. K. Yamawaki, in *Proceedings of the 1990 International Workshop on Strong Coupling Gauge Theories and Beyond*, eds. T. Muta and K. Yamawaki (World Scientific Publishing Co., Singapore, 1991).

33. T. Elliot and S. King, *Z. Phys.* **C58** (1993) 609.

34. G. 't Hooft, *Phys. Rev. Lett.* **37** (1976) 8.

35. S. Coleman, *Aspects of Symmetry* (Cambridge University Press, Camridge, 1985).

36. R. R. Mendel and V. A. Miransky *Phys. Lett.* **B268** (1991) 384.

37. V. A. Miransky, *Phys. Rev. Lett.* **69** (1992) 1022.

38. S. Martin, *Nucl. Phys.* **B398** (1993) 359.

39. Y. Nagoshi, K. Nakanishi and S. Tanaka, *Prog. Theor. Phys.* **85** (1991) 131.

40. M. Bando, T. Kugo and K. Suehiro, *Prog. Theor. Phys.* **85** (1991) 1299.

41. Particle Data Group, K. Hikasa et al., *Phys. Rev.* **D45** PartII (1992) S1.

42. C. T. Hill and E. A. Paschos, *Phys. Lett.* **B241** (1990) 96; C. T. Hill, M. Luty and E. A. Paschos, *Phys. Rev.* **D43** (1991) 3011.

43. S. F. King, *Phys. Rev.* **D45** (1992) 990.

44. K. S. Babu and R. N. Mohapatra, *Phys. Rev. Lett.* **66** (1991) 556.

45. Y. Achiman and A. Lukas, *Nucl. Phys.* **B384** (1992) 78.

Appendix A

1. S. S. Schweber, *An Introduction to Relativistic Quantum Field Theory* (Row, Peterson and Co., Evanston, Illinnois, 1961).

2. J. D. Bjorken and S. D. Drell, *Relativisitic Quantum Fields* (McGraw-Hill Book Co., New York, 1965).

3. E. P. Wigner, *Group Theory* (Academic Press, New York, 1959).

4. R. Streater and A. Wightman, *PCT, Spin and Statistics, and All That* (W. A. Benjamin Inc., New York, 1964).

Appendix B

1. R. Slansky, *Phys. Rep.* **79** (1981) 1.

2. K. I. Aoki, M. Bando, T. Kugo, M. G. Mitchard and H. Nakatani, *Prog. Theor. Phys.* **84** (1990) 683.

Appendix C

1. C. G. Bollini and J. J. Giambiagi, *Phys. Lett.* **40** (1972) 566; G. 't Hooft and M. Veltman, *Nucl. Phys.* **B44** (1972) 189.

2. T. Muta, *Foundations of Quantum Chromodynamics* (World Scientific Publishing Co., Singapore, 1987).

3. T. Kugo, *Quantum Gauge Field Theory Parts I, II* (in Japanese) (Baifukan, Tokyo, 1989).

4. G. 't Hooft, *Nucl. Phys.* **B61** (1973) 455.

5. W. A. Bardeen, A. J. Buras, D. W. Duke and T. Muta, *Phys. Rev.* **D18** (1978) 3998.

6. R. Feynman and A. Hibbs, *Quantum Mechanics and Path Integrals* (McGraw-Hill Book Co., New York, 1965).

7. C. Itzykson and J. Zuber, *Quantum Field Theory* (McGraw-Hill Book Co., New York, 1980).

8. S. Coleman, *Aspects of Symmetry* (Cambridge University Press, Cambridge, 1985).

9. F. A. Berezin, *The Method of Second Quantization* (Academic Press, New York, 1966).

Appendix D

1. A. I. Akhiezer and V. B. Berestetsky, *Quantum Electrodynamics* (Interscience, New York, 1965).

2. H. Bateman and A. Erdélyi, *Higher Transcendental Functions*, vol.1 (McGraw-Hill Book Co., New York, 1953).

Appendix E

1. I. J. R. Aitchison and C. M. Fraser, *Phys. Lett.* **B146** (1984) 63; L. H. Chan, *Phys. Rev. Lett.* **54** (1985) 1222.

2. V. P. Gusynin and V. A. Kushnir, *Sov. J. Nucl. Phys.* **51** (1990) 373.

Subject Index